Sperm Competition and Its Evolutionary Consequences in the Insects

MONOGRAPHS IN
BEHAVIOR AND ECOLOGY

Edited by John R. Krebs and Tim Clutton-Brock

Foraging Theory, *by David W. Stephens and John R. Krebs*

Dynamic Modeling in Behavioral Ecology, *by Marc Mangel and Colin W. Clark*

The Evolution of Parental Care, *by T. H. Clutton-Brock*

Parasitoids: Behavioral and Evolutionary Ecology, *by H.C.J. Godfray*

Sexual Selection, *by Malte Andersson*

Polygyny and Sexual Selection in Red-Winged Blackbirds, *by William A. Searcy and Ken Yasukawa*

Leks, *by Jacob Höglund and Rauno V. Alatalo*

Social Evolution in Ants, *by Andrew F. G. Bourke and Nigel R. Franks*

Female Control: Sexual Selection by Cryptic Female Choice, *by William G. Eberhard*

Sex, Color, and Mate Choice in Guppies, *by Anne E. Houde*

Foundations of Social Evolution, *by Steven A. Frank*

Parasites in Social Insects, *by Paul Schmid-Hempel*

Levels of Selection in Evolution, *edited by Laurent Keller*

Social Foraging Theory, *by Luc-Alain Giraldeau and Thomas Caraco*

Model Systems in Behavioral Ecology: Integrating Conceptual, Theoretical, and Empirical Approaches, *edited by Lee Alan Dugatkin*

Sperm Competition and Its Evolutionary Consequences in the Insects, *by Leigh W. Simmons*

Sperm Competition and Its Evolutionary Consequences in the Insects

LEIGH W. SIMMONS

Princeton University Press
Princeton and Oxford

Published by Princeton University Press, 41 William Street, Princeton, New Jersey 08540
In the United Kingdom: Princeton University Press, 3 Market Place, Woodstock, Oxfordshire OX20 1SY

Library of Congress Cataloging-in-Publication Data
Simmons, Leigh W., 1960–
Sperm competition and its evolutionary consequences in the insects / Leigh W. Simmons.
p. cm. — (Monographs in behavior and ecology)
Includes bibliographical references (p.).
ISBN 0-691-05987-X — ISBN 0-691-05988-8 (pbk.)
1. Insects—Evolution. 2. Sperm competition. 3. Insects—Reproduction. I. Title. II. Series.

QL468.7 .S56 2001
595.713′8—dc21 2001016377

British Library Cataloging-in-Publication Data is available

This book has been composed in Times Roman

Printed on acid-free paper. ∞

www.pup.princeton.edu

Printed in the United States of America

1 3 5 7 9 10 8 6 4 2

1 3 5 7 9 10 8 6 4 2
(Pbk.)

Contents

Preface

The image of cattle grazing in the fresh green summer meadows of northern Europe conveys a sense of peace and tranquility. But in reality the meadows are a battle ground on which a myriad of golden yellow flies can be found jostling for position on and around fresh droppings. At their peak, up to 1000 or more male flies can be found around a single dung pat. The focus of their frenzied search is the occasional arrival of drab olive females who travel to fresh droppings to deposit their eggs. Females fly from downwind, pass over the pat, and land in the grass upwind before walking back to the dung surface to lay their eggs. Males search on and upwind of droppings and seize females immediately they arrive. On capturing a female, males will begin to copulate immediately. Struggles for the possession of females are intense. Searching males will pounce upon copulating pairs, with the result that large balls of golden flies can be seen tumbling about the dung surface while the object of their desire is pushed and pulled in all directions; sometimes females are drowned in the dung surface or otherwise injured to the extent that they can no longer fly. When the density of males on and around pats is high, a male capturing an incoming female will carry her in flight to the surrounding grass to copulate before returning her to the dung to lay her eggs. During oviposition the male remains mounted upon the female and pairs separate only after a clutch of eggs is laid.

The yellow dung fly holds a unique place in the history of behavioral ecology, and the person responsible for placing it there was Geoff Parker. In the 1960s, Parker studied the behavior and ecology of the yellow dung fly within a conceptual framework of sexual selection operating at the level of the individual. At the time this was a controversial approach. Not only was he working within an era that propounded group-selectionist theories of evolution, but sexual selection was not then generally accepted as a significant driving force in evolution. Parker's work on the yellow dung fly's mating system represents one of the earliest quantitative demonstrations of the role of sexual selection in evolution. Moreover, it contributed directly to the development of some influential general theories in behavioral ecology, including the "ideal free distribution," the "marginal value theorem," and, the subject of this volume, sperm competition (see Parker 2001).

There were three behaviors in particular that Parker sought to explain. Why did males carry females away from droppings before copulating? Why did males remain with females during egg deposition? And why did copulation take so long, over half an hour, when in many other species copulation can be accomplished within a few seconds? As noted above, struggles for the possession of arriving females often occur on the dung surface. Parker had also observed that pairs returning to the dropping to deposit eggs could be subject to encounters with searching males that also led to struggles and takeovers. A successful attacking male would copulate with the female before she completed the deposition of her clutch of eggs. Parker used a technique that was widely used by applied entomologists studying biocontrol of insect populations. He produced sterile males by exposing them to ^{60}Co irradiation. Irradiation induces dominant lethal mutations in sperm such that eggs fertilized by sperm from irradiated males fail to hatch. In this way, he could determine the proportion of eggs that were sired by each of two males when one was sterilized, simply by determining the proportion of eggs that hatched. He found that the fertilization success of the last male to mate with a female increased with time spent copulating, following a curve of diminishing returns. Typically, males copulated for 36 minutes and obtained around 80% of fertilizations (Parker 1970f). Thus, by copulating for a long time, males were able to displace the sperm from previous males. The fact that males do not continue to copulate until they fertilize all of a female's eggs can be explained by considering the rate of fertilization gain, the costs of searching for and guarding a second female, and the maximization of fitness across all available females. Dung fly copula duration was thereby instrumental in the development of, and gave empirical evidence for, the marginal value theorem (Parker 1970f; Parker and Stuart 1976; Charnov 1976).

Given that females readily remate and that males can displace sperm from previous males, the reasons for postcopulatory guarding during oviposition and emigration from the dropping during copulation became apparent. A male who abandoned his mate following copulation would have a reduced fitness compared with a male who remained and protected his mate from searching males because of the risks of recopulation and sperm displacement (Parker 1970d). Moreover, emigration from the dropping during copulation reduced the risks of takeover and thus similarly increased male fitness (Parker 1971).

Parker drew two general principles from his observations of yellow dung fly paternity and reproductive behavior. First, selection should favor adaptations in males for the preemption of sperm stored from previous matings; and, second, selection should simultaneously favor adaptations that prevent the preemption of a male's sperm by future rivals. These general principles immediately made sense of numerous biological observations in the entomological literature; the widespread occurrence of "passive phases" in which males guard females either before or after copulation, the occurrence of mating plugs that were thought to block the female's reproductive tract after copulation, the incorporation into the ejaculate of chemical substances that influence future sexual receptivity of females, and the deposition of pheromones onto females that make them unattractive to future males. In 1970 Parker collected these observations, together with observations of last male sperm priority patterns prevalent in both the pure and applied entomological literature, and presented them in his review *Sperm Competition and Its Evolutionary Consequences in the Insects.* The review outlined a general theory of sexual selection via sperm competition (the temporal overlap of competing ejaculates) and stressed the widespread implications it held for an organism's reproductive behavior, physiology, and morphology.

Parker's review appeared ahead of its time. The 1970s saw an upheaval in evolutionary biology as group-selectionist thinking gave way to individual selection and modern behavioral ecology was born. It was also in the 1970s that the importance of sexual selection was finally recognized. The study of sperm competition thus had to await the settling of dust before it would itself come of age. It was in the late 1970s that evolutionary biologists studying insects realized the implications of selection via sperm competition and ten years after the publication of Parker's review Robert Smith organized a symposium on sperm competition from which arose his edited volume *Sperm Competition and the Evolution of Animal Mating Systems* published in 1984. Some 60% of the contributors to the book addressed sperm competition in insects or arachnids. Nevertheless, Smith recruited authors to discuss the possible implications of sperm competition in vertebrate taxa ranging from fish to humans in an attempt to stimulate a broadening of research effort on sperm competition.

Evolutionary biologists studying the behavior and reproductive biology of birds had noticed that apparently monogamous species often engaged in what has now become known as extrapair copulations. In the 1980s they came to realize that such behavior could result in sperm competition and favor adaptations similar to those found in insects. Sperm competition was soon recognized as being responsible for the occurrence of a series of behaviors that function as mate guarding during the female's fertile period (close following of the female, territorial defense, and frequent copulations) and favored increased male investment in sperm production. The literature on birds was reviewed by two key contributors, Tim Birkhead and Anders Møller in their 1991 volume *Sperm Competition in Birds.* Those studying mammals had also taken note of the evolutionary significance of sperm competition. Don Dewsbury had revealed multiple paternity, and thus the occurrence of sperm competition, in rodents (Dewsbury 1984), and Sandy Harcourt and his colleagues had examined patterns of testis size variation in relation to breeding behavior across primates. Consistent with sperm competition theory, primate taxa with promiscuous mating systems were found to have larger testes than those with relatively monandrous mating systems (Harcourt et al. 1981). Harcourt's comparative analysis suggests that humans have testes that are about average given human body size, suggesting that sperm competition is neither particularly prevalent nor absent. In a U.K. government survey, 13% of women reported copulating with two concurrent sexual partners (Wellings et al. 1994), suggesting that sperm competition can occur. In their controversial volume *Sperm Competition in Humans* published in 1995, Robin Baker and Mark Bellis suggested that sperm competition has been a significant evolutionary driving force in human sexual behavior and morphology. Nevertheless, some of their findings can be subject to alternative explanation (Gomendio et al. 1998) or have not proved robust to rigorous analysis (Moore et al. 1999).

Smith's volume was successful in its endeavor, and the late 1980s and the 1990s saw a surge of interest in the study of sperm competition. Birkhead and Møller's edited volume *Sperm Competition and Sexual Selection* published in 1998 shows that sperm competition has had significant widespread evolutionary consequences for the reproductive biology of almost every animal group examined, and its botanical equivalent, pollen competition, is equally significant. The study of insects, and in particular of the yellow dung fly, was instrumental in the empirical and theoretical development of the field of sperm competition which is now rightly recognized as a subdiscipline of sexual selection. Throughout the thirty years since Parker's review, research on insect sperm competition has increased almost exponentially. It is now time to reassess the entomological literature and its contribution to the discipline of sperm competition. With this volume I have therefore attempted to provide an update to Parker's original review.

I begin in chapter 1 by outlining the general theories of sexual selection before discussing more fully how sperm competition arises, its evolutionary implications, and how it forms a subset of sexual selection. In chapter 2 some general working definitions for terms that are used in the sperm competition literature are provided, together with a comprehensive review of the literature on patterns of sperm utilization in nonsocial insects. Typically, these patterns are assumed from the proportion of offspring sired by the second of two males to mate with a female. I discuss why an understanding of the mechanisms of sperm transfer and storage is essential for interpreting observed patterns of sperm utilization and important for the evolutionary interpretation of nonrandom paternity, and of traits believed to arise under sperm competition.

Chapters 3, 4, and 5 discuss morphological adaptations in genitalia, physiological adaptations in ejaculates, and behavioral adaptations of males, respectively, that are thought to have arisen under selection for the avoidance of sperm competition from previous or future males. In chapter 6 I examine how sperm competition has shaped male copulatory behavior, providing a more detailed analysis of copula duration in the yellow dung fly as well as reviewing theoretical and empirical studies that examine copula duration in other insect taxa. Chapter 7 reviews the evidence that, contrary to established dogma, males are limited in their ability to produce ejaculates. Over the last decade, Parker and others have developed a new theoretical base for predicting the evolutionary consequences of sperm competition and variation in mate quality, for male investment in the ejaculate with different females.

Chapter 7 reviews this theoretical approach and the empirical evidence available from studies of insect ejaculate size that either support or refute theoretical predictions.

In his original review, Parker noted that there is every reason that selection via sperm competition would also act on sperm morphology. Numerous authors, most notably John Sivinski, have explored this possibility, at least conceptually. The insects exhibit a staggering diversity of sperm morphologies and in chapter 8 I review the empirical evidence that is available for the role of sperm competition in the evolution of such aspects of morphology as sperm length and polymorphism. In chapter 9 I take up William Eberhard's (1996) plea to examine the influence of female interests in studies of sperm competition. I therefore examine the evidence for female manipulation of sperm transfer, storage, and utilization as mechanisms of female choice.

Multiple mating and sperm utilization raise special issues in social insects because of the effects of mixed paternity on mean offspring relatedness. Previously there has been little if any communication between researchers focused on nonsocial insects and those focused on social insects. Chapter 10 presents a review of multiple mating and sperm utilization in social insects in an effort to bring these two fields together. Finally, in chapter 11 I consider the broader implications of selection via sperm competition for insect life history evolution and speciation. I believe that insects continue to offer the greatest potential for studying the evolutionary consequences of sperm competition, simply because they exhibit an extraordinary diversity of sperm transfer and storage mechanisms that generate different types of selection pressures. Moreover, their abundance and the ease with which quantitative and molecular genetic approaches can be made allow us to penetrate to greater depths of analysis than is possible with vertebrates. The insects will therefore continue to provide insight and lead research into new and exciting areas of sperm competition.

Perth, Western Australia, March 2001

Dedication and Acknowledgments

I consider the most valuable period in my research career to have been the time I spent at Liverpool working with Geoff Parker. I am fortunate indeed to be able to call Geoff both a colleague and friend. I first met him in 1985 on a visit to Woodchester Park in Gloucestershire where he was involved in the running of an undergraduate field course. I had gone to present my work on field crickets and to discuss with him the possibility of moving to Liverpool as a postdoc. There, in a meadow in front of the old mansion, he introduced me to yellow dung flies. Of course I had learnt of his work as an undergraduate so I was delighted to witness events first hand with the very person who had made them so well known in behavioral ecology. Since that first meeting Geoff's personal and professional support has been total and unreserved. Working with Geoff has been educational and inspirational. He has been the finest of mentors. My hope is that this volume will in some way repay an ever-growing debt. This volume is for Geoff, after thirty years.

I would like to thank all those who have helped in the production of this work. I am particularly grateful to Martin Thompson for the artwork. For access to unpublished works, advice, discussions on insect sperm competition, and/or for reading previous drafts of the manuscript, I thank John Alcock, Göran Arnqvist, Winston Bailey, Tim Birkhead, Koos Boomsma, William Eberhard, Sam Elworthy, Else Fjerdingstad, Matthew Gage, Paco Garcia-Gonzalez, Mike Majerus, Geoff Parker, Andy Pervis, Mark Ransford, Dale Roberts, Ron Rutowski, Steve Simpson, Mike Siva-Jothy, Lotta Sundström, Stuart Warwick, and Nina Wedell. During the production of this book I was supported financially by the Australian Research Council. Finally, I must also thank my research officer, Julie Wernhem, and research colleagues, John Hunt, Lotta Kvarnemo, Janne Kotiaho, and Joseph Tomkins, for keeping my research alive while I have been library or desk bound; my friends and colleagues in the Department of Zoology at the University of Western Australia for support and assistance; and finally, and most of all, my family, Carol and Freddy.

Sperm Competition and Its Evolutionary Consequences in the Insects

1 Sexual Selection and Sperm Competition

1.1 Sexual Selection

In his first major treatise on organic evolution, Darwin (1859) made a clear distinction between natural and sexual selection. He argued that natural selection favored traits in individuals that enhanced their viability, while sexual selection favored traits that enhanced their success in reproduction. Darwin was seeking to explain the evolution of extravagant secondary sexual traits that appeared counter to his theories of natural selection. By their nature, extravagant secondary sexual traits can prove detrimental to the survival of individuals possessing them and so should be removed by natural selection. However, Darwin recognized that the process of evolutionary change would be facilitated by the differential transmission of parental characteristics to future generations through successful reproduction, so that traits that enhanced reproduction, even at the expense of decreased survival, would be favored. Darwin (1871) thus defined sexual selection as the advantage some individuals have over others of the same sex and species in exclusive relation to reproduction. He recognized that sexual selection could operate through two distinct, although not necessarily mutually exclusive, processes. On the one hand selection would favor traits that enhanced an individual's success in competition with members of the same sex for mating opportunities, while on the other there would be an advantage bestowed on those individuals that were more attractive to members of the opposite sex. Mating competition (intrasexual selection) can act on what might otherwise be considered naturally selected traits, such as visual acuity, which allow some individuals to locate mates more rapidly than their competitors. Moreover, selection can favour the evolution of elaborate secondary sexual traits such as weaponry used in competitive interactions over members of the opposite sex. Darwin thus invoked his theories of intrasexual selection to explain the origin of such structures as the horns of male beetles that are used in competition (Fig. 1.1). He also argued that mating preferences (intersexual selection) could simultaneously favor the evolution of weapons, if members of the opposite sex selected as mates those individuals who were "vigorous and well armed." Many secondary sexual traits appear ornamental, however, rather than serving a direct function during combat. Darwin thus invoked intersexual selection as a mechanism for the evolution of ornaments, such as

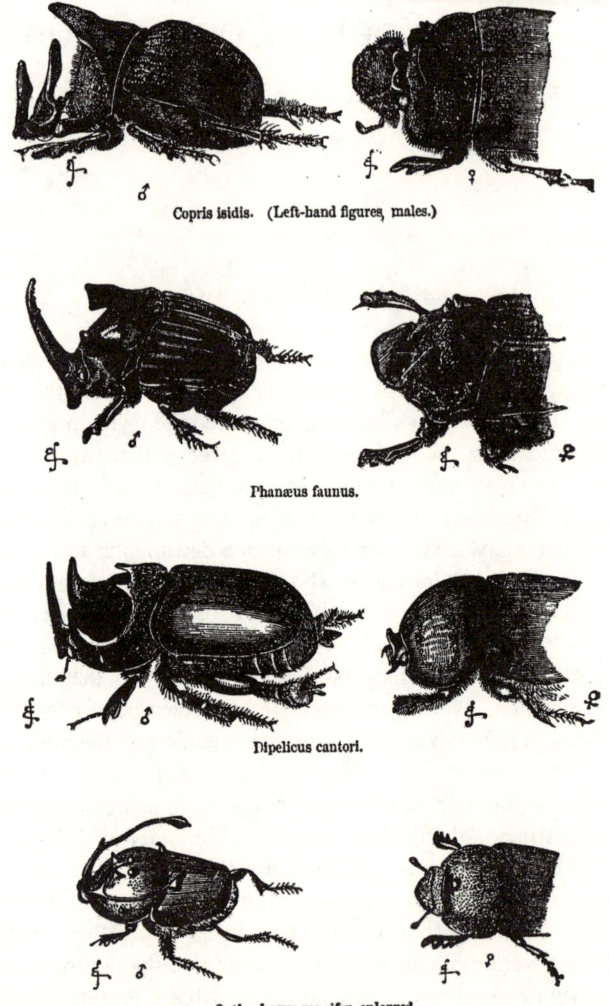

Figure 1.1 Darwin (1871) used the obvious sexual dimorphism in the horns of various species of beetle to illustrate his theories of sexual selection and the evolution of extravagant secondary sexual traits in males.

the elaborate plumage characteristics of birds or the songs produced by some insects, arguing that individuals of one sex preferred to mate with more attractive partners, thereby invoking a sense of beauty in the choosy sex. Darwin noted that it was generally males who competed for mating opportunities while females exercised mating preferences, so that it was males who were adorned with the products of sexual selection: exaggerated sensory structures, elaborate display ornaments, and weaponry.

Darwin's (1859, 1871) theories of sexual selection have fueled controversy since their inception. While it was clear that competition for mates could favor adaptations that enhanced success in the competitive race to reproduce, theorists were unable to envisage the selective advantage accrued through mating preferences (Wallace 1889, Huxley 1938). Furthermore, since intersexual competition is often very conspicuous and intense, ascribing nonrandom mating to female choice has proved difficult, particularly in cases where the traits reputed to be subject to female choice are also those responsible for success in competition (Partridge and Halliday 1984).

The 1980s and 1990s saw a rapid growth in the development of intersexual selection theory (Zahavi 1975, 1977; O'Donald 1980; Lande 1981; Hamilton and Zuk 1982; Kirkpatrick 1982; Pomiankowski 1987, 1988; Grafen 1990a,b; Iwasa et al. 1991), and innovative empirical investigations have clearly shown how female choice can be a potent force in evolution (Andersson 1982; Møller 1988; Basolo 1990; Houde and Endler 1990; Moore 1990; Houde 1994; Wilkinson and Reillo 1994; Basolo 1995a; see review of Andersson 1994). Controversy now centers not so much on whether mate choice occurs, but rather on whether secondary sexual traits that are subject to mating preferences represent arbitrary traits that have coevolved with the mating preference, perhaps due to what Darwin would have termed a sense of beauty, or are honest indicators of the underlying genetic quality of potential mates (Bradbury and Andersson 1987; Kirkpatrick and Ryan 1991; Maynard Smith 1991).

Arbitrary trait models of preference evolution were first proposed by Fisher (1915), who argued that traits in males and preferences in females would coevolve in a runaway process of ever-increasing exaggeration until checked by natural selection. The process relies on linkage disequilibrium between genes coding for the preference in females and the secondary sexual trait in males. Thus, females preferring males with exaggerated sexual traits produce sons that achieve a higher mating success and daughters with the preference. Formal quantitative models show that such a process is at least feasible (O'Donald 1980; Lande 1981; Kirkpatrick 1982; Pomiankowski et al. 1991) and recent empirical studies have demonstrated genetic correlations between traits and preferences both within and between populations (Houde and Endler 1990; Bakker 1993; Gilburn et al. 1993; Houde 1994; Wilkinson and Reillo 1994; Brooks and Couldridge 1999). However, such evidence is not in itself support for a Fisher process because genetic correlations are expected to arise under a number of preference evolution models, including those in which females use some arbitrary trait as a signal of the male's ability to provide resources or other material benefits to themselves and/or their offspring (Kirkpatrick 1982). Moreover, the Fisher process potentially explains how elaboration of a secondary sexual trait can occur, but not how it becomes established in the population. Fisher (1915) thus originally proposed that some fitness benefit might accrue to males possessing the trait that could be transmitted to offspring in order to initiate the process, essentially a "good genes" argument.

The so-called good genes models of sexual selection propose that the expression of the secondary sexual trait is an indication of the underlying genetic quality of the individual and that females choosing males with the most extravagant traits increase the viability of offspring produced (Zahavi 1975; Andersson 1986; Pomiankowski 1987; Grafen 1990a; Iwasa et al. 1991). One problem with good genes models of sexual selection is that directional selection is expected to erode fitness variation (Fisher 1930; Falconer and Mackay 1996). Where preferences are costly for females, they should not be maintained after fitness variation associated with the prefered trait in males is lost. Hamilton and Zuk (1982) suggested one mechanism for the maintenance of heritable fitness variation, proposing that the cyclical coadaptive nature of host-parasite relationships could maintain heritable resistance to infection. Since the expression of secondary sexual traits is often dependent on the general health and vigor of the individual, females could obtain genes conferring resistance to current parasite genotypes by choosing to mate with males whose secondary sexual traits indicate general health and vigor, and thus likely freedom from parasitic infection.

The Hamilton and Zuk hypothesis relies on the specific interaction between hosts and their parasites or pathogens. Fitness heritability can potentially be maintained under a variety of other conditions (Charlesworth 1987; Jones 1987; Rice 1988). For example it has been suggested that strong directional selection could potentially increase heritable variation in sexual traits if it favored an increase in the number of genes and the average contribution of each locus to phenotypic variance (Pomiankowski and Møller 1995). However, Rowe and Houle (1996) point out that sexually selected traits are unlikely to be subject to the strong directional selection required by Pomiankowski and Møller's (1995) model. Rather, sexual traits should reach some equilibrium value of expression where natural selection opposes sexual selection; at equilibrium sexual traits should be subject to stabilizing selection (Partridge and Endler 1987; Kirkpatrick and Ryan 1991). Rowe and Houle (1996) suggest that genetic variation in condition is likely to be high because it is a compound variable to which components of variance are contributed by a plethora of fitness-related traits. Secondary sexual traits that are condition dependent will thereby capture the genetic variance in condition and represent a reliable signal to females of underlying fitness benefits to mate choice. Certainly, there is a growing body of evidence from a broad range of taxa that mating preferences can have indirect fitness benefits for offspring (Watt et al. 1986; Crocker and Day 1987; Simmons 1987a; Reynolds and Gross 1992; Norris 1993; Moore 1994; Petrie 1994; Møller and Alatalo 1999).

An alternative view to the arbitrary traits and genetic benefit models of sexual selection is that male sexual displays arise because of perception biases in females that become established by natural selection operating on the female's sensory system (West-Eberhard 1984; Endler and McLellan 1988; Ryan and Rand 1993). For example, water mites respond to the vibratory signals of their copepod prey and the males of some species appear to mimic this signal in their courtship displays; males exploit the females predatory response to achieve insemination (Proctor 1992). The sensory exploitation hypothesis has been used to explain the paradoxical observation that the female's of some species can have preferences for display traits that are absent in conspecific males (Ryan et al. 1990; Basolo 1990, 1995b). These observations motivated Holland and Rice (1998) to view the coevolution of male sexual traits and female preferences from a different perspective. They suggested that once a sensory bias in females had resulted in the origin of a secondary sexual structure or display in males, females might be induced to mate suboptimally because the stimulus received from males results in them mating rather than performing an action appropriate to the sensory pathway that is exploited; in the case of the water mites, females become inseminated rather than capturing a prey item. Counterselection might then be expected to increase the female's sensory threshold so that the timing of mating or its frequency returns to the female's optimum; females become resistant to the male's sexual signal. Counterselection might then favor further elaboration of the male's signal to overcome the female's resistance. Under Holland and Rice's (1998) model, coevolution arises from chase-away selection where male traits chase increasing female resistance to mating. Holland and Rice's (1998) coevolutionary arms race is much the same as Parker's (1979) model, where he envisaged female resistance evolving not against a male signal but against male physical ability to overcome female resistance to mating. The chase-away model differs from Fisher's runaway selection where female preferences follow increased male trait size (see discussions of Getty 1999; Rice and Holland 1999; Rosenthal and Servedio 1999). Like runaway, chase-away selection is not predicted to generate a correlation between the success of individual males and the viability of their offspring.

1.2 Sexual Differences and the Evolution of Anisogamy

Although Darwin (1871) noted that it was generally males who competed for access to choosy females, it was nearly a century before Williams (1966) laid down the theoretical basis for these apparent sexual differences in reproductive behavior. Expanding on the early arguments of Bateman (1948), Williams (1966) argued that the ultimate factor controlling the sex on which selection operates was the relative parental expenditure of the sexes. For a given reproductive expenditure, males can produce many more gametes than can females. Thus, Bateman (1948) argued that males should be limited in their reproductive potential only by the number of females they can inseminate, so that we should expect to see competition among them for access to multiple females. Williams (1966) and later Trivers (1972) realized that expenditure on producing offspring often went further than simply producing gametes, and Trivers (1972) included such parameters as the care and protection of developing young as parental investment. Trivers' (1972) formal definition of parental investment was very precise, requiring that investment made by parents increased the offspring's chances of survival and reproduction. However, it is only the extent to which investments made by parents reduce their ability to invest in additional offspring that is relevant for the control of sexual selection, because it is only this cost of reproduction that will influence the relative availability of males and females for reproduction. Dewsbury (1982) pointed out that gametes are rarely transferred on a one-for-one basis. Rather, they are packaged in their thousands into ejaculates that contain nourishment for the sperm during transport to the site of fertilization and additional materials necessary for successful insemination of the female. The costs of ejaculate production are rarely trivial so that males may be limited in their potential to produce ejaculates. Nevertheless, sperm production has traditionally not been recognized as an important component of male reproductive investment. At the extreme, if ejaculate production represented a greater cost than egg production and maternal care, we should expect males to be a limited resource over which females competed.

A quantitative formulation for the influence of reproductive investment on the control of sexual selection was provided by Clutton-Brock and Parker (1992) and Parker and Simmons (1996) in their analyses of potential reproductive rates. Typically, males invest least in reproduction so that the time cost associated with producing a batch of offspring is lower for males than for females. Thus males have the higher potential reproductive rate and so are in a position to reproduce more often than females, leading to a bias in the operational sex ratio (OSR), the ratio of fertilizable females to sexually active males (Emlen and Oring 1977). When the OSR is biased toward an excess of males, males will compete for access to the limited supply of females. A corollary of the relatively greater potential reproductive rate of males is that the costs of mate rejection for males are high while the costs for females are low; females that reject a mating partner are likely to encounter an alternative mate sooner than would a male (Johnstone et al. 1996). Thus, sexual differences in both mating competition and mate choice arise because of differences in the relative time costs associated with reproductive investment by the sexes, in both their gametes and other forms of parental expenditure.

Sexual selection on males can therefore stem from a fundamental difference between the sexes in the costs associated with the production of gametes. Sexual reproduction in general is characterized by differences in the sizes of male and female gametes, termed anisogamy, which is believed to have arisen from the primitive state of isogamy (Parker et al. 1972). Parker et al. (1972) envisaged two distinct selective pressures related to gamete size, numerical productivity (the number of gametes produced per unit time) and zygote fitness (the probability that the zygote will survive to adulthood and reproduce). Numerical productivity would be favored by selection if this increased the reproductive rate of the parent. However, increased gamete size would be favored if additional resources laid down in gamete production had the effect of increasing the growth rate and survival of the resultant zygote. Parker et al. (1972) modeled the effects of these two selective forces on a hypothetical ancestral population of externally fertilizing organisms in which gamete fusion occurred at random. They assumed that there was natural variation in gamete size and that all gametes could fuse; there were no "male" or "female" gametes. Individuals with the highest productivity would be favored because they would have the higher reproductive rate, leading to the evolution of increasingly small gametes (curves 1–3 in fig. 1.2). At some stage (curve 3 in fig. 1.2) disruptive selection can occur because the effects of gamete size on zygote fitness become important; individuals producing the largest gametes will be favored because of the fitness of resultant zygotes, while those producing the smallest gametes continue to be favored because of their productivity (curves 4 and 5 in fig. 1.2). Once individuals producing either small or large gametes occur in the population, selection should then favor disassortative fusion. Because of the disadvantages associated with small zygotes, there will be strong selection against small gametes that fuse with other small gametes because the resulting zygotes will not survive without provisions for development. Large gametes that fuse with large gametes will not be disfavored and could possibly be favored. However, it is expected that there will be some optimal level of resources beyond which the zygote will gain no additional benefit, so that selection for assortative fusion among large gametes may not be as strong as selection favoring disassortative fusion by small gametes. Moreover, because of the higher productivity of individuals producing small gametes, the rate of evolutionary change toward disassortative fusion by small gametes will be greater than the rate of change toward assortative fusion by large gametes. It is easy to envisage an evolutionary conflict between large and small gametes because in one sense small gametes essentially parasitize the resources of large gametes. However, any evolutionary change in large gametes to prevent fusion by small gametes could be rapidly countered by changes in small gametes because of their greater productivity and thus evolutionary potential.

Once anisogamy is established in the population, additional selection pressures will arise that both maintain and further promote anisogamy. The numerical predominance of protosperm coupled with selection for disassortative fusion will result in intense competition among protosperm for access to the limited supply of protoeggs. This competition will favor the evolution of traits in protosperm that increase their success in encountering and penetrating the limited supply of protoeggs, traits such as tails that facilitate motility

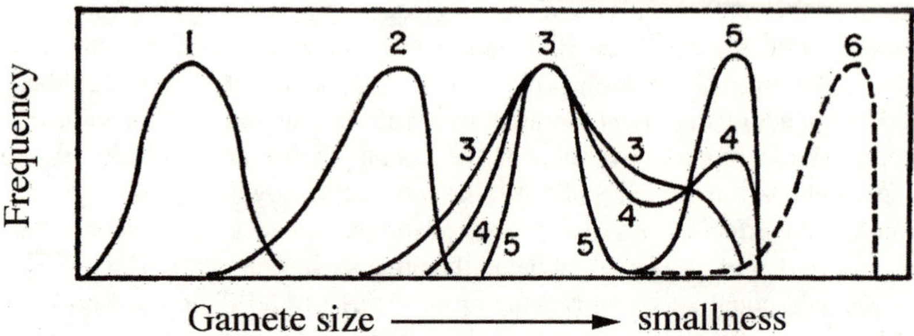

Figure 1.2 Stages in the evolution of anisogamy starting from an initial distribution of isogamete-producing variants (curve 1). There is drive toward greater productivity and a subsequent decrease in isogamete size (curve 2), but as zygote size becomes relatively important for fitness, disruptive selection begins to favor establishment of large- and small-producing morphs in stable frequency. Curves 3, 4, and 5 show progressive stages in the establishment of sperm producers and in the stabilization of ovum producers (bimodal distribution 5). Further reductions in sperm size (broken curve 6) would occur later so that sperm-sperm fusions may be lethal. (From Parker et al. 1972)

and acrosome reactions that facilitate rapid penetration of the exterior walls of protoeggs. Moreover, individuals that produce more protosperm will be favored because they will have an increased probability that some of their protosperm will encounter available protoeggs. Given that there is likely to be a tradeoff between gamete size and number, competition between proto sperm will lead to the evolution of increasingly small and numerous sperm. Parker et al. (1972) argued that for eggs motility would become redundant because the time taken to achieve fusion would be decreased by the strong selection on sperm to achieve fusion.

Parker (1982) extended his models to a system in which the mode of reproduction shifted from external fertilization to internal fertilization and showed how even a weak degree of competition between the sperm of different males could maintain and promote anisogamy. Imagine a population of organisms that evolved internal fertilization so that sperm were no longer faced with competition from other sperm in the fertilization environment. In the complete absence of competition, selection could favor an increase in sperm size so that sperm began to contribute some nourishment to zygotes. However, Parker (1982) showed that even with a low incidence of multiple mating by females, so that the sperm of different males were sometimes in competition, selection would prevent increased sperm size because the production of larger sperm would compromise productivity and thus the probability of winning fertilizations. Thus, sperm should be maintained small and numerous because the more sperm ejaculated the greater the chance of achieving fertilizations in competition with other sperm (Parker 1982).

Alternative selection pressures favoring small sperm size have been proposed, such as the avoidance of cytoplasmic organelle and/or parasite transmission (Hurst 1990; Hurst and Hamilton 1992; Randerson and Hurst 1999). Recent work by Levitan (1996) suggests that in free spawners, the ancestral state envisaged by Parker et al. (1972), females can be sperm limited despite the greater gamete productivity of males (see also Levitan and Petersen 1995; Yund 2000). Under sperm limitation, there can be strong selection for increased egg size because individuals with larger eggs have a higher probability of fertilization. Thus, selection appears to favor increasingly large and few eggs. Levitan (1996) suggests that sperm competition could be a derived selection pressure, resulting from the evolution of internal fertilization in response to sperm limitation. However, the selective pressures generated by sperm limitation and sperm competition are more likely extremes of a continuum, the importance of each depending on habitat and spawning conditions.

Alternative arguments for why sperm are so numerous have also been made. During the meiotic phase of spermatogenesis, homologous chromosomes align to facilitate the exchange of DNA during crossover and recombination (chiasmata). Cohen (1967, 1969, 1973) argued that errors occuring during chiasmata may result in inviable sperm. Species in which males have high chiasma frequencies would have to produce large numbers of sperm because only a small proportion of them would be viable. He pointed out that there is a positive association between chiasma frequency and gamete redundancy (number of sperm ejaculated per egg fertilized) that is not predicted by selection through sperm competition. However, Manning and Chamberlain (1994) show how such a relationship can be reconciled with sperm competition theory. They assumed that within ejaculates sperm vary in their competitiveness; essentially there is interejaculate competition between sperm that is akin to sib competition. Further, they assume that deleterious mutations arising during sperm production reduce an individual sperm's competitiveness. When there is sperm competition between males, the male with the greatest proportion of highly competitive sperm within his ejaculate will be favored. Chiasmate males should be favored over achiasmate males because they should have greater variation in sperm competitiveness so that, on average, there will be a higher probability that their ejaculate will contain one or more highly competitive sperm. Thus recombination should be favored by sperm competition. Furthermore, in order to produce one or more highly competitive sperm, the minimum number of sperm actually produced will be an increasing function of the number of genes that affect sperm competitiveness and the recombination rate of these genes; thus we should expect to see a relationship between chiasmata frequency and gamete redundancy under sperm competition (Manning and Chamberlain 1994).

1.3 Sperm Competition and Sexual Selection

Darwin (1871) viewed females as being fundamentally monogamous, so that the process of sexual selection acted prior to pairing. Smith (1984) suggests that this belief led Darwin to overlook the potent selective forces of sexual selection that arise due to competition among sperm after insemination. Birkhead (1995) suggests that Darwin (1871) may have been aware of the potential for sperm competition since he noted an example of mixed paternity within the brood of a pair of geese kept by his cousin, concluding that an extrapair male had "prepotent powers" over the other because of a bias in paternity to that male. Birkhead (1995) suggests that the social mores of the late 1800s prevented Darwin from pursuing the notion of sexual selection continuing after pairing. It was Parker (1970e) who first conceptualized sperm competition as a process of sexual selection, defining sperm competition as the competition between sperm of two or more males for the fertilization of ova. Parker (1970e) argued that copulation with internal fertilization may have originated via a selective advantage to males better able to position their gametes closer to the site of fertilization than other males. This view contrasts with that of Levitan (1996), who suggests that internal fertilization would be naturally selected due to sperm limitation in ancestral external fertilizers. Nevertheless, Parker (1970e) explicitly stated that sexual selection would not preclude a naturally selected advantage to internal fertilization, only that the sexual selection pressure of competition between spawning males was likely to be stronger. Moreover, he noted that copulation could not remove sexual selection via sperm competition because, when a female mates with a second male before sperm from her first mate are exhausted, the sperm from the two males will still be in direct competition for the fertilization of ova.

Multiple Mating by Females

A prerequisite for sexual selection via sperm competition, then, is multiple mating by females, the phenomenon largely ignored by Darwin (1871). In reality, multiple mating by females is remarkably common and, while the selective advantage of multiple mating for males is obvious, the benefits for females are not; male fecundity is a function of the number of mates he acquires while female fecundity need not be so (Bateman 1948; for a review on insects see Ridley 1988; Arnold and Duvall 1994). A number of hypotheses, which need not be mutually exclusive, have been proposed to explain the evolution of multiple mating by females (table 1.1).

Perhaps the most obvious selection pressure for female remating is the necessity to acquire viable sperm for fertilization. Unless females store sperm within their reproductive tracts, they will need to remate at each re-

Table 1.1
Hypotheses for the occurrence of multiple mating by females.

1. *Sperm replenishment:* Females remate to top up sperm stores depleted by previous ovipositions, replace inviable sperm, or otherwise ensure fertility.
2. *Material benefits:* Females remate to acquire resources controlled by males, such as nesting sites, food resources, or protection from conspecifics and/or heterospecifics.
3. *Genetic benefits:* Females replace sperm of previous mate with sperm from a genetically superior mate, encourage competition between sperm to ensure fertilization by sperm of high quality, or ensure genetic diversity in offspring.
4. *Convenience:* Females remate to minimize costs of harassment from males.
5. *Correlated evolution:* Females remate because of a correlated response to sexual selection on multiple mating by males.

Sources: 1. Thornhill and Alcock 1983; Gromko et al. 1984a; Sheldon 1994; 2. Borgia 1981; McLain 1981; Kaitala and Wiklund 1994; Tsubaki et al. 1994; 3. Williams 1978; Walker 1980; Madsen et al. 1992; Keller and Reeve 1995; Zeh and Zeh 1996, 1997; 4. Thornhill and Alcock 1983; Parker 1984; Rowe 1992; 5. Halliday and Arnold 1987.

productive event. Of course sperm storage in insects is almost ubiquitous, and it is often stated that females need only mate a single time in order to obtain enough sperm to fertilize the ova they will produce during their lifespan. While this may sometimes be true, it is not generally so. In Ridley's (1988) comparative analysis of mating frequency and fecundity, of 48 species for which there were data, the females of 58% ran out of sperm if not allowed to remate. Moreover, in almost all insects studied, fecundity was increased through multiple mating. It may be that sperm degrade during long-term storage. It is also the case that males may often be impotent, either because they are inefficient at insemination or because they have inviable sperm, so that multiple mating by females will be favored in order to guard against infertility (e.g., Labine 1966; Loher and Edson 1973; Simmons 1988b; see review of Ridley 1988;). Female fecundity may also be enhanced through multiple mating because of material benefits donated by males. For example, in many insects males transfer nutrients at mating, either with the ejaculate (Gwynne 1984a; Simmons and Gwynne 1991; Kaitala and Wiklund 1994) or as a prey item during copulation (Thornhill 1976b; Svensson and Petersson 1987), which can provide direct benefits to the female (see review by Vahed 1998b). Mating can even enhance female foraging efficiency (Rubenstein 1984; Wilcox 1984) or generally reduce harassment by other males, thereby facilitating female reproductive activities such as oviposition (Borgia 1981; Tsubaki et al. 1994).

Less obvious benefits for multiple mating have been proposed in the form of genetic benefits derived from the acquisition of sperm from males superior in quality to a female's previous mating partner; essentially females remate in order to gain indirect benefits of mate choice (e.g., Walker 1980;

Thornhill and Alcock 1983). A number of studies of birds have shown that females paired with males having inferior expression of secondary sexual traits are more likely to engage in extrapair copulations so that their nests contain more extrapair young. Males that obtain extrapair young tend to have superior secondary sexual traits. Moreover, extrapair young have been shown to be of greater viability, indicating that by remating with more attractive males females gain indirect fitness benefits for their offspring (e.g., Kempenaers et al. 1992; Graves et al. 1993; Sundberg and Dixon 1996; see review by Møller 1998). Females might also benefit from multiple mating by producing genetically diverse offspring because this would encourage sib competition (Pease 1968; Maynard Smith 1978; Williams 1978; Walker 1980). Madsen et al. (1992) found that multiple mating by female adders can reduce the incidence of stillborn young and suggest that under a regime of multiple mating, sperm competition would prevent all but the most viable sperm from gaining fertilizations. If sperm viability were positively associated with offspring viability then multiple mating could enhance the genetic quality of offspring, although such evidence is currently lacking (Parker 1992b). Zeh and Zeh (1996; 1997) have recently argued that female reproductive success is likely to be under considerable threat from genetic incompatibility, arising from a variety of factors such as cellular endosymbionts or maternal effect lethals. They argue that multiple mating by females may in general serve to increase the genetic diversity of gametes available for fertilization and potentially minimize the risks of fertilization by genetically incompatible sperm.

Alternatively, females may remate simply because the costs of repeated harassment from males outweigh the costs of additional matings (Parker 1984). Such convenience polyandry (Thornhill and Alcock 1983) has been noted in water striders where females generally refuse copulations from mounted males because mating incurs costs in the form of reduced foraging speeds and increased risk of predation (Rowe 1992, 1994). Nevertheless, when the density of males in the environment is high, the costs associated with continual mate rejection for females outweigh the costs of mating so that females become less reluctant to mate (Rowe et al. 1994). In general, the resolution of sexual conflict over mating is likely to reflect the relative costs and benefits of remating for females, since male interests will always be best served by copulation.

Finally, it has been proposed that female remating may simply be a nonadaptive by-product of selection for multiple mating in males (Halliday and Arnold 1987). The hypothesis predicts a genetic correlation between mating frequency of males and females. Female remating frequency has been shown to have a genetic basis in the field cricket *Gryllus integer* (Solymar and Cade 1990) and in *Drosophila melanogaster* (Gromko 1992). In *D. melanogaster*, a correlated response in males due to selection for female remating speed has been documented for some populations but not others. Nevertheless, female *D. melanogaster* are known to gain direct benefits from multiple mating, and a genetic correlation between male and female remating frequency is expected under both adaptive and nonadaptive hypotheses (Gromko 1992). Selection experiments on birds (Cheng and Siegel 1990) have failed to provide support for the genetic correlation between male and female mating frequency envisaged by Halliday and Arnold (1987).

It is often implicit in discussions of sperm competition that selection acts on males as a consequence of the evolution of multiple mating by females. Although multiple mating is a prerequisite for sperm competition, each of the hypotheses for multiple mating outlined in table 1.1 holds different implications for sperm competition. Sperm competition is not expected to arise where selection favors the evolution of multiple mating by females in the context of sperm replenishment. If a female refrains from copulating with a second male until such time as her sperm stores are exhausted or inviable, then neither the first nor a subsequent male copulating will be in competition for available ova. For sperm competition to occur, the sperm of two or more males must coexist within the reproductive tract of the female. Sperm competition is more likely to occur due to convenience polyandry or when females remate for material benefits, such as access to oviposition or nesting sites, and where sperm are not depleted between bouts of reproduction. The interplay between sperm competition, material donations by males, and female multiple mating may, however, be more complex than it would first seem. For significant male investment in reproduction is expected to arise only when there is a high probability that the investing male will father the offspring to which he contributes (Xia 1992; Westneat and Sherman 1993; see review by Wright 1998). Thus, material benefits obtained by females may arise subsequent to the evolution of mechanisms of paternity assurance that reduce sperm competition and/or female remating. The genetic benefits hypotheses propose that female remating is favored specifically because sperm competition occurs, predicting that sperm competition and female multiple mating should coevolve.

Sperm Storage

The second prerequisite for sexual selection via sperm competition is that the sperm of two or more males coexist within the reproductive tract of the female at the time of fertilization (Parker 1970e). Insects thus sustain high levels of sperm competition, because of their tendency to store and maintain sperm internally, in specially adapted storage organs, often referred to as spermathecae. The morphology of spermathecae varies widely across species (see fig. 1.3). In some, the spermatheca consists of a simple blind tube (Gregory 1965) while in others it is a spheroid sac capable of a considerable degree of expansion (Simmons 1986). In many Coleopterans and Dipterans, the spermathecae are heavily sclerotized and thus of fixed capacity (Parker et

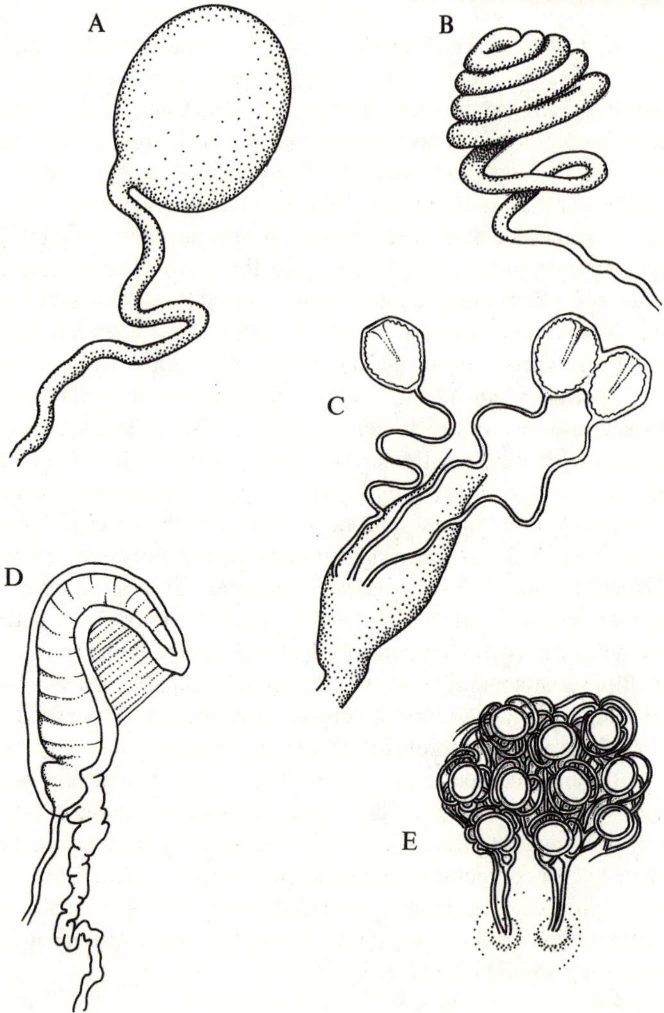

Figure 1.3 The diversity of sperm storage organs in insects: (A) the single membranous spermathecal sac of the field cricket *Gryllus bimaculatus* (Simmons 1986); (B) the coiled tubular spermatheca of the locust *Locusta migratoria migratorioides* (Gregory 1965); (C) the reproductive tract of the yellow dungfly *Scatophaga stercoraria*, in which three chitinous spermathecae (one singlet and one doublet) are connected to the bursa copulatrix via independent spermathecal ducts (Hosken et al. 1999); (D) the chitinous spermatheca of the beetle *Chelymorpha alternans* with its spermathecal muscle (Rodríguez 1994); (E) ten spermathecae of the earwig *Diplatys macrocephalus*, which share a complex of four interconnecting spermathecal ducts (Popham 1965).

al. 1990; Eady 1994b). Some species have a single spermatheca and others three separate spermathecae, as is typical in the Diptera (fig. 1.3). Some earwigs can have as many ten separate spermathecae, which share four spermathecal ducts that serve as entrance from and exit to the reproductive tract (Popham 1965). Often, sperm are stored in sites other than the spermathecae. In fruit flies, for example, sperm are stored in both the spermathecae and the seminal receptacle (Fowler 1973), while in damselflies the bursa copulatrix serves as a secondary site for sperm storage (Siva-Jothy and Hooper 1995). Nevertheless, it seems that the spermathecae are the sites where sperm are actively maintained by the female. Specialized cells provide both nutrition and a constant ionic environment for the sperm (Davey 1985a). Glandular tissue is closely associated with the spermatheca and can either consist of a layer of cells surrounding the spermatheca itself, or may be more highly differentiated into a spermathecal accessory gland (Davey 1985a). These glandular cells secrete a mucopolysaccharide that appears essential for the survival of sperm in the spermatheca. Experimental removal of accessory glands, or inhibition of secretory cell function, results in the gradual death of stored sperm (Davey and Webster 1967; Villavaso 1975b). Thus, sperm are maintained in a fertile state for considerable periods, as long as thirty years in some ants (Pamilo 1991d).

The numbers of sperm that can be stored by females varies across species and is likely to have direct implications for sexual selection via sperm competition. Where the sperm storage organs have a fixed storage capacity, the probability of temporal overlap between the sperm of two or more males should be dependent on the numbers of sperm transferred by the male and stored by the female at copulation. Eberhard (1996) reviewed the literature on ejaculate size and sperm storage capacities for species in which the spermathecae are sclerotized structures of fixed capacity. Of the 39 species of insect for which there were data, in 78% the male delivered an ejaculate that contained fewer sperm than the female could potentially store so that there would be a strong probability of the coexistence of ejaculates from different males and thus sperm competition following remating by the female. The remaining 22% of species had ejaculate sizes that completely filled the female's spermathecae, perhaps reducing the potential for overlap of ejaculates and sperm competition. Insect spermathecae are not uniformly of fixed capacity. Some insects have expandable spermathecae that can store ejaculates from multiple matings (e.g., Simmons 1986). The storage of multiple ejaculates will greatly increase the selective pressures arising from sperm competition. These simplified observations are made to stress the importance of the coexistence of successive ejaculates in generating selection via sperm competition. In reality, the morphology of the female's spermathecae may have more complex implications for the products of selection via sperm competition, and these will be discussed again in chapter 2 of this volume.

Sexual Selection via Sperm Competition

Parker (1970e) recognized two opposing selection pressures on males that would arise as a direct consequence of sperm competition. On the one hand, selection should favor traits in males that allow them to gain the majority of fertilizations when mating with previously mated females. On the other hand, when the probability of female remating is high, selection should favor adaptations in males that enable them to avoid or reduce subsequent competition from the sperm of another male. Thus, adaptations should arise for both engagement in and avoidance of sperm competition. Sexual selection via sperm competition has the potential to act at all levels of sexual interaction, such as behavioral interactions between males and females, the genitalic and secondary sexual morphology of males and females, the physiology of male seminal secretions and female reproduction, and the behavior and morphology of gametes themselves.

Prout and Bundgaard (1977) provided the first population genetic models of selection via sperm competition. Using a one-locus, two-allele model in which three diploid male genotypes conferred different abilities to preempt the sperm stored from previous males, they showed how the degree of directional selection observed in natural populations of *D. melanogaster* could drive alleles conferring superior preemption abilities to near fixation in as little as 40 generations. Nevertheless, empirical data showed that variability was maintained in the population, suggesting that some form of balancing selection must oppose directional selection for increased sperm preemption. Prout and Bungaard (1977) noted that selection due to sperm competition gave rise to frequency-dependent selection in their model and that balanced heterosis could maintain variation in sperm preemption ability. They concluded that it was entirely possible that the sperm preemption characteristics of a given species would be shaped by competition among males as envisaged by Parker (1970e). More recently, Prout and Clark (1996) have extended the original models of Prout and Bungaard (1977), including antagonistic pleiotropic effects of alleles conferring superior sperm preemption abilities on male mating success and female fecundity. The new models confirm that stable polymorphisms in sperm competition abilities can exist given the frequency-dependent properties arising from the interactions of different male genotypes.

It is clear that adaptations arising from sperm competition will be in conflict; there will be an evolutionary arms race involving adaptations for preempting stored sperm and for the avoidance of preemption by other males (Parker 1984). Where an adaptation arises in males for successfully avoiding sperm preemption, selection should favor mating strategies in which males reject matings with previously mated females. Selection for the avoidance of sperm preemption will in turn be relaxed so that, where costly, sperm preemption avoidance mechanisms will again disappear, making previously mated females once more a valuable reproductive resource. Parker (1984) used a game theory analysis to determine whether there could be an evolutionarily stable level of armaments in sperm preemption and its avoidance. He concluded that the only evolutionarily stable strategy (ESS) would be a pair of strategies in which males invest in both sperm preemption and its avoidance, with the magnitude of investment in each adaptation depending on their relative cost/benefit functions (Parker 1984). In fact, this is exactly what we see across animal taxa (Smith 1984; Birkhead and Møller 1992, 1998a).

Sexual selection via sperm competition has led to a rapid and divergent evolution of traits that function in sperm competition and its avoidance. Parker's (1970e) original exposé of sexual selection via sperm competition noted the occurrence of varying degrees of sperm preemption across insect taxa and listed a number of reproductive traits that he believed to be adaptations for the avoidance of sperm preemption (table 1.2). Here, it is sufficient to note that selection is proposed to act on variation in male fertilization and thus reproductive success. Any variation in behavior (such as the tendency to remain with the female after copulation), physiology (such as the chemical composition of the ejaculate that enhances the success of an individual male in gaining fertilizations), or morphology (such as genitalic structures that deliver sperm closer to the site of fertilization or manipulate previously stored sperm of rivals) will become subject to selection via sperm competition. Genes responsible for favored traits will rapidly spread to fixation and the behavioral, physiological, or morphological trait will become established in the population. Parker (1970e) thus viewed traits such as copulation, mate guarding, and chemicals in the seminal fluids that inhibit female sexual activity as adaptations that allow males to monopolize females, or, more specifically, their eggs (table 1.2). Parker (1970e) also noted that sexual selection via sperm competition would act on individual sperm; those that can outcompete others for access to ova, for example through greater motility or longevity, should confer an advantage on the male producing them so that adaptations in sperm morphology and/or behavior will also spread in the population. Such traits are best viewed as adaptations for the engagement in sperm competition, and distinct from those that function in the avoidance of sperm competition (table 1.2). Selection for enhanced competitive ability of ejaculates will not arise unless ejaculates coexist within the reproductive tract of females and compete directly for fertilizations. When adaptations arise that successfully avoid sperm competition, selection on the competitive ability of ejaculates will be relaxed or absent (Simmons and Siva-Jothy 1998). For heuristic purposes, I shall recognize two categories of adaptation in this volume, those for the avoidance of sperm competition and those for the engagement in sperm competition. Nevertheless, it may sometimes be difficult to classify traits in terms of such "defensive" or "offensive" functions; it is perhaps better to recognise a continuum (table 1.2). At one ex-

Table 1.2
Adaptations arising for the avoidance of and engagement in sperm competition.

Avoidance of sperm competition (defensive)
1. Physical barriers to female remating
2. Chemical barriers to female remating
3. Pre- and postcopulatory mate guarding
4. Avoidance of takeovers
5. Genitalic morphology
6. Copulation duration
7. Copulation frequency
8. Testis size and sperm production
9. Sperm characteristics
Engagement in sperm competition (offensive)

treme, traits are clearly defensive; traits such as copulatory mate guarding and sperm removing structures on the genitalia prevent the overlap of ejaculates and thus sperm competition. At the other extreme, traits such as sperm motility and longevity are clearly adaptations for gaining competitive superiority in the race for fertilizations. Some traits, however, such as testis size, copulation frequency, or copulation duration, increase the numbers of sperm transferred to the female. Increased sperm numbers could be viewed as an adaptation for the engagement in sperm competition if males with the greatest numbers of sperm at the site of fertilization have a competitive advantage. They could also be viewed as adaptations for the avoidance of sperm competition if the sperm from one male were in such high numbers relative to rival sperm that rival sperm were unable to compete for fertilizations. The interpretation of such traits requires knowledge of the patterns of sperm transfer and storage (Simmons and Siva-Jothy 1998).

The forms of adaptation to sperm competition described above might previously have been viewed as primary rather than secondary sexual characteristics. Recently, however, it has become clear that structures such as male genitalia can be subject to elaboration through sexual selection beyond their primary mating function (Waage 1979b; Eberhard 1985; Arnqvist 1997a), so that the distinction between primary and secondary sexual traits might seem redundant (Andersson 1994). Darwin (1871) himself noted that sexual selection would often lead to the exaggeration of traits, such as eyes or ears, that aid in finding members of the opposite sex. Although his principal reason for distinguishing sexual selection from natural selection was to explain the evolution of elaborate ornaments that seemed unnecessary for reproduction per se, elaboration of any trait beyond its naturally selected optimum can arise under sexual selection (Richards 1927; Thornhill and Alcock 1983; Gwynne and Bailey 1999).

Simmons and Parker (1996) formally examined whether sperm competition could influence the intensity of sexual selection acting on male secondary sexual traits. As outlined in section 1.2, the principal factor controlling the operation of sexual selection is the time cost associated with reproduction; selection via competition and mate choice will act predominantly on the sex that has the lower cost of reproduction (Clutton-Brock and Parker 1992; Johnstone et al. 1996; Parker and Simmons 1996). Simmons and Parker (1996) showed that sperm competition could not directly influence sexual selection operating on secondary sexual traits because whatever form it takes males must pay the time costs associated with mating in order to engage in sperm competition and thus reproduction, and it is this time cost that will determine the ratio of sexually active males to females (the OSR) and the intensity of sexual selection (Clutton-Brock and Parker 1992; Parker and Simmons 1996). However, Simmons and Parker (1996) noted that sperm competition could influence sexual selection indirectly through adaptive changes in male investment. First, as sperm competition risk increases, males are predicted to increase their investment in sperm production and thus competition (Parker 1998), which must be paid for by an increase in time out from mating activity for sperm replenishment. Thus, adaptive changes in investment in sperm competition should be accompanied by reductions in the ratio of the number of sexually active males to females, leading to a reduction in the intensity of sexual selection. A second adaptive change generated by sperm competition can arise when male time costs relate mainly to paternal care, rather than sperm replenishment. It is predicted that decreased confidence of paternity should be associated with decreased levels of paternal care (Wright 1998). Thus, increased risk of sperm competition should be associated with both an increase in investment in sperm production as noted above, and a concomitant decrease in levels of paternal care that will increase the amount of time available for additional mating activities. In paternally caring species, therefore, the two effects act in opposite directions. Nevertheless, the net effect should be a decrease in time costs of reproduction and an increase in the number of sexually active males, because time costs of paternal care are likely to represent a substantially higher proportion of the male's time budget than sperm replenishment. Thus, across species of birds, we find a negative association between the degree of extrapair paternity (an index of risk of sperm competition) and the contribution of males to the feeding of offspring (Møller and Birkhead 1993; Møller and Cuervo 2000), and a positive association between extrapair paternity and plumage brightness (Møller and Birkhead 1994), as might be expected from the predicted increase in intensity of sexual selection that will accompany reductions in paternal care.

Parker (1970e) viewed sperm competition as an extension of sexual selection via intermale competition. However, it is becoming increasingly clear that females have the potential to exploit processes of sperm transfer, storage, and utilization to bias paternity of offspring in favor of preferred males

(Lloyd 1979; Thornhill 1983; Simmons 1987b; Eberhard 1996). On the one hand females may copulate more frequently with preferred males or time their copulation behavior with different males so that preferred individuals have a greater probability of fertilizing eggs. In this case, it is not sperm competition per se that influences a male's fertilization and reproductive success. Males with more attractive secondary sexual traits need not be superior competitors in sperm competition. Rather, it is the behavior of the female that allows the sperm from males with more attractive secondary sexual traits to be in the right place at the right time (see review by Møller 1998). In reality, by adopting such behaviors, females circumvent sperm competition between males because the sperm of different individuals are not allowed to coexist in the reproductive tract at the time of fertilization. On the other hand, females may encourage sperm competition through multiple mating so that they have a mixture of sperm in their sperm storage organs, allowing them to choose among potential sires, or allowing competition among the sperm of potential sires, to ensure that only superior males fertilize eggs (Keller and Reeve 1995; Eberhard 1996).

Sperm competition can thus prove a potent sexually selective force acting on male reproductive biology. Insects perhaps sustain higher levels of sperm competition than any other taxa because of their propensity to remate, and because females are preadapted for the storage and maintenance of multiple ejaculates. Parker (1970e) thus used the insects as a model taxon to demonstrate the occurrence of sperm competition and its evolutionary consequences. Since Parker's influential paper, the importance of sperm competition as a sexually selective process has been recognized in most animal taxa, although the insects remain unparalleled in both the diversity of mechanisms involved in sperm competition and its evolutionary outcomes. Insects thus offer our best opportunity to reveal the full extent to which sperm competition can drive evolutionary change. Nevertheless, there is often confusion in the literature concerning what constitutes sperm competition, the mechanisms by which it is brought about, and the selective pressures that can arise from it. This volume offers a synthesis of our current knowledge of sperm competition in the insects. My aim has been to provide a conceptual framework for the evolutionary interpretation of observed processes of reproduction. Although, like Parker (1970e), I use insects as a model, my objectives are not taxon specific. The insects have provided, and will continue to provide, valuable insight into the evolutionary process and I use them to provide a general overview of sexual selection via sperm competition.

1.4 Summary

Darwin recognized that selection could operate through the differential reproductive success of individuals, so favoring traits that are otherwise costly in terms of reduced survival. Sexual selection can act through mating competition and/or female choice and favor the evolution of exaggerated secondary sexual traits. Males and females differ fundamentally in the number and size of gametes produced, so-called anisogamy. An inevitable consequence of anisogamy is sperm competition because small gametes must compete to fertilize the limited supply of large gametes. Whenever the sperm from two or more males overlap at the site of fertilization, there will be sperm competition, which will both maintain anisogamy and generate sexual selection on males for adaptations that increase the probability of fertilization. Multiple mating and sperm storage are thus prerequisites for sperm competition in internally fertilizing species. They are not, however, evidence for sperm competition since sperm storage and remating need not result in the coexistence of sperm from different males within the reproductive tract. Sperm competition is expected to favor opposing adaptations in males that prevent future males from mating with a female or preempt the sperm stored by females from previous matings. When sperm from different males temporally and spatially overlap at the site of fertilization, adaptations, such as variations in sperm morphology or longevity, can arise in sperm so as to enhance their success in competition for ova. It is clear that sperm competition can also be a potent sexually selective pressure acting on male reproductive morphology, physiology, and behavior. Furthermore, sperm competition can indirectly influence sexual selection acting on male secondary sexual traits, through adaptive changes in male investment in sperm production and parental care. Sperm competition is generally envisaged as an extension of intrasexual selection. Nevertheless, female behavior can bias male fertilization success so that intersexual selection can also favor adaptations in males in the context of sperm competition. Insects sustain high levels of sperm competition because of the strong remating tendencies of females and, perhaps more importantly, because females store and maintain sperm internally within specially adapted sperm storage organs. The considerable evolutionary diversity in mechanisms of sperm transfer, storage, and utilization in insects makes them unique models for exploring the evolutionary consequences of sperm competition.

2 Sperm Utilization: Concepts, Patterns, and Processes

2.1 Introduction

The first observations of multiple paternity in insects were those of Nonidez (1920) who noted that, when female *Drosophila melanogaster* mated with two males in close succession, they produced offspring that were sired by both males. At the time of his review, Parker (1970e) presented evidence of multiple paternity, and thus the potential for sexual selection via sperm competition, in sixteen species of insect from just five orders. We have since seen an exponential growth in studies of paternity outcomes in insects, perhaps not surprisingly given that paternity outcome is the trait most closely associated with male fitness. A recent review gave data on paternity outcomes following multiple matings for 109 species of insect from ten orders (Simmons and Siva-Jothy 1998). Many species have been studied because they represent insects of economic importance, and the realization that sterile male release could be an effective biocontrol has resulted in studies of paternity outcome when females mate with males that vary in their sterility status. Paternity is generally determined empirically as the proportion of offspring sired by the last male to mate with the female. In most studies, this is equal to the proportion of offspring sired by the second male to mate in controlled double mating trials, hence the widely used statistic P_2 (Boorman and Parker 1976).

Like most branches of science, the study of sperm competition has accumulated a glossary of terms that have never been formally defined. The common interests in paternity outcome of evolutionary biologists and applied scientists, coupled with a lack of clear definitions, has frequently led to confusing and incorrect usage of terminology throughout the sperm competition literature. Terms such as sperm precedence and sperm displacement, for example, are often used synonymously to describe paternity outcomes when observed values of P_2 are high. Apart from implying very different mechanisms of sperm transfer, storage, and utilization (see section 2.2), associating observed values of P_2 with specific mechanistic processes can be misguided. Sperm precedence can be reflected by either very high or very low values of P_2, depending on which of the two males gains precedence. More importantly, while high values of P_2 may be indicative of sperm displacement, they may equally reflect low rates of sperm mixing within the female sperm storage organ(s), and thus sperm precedence, or they may reflect the fact that the female has no sperm remaining in storage from her previous mating partner at the time of the second male's copulation. In general, P_2 is a greatly overinterpreted statistic (Simmons and Siva-Jothy 1998).

This chapter opens with a series of working definitions for terms that have arisen in the study of sperm competition, followed by a critical appraisal of the methods used for quantifying sperm utilization. It then provides a comprehensive review of the occurrence and degree of mixed paternity following multiple mating in nonsocial insects, before offering a mechanistic approach for the elucidation of processes by which observed patterns of paternity can arise following copulations with previously mated females. The chapter concludes with a discussion of the importance of understanding the mechanisms of sperm transfer, storage, and utilization for the evolutionary interpretation of nonrandom paternity and of traits believed to be the products of selection via sperm competition.

2.2 Classification and Definition of Terms Used in the Sperm Competition Literature

Sperm Competition

Sperm competition is a term that has often been stretched beyond its conceptual boundaries. Sperm competition, for internal fertilizers such as insects, is defined as "the competition within a single female between the sperm from two or more males for the fertilization of the ova" (Parker 1970e). To encompass external fertilizers, this definition has since been rewritten as "the competition between the sperm from two or more males for the fertilization of a given set of ova" (Parker 1998).

In its strictest sense, sperm competition does not occur unless there is a temporal and spatial overlap of ejaculates from competing males. Nevertheless, sperm competition is implicitly, and sometimes explicitly, linked to the value of P_2; when P_2 is high, sperm competition is said to be intense. However, P_2 will be high whenever sperm from previous matings are lost from the female's sperm storage organ(s), due to usage, inefficient storage, or digestion by the female. In these situations, there is no temporal overlap of ejaculates from competing males and no sperm competition. That sperm depletion/loss between matings is common in insects was clearly demonstrated in Ridley's (1988) comparative analysis of mating frequency and fecundity. The work of Tsubaki and Yamagishi (1991) and of Yamagishi et al. (1992) shows quite clearly the error in concluding that high values of P_2 are necessarily associated with sperm competition. They examined the patterns of sperm storage and utilization in the melon fly *Dacus cucurbitae*. Tsubaki and Yamagishi (1991) found that the number of sperm contained within the sperm storage organ of female *D. cucurbitae* decreased exponentially with time since mating. In double-mating trials, P_2 was found to increase with the interval between first and second matings, yielding values of paternity for the second male to mate that were not significantly different from those predicted, given that the first male would have had an increasingly reduced number of sperm in the female sperm stores, relative to the second male, due to sperm loss. Thus, increasing values of P_2 with increasing remating interval actually reflect a decrease in the temporal overlap of sperm from two males, and thus reduced sperm competition. Sperm competition will be at its greatest when there is complete overlap of competing ejaculates; where, on average, intermediate values of P_2 are observed. Differential sperm viability has been linked to male genotype in *D. melanogaster* where P_2 similarly increases with remating interval (Gromko et al. 1984a). Loss of first male sperm from the female reproductive tract appears to be the most important parameter generating nonrandom paternity in birds (Lessells and Birkhead 1990; Birkhead et al. 1995; Colegrave et al. 1995). In cases such as these, the occurrence and intensity of sperm competition will depend very much on the timing of copulation and oviposition events.

Boorman and Parker (1976) noted that competition would occur between the sperm contained within a given male's ejaculate, as well as between the ejaculates of different males, and that both phenomena could rightly be considered as sperm competition. They noted, however, that the evolutionary consequences of inter- and intraejaculate sperm competition would differ, and stressed that it was interejaculate sperm competition that was directly relevant to sexual selection. The terms ejaculate and sperm are often used to describe the same process of competition between males although the ejaculate contains both sperm and other components, such as the seminal fluids, that may sometimes have an indirect influence on sperm competition.

Finally, in a more general sense the term sperm competition can be used to identify a specific evolutionary pressure. Thus, adaptations that currently serve to avoid sperm competition (see table 1.2) may have arisen due to the selective pressure of sperm competition in the evolutionary past, even though sperm competition may no longer occur.

Sperm Mixing

This is the mixing of sperm from two or more males within a single female, usually within her sperm storage organs, that results in the conditions necessary for sperm competition. The degree of sperm competition will depend largely on the degree of sperm mixing. Sperm competition will be at its strongest when there is rapid and random mixing of sperm derived from different males in the female's sperm storage organ(s), and at its weakest when sperm mix slowly or nonrandomly so that the sperm from different males remain clumped. All things being equal, the observed outcome should be that when two males copulate, on average, each male gains 50% of the fertilizations. This scenario has been termed a "fair raffle" (Parker et al. 1990). However, in the case of random mixing, sperm competition is likely to favor males that transfer sperm of greater competitive ability so that other factors may generate what Parker (1990) called a "loaded raffle" in which for the same numbers of sperm, some males obtain more fertilizations than others.

Sperm Precedence

This is the nonrandom utilization of sperm from a particular male when the sperm of two or more males overlap within a single female, also termed "sperm predominance" (Gromko et al. 1984a). The last sperm to enter the female's sperm storage organ(s) may be the first to leave, yielding a high value of P_2. Alternatively, it may be the case that the first sperm to enter the sperm storage organ(s) are the first to leave, yielding a low value of P_2. It is important to note, however, that nonrandom fertilization following multiple matings does not indicate the occurrence of sperm precedence (section 1.1). When mechanisms of sperm precedence exist, the spatial overlap of ejaculates will be minimal so that there will be little or no sperm competition. Short-term sperm precedence can result from slow and/or nonrandom mixing of sperm within the female sperm storage organ(s). Thus, while on average each of two males may gain 50% of fertilizations in the long term, at single bouts of fertilization and oviposition, either male may gain complete paternity if his sperm are the closest to the site of fertilization and so take precedence (see, for example, Siva-Jothy and Tsubaki 1989a). There are a variety of mechanisms by which long-term sperm precedence may arise.

Sperm Stratification

Certain sperm can obtain a positional advantage within the storage organ(s) so that they are the first to leave during fertilization. Often, it is the last sperm to enter that are the first to leave so that when two males copulate P_2 approximates to 1.0. Sperm stratification may be a direct consequence of sperm repositioning (e.g., Waage 1984), or the passive consequence of mating order (e.g., Schlager 1960). It refers to the layered disposition of the sperm of several males within a female's sperm storage organ(s). Consequently, sperm precedence can break down if mixing of the sperm storage organ(s) contents occurs with time so that sperm competition can become increasingly important.

SPERM LOADING

The term "sperm loading" was first coined by Dickinson (1986) to describe the phenomenon where males increase the numerical representation of their own sperm, relative to rival sperm, in order to gain an advantage at fertilization. In this case the competitive superiority of one male relies on random mixing of sperm within the sperm storage organ(s) of the female. As such, the term is not distinct from sperm mixing.

SPERM INCAPACITATION

Sperm incapacitation as a mechanism of gaining sperm precedence was first proposed by Silberglied et al. (1984). Sperm incapacitation is the killing and/or inhibition of function of sperm from a female's previous mate by physiological adaptations in the seminal fluids and/or morphological adaptations in the sperm of the copulating male (sperm incapacitation via sperm-sperm interactions has also been termed the Kamikaze sperm hypothesis by Baker and Bellis 1988). Sperm incapacitation would require an incapacitating agent that either was short lived and delivered prior to insemination or had self-recognition capabilities. The latter is particularly unlikely in insects given their poor graft recognition abilities (see Lackie 1986). Nevertheless, recent work with *Drosophila* seminal fluids suggests that sperm incapacitation may be important in generating last-male sperm precedence (Price et al. 1999).

SPERM SELECTION

Sperm precedence can arise due to the nonrandom usage of sperm derived from a particular male when the female has sperm stored from a variety of potential sires. Sperm selection is the extension of female choice at the gametic level and has also been termed female sperm choice (Birkhead 1998). The mechanisms by which females could actively select particular sperm to fertilize their eggs are unclear, and indeed the occurrence of sperm selection itself is equivocal (Simmons and Siva-Jothy 1998). Nevertheless, females have ultimate control over the utilization of sperm within their bodies so that observed patterns of sperm precedence could at least potentially reflect adaptive fertilization strategies of females (Eberhard 1996). Sperm selection is fundamentally different from sperm competition in that sperm selection requires the evolution of traits by which females can circumvent the outcome of sperm competition when there is sexual conflict over which sperm are to be utilized. If there were no sexual conflict over sperm utilization, sperm selection strategies in females would not arise. Sperm selection is also distinct from behavioral means by which females ensure that the sperm of certain males are present at the site of fertilization. Like male combat for access to mates, such behavioral mechanisms are best viewed as conventional tactics of female choice. Rather, sperm selection, like sperm competition, is a continuation of the sexual selective process after insemination has occurred.

Sperm Displacement

This is the spatial displacement of sperm derived from a female's previous mate by the copulating male with the consequence that self sperm need not compete to fertilize ova. Previously, the term sperm displacement has been used to describe two rather different processes; the complete removal of rival sperm from the female sperm storage organ(s) by the copulating male, and the repositioning of rival sperm to a place where they cannot gain access to ova. This dual usage of the term is particularly true in the odonate literature where both processes occur within the same taxonomic group. The two processes do differ, however. Sperm repositioning is a mechanism whereby males achieve sperm precedence over their rivals through the stratification of ejaculates within the sperm storage organ(s) of the female (see above). Sperm competition may ensue with long-term mixing of sperm in storage (Schlager 1960; Siva-Jothy and Tsubaki 1989a). In contrast, sperm removal avoids the overlap of sperm from different males and thus the possibility of sperm competition. For clarity, therefore, I recommend restricting the term "sperm repositioning" to describe the mechanism by which males achieve sperm precedence (see above). The term "sperm displacement" was first coined by Lefevre and Jonsson (1962) to describe the removal of rival sperm from the female sperm storage organs during copulation by male *D. melanogaster*, giving priority for its usage in this context. There are two mechanisms of sperm displacement.

SPERM REMOVAL

Sperm removal involves the direct removal of rival sperm from the female's sperm storage organ(s) prior to ejaculation, facilitated by anatomical adaptations of the males genitalia (e.g. Waage 1979b).

SPERM FLUSHING

This is the indirect removal of previous sperm from the female's sperm storage organ(s) by the incoming ejaculate of the copulating male (e.g., Lefevre and Jonsson 1962; Ono et al. 1989; Parker and Simmons 1991). Sperm flushing will be facilitated largely by the female anatomy; where a single copulation is sufficient to fill the sperm storage organ(s), subsequent ejaculates entering the storage organ(s) will force previously stored sperm to be expelled (section 2.5). Again, the direct consequence of sperm flushing is that there is little overlap of sperm from different males and so little sperm competition.

Sperm Competition Risk

Recent literature has sought to identify traits in males that may have arisen due to sperm competition. Such studies require some objective measure of the relative occurrence of sperm competition and this has been termed, sperm competition risk. Risk is formally defined as the probability (between zero and 1) that females will engage in promiscuous mating activity that will result in the temporal and spatial overlap of the ejaculates from two or more males (Parker et al. 1997; Parker 1998). Thus sperm competition risk is the probability that males, within a given population or species, will engage in sperm competition. Where there are adaptations for the avoidance of sperm competition, such as mechanisms of sperm displacement, promiscuous mating activity by females can often result in complete loss of paternity, rather than increased risk of sperm competition.

Sperm Competition Intensity

The intensity of sperm competition is determined largely by the absolute number of different males engaged in competition for the ova of a single female (Parker et al. 1996; Parker 1998). Thus, once there is female remating and sperm competition, the intensity of that competition will increase as the number of mating partners increases beyond 2. Sperm competition intensity is analogous to the operational sex ratio (Emlen and Oring 1977) at the gametic level; the greater the excess of gametes from different males, the greater the intensity of competition for the limited supply of the female's gametes. The intensity of sperm competition will also depend on the degree of overlap between ejaculates. With sperm displacement, for example, the intensity of sperm competition will be a decreasing function of the proportion of rival sperm displaced prior to, or during, insemination. Likewise, the intensity of sperm competition will decrease with increased loss of sperm from the female's sperm storage organ(s) between matings.

Paternity Assurance

The term paternity assurance was first introduced by Smith (1979) to describe a mechanism by which copulating males ensured that the eggs laid by their mate were fertilized with their sperm. Paternity assurance refers broadly to any adaptation, such as mate guarding or sperm displacement, that arises in males for the avoidance of sperm competition from previous and/or future males.

Confidence of Paternity

Confidence of paternity describes a males average probability of siring offspring following copulation with a given female (Alexander 1974). It is thus congruent with the species average value of P_2 following a double mating. Confidence of paternity has been used as a key variable in understanding the evolution of male parental care (Maynard Smith 1977; Zeh and Smith 1985;

Xia 1992; Westneat and Sherman 1993; Wright 1998). Like sperm competition, confidence of paternity is a term that has often been stretched beyond its conceptual boundaries with the result that studies of the interaction between sperm competition and paternal care are often poorly designed and subject to misinterpretation (Schwagmeyer and Mock 1993; Wright 1998). For example, it has been used to describe a male's "perception" of his likelihood of fertilizing ova, a variable that the empirical worker is unlikely to be able to predict and/or manipulate with any accuracy (see also Kempenaers and Sheldon 1997).

2.3 Quantifying Paternity

Morphological Markers

Quantifying patterns of sperm utilization by females following multiple mating requires genetic markers that can be used to unambiguously assign paternity of offspring to individual males. One such method is the use of morphological markers, such as body or eye color, that are inherited via simple Mendelian ratios. Offspring phenotype can be used to accurately determine which of two males' sperm have been successful in fertilization when homozygous recessive females copulate with both a recessive and a dominant homozygous male.

There are a number of problems associated with the use of genetically determined morphological markers for quantifying sperm utilization in sperm competition studies. Sperm carrying different genotypes can have different fertilizing capacities. At one level, differential fertilizing capacity is of fundamental importance in the study of sperm competition since individual variation in fertilization ability in natural populations will be the focus of selection via sperm competition. However, differential fertilizing capacities between strains or genotypes can arise due to artificial selection imposed during the isolation of the marker. Thus, any patterns of sperm utilization observed after inter- and intrastrain matings may not reflect true patterns of sperm utilization occurring in natural populations of the species in question. Moreover, sperm carrying different genotypes may give rise to zygotes with different egg-to-adult viabilities. Morphological markers are often expressed in adults so that estimates of nonrandom paternity are based on adult progeny counts. Apparent biases in paternity may therefore reflect random fertilization followed by differential offspring survival, rather than nonrandom sperm utilization. This problem will be less acute when markers are phenotypically expressed in early stages of development, but even then differential viability of zygotes, prior to hatching, can still bias sperm utilization estimates.

Studies of sperm utilization in *D. melanogaster* have made use of the many morphological mutants that are available for this species. The data clearly show that values of P_2 obtained with wild-type flies are greater than those obtained using flies with morphological markers; when wild-type males compete against mutant males, wild-type males generally have superior fertilizing capacity (fig. 2.1). The data in fig. 2.1 control for any additional confounding influence of differential zygote mortality (Gromko et al. 1984). They also reveal that reduced levels of last-male sperm use are associated with brood effects, that is, changes in P_2 with successive egg batches, indicating that failure to achieve high values of P_2 most likely results from an inability of mutant males to displace sperm from the female sperm storage organs. There are fifteen other species of insect in which morphological markers have been used to assess patterns of sperm utilization and which also provide data sufficient to assess order effects of wild-type and mutant males (table 2.1). Of these 60% show the same bias toward a wild-type advantage seen in studies of *D. melanogaster*. Most frequently, studies using morphological markers use females of a single genotype, typically homozygous recessives. Such practices ignore the possibility of a female influence over sperm utilization (Eberhard 1996). Indeed, studies of *Tribolium confusum* and *T. castaneum* have examined sperm utilization using both wild-type and mutant females and both show interactions between male and female genotypes in determining paternity (table 2.1) that potentially confound inferences concerning the nature and consequences of sperm competition in natural populations.

The potentially confounding effects of differential mortality are less clearly assessed, and indeed have been largely ignored. Studies of *Oncopeltus fasciatus* and *Plodia interpunctella* both found that mutant strains had lower egg-to-adult viability so that we should expect to see paternity based on adult progeny counts biased toward wild-type males. However, in both studies paternity was biased toward mutant males. The lower mutant viability would suggest that the observed fertilization bias toward sperm from mutant males is in fact underestimated by progeny counts. A bias toward mutant sperm was also apparent in a study of the wasps *Dahlbominus fuscipennis* and *D. littoralis* (table 2.1). It is interesting to note that in each of these cases females used in the mating trials were homozygous mutants, suggesting some form of assortative fertilization or homogamy (chapter 9). The problems associated with differential viability were highlighted in a recent study of the genetics of sperm utilization. Gilchrist and Partridge (1997) used an adult body color mutant to examine paternity in *D. melanogaster*. They found that homozygous mutants had a lower heritable mean preadult viability that resulted in a positive association between the paternity success of individual males and the preadult viability of their offspring. This relationship could thereby lead to the erroneous conclusion that there are heritable differences in sperm displacement ability.

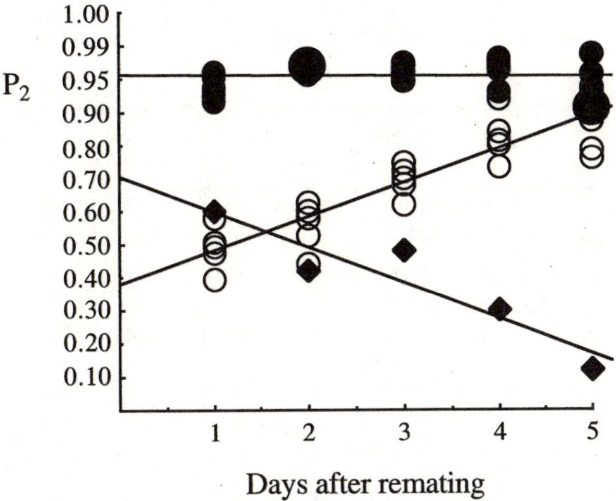

Figure 2.1 The mean proportion of offspring sired by the second male to mate (P_2) with a doubly mated female *Drosophila melanogaster* as a function of time since the second mating. Paternity was determined using morphological markers. Three data sets are presented. The first (solid circles) are for wild-type second males, the second (open circles) are for the mutant *forked* second males, and the third (solid diamonds) are for the mutant *carnation* second males. P_2 values are greater when the last male to mate is a wild-type male. Significant brood effects occur when mutant males are the second to mate. In the case of forked males, P_2 appears to rise perhaps as wild-type sperm become depleted. In the case of carnation males, P_2 declines, possibly due to a decline in sperm function. (Redrawn from Gromko et al. 1984a)

Allozyme Markers

Genetic markers that are less susceptible to the types of problem outlined above can be obtained from phenotypic variation manifest at the molecular level (Avise 1993). Allozymes are genetically polymorphic enzyme systems that catalyze biochemical pathways (Avise 1993). Natural variants can be scored using electrophoretic techniques because active proteins migrate at different rates dependent on their charge characteristics, size, and shape. Like morphological mutants, allozyme variants are often inherited in simple Mendelian ratios. They offer two possible means for assigning paternity. If prospective parents are first screened for allozyme phenotype, they can be used to unambiguously assign paternity in staged mating trials using an experimental protocol similar to that adopted with morphological markers. Given the existence of allozyme variants that differ in their speed of electrophoretic migration, females homozygous for one allele can be alternately mated to males that carry different alleles so that offspring can be uniquely assigned to each male. Allozyme markers do not appear to be subject to the same confounding problems that are encountered with morphological markers. Offspring need not be reared to adulthood before being scored,

Table 2.1
P_2 values obtained from reciprocal mating crosses in which morphological markers were used to assess sperm utilization.

Species	*Female*	*Males*		*Reference*
		m/w	*w/m*	
Blatella germanica	*m*	0.69	0.33	Cochran 1979
Abedus herberti	*w*	0.94	1.00	Smith 1979
Oncopeltus fasciatus	*m*	0.39	0.61	Economorpoulos and Gordon 1972
Henosepilachna pustulosa	*w*	0.65	0.50	Nakano 1985
	m	0.55	0.69	
Tribolium confusum	*w*	0.79	0.67	Vardell and Brower 1978
	m	0.92	0.91	
Tribolium castaneum	*w*	0.62	0.34	Lewis and Austad 1990
	m	0.69	0.65	
Anthonomus grandis	*m*	0.70	0.30	Bartlett et al. 1968
Callosobruchus maculatus	*w*	0.87	0.83	Eady 1991
Adalia decempunctata	*m*	0.74	0.69	Ransford 1997
Drosophila hydei	*m*	0.72	0.23	Markow 1985
Drosophila mojavensis	*m*	0.68	0.64	Markow 1988
Drosophila simulans	*m*	0.74	0.74	Price 1997
Drosophila mauritiana	*m*	0.95	0.82	Price 1997
Drosophila sechellia	*m*	0.84	0.75	Price 1997
Drosophila littoralis	*m*	0.63	0.85	Aspi 1992
Drosophila montana	*m*	0.67	0.70	Aspi 1992
Dahlbominus fuscipennis	*m*	0.24	0.76	Wilkes 1966
Nasonia vitripennis	*m*	0.92	0.42	Beukeboom 1994
Trichoplusia ni	*m*	1.00	0.84	North and Holt 1968
Plodia interpunctella	*m*	0.67	0.69	Brower 1975

Note: *w* = wild type, *m* = mutant.

alleviating potential problems with biased mortality. In general, those studies that have used allozyme variation in staged mating trials have failed to find any order effects that are related to male genotype (Boggs and Watt 1981; Gwynne 1988b; Allen et al. 1994; Sakaluk and Eggert 1996; Ransford 1997; Oberhauser et al. unpublished). There are two exceptions that are informative. A study of *Conotrachelus nenuphar* (Huettel et al. 1972) used a laboratory strain in which an allozyme morph had been increased by artificial selection through inbreeding and sib selection, much the same process by which morphological markers are isolated. Here, there was a sex by genotype interaction, similar to that observed in the morphological marker studies with *Tribolium castaneum* (Lewis and Austad 1990), whereby males from the unselected population had superior fertilizing capacity when competing within females from the unselected population. A study of *D. teissieri* used allozyme morphs from geographically isolated populations (Joly et al. 1991b). Again, there was a sex by genotype interaction in that fertilization was biased toward conspecific males. Thus, unlike morphological markers, allozyme markers appear to offer a genetic marker system for reliably quantifying sperm utilization, provided that the markers used represent natural variation present within the population being studied.

The second advantage of allozyme markers in sperm competition studies is the ability to assign paternity following natural matings, without prior knowledge of male genotypes. By scoring offspring at a number of different polymorphic loci, it is possible to assign paternity from maximum likelihood estimates based on the expected Mendelian ratios of electrophoretic variants (e.g., Dickinson 1988). Moreover, by screening multiple enzyme systems, it is possible to examine the extent of naturally occurring multiple paternity in field-collected females (LaMunyon 1994) or to perform experiments that mimic natural sperm competition situations, where three or more males mate with a given female.

DNA Markers

Recent advances in molecular techniques have provided a variety of means by which individuals can be recognized, and thus parenthood determined, from their DNA. Molecular markers can be derived from DNA by incubation with restriction enzymes that cleave DNA at specific oligonucleotide sequences. Single-strand DNA fragments are then subjected to electrophoresis. Fragments vary in their rate of electrophoretic migration due to differences in the number of oligonucleotides they contain. DNA markers are viewed by the addition of specific, previously isolated, purified, and radioactively labeled DNA sequences or "probes" (Avise 1993). The single-strand probes hybridize with homologous sequences which can then be viewed by autoradiography. Banding sequences are typically unique to each individual, hence the term DNA fingerprinting, and since each band in an individual's DNA fingerprint must be derived from its parent, paternity can be assigned based on shared banding patterns. There are a number of molecular DNA techniques now available, differing, primarily in the size of DNA fragments examined (for an overview of their relative utilities see Burke 1989; Avise 1993; Queller et al. 1993a).

DNA fingerprints have only recently been used in the study of insect paternity (Achmann et al. 1992; Hadrys et al. 1993; Cooper et al. 1996b; Hooper and Siva-Jothy 1996; Simmons and Achmann 2000). Like allozymes, they offer the benefit of not having to know the genotype of potential fathers prior to multiple matings, and they can allow the assignment of paternity following multiple matings with more than two males. However, DNA fingerprinting requires considerably greater effort because of the need to first isolate and purify suitable probes. One problem inherent in both allozyme and DNA marker techniques is nondetection of paternity due to poor sampling efficiency (Gromko et al. 1984a; Boomsma and Ratnieks 1996). The problem is particularly acute when the proportion of offspring sired by a particular male is low, for example, when sperm precedence is high. The probability of sampling the disfavored male's offspring increases with the total number of offspring sampled, so that very large samples are required for accurate assessment of paternity when precedence is high. Sampling efficiency is likely to be particularly relevant when time and cost considerations associated with DNA techniques demand small sample sizes. The biggest contribution of DNA fingerprinting to the study of sperm competition comes from the ability to examine patterns of paternity and reproductive success in field populations because, unlike allozyme analysis, it is possible to genotype individuals from very small samples of tissue, thereby allowing nondestructive sampling. Moreover, DNA samples are more easily collected in the field since, unlike allozyme samples, they do not need to be stored at extreme low temperatures. Because only small samples are required, DNA markers may also provide the means for examining mechanisms behind nonrandom paternity. Recent studies have examined the contribution of different males to the sperm storage organ(s) of the female using DNA fingerprints obtained from the sperm, contributions that are congruent with observed patterns of paternity (Achmann et al. 1992; Siva-Jothy and Hooper 1995, 1996; Chapuisat 1998).

Sterile Male Technique

The most widely adopted method for estimating paternity is the use of males sterilized by sublethal doses of radiation (Boorman and Parker 1976). The technique relies on the fact that sublethal doses of radiation induce chromosomal mutations in sperm that result in early embryonic death. The irradiation dose that provides the optimal level of embryonic mortality while maintaining sperm viability and fertilizing capacity must be determined prior to experimental matings. Females can then be mated with both a normal and an irradiated male and paternity assigned on the basis of egg viability. The sterile male technique gained favor because of its ease of application and because paternity biases toward particular males arising from differences in zygote viability are more easily controlled. In contrast to studies using morphological markers, differential zygote viability is directly examined within the context of the experimental protocol, and is the means by which paternity is assigned. Control treatments are established in which the viability of eggs fertilized by normal and irradiated males is assessed and paternity estimates from double matings with both normal and irradiated males take into account the differences in viability that are expected for each male type (Boorman and Parker 1976). However, differential fertilizing capacity can also be induced through irradiation. In general, studies using the irradiated

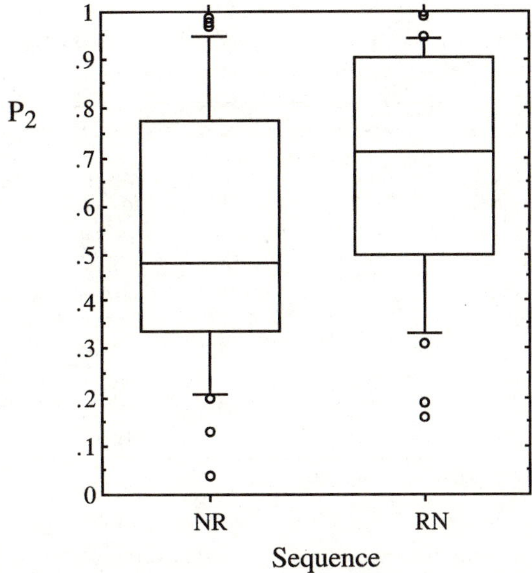

Figure 2.2 Differential fertilizing capacity of sperm from normal and irradiated males. Data are box plots of P_2 values obtained using the irradiated male technique for twenty-nine species of insect, plotted in relation to mating order. The horizontal line represents the median value for P_2. The upper and lower boundaries of the box are the 75th and 25th percentiles and the bars represent the 90th and 10th percentiles. Open symbols give the upper and lower 10% values. P_2 values are consistently higher when the normal male is the second to mate with the female (Wilcoxan-matched pairs, $z = 3.83$, $N = 29$, $P < 0.001$). (Data from studies cited in table 2.1)

male technique have found that irradiated males have a reduced fertilizing capacity compared with normal males; normal males obtain on average 13% more fertilizations as second mates than do irradiated males (fig. 2.2). Moreover, values tend to be skewed toward zero when the second male is irradiated and toward 1 when the second male is normal. There are exceptions to the general pattern. In studies of the armyworm *Pseudaletia separata* and the melon fly *Dacus cucurbitae* irradiated males had a higher fertilizing capacity than did normal males (Yamagishi et al. 1992; He et al. 1995b). It is not clear how irradiation influences fertilizing capacity, although most likely it affects sperm numbers, motility, and/or viability. In their study of the dung fly *Scatophaga stercoraria*, Simmons et al. (1996) found that the fertilization capacity of irradiated sperm declined over a period of seven days. Since sperm generally remain viable for several weeks (Parker 1970f), these data suggest that irradiation reduces sperm viability. Nevertheless, the problems associated with differential fertilizing capacity can be overcome by randomization of the mating sequence. The sequence normal-irradiated will tend to underestimate while the sequence irradiated-normal will tend to overestimate

Table 2.2
Comparison between P_2 values obtained using the irradiated male and genetic marker techniques.

Species	*Irradiated Male*	*Genetic Marker*	*Source*
Callosobruchus maculatus	0.83	0.85^{P}	Eady 1991
Rhagoletis pomenella	0.80	0.83^{A}	Myers et al. 1976; Opp et al. 1990
Plodia interpunctella	0.70	0.68^{P}	Brower 1975; Cook et al. 1997
Requena verticalis	≈0.00	$\approx 0.00^{AM}$	Gwynne 1988b; Gwynne and Snedden 1995; Simmons and Achmann 2000
Grylloides sigillatus	0.38	0.42^{A}	Sakaluk 1986; Sakaluk & Eggart 1996

Note: P, phenotypic; A, allozyme; M, molecular.

the potential paternity of second males. Thus, averaging an equal number of reciprocal crosses will provide a balanced estimate of P_2. Indeed, Eady (1991) assessed the accuracy of the technique in estimating P_2 by comparing the results of an irradiated male experiment with those obtained using a morphological marker in the bean weevil *Callosobruchus maculatus*. There were no viability or differential fertilizing capacity problems associated with the morphological marker in this species. The two methods provided quantitatively similar results and the data in table 2.2 suggest that this is generally the case. The principal drawback of the sterile male technique is that it allows paternity to be determined for only one male in a multiple mating situation, or for two males in a double-mating situation. Thus, the technique does not allow an estimation of the patterns of sperm utilization by multiply mated females under field conditions.

Direct Observation of Sperm

A number of studies have inferred paternity on the basis of direct observation of the displacement and/or mixing of sperm in the female sperm storage organ(s). In some cases this can give a reliable quantitative indication of order effects. For example, because the males of some odonates remove sperm stored in the female sperm storage organ(s) before delivering their own ejaculate (see chapter 3), it is possible to quantify the proportion of sperm removed by interrupting copulations after varying periods of time and counting the numbers of sperm present within the sperm storage organ(s) (Waage 1979b, 1984, 1986; Ono et al. 1989; von Helversen and von Helversen 1991). Assuming random mixing of the remaining sperm, the proportion of offspring sired by the last male to mate can be inferred. Studies that have incorporated both observations of sperm removal and other means of assigning paternity suggest that estimates based on observations of sperm removal can be quantitatively accurate (Fincke 1984; Siva-Jothy and Tsubaki 1989a; Cordero and Miller 1992). Nevertheless, in some cases observational data may only be sufficient to provide qualitative insight into sperm utilization patterns. In her studies of rabbit fleas *Spilopsyllus cuniculi*, Rothschild (1991) used histological sections to show that copulations by immature females resulted in the storage of relatively few sperm but that these sperm were stored in a site close to the exit of the spermatheca, perhaps giving them a positional advantage. Copulations by mature females resulted in the storage of many more sperm albeit at a site further back in the spermatheca, away from the exit. However, at the time of fertilization, Rothschild (1991) showed that the contraction of spermathecal muscles resulted in a mixing of sperm from both sites, so that the last male to mate may be advantaged given that females store more sperm from their most recent mates. Paternity data are required to assess how these patterns of sperm storage determine the relative contribution of different males at fertilization.

2.4 Patterns of Sperm Utilization: An Overview

There is now information on the patterns of sperm utilization following double matings by females from 133 species of nonsocial insect from eleven orders (see table 2.3 at end of chapter). Because the sterile male technique has been the most widely used method for examining sperm utilization in insects, these studies typically report the proportion of offspring sired by the second male to mate (P_2) when a female is mated twice. Typically, patterns of sperm utilization are reported as species-specific mean values of P_2. However, intraspecific variation in P_2 can provide a greater insight into the patterns and processes behind sperm utilization than the mean values themselves (Lewis and Austad 1990; Simmons and Siva-Jothy 1998), a point that will be exemplified throughout this volume. Unfortunately, until recently variance in P_2 has been largely ignored. Nevertheless, some published studies provide an indication of the variance, either as a standard deviation or standard error about the mean value of P_2, or simply as the range of values obtained. In some cases, the actual data are provided, allowing the variance to be calculated directly. Data are provided on the mean and variance in P_2 for many of the studies of sperm competition in table 2.3. Patterns of sperm utilization in social insects are discussed later in this volume (see chapter 10). The standard deviation is used as a measure of variance wherever possible but the ranges are also provided. The standard deviation is a more realistic measure of variability than the range because single values of 0 or 1 may represent outliers in an otherwise invariant distribution. For example, P_2 has a full range of 1 in the dragonfly, *Sympetrum danae*, but the

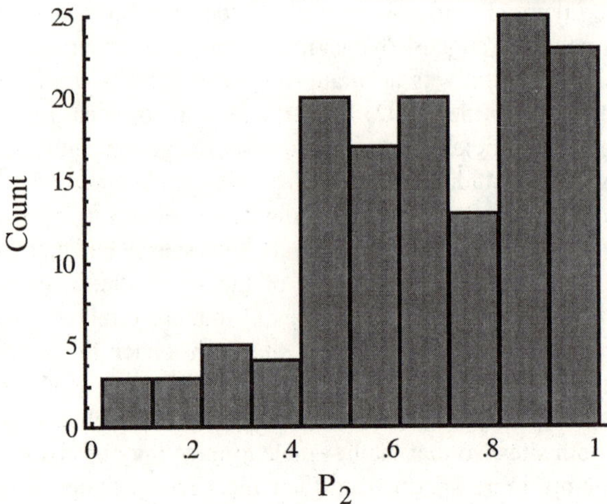

Figure 2.3 The distribution of species-specific mean values of P_2 for the 133 species of insect reported in table 2.3. The distribution has modes at intermediate values of P_2 indicative of mixed sperm utilization, and at extreme high values of P_2, indicative of sperm loss prior to remating or mechanisms of sperm precedence or displacement.

variance is only 0.05. In contrast, *Drosophila hydei* has an order of magnitude higher variance even though the range is lower, at only 0.75 (table 2.3).

The interspecific distribution of mean P_2 values exhibits a distinct bimodality (fig. 2.3), suggesting that there are two predominant patterns of sperm utilization in insects. The first mode occurs at values of P_2 between 0.4 and 0.7, indicative of species in which sperm from the two males mix within the female sperm storage organ(s) giving rise to sperm competition. The second mode occurs at values of $P_2 > 0.80$, indicative of species in which either sperm are lost prior to the second mating, or mechanisms of second-male sperm precedence or sperm displacement operate. Values of $P_2 < 0.4$ are rare. Thus, in general 43% of insect species show evidence of mixed sperm utilization, 44% of species show nonrandom fertilization favoring the second male to copulate, and 12% of species show nonrandom fertilization favoring the first male to copulate.

Intraspecific variance in P_2 is as high as interspecific variance; across the 133 species in table 2.3 for which data are available, the standard deviation in P_2 is 0.23 compared with the mean intraspecific value of 0.24 ± 0.14 (data for 69 species). However, there does appear to be a clear pattern in the data. In general, variation in P_2 is negatively associated with the species-specific mean; species in which paternity is biased toward the last male to copulate show little variance in paternity outcome while those showing sperm mixing have high variance in paternity outcome (fig. 2.4). This pattern appears to be robust with respect to phylogenetic associations, and the mea-

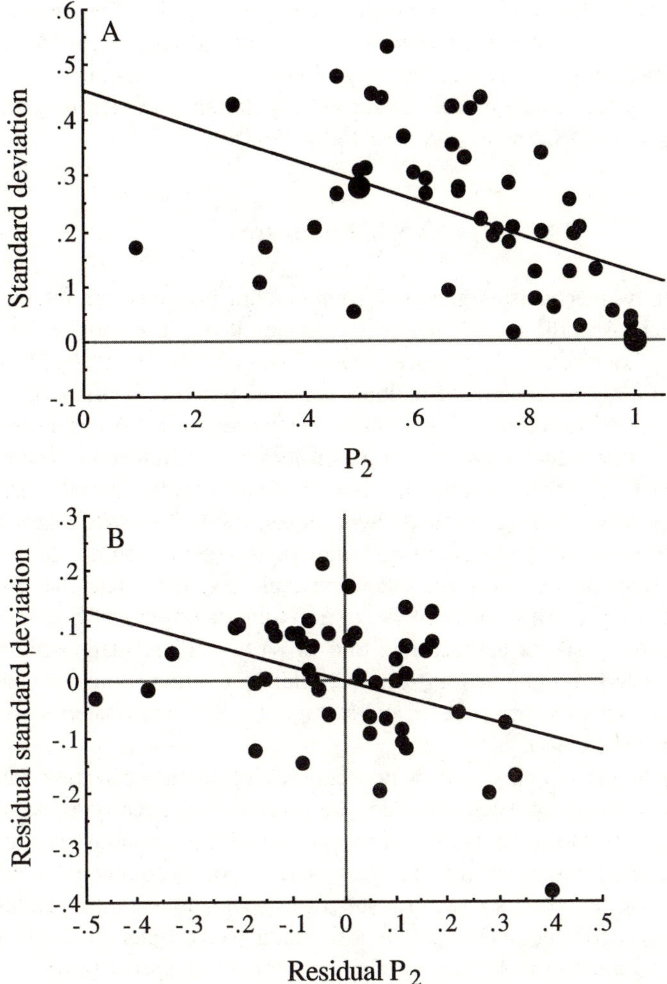

Figure 2.4 Variation across species of insects in the proportion of offspring sired by the second male to copulate appears to be negatively associated with the species mean. This pattern is true for species comparisons (A) and is maintained after controlling for common ancestry using Stearns's (1983) method of phylogenetic subtraction (B). (Redrawn from Simmons and Siva-Jothy 1998)

sure of variability used (Simmons and Siva-Jothy 1998). The relationship is also apparent within species. For example, many of the species reported in table 2.3 show initially high values of P_2 that decline with time, either between copulations or after the second copulation. Increases in the variance in P_2 are associated with decreases in the mean value of P_2. Thus, in general it seems that variation in fertilization success is restricted to species in which sperm mix in the female's sperm storage organ(s) so that they must compete for fertilizations. Where mechanisms of sperm precedence and/or displacement function to reduce the intensity of sperm competition, or where sperm are lost by the female before the second copulation, variation in fertilization success is low (Simmons and Siva-Jothy 1998).

2.5 Mechanisms

Frequently, mechanisms generating nonrandom paternity are inferred from the species-specific mean value of P_2. As outlined in the introduction to this chapter, however, such practices are misguided. Values of P_2 provide an indication of the outcome of multiple matings, in terms of offspring sired by individual males, but tell us nothing of the processes that occur during copulation and/or between copulation and paternity determination. There are at least six factors that can influence which of two copulating males are represented in the offspring produced by a female (table 2.4). While sperm competition and/or adaptations arising from sperm competition are determinants of paternity in the first three factors in table 2.4, it is clear that the same patterns of paternity can arise for reasons unrelated to sperm competition. Documenting sperm competition, and in particular elucidating the mechanisms underlying nonrandom paternity, clearly requires an understanding of the processes of sperm transfer and storage, in addition to paternity outcome (Parker et al. 1990).

Despite the ubiquity with which patterns of sperm utilization following double matings has been assessed, the mechanisms underlying nonrandom paternity are largely unknown. There are notable exceptions. In particular, the bias in paternity toward the last male to mate, common for odonates (table 2.3), has been shown to result from adaptations of the genitalia that facilitate sperm removal and/or repositioning (see chapter 3). These studies have examined directly the transfer and storage of sperm from successive males, as well as sperm utilization. Wilkes (1966) examined the process of sperm transfer and storage in the wasp *D. fuscipennis*. The spermatheca is a spherical structure with a chitinous lining that resists any change in its storage capacity (fig. 2.5; Wilkes and Lee 1965). Contraction of the spermathecal duct musculature straightens the U-shaped section of the duct, thereby controlling the entry and exit of sperm to the spermatheca. During a female's first copulation, sperm enter the spermatheca rapidly and come to rest with their heads protruding into the opening of the spermatheca to the spermathecal duct, their exit being hindered by incoming sperm. With the accumulation of motile sperm at the opening of the spermatheca, entry of new sperm is impeded. With successive matings, entry of new sperm appears to become progressively slower so that the total number of sperm stored following a second copulation is increased by only c. 30% and after a third copulation by

Table 2.4
Some factors that may determine the patterns of paternity observed in offspring following multiple mating.

Process	*Expected Paternity for Second Male*
1. Sperm from both males mix in females sperm storage organ(s) resulting in competition for available ova.	Variable, favoring superior competitor
2. Sperm from both males mix in females sperm storage organ(s) allowing them to select among available sires.	Variable, favoring preferred sire
3. Sperm from first male are displaced from the sperm storage organ(s) or precluded from fertilization site so that one male avoids sperm competition and/or selection.	High with displacement, high or low with precedence
4. Sperm from first male lost from storage prior to second copulation.	High
5. Sperm from first or second male not stored by female, either through preference or through failed insemination.	High or low
6. Eggs fertilized by one male fail to develop	Variable, favoring viable genetic combinations

only c. 17% (fig. 2.5). Thus, if there were no displacement of sperm, the representation of sperm in the female spermatheca would be 3:1 in favor of the first male when two males copulate, or 4.3:1.3:1 when three males copulate. The patterns of paternity observed following double and triple matings suggest that sperm are used in relation to their representation in the spermatheca (after double matings paternity is shared 2:1 and after triple matings 2.8:2.5:1; fig. 2.5) although the increasing underrepresentation of first males following second and third matings, and the overrepresentation of second males following a third mating, strongly suggest that first-male sperm are subject to an increasing degree of displacement during subsequent inseminations. Thus, Wilkes' (1965, 1966) observations of sperm transfer, storage, and utilization are congruent with a mechanism generating nonrandom paternity in *D. fuscipennis* in which there is partial displacement of sperm from the spermatheca during insemination followed by random utilization of sperm at the time of fertilization. The inelastic nature of the female spermatheca in this and many other species is likely to be a causal factor driving the displacement of sperm from storage as the spermatheca approaches its carrying capacity (see below).

Parker et al. (1990) advocated a theoretical approach to objectively and quantitatively compare observed patterns of paternity with those predicted

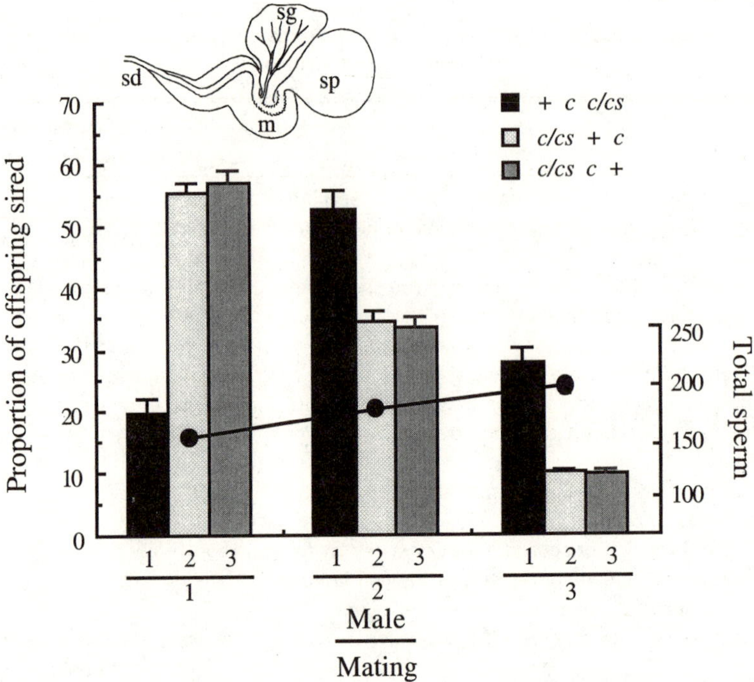

Figure 2.5 Sperm transfer, storage, and utilization in the wasp *Dahlbominus fuscipennis* after three successive copulations. The total number of sperm stored in the spermatheca increases by only 30% and 17% after second and third matings, respectively. Sperm utilization was assessed using wild-type males and males carrying the morphological marker *carmine c*, and the yellow eyed double recessive *c/cs*. All females used were homozygous *carmine*, and three experiments were performed in which the sequence of males used differed. Although there appears to be a disadvantage for males bearing wild-type sperm (+) when they are the first to copulate, on average across all experiments the first male obtains the majority of fertilizations, as might be expected given that the majority of sperm stored by the female come from her first copulation (from Wilkes 1965, 1966). Vignette shows the morphology of the female reproductive tract: sd, spermathecal duct; sg, spermathecal gland; sp, spermatheca; m, muscle (modified from Eberhard 1996).

from a variety of potential mechanisms of sperm utilization. The approach requires knowledge of sperm utilization patterns and quantitative data on the numbers of sperm transferred by males and stored by females, data such as those reported by Wilkes (1965, 1966). Although they had no data on sperm transfer and storage to support their arguments, Economorpoulos and Gordon (1972) used a remarkably similar approach in interpreting the patterns of paternity they observed following double matings of the hemipteran *Oncopeltus fasciatus*. Lessells and Birkhead (1990) simultaneously developed a series of models specific to birds. Parker et al.'s (1990) purpose was to outline a general philosophy for the interpretation of sperm competition data rather than to claim the existence of specific mechanisms of sperm storage and utilization. Thus, they broadly outlined two mechanisms of sperm utilization, the raffle, in which sperm from both males were present in the female sperm storage organ(s), and displacement, in which the second male displaces sperm stored from the first male during copulation. Likewise, they outlined two forms of displacement that differed in the time at which sperm from competing males mixed in the sperm storage organ(s), either instantaneously during insemination or after insemination and displacement were completed (box 2.1). The categorization of mechanisms into mixing and displacement obfuscates reality. These mechanisms will be extremes of two continua from zero to complete mixing and from zero to complete displacement. The realized P_2 will thus reflect the combination of displacement and mixing. For example, in species such as *D. fuscipennis*, second males may only partially displace sperm stored from the first male so that the actual mechanism is a raffle following partial displacement. The rate of sperm mixing is also likely to vary between species, so that a mechanism with properties expected from both the instantaneous mixing and no mixing models will more accurately describe observed paternity. Thus, Parker et al. (1990) described their models as prospective in the hope that they would be either refined to suit different systems and/or replaced with new models that were more appropriate to the system in question.

A number of studies have provided strong insight into the mechanisms generating nonrandom paternity utilizing Parker et al.'s (1990) approach. Parker et al. (1990) themselves used data for the field cricket *Gryllus bimaculatus*, showing that observed values of P_2 best fit the fair raffle model of sperm competition, while data for the dung fly *Scatophaga stercoraria* best fit a model of volumetric sperm displacement with instantaneous mixing (table 2.5; see also Parker and Simmons 1991). Eady (1994b) extended Parker et al.'s (1991) original models to analyze P_2 data obtained in a study of bean weevils, *Callosobruchus maculatus*. He found that sperm were lost from the fertilization set at a constant (exponential) rate following copulation so that 24 h after the first copulation only 58% of those sperm entering the fertilization set remained. Eady (1994b) thus extended the sperm displacement models (equations [2.6] and [2.8] in box 2.1) to account for sperm loss between copulations. Neither version of the extended displacement model accurately explained the observed variation in P_2. However, Eady (1994b) found that of the average 46000 sperm inseminated into the bursa copulatrix by the male, only 6468 were transferred to the fertilization set (the spermathecae) by the female. Thus, the parameter p, the probability that sperm enter the fertilization set, in the models described in box 2.1 is only 0.16. Eady (1994b) assumed $p = 1.0$ when fitting Parker et al.'s (1990) models so that neither would be expected to describe the observed variation in P_2. Using the mean value of P_2 observed and solving for p in the sperm-for-

Box 2.1
The Logic of Models for Predicting Mechanisms of Nonrandom Paternity

Parker et al. (1990) developed a series of prospective models for predicting the mechanisms underlying observed patterns of paternity based on an understanding of the processes of sperm transfer and storage. The models are based on the parsimonious assumption that there is a "fertilization set" of sperm from which sperm are drawn at random to fertilize ova. At some time between insemination and fertilization there is random mixing of sperm within the fertilization set so that sperm are used in direct proportion to their numerical representation in the set; when considering just two males mating, the probability of successful fertilization by male 2 in competition with male 1 is simply

$$P_2 = S_2 / (S_1 + S_2), \qquad [2.1]$$

where S_1 is the number of sperm transferred to the fertilization set by male 1, and S_2 is the number of sperm transferred by male 2. This simple model was referred to as the *fair raffle*, since winning paternity resembles a true raffle or lottery in which each sperm is a fertilization ticket (Parker 1982; Parker 1984). When ejaculate sizes, or sperm numbers, in the fertilization set from males 1 and 2 are shown to be equal, the observed paternity P_1 or P_2 can be compared statistically with the predicted paternity; in the case of two males this would be 0.50. A more sensitive test of the model, however, would be to generate unequal sperm numbers from the two males by manipulating copulations and then fit the data to a linear version of the model, where

$$1/P_2 = (S_1/S_2) + 1. \qquad [2.2]$$

A plot of $1/P_2$ against S_1/S_2 should yield a slope and intercept that do not differ significantly from +1.0. The purpose of linearization was to make testing of the models relatively simple. However, it would now be more accurate to use nonlinear curve-fitting packages that have since been developed. Deviations from the model might indicate that the raffle is "loaded"; one male's sperm are more likely to enter the fertilization set, or more likely to be drawn from the fertilization set, due to some competitive superiority of that male's sperm, female preference, or nonrandom mixing. Parker et al. (1990) provided a basic model for a loaded raffle in which the probability of entering the fertilization set for each male was expressed as p_1 and p_2; the ratio of the two probabilities, $r = p_1/p_2$, gives the degree of loading in the raffle so that the linearized version becomes

$$1/P_2 = r\,(S_1/S_2) + 1. \qquad [2.3]$$

The slope of the relationship between $1/P_2$ and S_1/S_2 thus gives an indication of the degree of unfairness.

(*continued on next page*)

Box 2.1 (*cont.*)

In many cases, however, there will be an upper limit to the amount of sperm a female can store in her sperm storage organ(s) so that sperm entering the fertilization set do so by displacing sperm that are currently stored. Thus, when the fertilization set reaches its total storage capacity $\boldsymbol{S}$ following a single copulation, insemination by male 2 will result in an increase in s_2 (where s_i is sperm in the sperm store from male i as opposed to S_i, which is total sperm transferred by male i) and an equivalent decrease in s_1, so that at any time during the second copulation $\boldsymbol{S} = s_1 + s_2$. Parker et al. (1990) developed two models of sperm displacement that differed in the time at which sperm mixing within the sperm storage organ(s) occurred.

With instantaneous mixing, the sperm from the second male comes to occupy a greater proportion of the total sperm stored, so he will become increasingly likely to displace his own sperm with continued insemination. To predict the value of P_2 following sperm displacement, we need to know how many sperm from male 2 remain in the fertilization set, s_2, following both the displacement of sperm from male 1 and the inevitable displacement of self's sperm. With a total of S_2 sperm transferred, pS_2 enter but only $s_2(S_2)$ remain in the fertilization set, so that the rate of change of s_2 with respect to S_2 is

$$\mathrm{d}s_2/\mathrm{d}S_2 = p - p\, s_2(S_2)/\boldsymbol{S}. \qquad [2.4]$$

The appropriate function for $s_2(S_2)$ is $\boldsymbol{S}[1-\exp(-pS_2/\boldsymbol{S})]$, so that

$$\mathrm{d}s_2/\mathrm{d}S_2 = p\exp(-pS_2/\boldsymbol{S}). \qquad [2.5]$$

With random utilization of sperm from the fertilization set, after transferring S_2 sperm, $P_2 = s_2(S_2)/\boldsymbol{S} = [1 - \exp(-pS_2/\boldsymbol{S})]$ and $P_1 = 1 - P_2 = \exp(-pS_2/\boldsymbol{S})$. The latter can be simplified as

$$-\ln(P_1) = pS_2/\boldsymbol{S}, \qquad [2.6]$$

so that plotting $-\ln(P_1)$ against $S_2/\boldsymbol{S}$ should give an intercept of zero and a slope of $+p$, the probability that a given sperm transferred will enter the fertilization set. If absolute sperm numbers are unknown but the relation between copulation duration and sperm transfer is known to be a linear function of time, then

$$-\ln(P_1) = zt/\boldsymbol{S}, \qquad [2.7]$$

where z is the number of sperm displaced per unit time and t is the duration of copulation. Thus plotting $-\ln(P_1)$ against t will give an intercept of zero and a slope $+z/\boldsymbol{S}$.

(*continued on next page*)

Box 2.1 *(cont.)*

The above assumes that sperm are replaced on a one-for-one basis. A biologically more realistic assumption is that there is a volumetric displacement of fluid from the sperm storage organ(s), so that during the female's first copulation the density of sperm increases to its maximum value. Thus, Parker and Simmons (1991) extended the model to show how the proportion of offspring sired by the second male could be predicted by

$$P_2 = [1 - \exp(-z_2 t_2/\mathbf{S})]/[1 - \exp(-(z_1 t_1 + z_2 t_2)/\mathbf{S})]. \quad [2.8]$$

If the total sperm input $(z_1 t_1 + z_2 t_2)$ is high relative to $\mathbf{S}$, the denominator approaches 1.0 so that $P_2 \approx [1 - \exp(-z_2 t_2/\mathbf{S})]$, which is equivalent to [2.7] above. Note that now the sperm transferred by male 1 will have no bearing on male 2's paternity.

With no mixing until after sperm displacement so that male 2 never displaces self sperm, the model is greatly simplified because $s_2(S_2) = pS_2$ and

$$P_2 = pS_2/\mathbf{S}, \quad [2.9]$$

so that plotting P_2 against $S_2/\mathbf{S}$ should have an intercept of zero and a slope of $+p$, the proportion of male 2 sperm entering the fertilization set. If sperm are inseminated directly into the female's sperm stores, $p = 1.0$. The analogous model to [2.7] would be

$$P_2 = zt/\mathbf{S}, \quad [2.10]$$

so that male 2's paternity should increase linearly with time in copulation.

sperm and volumetric displacement models of Eady (1994b) yielded values for p of 0.23 and 0.14, respectively, the latter giving a remarkably close fit to the observed value of 0.16. Thus, for *C. maculatus*, the data support a mechanism of sperm displacement in which only 16% of inseminated sperm enter the fertilization set to effect displacement. Sakaluk and Eggert's (1996) study of the cricket *Gryllodes sigillatus* reported a significant fit to the sperm displacement model with instantaneous mixing (table 2.5). However, the assumption of this model, that there is a fixed capacity for sperm storage, $\mathbf{S}$, was violated; it is known that the females can store sperm from several copulations (Sakaluk 1986) but Sakaluk and Eggert (1996) assumed that $\mathbf{S} = S_1$, making the observed fit biologically unrealistic. The data for *G. sigillatus* also fitted the model in which there was no mixing until after displacement (table 2.5). Sakaluk and Eggert (1996) rejected this model on

Table 2.5
Studies in which theoretical models have been used to evaluate alternative mechanisms underlying observed patterns of paternity following double matings

Species[a]		*Lottery*	*Displacement*		*Extension*	*Source*
			Instantaneous mixing	Mixing after displacement		
		$1/P_2 = r(S_1/S_2) + a$	$-\ln(P_1) = S_2/\boldsymbol{S}$	$P_2 = S_2/\boldsymbol{S}$		
Gryllus bimaculatus	*r*	$1(0.80 \pm 0.28)^{\dagger}$	$0(0.15 \pm 0.09)$	—		Parker et al. 1990
	a	$1(1.44 \pm 0.86)$	$0(0.002 \pm 0.001)$	—		
Scatophaga stercoraria	*r*	$0(0.02 \pm 20.05)$	$1(1.38 \pm 0.49)^{\dagger}$	$>0(0.71 \pm 0.06)$	$P_2 = [1 - \exp(-z_2 t_2/\boldsymbol{S})]^{b\dagger}$	Parker et al. 1990
	a	$>1(1.17 \pm 0.08)$	$0(0.88 \pm 0.49)$	$<1(0.15 \pm 0.06)$		Parker and Simmons 1991
Adalia bipunctata	*r*	$0(6.35 \pm 4.26)$	$1(1.43 \pm 0.36)^{\dagger}$	$<1(0.46 \pm 0.14)$		Ransford 1997
Gryllodes sigillatus	*a*	$0(-2.91 \pm 4.90)$	$0(-0.47 \pm 0.42)$	$0(0.05 \pm 0.16)$		
	r	$0(0.64 \pm 0.50)$	$1(1.02 \pm 0.34)$	$>0(0.50 \pm 0.15)$	$1/P_2 + b = (S_1 + S_2)/S_2{}^{c}$	Sakaluk and Eggert 1996
	a	$1(2.56 \pm 1.54)$	$0(0.08 \pm 0.21)$	$0(0.15 \pm 0.09)$	$P_1{}^{b-1} = [(1-b)/S_1]S_2 + 1^{d\dagger}$	
Harmonia axyridis	*r*	$0(0.46)$	—[e]	—[e]		Ueno 1992
	a	$1(1.82)$	—	—		

[a]Because he was unable to manipulate S_2 in his study of *Callosobruchus maculatus*, Eady (1994) plotted the values of P_2 predicted against number of sperm inseminated by the second male and contrasted the mean observed P_2 against that predicted for the mean number of sperm inseminated (see text for more details).
[b]Volumetric sperm displacement with instantaneous mixing.
[c]Partial sperm displacement with no mixing until after displacement, where *b* is the displacement efficacy.
[d]Partial sperm displacement with instantaneous mixing.
[e]The fit to these models cannot be evaluated because S_2 was not divided by $\boldsymbol{S}$, the sperm storage capacity.
[†]The model that best predicts the observed values of P_2.

the grounds that the slope of the regression was significantly different from 1.0. However, a slope of 1.0 is expected only if all sperm enter the fertilization set (where $p = 1.0$ see equation [2.9] in box 2.1). The model requires only that the slope is positive and significant, the slope providing an estimate of the proportion of second-male sperm entering the fertilization set. In the case of *G. sigillatus* the model suggested that only 50% of the second males' sperm entered the fertilization set. Nevertheless, again, this model should not have been used since $S \neq S_1$. The paternity data did not fit either the fair raffle or the loaded raffle models. Sakaluk and Eggert (1996) thus extended the displacement models [2.6] and [2.9] in box 2.1 to take into account the possibility that an imperfectly elastic spermatheca could result in only partial sperm displacement. They incorporated a parameter b, the displacement efficacy, which they estimated to be around 0.50 from the change in volume of the spermatheca following a second copulation. The data for *G. sigillatus* then gave an almost perfect fit to a model of partial sperm displacement with instantaneous mixing (table 2.5).

Physical Influences

The theoretical models of sperm utilization outlined above assume variation in certain physical properties of the sperm transfer and storage processes that give rise to variation in the extent of sperm displacement and mixing. Walker (1980) was the first to posit the notion that physical characteristics of the female sperm storage organ(s) could influence the extent of nonrandom paternity. He argued that the extent of sperm mixing within the female sperm storage organ(s) would depend on spermathecal shape. Species with complex, elongated, or tubular spermathecae should afford a low potential for sperm mixing so that the last sperm to enter the female storage organ(s) would be the first to leave, resulting in last-male sperm precedence. In contrast, species with simple, spherical spermathecae should have a high potential for sperm mixing, resulting in mixed paternity. Ridley (1989) tested Walker's hypothesis using outgroup comparisons to recognize independent evolutionary events of character evolution, but found no support for the hypothesis. The problem with Ridley's analysis, however, was that he used the value of P_2 as a measure of sperm precedence. As outlined above, high values of P_2 do not necessarily reflect second-male sperm precedence so that Ridley's analysis was not a valid test of Walker's hypothesis (see also section 2.8). Nevertheless, it is clear that species with spherical spermathecae can have strong last-male paternity because of sperm removal or displacement (Parker et al. 1999) rather than a last-in first-out mechanism of sperm precedence. Equally, species with tubular spermathecae can show almost complete mixed paternity (Vardell and Brower 1978; Dickinson 1988). That is not to say that spermathecal morphology has no role to play in influencing the patterns of sperm utilization, rather that Walker's (1980) reasoning was too simplistic. As well as varying in shape, sperm storage organ(s) vary in their physical properties, such as their initial volume and elasticity, which can significantly influence sperm competition mechanisms; small structures of fixed volume may afford the greatest potential for sperm displacement while large and elastic structures may afford the greatest potential for mixing.

The influence of storage capacity on mechanisms of sperm utilization is perhaps best illustrated by studies of the yellow dung fly. Sperm counts from females mating for the first time have demonstrated that sperm numbers increase with copula duration to an asymptotic level that is maintained irrespective of the number of successive copulations (Parker et al. 1990; Parker and Simmons 1991; Ward 1993). The sperm storage organs consist of three chitinous spermathecae of fixed volume (see fig. 1.3); the maximum storage capacity of the three spermathecae combined appears to be around 950 sperm, which is reached during a single normal copulation of 35 minutes. Parker et al.'s (1991) model of constant random sperm displacement (equation [2.8], box 2.1) assumes that displacement proceeds because the sperm stores have reached their asymptotic sperm density. Consequently, one unit of fluid entering the sperm stores must equal one unit of fluid leaving the sperm stores. With instantaneous mixing of ejaculates the density of the copulating male's sperm in the spermathecae will increase with exponentially diminishing returns because, as his representation in the spermathecae increases, so "self-displacement" becomes increasingly likely. In mathematical terms, when the female's sperm stores are filled to capacity before the second male copulates, the denominator in equation [2.8] approaches 1.0 and the contribution of the previous male to the spermathecae has little bearing on the copulating male's fertilization success (box 2.1). Rather, P_2 rises asymptotically with copula duration of the second male (fig. 2.6). Explicit in this model of sperm displacement is that, when the spermathecae are not filled to capacity at the time of the second male's copulation, sperm displacement will not occur, so that the second male's fertilization success should depend on his contribution to the fertilization set relative to the contribution of the first male. Mathematically, as the total sperm input ($z_1t_1 + z_2t_2$) decreases relative to S, the denominator in equation [2.8] should become increasingly important for predicting P_2. Thus, for very low total sperm inputs the system should conform to a simple raffle. Simmons and Stockley (unpublished data) found that, by manipulating copulations of first and second males so that the total copulation duration of both males was just 10 min (30% of that required to reach asymptotic sperm densities), P_2 increased in direct proportion to copula duration of the second male relative to the first, conforming exactly to a mechanism of random sperm mixing (fig. 2.6).

The data for *Scatophaga stercoraria* thus show how physical properties of the female sperm storage organ(s) can determine the mechanisms of sperm displacement. In the case of *S. stercoraria* spermathecal capacity can be

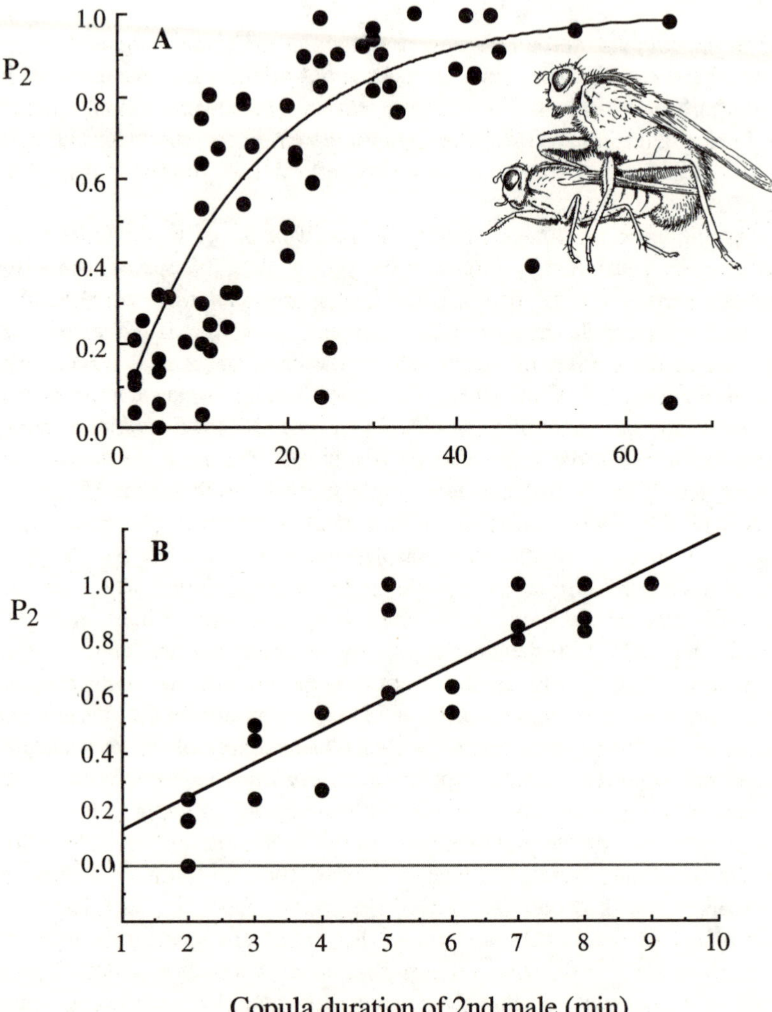

Fig. 2.6 Increase in fertilization success for male dung flies *Scatophaga stercoraria*, with increasing copula duration (A) when the first male has filled the female's sperm storage organs, and (B) when the total copulation duration for first and second males is 10 min so that the sperm storage organs have reached only 30% of their storage capacity. In (A) the data fit a model of constant random sperm displacement so that the second male becomes increasingly likely to displace his own sperm (see box 2.1; table 2.5). The slope is described by $P_2 = 1 - e^{-0.061t}$, where t is the copula duration of the second male (combined data set from Parker and Simmons 1991). On average, one-sixteenth of the female's sperm stores are displaced per minute of copulation. In (B) the sperm storage organs have not reached their asymptotic sperm density so that sperm displacement cannot occur. Second males gain fertilizations in direct proportion to the representation of their sperm in the female's sperm stores; the line is described by the equation $P_2 = 0.11t + 0.01$ ($r^2 = 0.76$, $F_{(1,\ 20)} = 64.94$, $P < 0.001$) so that for a relative increase of 1 min of copulation, second males have a 10% increase in fertilization success (Simmons and Stockley, unpublished). Thus, the data best fit a model of random sperm mixing (plotting $1/P^2$ against S_1/S_2 yields a significant regression $F_{(1,\ 20)} = 63.72$, $r^2 = 0.77$, $P < 0.001$, with a slope that does not differ from 1.0 (1.05 ± 0.13) and an intercept that does not differ from 1.0 (0.74 ± 0.22) (see box 2.1).

reached rapidly within a single copulation. In other species, the capacity of the female sperm storage organ(s) may be much greater so that it is not reached without prolonged copulation or until the female has engaged in several copulations. Males of the milkweed leaf beetle *Labidomera clivicollis clivicollis* copulate for periods up to 42 h in the field. Dickinson (1986) found that sperm numbers increased linearly over the first 24 h of copulation but did not increase when females mated for 45 h were remated for a further 45 h with a second male (Dickinson 1988); the female sperm storage organ(s) thus appear to be filled after a period of 45 h of copulation. Sperm mixing was strong when two males copulated for a total of 30 h with paternity biased toward the male with the relatively longer copulation (Dickinson 1986). However, last-male sperm precedence was strong when both first and second males were allowed to copulate for 45 h so that the total copulation duration was 90 h (Dickinson 1988). These changes in the pattern of sperm utilization with total copula duration are consistent with those seen in *S. stercoraria* and thus similarly suggest a mechanism of instantaneous random sperm displacement in *L. clivicollis* that arises due to the sperm storage capacity of the spermatheca.

In contrast, the spermatheca of gryllids is an elastic sac that expands to accommodate successive ejaculates. For *Gryllus bimaculatus* the spermatheca appears to be perfectly elastic so that all sperm are stored by the female and random utilization of sperm results in equal paternity for each male in competition (Simmons 1986). The spermatheca of *G. sigillatus* however, appears to be less elastic than that of *G. bimaculatus* so that first male sperm suffer a small degree of displacement during insemination by the second male, thus biasing paternity (Sakaluk and Eggert 1996).

As well as variation in displacement potential, there will be variation in the mixing potential of ejaculates that will influence sperm utilization (box 2.1). Mixing potential is likely to be determined by a variety of physical characteristics, such as the viscosity of the ejaculate, the motility of sperm, or the space available within the female sperm storage organ(s) for sperm movement. With low mixing potential, males will periodically gain sperm precedence at fertilization, depending on the relative positions of sperm clumps within the sperm storage organ(s), so that for any given clutch of eggs one male may sire more offspring than others (see fig. 2.7). High mixing potential will be manifest by random and invariant paternity within clutches. Harvey and Parker (2000) used mathematical simulations to show how varying degrees of sperm mixing could influence within-species variance in P_2. They used simulation models to vary the number of clumps or "packets" of sperm from two different males; at one extreme sperm from different males could remain in two discrete clumps or they could break up into many smaller clumps which mixed randomly in the sperm storage organ(s) of the female. These models generated distributions in P_2 values identical to those seen across insects, ranging from unimodal distributions when sperm

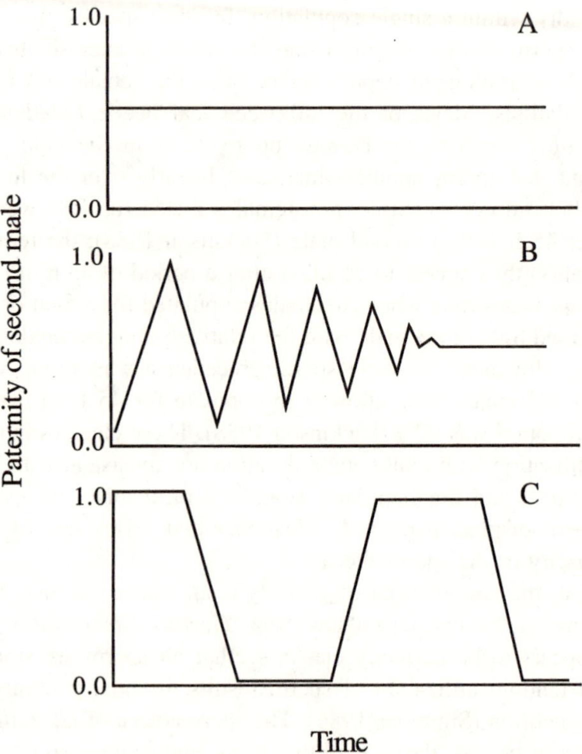

Figure 2.7 Effect of mixing potential on the proportion of eggs fertilized by the second male to mate with a doubly mated female. In (A), there is a high mixing potential and male 2 obtains an equal and invariant proportion of fertilizations relative to male 1 (e.g., Simmons 1987b). In (B), sperm mixing is slow so that initially the second male gains high paternity in some clutches and low paternity in others, depending on the position of his sperm clump relative to that of the first male. With time, complete mixing is achieved and both males obtain equal paternity (e.g., Siva-Jothy and Tsubaki 1989a). In (C), mixing potential is low. Sperm clumps remain separate and movement within the sperm storage organs is limited, resulting in periods of complete first-male or second-male paternity. Low mixing potential may underlie the apparent bimodality in sperm utilization patterns seen in many Lepidoptera (table 2.3).

break into multiple small packets, to strictly bimodal distributions, with modes at zero and 1, when sperm from different males remained in single discrete packets. Harvey and Parker's (2000) analysis suggests that variation in the degree of mixing can explain much of the widespread variation in P_2 seen in table 2.3.

It should be remembered that mixing potential, and indeed displacement potential, can be influenced by active biological processes within the sperm storage organ(s) of the female. In a number of species, females have been shown to mix sperm within their sperm storage organ(s) by contraction of spermathecal musculature or the use of specially adapted pumps that circulate sperm (e.g., Heming–Van Battum and Heming 1986; Rothschild 1991; Kaufmann 1996). In general the mechanistic approach to understanding observed patterns of sperm utilization outlined here does not intend to advocate that females are passive receptacles into which males ejaculate. Females are known to have considerable control over the transport and storage of sperm within their reproductive tracts (reviewed by Eberhard 1996) so that the realized P_2 for an individual male may often reflect the resolution of a conflict of interests between the male and female. Rather, the approach seeks to best explain the *average* pattern of sperm utilization seen for a particular species and thereby provide insight into mechanism. Variance about the mean species value of P_2 can arise due to random variation in the degree of sperm mixing (Harvey and Parker 2000) or to the mechanism of sperm utilization itself (Simmons and Siva-Jothy 1998), and indeed can be used to predict mechanisms of sperm competition (Cook et al. 1997; Simmons and Siva-Jothy 1998). Alternatively, variance may be due to differences among males in their ability to compete for fertilizations and/or reflect female assessment of individual males as potential sires.

2.6 Sperm Utilization and Multiple Mating

It is clear from the considerations in section 2.5 that the mechanisms underlying nonrandom sperm utilization can result in changes in the patterns of paternity as female sperm storage organ(s) become filled. Consequently, it is possible that the patterns of sperm utilization may change as females mate with increasing numbers of males. Unfortunately, however, documentation of patterns of sperm utilization has been constrained by the methods available for distinguishing between putative sires. Observations of sperm utilization have thus generally focused on situations where the female mates with just two males (table 2.3). The question arises as to whether the patterns of fertilization observed in laboratory experiments actually reflect those occurring in natural populations of insects, where females frequently mate with many more than two males. In their study of sperm utilization in pseudoscorpions, Zeh and Zeh (1994) found that when females mated just twice there were strong order effects, with second males obtaining 98% of fertilizations when they copulated within 24 hours of the first male. However, by using minisatellite probes, they were also able to assign paternity to three males. When females mated with a third male, mating order effects disappeared, so that paternity was distributed randomly with each male gaining a proportion of fertilizations that did not differ from 33%. Zeh and Zeh (1994) argued that stratification of sperm occurs within the tubular spermatheca until it is filled to capacity but that further ejaculates generate sufficient pressure to cause sperm mixing.

Table 2.6
Comparison of patterns of sperm utilization following double (P_2) and multiple (P_n) matings and estimates of last-male sperm utilization in the field.

Species	*Last-Male Paternity* Laboratory	Field	*Source*
Teleogryllus oceanicus	$0.46P_2 : 0.20P_4$		Simmons, in press
Tribolium castaneum	$0.75P_2 : 0.67P_3$		
Labidomera clivicollis	0.65–0.87	0.26–0.79	Dickinson 1988
Callosobruchus maculatus	$0.78P_2 : 0.83P_3$		Eady and Tubman 1996
Chelymorpha alternans	$0.50P_2 : 0.30P_3$		Rodriguez 1994a
Adalia bipunctata	$0.57P_2$	0.55	Ransford 1997
Gerris lacustris	$0.68P_2 : 0.22P_3$		
Drosophila melanogaster	$0.93P_2$	0.71, 0.83	Gromko et al. 1984a; Marks et al. 1988; Harshman and Clark 1998
Drosophila pseudoobscura	$0.82P_2$	0.82	Cobbs 1977; Turner and Anderson 1984
Drosophila hydei	$0.48P_2 : 0.25P_3$		Markow 1985
Scatophaga stercoraria	$0.81P_2 : 0.84P_3$ $0.86P_5 : 0.77P_7$		Parker 1970f
Dahlbominus fuscipennis	$0.32P_2 : 0.16P_3$		Wilkes 1966
Utetheisa ornatrix	$0.52P_2$	0.36	LaMunyon and Eisner 1993; LaMunyon 1994

There are a number of studies that can assess the validity of extrapolating patterns of paternity from two male studies to natural situations (table 2.6). In general, observations from two-male and multimale matings reveal consistent patterns of sperm utilization. The studies in table 2.6 fall into two categories, those that suggest order effects and those that conform to a model of random sperm mixing. For species with order effects last-male advantages at fertilization appear to be maintained irrespective of the number of times the female mates prior to the last male; thus, for *C. maculatus*, *S. stercoraria*, and *T. castaneum*, the paternity achieved by the last male does not change significantly with the number of previous matings (table 2.6). Similarly, there is only a slight decrease in the superiority of first males when female *D. fuscipennis* have three rather than two copulations (table 2.6). In contrast, patterns of sperm utilization in species such as *T. oceanicus, C. alternans, D. hydei*, and *G. lacustris* conform to a model of random mixing of sperm that predicts a decrease in the last male's paternity as the number of males in competition increases, a prediction that is supported by observed decreases in last-male paternity following multiple matings (table 2.6). Thus, the available data suggest that the changes in the pattern of sperm utilization following multiple matings reported by Zeh and Zeh (1994) are not ubiquitous. Nevertheless, the data in table 2.3 show that paternity can vary widely dependent on the experimental protocol adopted; parameters such as the interval between matings can either increase or decrease the second male's paternity expectation. Thus it is important to know what the values of such parameters are in natural populations before one can extrapolate findings from laboratory studies of sperm utilization to the field and thus draw conclusions regarding selection operating via differential fertilization success.

2.7 Sperm Utilization in Natural Populations

Although a number of studies have documented the occurrence of mixed paternity from broods of field-collected females (Allen et al. 1994; López-León et al. 1995; Gregory and Howard 1996; Imhof et al. 1998), studies that examine patterns of sperm utilization in natural populations of nonsocial insects are rare. Perhaps because of the practical difficulties involved in performing controlled matings in odonate systems, this taxa provide some of the best information on patterns of sperm utilization in the field. The last-male advantage at fertilization seen in *Mnais pruinosa, Leucorrhinia intacta*, and *Erythemis simplicollis* was demonstrated by releasing sterile males into field populations and observing copulations between these males and females that arrived at ponds to oviposit (McVey and Smittle 1984; Siva-Jothy and Tsubaki 1989a; Wolf et al. 1989). These estimates thus gave patterns of sperm utilization for females that would have mated an indeterminate number of times previously, but who were representative of females within natural populations. Similarly, the recent availability of molecular genetic markers has allowed the extent of last-male paternity to be established in natural populations of *Anax parthenope, Orthetrum coerulescens*, and *Calopteryx splendens* (Hadrys et al. 1993; Hooper and Siva-Jothy 1996). The data from these studies provide unequivocal evidence for a last-male fertilization advantage in natural populations of odonates.

There are a number of studies available that provide comparative data on patterns of sperm utilization in both laboratory and field situations (table 2.6). One method for estimating last-male paternity from field populations is to collect mating pairs and attempt to assign paternity to the putative father using allozyme or DNA markers. Studies of natural populations of *Drosophila* have shown that the observed patterns of last-male paternity following field matings are very close to those documented in double-mating trials in the laboratory (table 2.6); although the field estimate of last-male paternity in *D. melanogaster* appears 23% lower than the average for laboratory studies, it is within the normal range of estimates obtained from different laboratory strains of this species (Gromko et al. 1984a). Dickinson's (1988) study of the milkweed beetle similarly found upper and lower bounds of last-male paternity in field populations that were within the range expected, given the variance in last-male paternity that is generated from variation in the duration of mating associations in the laboratory. Thus, it would seem that laboratory estimates of the average values of P_2 for a species can often provide a reliable indication of the general pattern of sperm utilization in natural populations. However, there may be exceptions. In two-male mating trials of the moth *Utetheisa ornatrix*, LaMunyon and Eisner (1993) noted that values of paternity were bimodally distributed, with the majority of males gaining all or none of the fertilizations; cases of mixed paternity were rare (fig. 2.8). This pattern of bimodality in sperm utilization for first and second males is commonly found in the Lepidoptera (table. 2.3) although there has been little attempt to explain the phenomenon (but see Simmons and Siva-Jothy 1998; Harvey and Parker 2000). LaMunyon (1994) examined last-male paternity in a field population of *U. ornatrix*, where he used spermatophore counts (see chapter 4) to show that females had mated as many as thirteen times, and on average five times prior to the observed copulation. The mean estimate for last-male paternity was lower than that observed in controlled double-mating experiments (table 2.6). Moreover, the incidence of mixed male paternity was significantly greater in the field (18 of 33 females) than in controlled double mating trials (17 of 53 females; $\chi^2 = 4.255$, $P = 0.042$) (fig. 2.8). This change in the pattern of sperm utilization is reminiscent of that reported for pseudoscorpions by Zeh and Zeh (1994), and may similarly reflect the mixing potential for the species. If mixing potential is low then one might expect to find both strong first-male and strong last-male sperm precedence following a double mating, dependent on the relative positions of each male's sperm mass at the time when sperm are drawn from the storage organ for fertilization (see fig. 2.7 and Harvey and Parker 2000). With increasing numbers of matings, and thus inseminations, sperm mixing may become more pronounced as the pressure within the sperm storage organ, generated by additional ejaculates, increases (see also Zeh and Zeh 1994). Nevertheless, sperm mixing in *U. ornatrix* does not appear to be random, even after high numbers of inseminations, since the average paternity for the last male to copulate did not decline with increasing numbers of previous matings (LaMunyon 1994). Rather, last males consistently gained an average of 40% of fertilizations with the remaining 60% partitioned among previous males in an unknown ratio. This pattern would suggest that some combination of nonrandom sperm mixing and sperm precedence mechanisms determines the pattern of sperm utilization in this system.

2.8 Mechanisms and the Potential for Selection

An understanding of the precise mechanisms underlying sperm utilization is important for interpreting nonrandom patterns of paternity and predicting the types of adaptation that sperm competition can generate. Since Parker

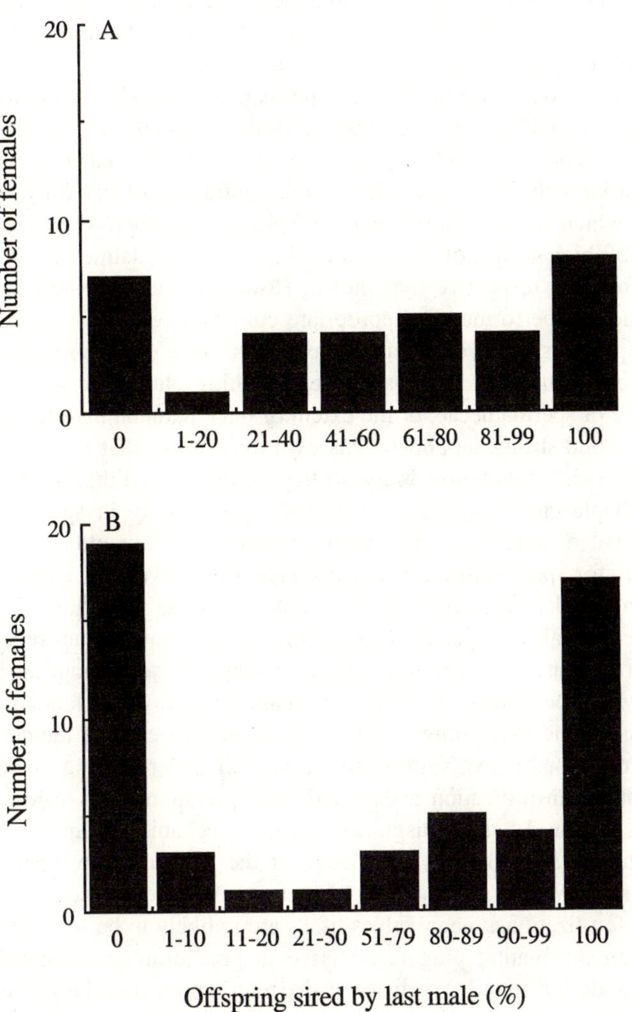

Figure 2.8 Variation in the proportion of offspring sired by the last male to copulate with a female in natural populations (A) and in laboratory double-mating trials (B) of the arctiid moth *Utetheisa ornatrix.* Females in natural populations mated on average five times prior to their final mating and had a higher incidence of mixed paternity than females who were only mated twice in the laboratory. (Data from LaMunyon and Eisner 1993 and LaMunyon 1994)

(1970e) there have been four reviews of insect sperm competition, each attempting to draw associations between morphological or behavioral adaptations and patterns of sperm utilization. In their update, Boorman and Parker (1976) argued that mating plugs arise as an adaptation for the avoidance of sperm competition. Thus, they predicted that species with mating plugs should have low values of P_2. Walker (1980) attempted to establish a mechanism for sperm precedence based on spermathecal shape (see section 2.5); species with complex and elongate spermathecae should have a lower mixing potential and thus higher values of P_2 than species with simple spherical structures. Gwynne (1984b) predicted that, because paternal investment should arise only in species with a high confidence of paternity, those species in which males contribute to offspring production should be characterized by high values of P_2. Each of these reviews claimed to provide evidence for their respective hypotheses. However, Ridley (1989) pointed out that none had performed an appropriate comparative analysis, and he tested each hypothesis using outgroup comparisons. Ridley (1989) found no association between P_2 and the existence of mating plugs (see fig. 4.7 below), the shape of spermathecae, or the extent of male parental investment. He did, however, find strong support for his own hypothesis, that P_2 should be positively associated with polyandry. Ridley (1989) argued that mechanisms for sperm displacement would be expected to arise only in species were the female mated more than once; with monandry there would be no selection on males for sperm displacement mechanisms. Conversely, the preexistence of sperm displacement mechanisms could favor the evolution of polyandry because the males of species with sperm displacement mechanisms would be selected to mate with nonvirgin females while those without such mechanisms would be selected to mate preferentially with virgin females.

Simmons and Siva-Jothy (1998) pointed out that each of the comparative studies described above suffers from a lack of understanding of the mechanisms of sperm utilization and/or a misinterpretation of P_2 values. As outlined in section 2.1, it is misguided to infer mechanism from the mean species value of P_2 without a knowledge of the mechanism of sperm transfer and storage. Where mating plugs function to prevent second males from gaining fertilizations, mean values of P_2 are unlikely to be low; they may be low when the mating plug is effective in preventing insemination by the second male but may be high when the plug's protective barrier is circumvented by the second male, assuming of course that females store all of the sperm that males inseminate. The average values of P_2 for such species are more likely to be intermediate but with high variance, so that either the first or second males gain fertilizations depending on the efficacy of the plug (see section 4.2). Walker's (1980) hypothesis assumes that females store all sperm from successive males and sperm precedence arises because of stratification. Assumptions concerning sperm precedence cannot be made based on the observed value of P_2; high values of P_2 can equally reflect sperm displacement or passive sperm loss (section 2.2), neither of which need be associated with spermathecal shape. Although the value of P_2 may be a reliable measure of confidence of paternity in the context of the evolution of male parental investment, hypotheses based on P_2 alone are oversimplistic.

One could argue that P_2 in paternally investing species should be low. That is, males invest paternally because there is little risk that females will utilize sperm from future males. Indeed, there is evidence for just such a relationship in the tettigoniid, *Requena verticalis* (Simmons and Achmann 2000). The evolution of paternal investment will depend on additional reproductive variables such as the interval between matings and the rate at which females utilize male contributions (Simmons and Parker 1989). Moreover, Ridley (1989) included in his analysis male investments such as postcopulatory guarding. However, unlike paternal investment, mate guarding arises in response to increased risks of sperm competition (Parker 1970e; chapter 5). Finally, Ridley's finding that polyandrous species have higher values of P_2 than monandrous species need not support his argument that sperm displacement mechanisms have coevolved with female multiple mating. Passive sperm loss between successive matings results in high values of P_2 (section 2.2), and Ridley (1988) has shown elsewhere that mating frequency in insects is associated with sperm depletion and the maintenance of fertility. Thus, polyandrous species would be expected to have high values of P_2 due to the depletion of sperm from their previous mates. These studies illustrate the pitfalls of using P_2 values alone for predicting selection arising from sperm competition.

Over the last decade a series of sperm competition game models have been developed to predict the types of adaptation that should be expected from selection via sperm competition (reviewed by Parker 1998). These theoretical models are based on the assumption that sperm competition conforms to a raffle (box 2.1). For the majority of vertebrate taxa this seems a reasonable assumption (Gomendio and Roldan 1993; Birkhead and Biggins 1998) and comparative analyses appear to support predictions made by the models. Increased sperm competition risk is predicted to favor increased investment in sperm and thus gametic tissue (Parker et al. 1997). Across species of birds (Møller 1991; Birkhead and Møller 1992), mammals (Short 1979; Harcourt et al. 1981; Ginsberg and Rubenstein 1990; Hosken 1997; Kappeler 1997; Rose et al. 1997; Hosken 1998), and fish (Stockley et al. 1997), testis size does appear to increase with sperm competition risk, as estimated from aspects of female mating frequency. However, in the insects it is clear that sperm competition mechanisms can be more complex than the simple raffle (section 2.2). In species with sperm repositioning or mechanical displacement there will be no sperm competition at the time of fertilization so that sperm numbers should be unimportant for male fitness. Thus, we might not expect selection to favor increased investment in gametic tissue and so might not expect to find a relationship across species between female mating frequency and testis size. Nevertheless, in some species males use their own ejaculates to displace sperm stored from previous matings (Ono et al. 1989; Otronen and Siva-Jothy 1991; Parker and Simmons 1991) so that selection may favor increased testis size for the avoidance of rather than engagement in sperm competition (Hosken and Ward 2001). Selection for increased testis size could therefore be characteristic of species with intermediate values of P_2 (where sperm competition is intense) or very high values of P_2 (were there is little or no sperm competition). Large testes might even be characteristic of species with very low values of P_2 if, for example, males were selected to fill the female sperm storage organ(s) as a means of preventing the storage of sperm from future males (Simmons and Siva-Jothy 1998; Simmons and Achmann 2000). Recent sperm competition game models have begun to examine the possible responses to selection in sperm morphology and behavior (Parker 1993; Parker and Begon 1993; Kura and Nakashima 2000). Thus, it is argued that sperm size, longevity, and swimming speed could be the focus of selection via sperm competition (Sivinski 1984; Baker and Bellis 1988; Gomendio and Roldan 1991). However, looking for such responses in taxa where males have evolved mechanisms for the avoidance of sperm competition is unlikely to be enlightening. Likewise, the potential for sperm selection will depend on the availability of sperm from multiple mating partners so that sperm selection potential will be inversely related to the degree of sperm displacement (Simmons and Siva-Jothy 1998). In short, for any analysis of traits believed to be selected in the context of sperm competition, it is essential to have a knowledge of the mechanisms of sperm transfer, storage, and utilization. Unfortunately, with few exceptions, such data are currently lacking.

2.9 Summary

The utilization of sperm derived from different males by polyandrous female insects has been widely examined using a variety of techniques to assign paternity. Because of the limitations of early methods, most studies document the proportion of offspring sired by the second male to mate with a doubly mated female, hence the statistic P_2. Although there is a continuum, the data suggest that insect species can be broadly categorized into two groups, those with mixed sperm utilization and those in which the last male dominates at fertilization. However, despite the ubiquity with which it has been measured, the mechanisms underlying sperm utilization are largely unknown. Process is frequently inferred from pattern, a practice that has led to erroneous conclusions. A knowledge of the processes of sperm transfer and storage are essential for the interpretation of P_2 data in the context of sperm competition theory. The P_2 statistic is not the only term that has been subject to misuse in the sperm competition literature and I provide a series of working definitions for specialist terms that are aimed at clarifying patterns and processes. Quantification of the transfer and storage of sperm can be used to elucidate the mechanisms behind sperm utilization following a theoretical approach outlined by Parker and his colleagues. Such approaches have revealed a number of factors, such as the displacement potential and the mixing potential of the system, that may act as physical determinants of sperm utilization patterns. Despite the artificiality of two-male mating trials, patterns of sperm utilization appear to be consistent following multiple matings by females, although there are exceptions. Nevertheless, paternity outcomes can vary considerably with factors such as remating interval or the time between mating and oviposition. Estimates of the values of such parameters are required from field populations before conclusions based on paternity outcome measured in the laboratory can be extended to predictions of selection operating via sperm competition in natural populations. In general, a knowledge of the mechanisms of sperm transfer, storage, and utilization are required for the evolutionary interpretation of adaptations that are believed to have arisen in the context of sperm competition, and for predicting the nature of selection imposed on males via female polyandrous mating.

Table 2.3
Patterns of sperm utilization in nonsocial insects.[a]

Species	*Mean* P_2	*Range*	*SD*	*Method*	*Source of Variance*
ODONATA					
Zygoptera					
Argia moesta[1]	0.93			S	
Argia sedula[1]	0.71			S	
Calopteryx dimidiata[2]	0.98			S	
Calopteryx maculata[3]	0.90			S	
Calopteryx splendens[4]	0.98			M	
Ceriagrion tenellum[5]	0.87/1.00		0.21/0.11	R	Copula duration of second male (+)
Enallagma hageni[6]	0.78	0.44–0.95	0.21	R	
Ischnura elegans[7]	0.82	0.44–1.00	0.19	M	
Ischnura graellsi[8]	0.99	0.92–1.00	0.03	P	Time since final copulation (−)
Ischnura ramburi[1]	0.82			S	
Ischnura senegalensis[9]	1.00			S	
Mnais pruinosa pruinosa[10]	1.00–0.50		0.00–0.40	R	Time since final copulation (−); copula duration of second male (+)
Anisoptera					
Anax parthenope[11]	1.00		0.00	M	
Erythemis simplicollis[12]	0.99–0.65		0.00–0.13	R	
Lestes vigilax[13]	0.75			S	
Leucorrhinia intacta[14]	0.90	0.00–1.00	0.21	R	Copula duration of second male (+ asymptotic)
Nannophya pygmaea[15]	0.98–0.72		0.02–0.10	R	Time since final copulation (−)
Orthetrum coerulescens[11]	1.00–0.77	0.19–1.00	0.00–0.34	M	Copula duration of second male (+); time since final copulation (−)
Sympetrum danae[17]	0.96	0.00–1.00	0.05	R	Copula duration of second male (+ asymptotic); time since final copulation (+)

ORTHOPTERA					
Grylloidea					
Allenomobius fasciatus[17]	0.62			A	
Allenomobius socius[17]	0.43			A	
Gryllodes sigillatus[18]	0.42	0.04–0.88	0.20	RA	Relative spermatophore attachment times
Gryllus bimaculatus[19]	0.33/0.45/0.68	0.00–0.80	0.06/0.03/0.06	R	Relative spermatophore attachment times; relative number of copulations
Gryllus integer[20]	0.72	0.08–1.00		R	
Teleogryllus oceanicus[21]	0.46	0.05–0.86	0.24	A	
Truljalia hibinonis[22]	0.88		0.12	S	
Tettigonioidea					
Decticus verrucivorus[23]	0.50	0.03–1.00	0.31	R	Relative spermatophore size
Kawanaphila nartee[24]	0.69	0.13–1.00	0.33	R	
Metaplastes ornatus[25]	0.85			S	
Poecilimon veluchianus[26]	0.90	0.87–0.93	0.03	M	
Requena verticalis[27]	0.00–0.19	0.00–0.94	0.17	ARM	Spermatophore size of first male (−); remating interval (+)
Steropleurus stali[28]	0.95	0.08–1.00		R	
Acridoidea					
Chorthippus parallelus[29]	0.60	0.35–0.84		P	Homogamy
Eyprepocnemis plorans[30]	0.92	0.40–1.00		P	Relative number of copulations
Locusta migratoria[31]	0.38–0.86	0.00–1.00		R	Remating interval (+)
Paratettix texanus[32]	0.58	0.00–0.91	0.37	P	
Podisma pedestris[33]	0.39	0.00–1.00		P	Homogamy
Schistocerca gregaria[34]	1.00			P	
BLATTODEA					
Blatella germanica[35]	<0.37[b]	0.00–1.00		P	Remating interval (+)
Diploptera punctata[36]	0.68	0.00–1.00	0.28	R	

Table 2.3 (*cont.*)

Species	Mean P_2	Range	SD	Method	Source of Variance
PHASMATODEA					
Baculum sp 1.[37]	0.99	0.80–1.00	0.04	R	
Extatosoma tiaratum[38]	0.98	0.83–1.00	0.08	R	Number of copulations (+)
MANTODEA					
Hierodula membranacea[39]	0.24	0.00–1.00		R	
Hierodula patellifera[39]	0.85	0.00–1.00		R	
HEMIPTERA					
Abedus herberti[40]	0.99	0.97–1.00	0.01	P	
Dysdercus koenigii[41]	0.66			R	
Eysarcoris lewisi[42]	0.44/1.00			R	Time since final copulation (+)
Gerris lacustris[43]	0.68	0.00–1.00[c]	0.49	R	Remating interval (−); male genital morphology
Gerris lateralis[44]	0.81	0.10–1.00		R	Copula duration (+); male genital morphology
Gerris remigis[45]	0.65	0.28–1.00		R	Relative copula duration
Jadera haematoloma[46]	0.62	0.05–0.95	0.30	R	
Lygaeus equestris[47]	0.92			R	
Neacoryphus bicrucis[48]	0.78		0.01	R	Copula duration of second male (+ asymptotic)
Nezara viridula[49]	0.51	0.00–1.00	0.31	P	Size of first male (−); remating interval (+)
Oncopeltus fasciatus[50]	0.50	0.03–0.63	0.23	P	
COLEOPTERA					
Adalia bipunctata[51]	0.60	0.00–1.00	0.40	PA	Copula duration (+)
Adalia decempunctata[52]	0.71	0.00–1.00	0.36	M	Number of inseminations by first male (−)
Aleochara curtula[53]	0.87	0.00–1.00		M	
Anthonomus grandis[54]	0.52–0.90	0.10–0.90		P	Remating interval (+)
Bolitotheros cornutus[55]	0.67	0.00–1.00	0.34	A	
Callosobruchus maculatus[56]	0.83	0.55–1.00	0.34	PR	Number of sperm transferred by second male (+)

Chelymorpha alternans[57]	0.50			P	Copula duration (+); genital flagellum length (+)
Conotrachelus nenuphar[58]	0.50	0.02–0.90	0.28	A	
Dermestes maculatus[59]	0.58	0.00–1.00	0.56	R	
Epilachna varivestis[60]	0.70			C	
Harmonia axyridis[61]	0.55	0.16–1.00	0.53	P	Copula duration of second male (+)
Henosepilachna pustulosa[62]	0.60	0.00–0.97	0.31	P	
Ips pini[63]	0.58	0.00–1.00	0.34	R	Mating access (+)
Labidomera clivicollis[64]	0.65/0.78/0.87	0.29–1.00	0.27/0.32/0.26	A	Copula duration of second male (+ asymptotic)
Lasioderma serricorne[65]	0.84			R	
Leptinotarsa decemlineata[66]	0.32–0.53			P	Relative copulation frequency
Necrophorus orbicollis[67]	0.94			P	
Necrophorus vespilloides[68]	0.11–0.92	0.00–1.00		P	Number of copulations by second male (+)
Onthophagus binodis[69]	0.55	0.00–1.00	0.27	R	
Onthophagus taurus[69]	0.58	0.00–1.00	0.25	R	
Onymacris unguicularis[70]	0.82			R	
Popillia japonica[71]	0.85			C	
Tenebrio molitor[72]	0.91			R	Time since final copulation (−)
Tetraopes tetraophthalmus[73]	0.72	0.33–1.00	0.22	A	
Tribolium castaneum[74]	0.62	0.40–0.86	0.27	P	Relative male size; females; time since final copulation (−); homogamy; parasites
Tribolium confusum[75]	0.82			P	
Trogoderma inclusum[76]	0.52			R	
MECOPTERA					
Panorpa vulgaris[77]	0.46	0.02–0.95	0.27	A	Relative copula duration
DIPTERA					
Aedes aegypti[78]	0.15	0.08–0.83		R	
Anopheles albimanus[79]	0.49[b]			A	
Anopheles gambiae[80]	0.02			R	
Ceratitis capitata[81]	0.68	0.16–0.99	0.27	P	
Culex pipiens[82]	0.11–1.00			C	Remating interval (+)
Culicoides mellitus[83]	0.29			R	

Table 2.3 (*cont.*)

Species	*Mean P_2*	*Range*	*SD*	*Method*	*Source of Variance*
Cyrtodiopsis whitei[84]	0.48	0.00–1.00		R	
Dacus cucurbitae[85]	0.42–0.74	0.35–1.00	0.19	R	Relative copula duration; remating interval (+)
Dacus oleae[86]	0.46–0.52			R	Age at irradiation
Drosophila hydei[87]	0.50	0.25–1.00	0.28	P	Remating interval (+)
Drosophila littoralis[88]	0.72		0.35	P	
Drosophila mauritiana[89]	0.83			P	
Drosophila melanogaster[90]	0.93	0.31–1.00	0.13	RAP	Number of sperm transferred by second male (+); resistance by first male (−); genetic, remating interval (+)
Drosophila mojavensis[91]	0.66	0.52–0.81	0.09	P	Remating interval (+); time since final copulation (−)
Drosophila montana[88]	0.68		0.38	P	Remating interval (+)
Drosophila pseudoobscura[92]	0.82	0.00–1.00	0.07	P	Resistance by first male (−)
Drosophila sechellia[89]	0.79			P	
Drosophila simulans[89]	0.74			P	
Drosophila teissieri[93]	0.77	0.30–1.00	0.18	A	Homogamy; sperm polymorphism
Dryomyza anilis[94]	0.33 0.18–0.75 0.45–0.85	0.05–0.73	0.17	R	Size of second male (+); eggs laid (−); mating situation; number of tapping sequences (+ asymptotic); time since final copulation (−); number of copulations by second male (+); male genital morpholgy
Glossina austeni[95]	0.29			R	Time since final copulation (−)
Glossina morsitans[96]	0.45–0.11			R	Remating interval (−)
Rhagoletis pomonella[97]	0.83	0.31–1.00	0.20	AR	
Scatophaga stercoraria[98]	0.88	0.02–1.00	0.26	R	Copula duration of second male (+ asymptotic); size of second male (+); male nutrient reserves

HYMENOPTERA					
Aphytis melinus[99]	0.14	0.00–0.50		A	Remating interval (−)
Dahlbominus fuscipennis[100]	0.32	0.20–0.76	0.11	P	Time since final copulation (+); remating interval (+/−)
Diachasmimorpha longicaudata[101]	0.49			P	Remating interval (+)
Diadromus pulchellus[102]	0.06	0.00–0.77	0.19	P	
Lariophagus distinguendus[103]	0.07			P	
Nasonia vitripennis[104]	0.02–0.65	0.01–1.00		P	First male mating history (+)
LEPIDOPTERA					
Bombyx mori[105]	0.95–0.06	0.00–1.00[c]	0.05	P	Remating interval (−)
Carpocapsa pomonella[106]	0.56			R	
Choristoneura fumiferana[107]	0.46	0.00–1.00[c]	0.48	P	Remating interval (−)
Helicoverpa zea[108]	0.71	0.00–1.00		R	
Heliothis virescens[109]	0.47	0.00–1.00[c]		R	
Phthorimaea operculella[110]	0.94			R	
Plodia interpunctella[111]	0.57–0.82	0.00–1.00[c]	0.55–0.29	R	Relative spermatophore size
Pseudaletia separata[112]	0.83			R	
Pseudaletia unipuncta[113]	0.47	0.00–1.00[c]		P	
Pseudoplusia includens[114]	0.27	0.00–1.00[c]	0.43	A	
Spodoptera frugiperda[115]	0.54	0.00–1.00[c]	0.44	C	
Spodoptera litura[116]	1.00			S	
Trichoplusia ni[117]	0.92			P	
Utetheisa ornatrix[118]	0.52	0.00–1.00[c]	0.45	A	Relative spermatophore size
Papilionoidea					
Pieris napi[119]	0.66	0.00–1.00[c]		R	Male size (+)
Pieris rapae[120]	0.73	0.00–1.00[c]	0.38	R	Male size (+); male mating history (+)
Colias erytheme[121]	1.00		0.00	A	
Euphydryas editha[122]	0.72	0.00–1.00[c]	0.44	R	
Papilio dardanus[123]	0.86			P	
Danaeus plexippus[124]	0.67	0.00–1.00	0.30	A	Male size (+); male mating history (+); time since final copulation (+/−)

Table 2.3 (*cont.*)

Sources: 1. Waage 1986; 2. Waage 1984; 3. Waage 1979b; 4. Hooper and Siva-Jothy 1996; 5. Andrés and Cordero Rivera 2000; 6. Fincke 1984; 7. Cooper et al. 1996; 8. Cordero and Miller 1992; 9. Sawada 1995; 10. Siva-Jothy and Tsubaki 1989a; 11. Hadrys et al. 1993; 12. McVey and Smittle 1984; 13. Waage 1982; 14. Wolf et al. 1989; 15. Siva-Jothy and Tsubaki 1994; 16. Michiels 1992; Nuyts and Michiels 1996; 17. Gregory and Howard 1994; 18. Sakaluk 1986; Sakaluk and Eggert 1996; Calos and Sakaluk 1998; 19. Simmons 1987b; 20. Backus and Cade 1986; 21. Simmons, in press; 22. Ono et al. 1989; 23. Wedell 1991; 24. Simmons 1995c; 25. von Helverson and von Helverson 1991; 26. Achmann et al. 1992; 27. Gwynne 1988b; Gwynne and Snedden 1995; Simmons and Achmann 2000; 28. Vahed 1998c; 29. Bella et al. 1992; 30. López-Leon et al. 1993; 31. Parker and Smith 1975; 32. Nabours 1927; 33. Hewitt et al. 1989; 34. Hunter-Jones 1960; 35. Cochran 1979; 36. Woodhead 1985; 37. Carlberg 1987a; 38. Carlberg 1987b; 39. Lawrence 1991; 40. Smith 1979; 41. Harwalker and Rahalkar 1973; 42. Ueno and Ito 1992; 43. Danielsson and Askenmo 1999; 44. Arnqvist 1988; Arnqvist and Danielsson 1999a; Arnqvist and Danielsson 1999b; 45. Rubenstein 1989; 46. Carroll 1991; 47. Sillen-Tullberg 1981; 48. McLain 1989; 49. McLain 1985; 50. Economorpoulos and Gordon 1972; 51. de Jong et al. 1993; Ransford 1997; de Jong et al. 1998; 52. Ransford 1997; 53. Benken et al. 1999; 54. Bartlett et al. 1968; 55. Conner 1995; 56. Eady 1991; Eady 1994a; 57. Rodriguez 1994a; 58. Huettel et al. 1972; 59. Archer and Elgar 1999; 60. Webb and Smith 1968; 61. Ueno 1994; 62. Nakano 1985; 63. Lissemore 1997; 64. Dickinson 1986; Dickinson 1988; 65. Coffelt 1975; 66. Boiteau 1988; 67. Trumbo and Fiore 1991; 68. Müller and Eggert 1989; 69. Tomkins and Simmons 2000; 70. De Villiers and Hanrahan 1991; 71. Ladd 1966; 72. Siva-Jothy et al. 1996; 73. McCauley and Reilly 1984; 74. Schlager 1960; Lewis and Austad 1990; Robinson et al. 1994; Yan and Stevens 1995; 75. Vardell and Brower 1978; 76. Vick et al. 1972; 77. Thornhill and Sauer 1991; Sauer et al. 1999; 78. George 1967; 79. Villarreal et al. 1994; 80. Bryan 1968; 81. Saul et al. 1988; Saul and McCombs 1993; 82. Bullini et al. 1976; 83. Linley 1975; 84. Lorch et al. 1993; 85. Tsubaki and Sokei 1988; Yamagishi et al. 1992; 86. Cavalloro and Delrio 1974; 87. Markow 1985; 88. Aspi 1992; 89. Price 1997; 90. Gromko et al. 1984a; Letsinger and Gromko 1985; Clark et al. 1995; Hughes 1997; 91. Markow 1988; 92. Turner and Anderson 1984; 93. Joly et al. 1991b; 94. Otronen 1990; Otronen 1994a; Otronen 1994b; 95. Curtis 1968; 96. Dame and Ford 1968; 97. Opp et al. 1990; 98. Parker 1970f; Simmons and Parker 1992; 99. Allen et al. 1994; 100. Wilkes 1966; 101. Martinez et al. 1993; 102. El Agoze et al. 1995; 103. van den Assem et al. 1989; 104. Holmes 1974; Beukeboom 1994; 105. Omura 1939; Suzuki et al. 1996; 106. Proverbs and Newton 1962; 107. Retnakaran 1974; 108. Carpenter 1992; 109. Flint and Kressin 1968; 110. Rananavare et al. 1990; 111. Cook et al. 1997; 112. He et al. 1995b; 113. Svärd and McNeil 1994; 114. Mason and Pashley 1991; 115. Snow et al. 1970; 116. Etman and Hooper 1979; 117. North and Holt 1968; 118. LaMunyon and Eisner 1993; LaMunyon and Eisner 1994; 119. Bissoondath and Wiklund 1997; 120. Wedell and Cook 1998; 121. Boggs and Watt 1981; 122. Labine 1966; 123. Clarke and Shepard 1962; 124. Oberhauser et al., unpublished manuscript.

Notes: Although many authors choose to exclude values of $P_2 = 0$ in their analyses, they are included in the calculation of means and variances here since they are genuine observations of P_2 after double matings.

+/− indicate direction of change in P_2 value with described variable. P: phenotypic markers; A: allozyme markers; M: molecular markers; R: irradiated male technique; C: chemosterilization; S: values based on direct observations of sperm removal and the assumption of random mixing and utilization of remaining sperm.

[a]Sperm utilization expressed as the proportion of offspring sired by the second male to mate, P_2, when a female is mated by two males. Estimates of variation in the species-specific value of P_2 provided in the form of the range and standard deviation (SD), and known causes of variation in P_2 noted. Species grouped within orders by commonly recognized major divisions.

[b]Values calculated from mixed broods only; in *Blatella germanica* 22% of females had mixed broods, 78% of females had $P_2 = 0$; in *Anopheles albimanus* only 0.88% of females remated and produced mixed broods.

[c]Bimodally distributed with modes at zero and one.

3 Avoidance of Sperm Competition I: Morphological Adaptations

3.1 Introduction

When females are polyandrous, selection will favor traits in conspecific males that (i) reduce the probability that their sperm will overlap spatially and/or temporally with those stored from previous males, and (ii) reduce the probability that their mate will engage in further matings. While such adaptations are "diametrically opposed" (Parker 1970e) the evolutionary functions of both are the same, to avoid sperm competition. Over the next three chapters I present a review of the literature on adaptations that are thought to have arisen to increase the probability that the copulating male fertilizes the ova subsequently laid by the female. Adaptations for the avoidance of sperm competition by males may often not be in the best interests of the female, and previous work has been criticized for adopting an androcentric view (Eberhard 1996). However, the role of the female in generating selection on males is implicit in that, without polyandry, there would be no sperm competition and no selection on males to avoid sperm competition. Moreover, despite specific adaptations, males may not be successful in avoiding sperm competition without the cooperation of females. While female influences are recognized here and throughout this volume, they will be discussed in detail in chapter 9.

3.2 Internal Fertilization

Parker (1970e) argued that internal fertilization most probably arose by sexual selection resulting from sperm competition. Males that can deliver their ejaculates closer to the site of fertilization should be at a selective advantage because they would incur a reduced risk and intensity of sperm competition. Moreover, gamete wastage would be greatly reduced compared with the delivery of ejaculates into the environment (Levitan 1996). Internal fertilization in the insects most likely originated in the production of spermatophores

(Alexander 1964; Hinton 1964). However, spermatophores appear to afford females considerable control over the insemination process (Simmons 1986; Sakaluk and Eggert 1996; see section 9.3) so that the selective pressures for the evolution of intromittent genitalia and direct insemination may also lie, to some extent, in a conflict of interests between males and females over insemination. Thornhill and Alcock (1983) provide a number of examples in which the aedeagus of male insects extends through the female genitalia to lie in a position where ejaculated sperm can be delivered directly to the site of storage and fertilization, often requiring the aedeagus to reach lengths in excess of the body itself (fig. 3.1). Extreme adaptations for delivering the ejaculate to the site of fertilization are seen in some groups of hemipteran bugs that practice traumatic insemination (Hinton 1964). In the bedbug *Cimex lectularis* and several anthocorid bugs, the male bypasses the usual route of insemination (the female's genital tract) and cuts through the body wall of the female's abdomen to ejaculate into the hemocoel. In some genera sperm migrate directly to the ovaries and fertilize eggs, thereby bypassing female storage of sperm and control of fertilization (Carayon 1966, 1974). With the exception of *Primicimex*, the females of all other cimicid genera have some form of special tissue or "spermalege" that receive male sperm. The function of the spermalege is currently unknown but Eberhard (1985) suggests that it may have arisen in females to regain control over the fertilization process. Traumatic insemination represents an extreme case of adaptation in males to achieve fertilization. In general, sperm competition does appear to have been important in the evolution of male genital morphology, and adaptations for the manipulation of rival sperm and/or the internal genitalia of females that increase the copulating male's fertilization success may be widespread in insects.

3.3 Sperm Removal and Repositioning

The importance of sperm competition as a selective pressure on male genitalia is most clearly recognized in the odonates, where genitalia appear to be adapted for the avoidance of sperm competition. The odonate penis is not associated with the primary genitalia, but rather is part of a secondary copulatory apparatus, itself modified from the ventral surface of the second and third abdominal segments (fig. 3.2). The male ejaculates into the sperm vesicle of the secondary copulatory apparatus after tandem formation but prior to copulation, a process known as sperm translocation. The zygopteran (damselfly) penis is a solid projection that is inserted into the female's genital opening prior to the transfer of sperm from the sperm vesicle to the bursa copulatrix. In contrast, the anisopteran (dragonfly) penis is an extension of the sperm vesicle, which may contain sperm at any time. The penis has a lumen through which sperm flow from the sperm vesicle into the bursa cop-

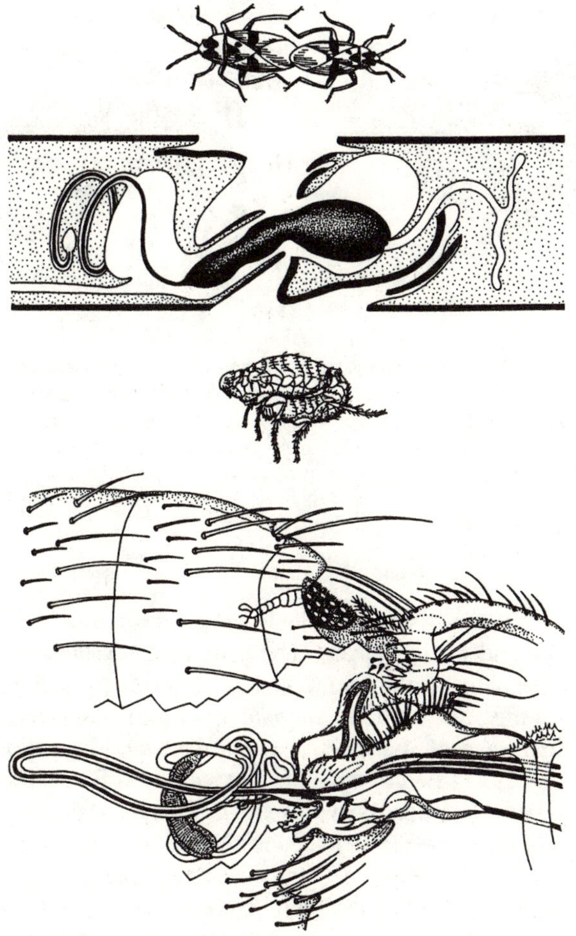

Figure 3.1 Intromittent genitalia may have arisen to facilitate the delivery of the ejaculate to the site of fertilization and/or sperm storage so that the risks of sperm competition could be reduced. In lygaeid bugs (*top*: from Weber 1930, reproduced by permission of Springer-Verlag) and ceratophylline fleas (*bottom*: from Holland 1955, reproduced by permission of the Royal Entomological Society) the aedeagus, shown in black, penetrates far into the female's spermathecal ducts to deliver sperm directly to the spermatheca.

ulatrix of the female. Moreover, the distal segment of the anisopteran penis is erectile, with a variety of processes evident only when in the erected state (fig. 3.3).

Waage's (1979) pioneering work with the damselfly *Calopteryx maculata* was the first of a series of studies showing that the zygopteran penis is adapted for the removal of rival sperm stored by the female (fig. 3.3). Sexual

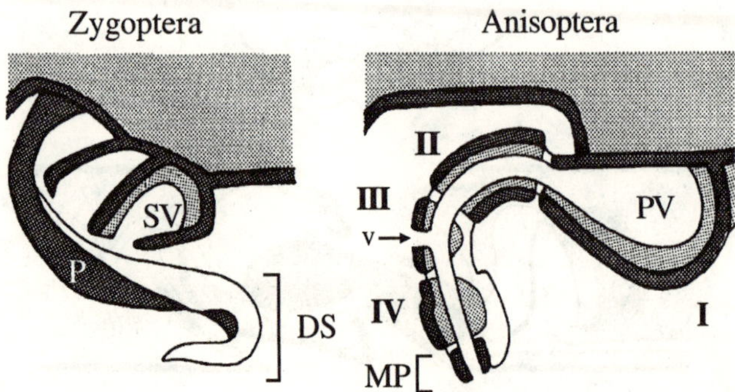

Figure 3.2 Schematic representation showing the morphological differences in the secondary genitalia between a male zygopteran (Coenagrionidae) and anisopteran (Libellulidae). Zygoptera: DS, distal segment; SV, seminal vesicle; P, penis shaft. Anisoptera: I–IV, segments of the penis; MP, medial process; v, valve through which sperm is translocated to the penis vesicle prior to copulation; PV, penis vesicle. Darkly shaded regions represent chitinous structures; lack of shading represents membranous structures. (From Waage 1984)

interactions are initiated when the male grasps the female's prothorax and the pair fly in tandem (precopula). The male then translocates sperm from the primary genitalia to the sperm vesicle. Copulation is initiated by the female who brings her abdomen forward to engage her genitalia with the copulatory apparatus of the male, the so-called wheel position (fig. 3.4). Copulation consists of two to three distinct stages (Miller and Miller 1981). During stage I of copulation the male exhibits a series of rhythmical abdominal flexions during which the abdomen is repeatedly arched and relaxed. The pair then enter a short period of quiescence that some authors recognize as stage II. During stage III, the rhythmic flexions continue but with increased intensity (Waage 1984; Siva-Jothy and Tsubaki 1989a; Cordero and Miller 1992). By interrupting copulations at various stages during pair formation and copulation, and determining the numbers of sperm contained within the sperm storage organ(s) of males and females, Waage (1979b) and others (see table 2.3) have shown that rival sperm are first removed from the female's sperm stores during stage I, before the male delivers his own ejaculate during stage III (see fig. 3.4). To aid the process of sperm removal, the penis and/or its projections are often endowed with backward facing spines or hairs that entrap stored sperm (fig. 3.3).

Morphological studies of zygopteran genitalia suggest that the avoidance of sperm competition by sperm removal is widespread in the taxon. Waage (1986) recognized four morphological types of zygopteran penis that differ in their degree of complexity (fig. 3.5). The simplest penis type (type I) appears to be ancestral within the calopterygid lineage and has neither appendages or spination. Increased complexity is associated with the presence

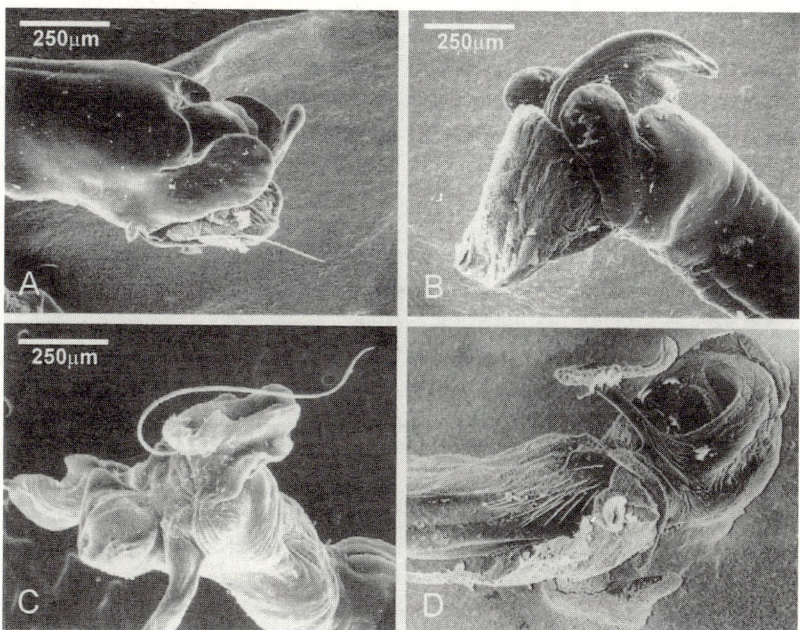

Figure 3.3 Scanning electron micrographs of the secondary genitalia for three species of Odonate. (A) is the uninflated and (B) the inflated penis of the dragonfly *Crocothemis erythraea*. The hornlike process in the inflated penis repositions sperm from rival males within the female's spermatheca. (C) is the inflated penis of the dragonfly *Orthetrum cancellatum*. The flagella carries barbs that remove sperm from the ducts of the spermatheca. (D) is the penis of the damselfly *Calopteryx maculata*. A sperm mass is attached to the ventral size by the stiff hairs, and a second sperm mass is attached to the backwardly facing barbs of the hornlike process that extends from the penis head. (A–C photos by M. Siva-Jothy. D from Waage, J. K. 1979. Dual function of the damselfly penis: sperm removal and transfer. *Science* 203: 916–918. Copyright 1979 American Association for the Advancement of Science)

of a recurved flap (type II), which can possess lateral horns (type III), or in the most complex penes (type IV) single or bifurcated flagella that are heavily spinated. Penis structures of type III appear to have had a single origin in the calopterygid lineage and a second convergent origin in the coenagrionid lineage. The bifurcation of the distal segment may also have multiple origins, independent of phylogenetic inertia, within the coenagrionid lineage (Waage 1986). Variation in penis morphology is associated with variation in the extent of sperm removal (Waage 1984, 1986). In species with type I and II penes, sperm are removed from the bursa copulatrix only (Miller and Miller 1981; Waage 1982). However, in species with type III or IV penes, the lateral horns and/or flagella can often enter the spermathecae, so that sperm are removed from both storage organs (Waage 1979b, 1986; Cordero and Miller 1992). This is not necessarily the case,

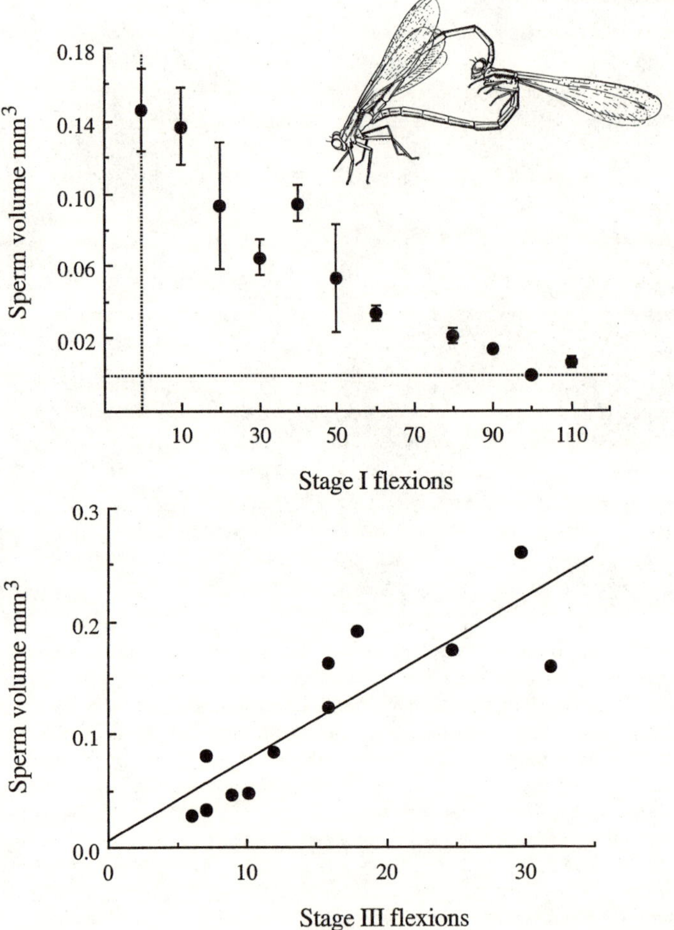

Figure 3.4 Changes in the volume of sperm stored in the sperm storage organ of the damselfly *Mnais pruinosa pruinosa* during copulation. During stage I of copulation (*top*), the volume of sperm declines with the number of abdominal flexions by the male until all sperm are removed. Following the quiescent stage II, the volume of sperm begins to rise (*bottom*) with the number of flexions in stage III as the copulating male delivers his own ejaculate. (Redrawn from Siva-Jothy and Tsubaki 1989a)

however, since in some species the horns and/or flagella do not appear long enough or have too large a girth to enter the spermathecae (e.g., Sawada 1995; Córdoba-Aguilar 1999). Moreover, sperm counts and molecular analysis of DNA derived from spermathecal sperm stores show that sperm may not be removed from the spermathecae during copulation (Siva-Jothy and Hooper 1995; Siva-Jothy and Hooper 1996).

By removing rival sperm from the female sperm storage organs, males

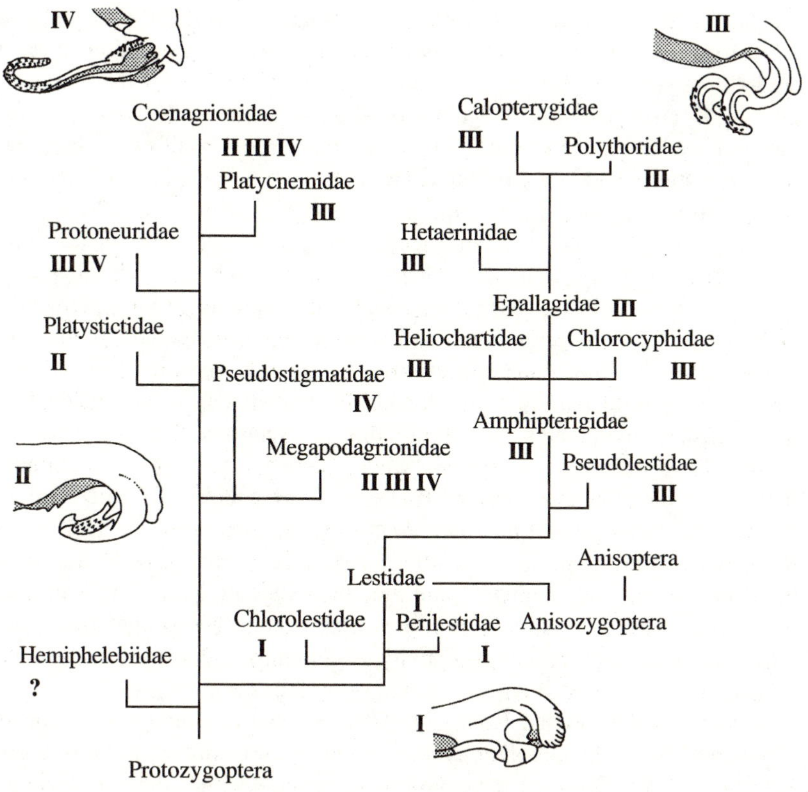

Figure 3.5 The distributions of four major types of penis morphology (I–IV) mapped onto a phylogeny of the Zygoptera (from Waage 1986). The four penis types shown are for I, *Lestes*; II, *Enellagma*; III, *Calopteryx*; IV, *Argia* (from Waage 1984). Type III genitalia appear to have had a single origin in the lineage leading to Calopterygidae and a second origin in the Coenagrionidae.

effectively avoid competition with rival males for the fertilization of ova. Studies of parentage in zygopterans have revealed that the last male to copulate with the female fertilizes the majority of ova (table 2.3). Moreover, because males greatly reduce the intensity of sperm competition by sperm removal, high paternity is also associated with low variance in paternity (table 2.3). Nevertheless, there is some variation in last-male fertilization success both within and between species, variation that appears to be associated primarily with the extent of sperm removal. For most species studied, immediate values of P_2 are uniformly high but in many P_2 declines with time since copulation. Decreased fertilization success for second males seems to occur as sperm mix in the sperm storage organs of the female. In their study of the calopterygid *Mnais pruinosa pruinosa*, Siva-Jothy and Tsubaki (1989a) found that the clutch laid immediately after copulation was fertilized exclusively by the last male to copulate. However, they found that paternity in subsequent clutches was dependent on the proportion of rival sperm removed from the female during copulation; thus males that removed all rival sperm maintained high fertilization success, while those that removed only half of the stored rival sperm obtained only half of the fertilizations in clutches laid 7–8 days later. These data are best explained by mixing of sperm within the sperm storage organs of females (Siva-Jothy and Tsubaki 1989a). Variation in the degree of sperm removal by males may represent a facultative behavior, dependent on the probability of female remating and on the time costs associated with removing rival sperm (see chapter 6). The same reasoning could account for interspecific variation. In species where females remate frequently, it may only be in the male's interest to ensure fertilization of the clutch laid immediately following copulation, because of the high probability that his sperm will be removed by future males. However, in species where female remating is uncommon, males will have an increased probability of being able to fertilize future clutches, so that it should then pay them to remove more of the rival sperm in order to ensure high paternity after sperm mixing. Interspecific variation in adaptations in male genitalia, and thus the extent of sperm removal, may thus reflect differences in selection pressures arising due to operational sex ratios at mating sites, female receptivity to remating, and/or the environmental potential for males to enforce remating by females.

Recently, Robinson and Novak (1997) attempted to examine variation in zygopteran genitalic morphology in a comparative context. Robinson and Allgeyer (1996) identified three groups of species within the coenagrionid genus *Ischnura* that differed in their ecology, life history, and behavior. They categorized species as polyandrous if published reports of mating females were common, the operational sex ratios at aquatic sites were male biased, and females exhibited color polymorphisms that persisted throughout reproductive life (color patterns in female zygopterans can influence female remating frequencies; see Cordero and Miller 1992; Forbes et al. 1995; Cordero et al. 1998). In contrast, monandrous species are rarely observed copulating, have balanced operational sex ratios at aquatic sites, and females are both monochromatic and exhibit pattern obscuring pruinescence that develops soon after adult emergence. Thus, Robinson and Allgeyer (1996) identified species of *Ischnura* that potentially differed in the risk of sperm displacement. Robinson and Novak (1997) conducted a multivariate analysis of eleven morphological measurements of the male penis, for a total of fifteen species of ischnid; the genus *Ischnura* is characterized by type IV morphology (fig. 3.5). Discriminant function analysis correctly assigned species to their mating system group, with the traits most useful for discrimination being the length and width of the terminal flagellae; monandrous species tend to have short and fat flagella while polyandrous species have long and thin flagella. Robinson and Novak (1997) also examined the distribution of microspination on the flagellum tips, concluding that monandrous species have, on average, less microspination than do polyandrous species. These data suggest that species in which females mate multiply have adaptations of the male genitalia for sperm removal, while those in which females mate only once do not. Congruent with the morphological data are data on copulation duration in these species. Robinson and Allgeyer (1996) found that species categorized as polyandrous copulated for significantly longer than do monandrous species, as might be expected given that sperm removal represents a significant time investment for copulating males (Siva-Jothy and Tsubaki 1989a; Siva-Jothy and Hooper 1995).

The comparative analyses of ischnid behavior and genitalia described above suffer from a number of problems that were recognized by the authors. First, the discriminant function analysis did not control for body size. Robinson and Novak (1997) did perform a principal component analysis with the data, yielding four principal components that collectively accounted for 76% of the variance in penis morphology. The first component, which is generally size related (Bookstein et al. 1985), accounted for 31% of the variance, leaving 48% unexplained. Such an analysis, while useful, cannot distinguish whether species fall into their mating system grouping in a discriminant function analysis based on penis morphology or differences in body size. In figure 3.6 I have presented the residual values from regressions of flagellum length and width on male body size using the data provided in Robinson and Novak (1997). Separation of groups based on flagellum length seems robust but not on flagellum width (fig. 3.6). Most importantly, Robinson and Novak's (1997) analysis was not performed within a phylogenetic context so that the association between penis morphology and mating system may be confounded by common ancestry. Unfortunately, given the lack of a suitable phylogeny for the ischnid group, this problem is currently insurmountable. Nevertheless, the observed association is an important first step in the study of genitalic evolution in this group and may prove robust to phylogenetic analysis, given that variation in zygopteran penis morphology does appear to be phylogenetically labile (fig. 3.5; Waage 1986). In this regard, it is interesting to note an apparent case of character loss in the coenagrionid lineage. Cordero et al. (1995) have shown that the penis of *Coenagrion scitulum* is devoid of spines and has shorter terminal horns than other coenagrionids. Perhaps not surprisingly, the males of this species are unable to remove sperm from the female sperm storage organs. Rather, they perform multiple copulation cycles during which they repeatedly inseminate the female so that the proportion of their sperm, relative to rival sperm, is increased. In this case, however, the simplified penis morphology does not appear to be associated with female monogamy.

The best comparative study of odonate genitalia is Miller's (1991) morphological analysis of 72 species from the anisopteran family Libellulidae. The libellulid lineage provides evidence consistent with the hypothesis that

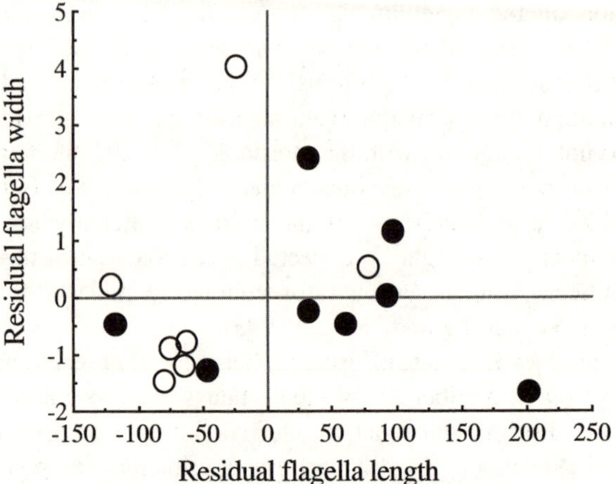

Figure 3.6 Variation in the length and width of the genital flagella across fifteen species of *Ishnura* after controlling for variation in body size. Relatively monandrous species are shown as open circles and polyandrous species as solid circles. Six of eight polyandrous species have relatively long flagella for their body size compared with one of seven monandrous species (exact probability from randomization test, $P = 0.0099 \pm 0.001$). In contrast, two of eight polyandrous species have wider flagella for their body size, compared with three of seven monandrous species ($P = 0.609 \pm 0.002$).

odonate genitalia are phylogenetically labile, and subject to selection via sperm competition. While the secondary genitalic apparatus differs morphologically from that seen in the zygoptera (fig. 3.2), its function in the avoidance of sperm competition appears the same. Copulatory behavior of anisopterans is similar to that of zygopterans, in that males first engage females in tandem before copulating in the wheel position (Waage 1984). They differ in that sperm may be present in the male's seminal vesicle at any time and copulations cannot be divided into discrete stages. Moreover, once the penis is inserted into the female's genital tract, movement of fluid from the first segment causes the distal process to inflate. Miller (1991) identified considerable variation in the presence and form of penis structures (the cornua, flagella, and inner lobes; fig. 3.7) associated with the medial process of the distal segment. He also documented considerable variation in female reproductive tract morphology, particularly in the volume and number of sperm storage organs (see also Siva-Jothy 1987a). One striking pattern to emerge from Miller's (1991) data was an association between penis structure and the volume of the female sperm stores. In species where females store a large volume of sperm in just one or two hypertrophied storage organs, the males have highly inflatable lobate penes (fig. 3.7). In contrast, in species where females store small volumes of sperm in several separate storage organs that connect with the bursa copulatrix via narrow ducts, the males possess flagella, or flagellalike inner lobes on the medial process. Again, Miller's

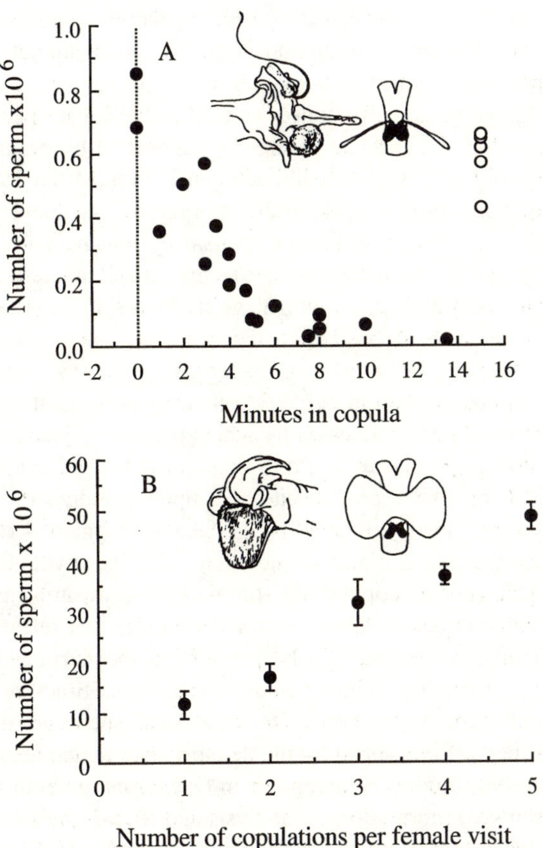

Figure 3.7 Sperm removal and repositioning in Libellulid dragonflies. (A) In *Orthetrum cancellatum*, females store small volumes of sperm in the bursa and spermathecae that connect with the bursa copulatrix via narrow ducts. The males possess flagella and flagellalike inner lobes on the medial process that enter the stores to remove sperm during copula. The solid circles show the decline in sperm number observed by interrupting pairs in copula. After removing rival sperm, the male immediately delivers his own ejaculate (open circles) (from Siva-Jothy 1987b). (B) In *Crocothemis erythraea*, females store large volumes of sperm in the bursa; the males possess large inflatable lobes that pack sperm to the rear of the sperm storage organ, giving their own sperm precedence when females oviposit immediately after copulation. The data show the increase in the mean (±SE) numbers of sperm stored by females following repeated copulations (from Siva-Jothy 1988). Vignettes show the male's inflated penis (*left*) and the female's sperm storage organs (*right*) (from Siva-Jothy 1984). Scanning electron micrographs of male genitalia are shown in figure 3.3.

(1991) analysis was not performed in a phylogenetic context so that he recommended caution in its interpretation. Nevertheless, examination of the distribution of penis type with taxonomic affiliation provides little evidence that penis type is phylogenetically conserved; inflatable lobate penes and those with flagella and/or flagella-like inner lobes appear to be evenly distributed across the ten subfamilies of Libellulidae considered (Miller 1991).

The apparent dichotomy in genitalic morphology in the Libellulidae is associated with differences in the mechanism by which males avoid sperm competition. Females in the genus *Orthetrum* are characterized by small-volume sperm stores and males with flagella. Siva-Jothy's (1987b) study of *Orthetrum cancellatum* showed that, like their zygopteran relatives, males remove all rival sperm from the female sperm storage organs over the course of a 15-min copulation, before delivering their own ejaculate (fig. 3.7). Miller (1990) proposed three processes by which the male's penis could remove sperm from storage. First, the size and shape of the flagellum is such that it can be inserted into the spermathecae and, upon withdrawal, entrap sperm under the distal proximally directed barbs. Second, segment 4 of the penis is pressed against the lateral plates of the vagina and the rhythmic movements of the males penis during copulations stimulate the campaniform sensilla and cause reflex contractions of the spermathecal muscle that release sperm into the bursa. Finally, males may flush sperm from the bursa with their own ejaculate, given that they transfer about twenty-four times the volume of sperm as can be stored in the bursa. On withdrawal, sperm are removed from the vagina as they are entrapped by the deflating lateral and apical lobes, and the medial process. Studies of parentage in *Orthetrum* are entirely consistent with Miller's (1990) morphological analysis and Siva-Jothy's (1987b) examination of female sperm stores; copulations in excess of 12 min result in complete paternity for the copulating male that persists across successive ovipositions (Hadrys et al. 1993). Shorter copulations result in the removal of fewer rival sperm (Siva-Jothy 1987b) so that gradual mixing of ejaculates results in reduced paternity in successive clutches (Hadrys et al. 1993). The long curved cornua of *Sympetrum danae* function in a convergent manner to the flagella of *Orthetrum* spp., entering the spermathecae to remove sperm from storage (Michiels and Dhondt 1988). Again, sperm are also flushed from the bursa by the male's own ejaculate (Michiels 1989) so that the last male to copulate gains up to 100% of the fertilizations, dependent on the time he spends copulating and thus the amount of rival sperm he removes (Michiels 1992).

In contrast, the inflatable lobate penes of species such as *Crocothemis erythraea* or *Nesciothemis farinosa*, appear to function as sperm packers rather than removers. That is, males reposition rival sperm to the back of the female's large sperm storage organ before delivering their own ejaculate (see also Waage 1984). Thus, the number of sperm stored by females increases linearly with the number of copulations (Miller 1984; Siva-Jothy 1988) (fig.

3.7). The resultant stratification of ejaculates within the female sperm store results in the last male gaining most, if not all, fertilizations at the oviposition following copulation. However, subsequent mixing of ejaculates results in a steady decline in last-male fertilization success with time after copulation, presumably to a level where each male sires offspring in proportion to the total number of males the female has copulated with (Siva-Jothy and Tsubaki 1994).

To claim that anisopterans fall into discrete categories of removers or packers is to obfuscate reality. Clearly, there is a continuum between species in which males remove all rival sperm (Siva-Jothy 1987b) and those in which males only reposition rival sperm (Siva-Jothy 1988). Between lie species such as *Erythemis simplicollis* that adopt both strategies, removing a proportion of rival sperm and repositioning those remaining (McVey and Smittle 1984). Likewise, there is a continuum in anisopteran morphology from simple penes through those with intermediate complexity in lobe structure to those with both inflatable lobes and additional elongate cornua and/or flagella (Miller 1991).

Convergent adaptations in genitalia that function in sperm removal have been reported from two other insect orders. Male longicorn beetles *Psacothea hilaris* have two discrete stages to copulation. In stage 1, males repeatedly insert and withdraw the penis, while in stage 2 the penis is inserted once and the pair remain in copula until mating ends (Yokoi 1990). The number of sperm in the female sperm storage organs is reduced by 98% during stage 1 of copulation, and the penis has a comb of microbristles and a scooplike structure at its distal end that removes sperm from the female (Yokoi 1990). The penis of *Tribolium castaneum* also has an array of chitinous spines which entrap rival sperm present in the female's bursa copulatrix. Although Haubruge et al. (1999) conclude that males gain the majority of fertilizations via the removal of rival sperm, immediate fertilization success for the second male is just 66% and declines with subsequent mixing of ejaculates to equal paternity for both males (Schlager 1960), indicating that sperm are not removed from the mixture (presumably those in the spermatheca) that is used for fertilization. Nevertheless, male postcopulatory courtship increases the number of sperm stored in the spermatheca and thus the copulating male's fertilization success (Edvardsson and Arnqvist 2000). Rival male sperm are clearly removed from the bursa copulatrix, as these can be transferred to subsequent mates of the sperm-removing male, resulting in rival males gaining fertilizations with females they do not mate with (Haubruge et al. 1999). Removal of rival sperm from the bursa copulatrix will increase the copulating male's chances of gaining fertilizations with his current mate, since this will prevent its translocation to the spermatheca by the female during the current male's postcopulatory courtship. Finally, in tettigoniids, insemination is indirect via an externally attached spermatophore (section 9.3). Although there is no penis, male *Metaplastes ornatus* have an enlarged spinated subgenital plate that is inserted into the females reproductive tract in phase 1 of copulation (von Helversen and von Helversen 1991). The structure appears to be used to stimulate release of sperm from the spermatheca and to evert the female's genital chamber, which is then cleared of previously stored sperm by the female herself. The process removes some 85% of stored sperm before the male transfers his own spermatophore (see fig. 9.6).

The process of sperm removal in *M. ornatus* results in considerable damage to the female's reproductive tract (von Helversen and von Helversen 1991). Crudgington and Siva-Jothy (2000) have similarly found that sclerotized spines on the aedeagus of *Callosobruchus maculatus* puncture the female's reproductive tract during copulation. Genital tract damage appears to reduce female lifespan so that there is sexual conflict over the duration of copula and the extent of damage sustained by the female; females terminate copula by kicking the male. It is not yet clear how males benefit from damaging the female genital tract. Crudgington and Siva-Jothy (2000) suggest that by doing so males may dissuade females from copulating again, and thereby avoid sperm competition from future males. However, Eady (1994) has shown that second males achieve high fertilization success via a mechanism involving sperm displacement, and mechanical stimulation of the female tract may facilitate displacement in a manner similar to that in *M. ornatus* (see also the discussion on sensory exploitation in section 9.4).

3.4 Alternative Explanations for Complex Genitalia

In general, male genitalia exhibit extreme patterns of evolutionary divergence (Eberhard 1985). Indeed, male genitalia have long been used as a taxonomic tool for distinguishing between species. Nevertheless, it is only recently that evolutionary biologists have begun to question the processes that have lead to the patterns of genitalic morphology that we see. The presence of complex male genitalia such as that typical of the odonates is by no means indicative of sperm removal and/or repositioning (see below). Three principal mechanisms have been proposed to explain genitalic evolution; the lock-and-key hypothesis, the pleiotropy hypothesis, and the sexual selection hypothesis.

The lock-and-key hypothesis is perhaps the most widely recognized, dating back to Dufour (1848). Species-specific genitalia are proposed to have evolved via selection for species isolation. Thus, hybrid matings are avoided because males must have the appropriate genitalia (the key) in order to achieve coupling with the genitalia of conspecific females (the lock). There are a variety of theoretical objections to the hypothesis (Eberhard 1985) and empirical studies, both across species (Eberhard 1985; Shapiro and Porter 1989) and within species (Eberhard 1992; Goulson 1993; Arnqvist et al.

1997), fail to provide support for it. Mayr (1963) suggested that because genitalia are internal structures, they should be less subject to natural selection and thus able to evolve in an arbitrary fashion due to pleiotropic effects of selection acting on general morphology. The pleiotropy hypothesis has been criticized for a number of reasons, including the lack of rapid and divergent evolution in the genitalia of species, such as the odonates, in which the primary genitalia are not used to inseminate the female (Eberhard 1985). Nevertheless, there is also some empirical support for the hypothesis (Arnqvist et al. 1997).

The sexual selection hypothesis is the most recently advanced explanation for the patterns of morphological variation seen in male genitalia (Eberhard 1985). Arnqvist (1997a) recognized three potential models for the evolution of genitalia under sexual selection. First, Eberhard (1985) proposed that male genitalia represent arbitrary traits that arise due to Fisherian runaway sexual selection (see section 1.1). The genitalia are proposed to function as internal courtship devices, with males best able to stimulate females siring more offspring. Females that favor males with elaborate genitalia will produce sons with elaborate genitalia, and daughters who require the appropriate stimulation during copulation, so that male genitalia evolve rapidly until checked by natural selection. Second, variation in male genitalic morphology may arise because of sexual conflict over fertilization events (Lloyd 1979; Alexander et al. 1997). Intromittent genitalia thus function to induce a female to receive more sperm, for example through prolonged copulations, and/or to use sperm from a particular male, even if it is not in her best interests to do so. Thus, as traits in females evolve to allow them to control insemination and/or fertilization events, traits in male genitalia evolve to circumvent female control, a process Parker (1979) termed an evolutionary chase, and Holland and Rice (1998) later termed chase-away sexual selection. Finally, male genitalic morphology can evolve under selection arising from sperm competition. Thus, males with adaptations to the genitalia that best remove and/or reposition rival sperm or are better able to deposit sperm directly into the sperm storage organ(s) will obtain more fertilizations so that directional selection under sperm competition will favor increased genitalic complexity (see section 3.2). I recognize a fourth model for the evolution of male genitalic complexity, based on direct competition among males for mates (see also Darwin 1871; Richards 1927; Thornhill and Alcock 1983). In many insects, matings occur on or around resources, such as feeding or oviposition sites. Competition for females is intense so that takeover attempts by searching males are common. Moreover, copulatory mate guarding is a common behavioral strategy for avoiding sperm competition (chapter 5). Thus, once coupled, any morphological trait that decreases the probability that a male will be displaced from the female will be favored by sexual selection. Complex genitalic morphology may thus represent holdfast devices that prevent the occurrence of takeovers.

Distinguishing between alternative hypotheses for genitalic evolution will be difficult. There are currently few studies that address this issue directly. Arnqvist (1997a) suggested a framework with which to distinguish between the three principal mechanisms (lock-and-key, pleiotropy, and sexual selection) based on patterns of selection, morphological variation, and the inheritance of morphological traits in single-species studies. Arnqvist's (1997) and Arnqvist and Thornhill's (1998) studies of the genitalia of water striders clearly refuted the lock-and-key hypothesis, but they found indirect selection on genitalic traits that arose due to selection acting on phenotypically correlated traits, consistent with the pleiotropy hypothesis. They also revealed patterns of genetic and phenotypic variation in genitalic morphology consistent with the sexual selection hypothesis. Distinguishing between models of sexual selection will prove more difficult. The model of direct male competition can be distinguished because it predicts that variation in male genitalia should be associated with the ability of males to avoid takeovers, and thus influence male mating success. Thus, Goulson (1993) was able to exclude direct male competition in his study of the butterfly *Maniola jurtina*. In contrast, Ferguson and Fairbairn (2000) found significant directional selection across two generations on the length of the external genitalia of the water strider *Gerris remigis*. The external genitalia act as clasping devices used to subdue females and to maintain contact with them during postcopulatory mate guarding and so are under direct sexual selection. The remaining three hypotheses (Fisherian runaway, sexual conflict, and sperm competition) all predict variation in male fertilization success that is attributable to variation in genitalic morphology. However, without knowledge of the precise mechanisms of sperm storage and utilization, it is not possible to determine the degree to which males and females control fertilization events, and thus attribute genitalic evolution to one particular model of sexual selection (Simmons et al. 1996; Arnqvist 1997a; Simmons and Siva-Jothy 1998).

The evidence for sperm competition selecting for odonate genitalic morphology seems strong (section 3.2). Adaptations of male genitalia function in removing and/or repositioning sperm in a manner that increases male fertilization success and thus fitness. Nevertheless, such evidence is also supportive of the sexual conflict model for genitalic evolution. In reality, the two models are not mutually exclusive, since the focus of sexual conflict, if it exists, may be over male interests in sperm competition (Knowlton and Greenwell 1984; Rice 1996; Arnqvist 1997a). Eberhard (1985) concluded that, in general terms, sperm competition was not sufficient to account for rapid and divergent evolution of male genitalia because males rarely gain direct access to sites of sperm storage. His conclusion was overstated, however, because the structures involved in manipulating sperm can have their influence via mechanical stimulation of the female reproductive tract (Miller 1990; von Helversen and von Helversen 1991; Eberhard 1996; Córdoba-Aguilar 1999) (see section 9.4). In this case males compete indirectly through differing abilities to stimulate the female, itself a process of cryptic female choice. Thus, as with models of sexual conflict, models of cryptic female choice and sperm competition are unlikely to be mutually exclusive.

The problems associated with assigning specific models of sexual selection to explain variation in male genitalic morphology are illustrated by recent studies. Males of the tenebrionid beetle *Tenebrio molitor* have a heavily spinated sheath covering the central shaft of the penis. When inserted into the bursa copulatrix of the female, the sheath rolls back, producing a "scouring" effect (Gage 1992). Gage (1992) found that the spines entrapped and removed sperm from the bursa copulatrix when copulation occurred soon after a female's previous mating, and suggested that the penis may function to reduce sperm competition from rival males. Siva-Jothy et al. (1996) tested the sperm competition hypothesis in detail, estimating both the amount of sperm removed by males, and the subsequent patterns of second-male paternity. They found that the amount of DNA (used as a measure of sperm removal) adhering to the sheath of the first male to copulate with an unmated female did not differ significantly from the amount adhering to the second male's sheath when females mated twice, irrespective of the interval between first and second matings. Thus, males also remove sperm from females who have not mated previously. Moreover, the proportion of offspring sired by the second male was not higher when a second male mated immediately, as might be expected if males remove recently deposited rival sperm from the female's bursa copulatrix. Siva-Jothy et al. (1996) concluded that sperm adhering to the spinated sheath most likely belong to the copulating male and leak back into the posterior part of the female's reproductive tract during insemination. Thus, despite appearances, genitalic morphology appears not to be related to sperm displacement in *T. molitor*. Rather, the spines on the penis sheath appear to maintain genital contact during takeover attempts by rival males (Siva-Jothy et al. 1996).

In support of Fisherian runaway selection on male genitalia, Eberhard (1985) presented evidence of an association between female remating and species specificity of male genitalia across 32 species of butterfly from the genus *Heliconius*; in monandrous species male genitalia were simple in form and differed little between species. Such a pattern, however, is consistent with all models of sexual selection on male genitalia. For example, sperm competition occurs only when females mate with more than one male so that genitalic adaptation for manipulating rival sperm is only expected in polyandrous species. On the other hand, takeover attempts by rival males will only occur if the captured female will be receptive to remating. Thus, holdfast devices associated with male genitalia should only be expected in species with polyandrous females. A recent examination of the genitalic morphology of apoid bees similarly found that complexity of the endophallus, an inflatable structure akin to the inflatable lobes of some anisopteran secondary genitalia, was associated with the mating system (Roig-Alsina 1993). As with the heliconids, monandrous species have simple endophalli while polyandrous species have complex endophalli. Roig-Alsina (1993) noted that this pattern was consistent with both female choice and sperm competition models of sexual selection. Nevertheless, the fact that, in general, the sclerotized parts of the apoid genital capsule form part of a clasping device suggests that the endophallus is also involved in clasping the female. For many bees, competition over females is intense, takeover attempts are frequent, and males often copulate in flight (reviewed in Roig-Alsina 1993). Adaptations that facilitate the maintenance of a strong genital hold on the female once intromission is achieved will be the focus of strong sexual selection (see also Thornhill and Alcock 1983).

A recurrent problem with the comparative studies described above is the lack of phylogenetic control. Recently, Arnqvist (1998) used the method of phylogenetic contrasts to compare the morphological divergence of male genitalia between insect clades that differed in their mating system and thus opportunity for sexual selection. He found that across four different orders of insect, male genital evolution was more than twice as divergent in polyandrous clades than in monandrous clades. This pattern was not found for other morphological traits, strongly supporting the hypothesis that sexual selection drives the evolution of male genitalia. The most detailed phylogenetic analysis of genitalic morphology currently available is Liebherr's (1992) study of 36 species of carabid beetle in the *Platynus degallieri* species group. Liebherr (1992) found that evolutionary changes in the morphology of the bursa copulatrix (specifically, the origin of a lobe on the dorsal surface of the bursa) preceded evolutionary changes in the morphology of the male penis (specifically, modifications to the apex of the aedeagal medial lobe). Only phylogenetic reconstructions of this type can enable us to assign a cause and effect relationship in the evolution of male and female reproductive tract morphologies. An understanding of such relationships is crucial in distinguishing between alternative models of sexual selection for genitalic evolution. Although a Fisherian process such as that envisaged by Eberhard (1985) predicts a genetic correlation between the trait in males (genitalia) and the preference in females (biased sperm use), it does not predict evolutionary changes in female reproductive tract morphology. Indeed, Eberhard (1985) argued that the relative simplicity of female reproductive tract morphology was evidence in favor of his model of female choice. In contrast, the sexual conflict or chase-away model of sexual selection does predict that adaptations in female genitalia should precede, and be countered by, adaptations in male genitalia. Although a Fisherian process may be ruled out as an explanation for genitalic evolution in the *P. degallieri* species group, changes in male genitalia may follow changes in female genitalia because of sperm competition and/or direct male-male competition in the absence of sexual conflict, if the initial changes in female morphology arise because of a naturally selected advantage associated with, for example, more efficient sperm storage or egg alignment during fertilization. The association between male genitalic complexity and female genitalic complexity in anisopteran odonates is similarly consistent with a sperm competition or combined sperm competition and sexual conflict model of sexual selection (Siva-Jothy 1987a).

The strongest evidence for sexual selection acting on male genital morphology will come from intraspecific studies, like those in odonates, that examine the influence of variation in male genital morphology on fertilization success. Tadler (1999) examined selection acting on the aedeagus processus of the lygaeid bug *Lygaeus simulans*. The aedeagal processus reaches through the length of the female's spermathecal ducts to deliver sperm into the spermathecae (see fig. 3.1). Using full sisters and randomly selected males, Tadler (1999) found stabilizing selection on aedeagal processus length by scoring males as having been either successful or unsuccessful in delivering sperm to the female's spermatheca during a single copulation. Unfortunately Tadler (1999) did not examine the critical prediction that variation in aedeagal process length was associated with variation in fertilization success. Nevertheless, Tadler's (1999) observations are consistent with Eberhard et al.'s (1998) examination of genital allometry in insects. Eberhard et al. (1998) concluded that, in general, male genitalia scale with body size with exponents less than 1 so that there is a single optimal genital size for all males. This argument and the stabilising selection noted by Tadler (1999) are consistent with the traditional lock-and-key hypothesis for genital evolution. Nevertheless, Eberhard (1998) argued that unless there were size assortative mating there could also be sexual selection on males for genitalia appropriate for the size of the most typical (average) female, because males with average genitalia could accurately align structures for stimulating or accurately position sperm for storage in the majority of females encountered. There are problems associated with both Tadler's (1999) and Eberhard et al.'s (1998) observations. First, Tadler (1999) used full sibling females that would have been genetically and phenotypically similar in their genital morphology. Thus, the stabilizing selection observed may have been specific to the size class of female used in Tadler's (1999) experiments. Indeed, in a second study in which females were assigned randomly to males, there was no evidence of selection acting on aedeagal processus length (Tadler and Nemeschkal 1999), suggesting that selection is unlikely in natural populations. Second, as noted by Green (1999), the true relationship between morphological traits is underestimated by the standard least-squares regression techniques used by Eberhard et al. (1998) so that in reality there may not be a single optimal size for male genitalia.

Three recent studies provide good evidence that male genital morphology can influence paternity. In the water striders *Gerris lacustris* and *G. lateralis*, the size and shape of the genital sclerites have a marked impact on male fertilization success (Arnqvist and Danielsson 1999a; Danielsson and Askenmo 1999) (see fig. 3.8). Moreover, different traits of male genitalia appear to serve different functions; the dorsal and ventral genital sclerites of

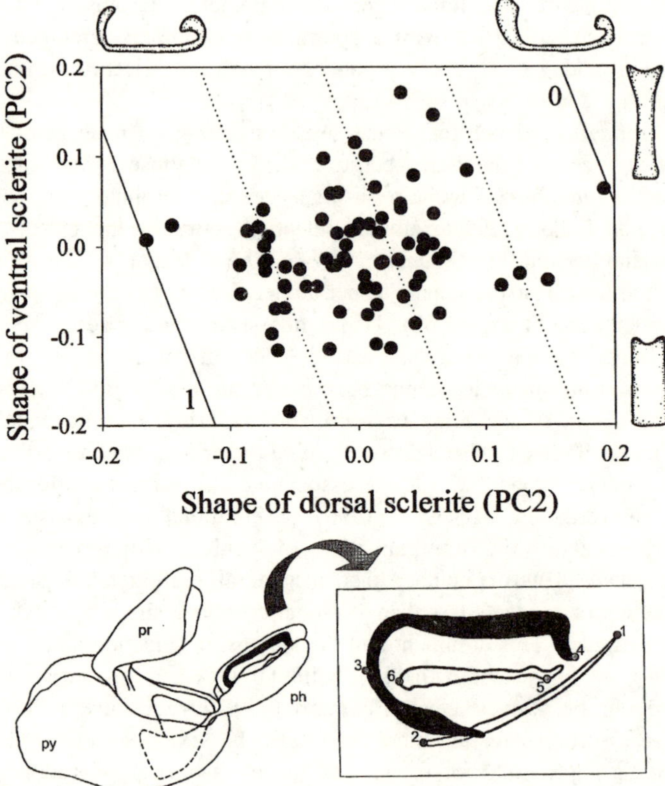

Figure 3.8 The influence of male genital morphology on fertilization success in the water strider *Gerris lateralis*. The contours represent the linear topography of fitness measured as the proportion of offspring sired by the second male, or P_2. The solid contour at the top right is for $P_2 = 0$ and at the bottom left for $P_2 = 1.00$. The positions of individual values of P_2 are indicated on the contour in relation to two measures of genital morphology, the shape of the dorsal and ventral sclerites. The second principal components are plotted because these were the shape dimensions that tended to be most related to fertilization success. Males with high paternity have stout ventral sclerites and narrow and apically incurved dorsal sclerites (from Arnqvist and Danielsson 1999a). The genital morphology for *G. lacustris* is shown below: *left*, the two posterior genital segments; py, phygophore; pr, proctiger; ph, phallus; and *right*, lateral view of the vesica with measurement landmarks. The genital sclerites are associated with the phallus, which is inflated during copulation (dorsal sclerite shown in black, lateral and ventral sclerites in white). (From Danielsson and Askenmo 1999, with permission of Springer-Verlag)

male *G. lateralis* influence the ability of males to obtain fertilizations with previously mated females, while the lateral genital sclerites influence a male's ability to gain fertilizations when his mate copulates with a second male. Males of the chrysomelid beetle *Chelymorpha alternans* have an elongated genital flagellum that is threaded into the spermathecal duct of the female. A spermatophore is then deposited into the bursa copulatrix of the female and sperm migrate up the flagellum to the spermatheca (Eberhard 1996). Rodriguez (1995) found that males with longer flagella had more sperm stored in the spermatheca and less sperm lost from the female's gonopore after copulation than males with shorter flagella. Consequently, the success of males in obtaining fertilizations was dependent on the relative lengths of their flagella (Rodriguez 1994a) (see also section 9.3). Although these examples clearly show that variation in male genital morphology can influence fertilization success, and thus be subject to sexual selection, it is difficult to distinguish between alternative models of sexual selection from the available data; an effect of genital morphology on fertilization success is expected from all sexual selection models. Nevertheless, female morphology is clearly an important determinant of male success in gaining fertilizations in *G. lateralis* (Arnqvist and Danielsson 1999a), a result not predicted from a purely sperm competition model. On the other hand, copulatory courtship had no influence on fertilization success in Arnqvist and Danielsson's (1999a) study, a result that is inconsistent with a purely Fisherian preference model.

It is naive to expect a single general model of sexual selection to explain the rapid and divergent evolution of male genitalia. As is true for classical secondary sexual ornaments (Evans and Hatchwell 1992a,b; Andersson 1994; Berglund et al. 1996), sexual selection on male genitalia is likely to arise through a multitude of channels so that our task will be to determine the relative importance of different selection processes. The key will lie in uncovering the mechanisms that govern sperm storage and utilization (Simmons and Siva-Jothy 1998). Only then will we be in a position to determine the extent to which males and females influence sperm storage and fertilization events. Currently, our knowledge of the mechanisms of sperm utilization is limited, although recent studies provide a protocol by which male and female influences can be assessed (Simmons et al. 1996; Simmons et al. 1999b; see chapter 9). To conclude that male genitalic morphology results from selection due to sperm competition alone requires a knowledge of the mechanism by which genitalic morphology influences sperm storage and the impact this has on male fertilization success. To exclude simultaneous processes of female choice and/or sexual conflict will require a demonstration that variation in male fertilization success is due predominantly to males, that any variation due to females is random with respect to male morphology, and that male and female reproductive interests coincide.

3.5 Summary

There is growing evidence that sexual selection generated through variation in fertilization success is an important process driving rapid and divergent evolution of male genitalic morphology. Adaptations in male genitalia can function directly in the mechanical removal and/or repositioning of rival sperm, so that males bearing genitalic traits that are best able to avoid the spatial and/or temporal overlap of their ejaculate with that of rivals will be favored by sexual selection. Additionally or alternatively, genitalic traits can function indirectly by stimulating the female reproductive tract to release sperm from storage, placing them at a site where the copulating male can dilute and/or flush away rival sperm using his own ejaculate. It is difficult to identify the precise nature of sexual selection operating on male genitalia. A pure sperm competition model would assume the female to be a passive vehicle within which males play out their competitive battles. However, females can be directly involved in the sperm displacement process, either assisting or resisting sperm displacement, so that the net action of sexual selection on male genitalia will involve elements of both male competition and female choice. Moreover, it is important to remember that in some cases aspects of male genitalia may evolve under classical preinsemination sexual selection, where complex genitalic structures serve as holdfast devices effective in avoiding takeovers from rival males or in subduing reluctant females during precopulatory and copulatory struggles. Despite the problems associated with identifying the exact process of sexual selection, it is clear that male genitalia have been elaborated by sexual selection in much the same way as male weapons and/or ornaments, obfuscating the dichotomy between primary and secondary sexual traits.

4 Avoidance of Sperm Competition II: Physiological Adaptations

4.1 Introduction

Multiple mating by females increases both the risk and intensity of sperm competition for males. In chapter 3 I discussed how adaptations in male genitalia can be favored by sexual selection if they reduce the temporal and/or spatial overlap of ejaculates present in the female's reproductive tract at the time of fertilization, and thus avoid sperm competition from a female's previous mates. Nevertheless, males are still faced with the risk of sperm competition from rivals that will mate in the future. Consequently, selection should favor any adaptation in males that reduces the probability that females will remate. In this chapter I examine physiological adaptations that appear to function in reducing female remating.

An examination of the reproductive tracts of male insects reveals a remarkable diversity of secondary glandular structures collectively referred to as accessory glands (Leopold 1976; Chen 1984; Gillott 1996). Like male genitalia, accessory glands and their products appear to have undergone rapid and divergent evolution. The number of glands can vary considerably across taxa; a single pair of accessory glands occur in *Drosophila melanogaster* (Gromko et al. 1984a) while in *Locusta migratoria migratorioides* there are 15 pairs of glands that are closely associated with the seminal vesicles (Gregory 1965; fig. 4.1). Moreover, male accessory glands show divergent evolutionary origins, being derived from the mesoderm, the ectoderm, or both (Davey 1985b), so that across taxa male accessory glands are not necessarily homologous structures (Leopold 1976). Nevertheless, there does appear to be functional convergence in that the secretions produced by these glands enter the ejaculatory duct and are transferred to the female during insemination. The biochemical properties of accessory gland products show an even greater divergence than gland morphology (Chen 1984). Again, there is a functional convergence in the principal effect male accessory gland products have on the female; they appear to be responsible for, among other things, a reduction in the probability of remating and thus the risk of sperm competition from rival males (table 4.1). Research on male accessory glands and their products has focused primarily on the biochemis-

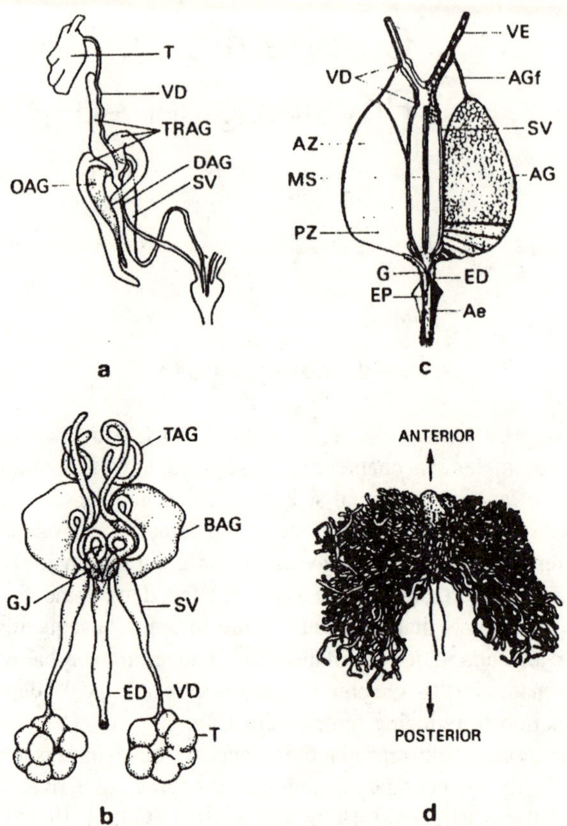

Figure 4.1 Morphological diversity of male accessory glands in insects. (a) *Rhodnius pulex*; (b) *Tenebrio molitor*; (c) *Aedes aegypti*; (d) *Acheta domestica*. Ae, aedeagus; AG, accessory gland; AGf, accessory gland filament; AZ, anterior zone; BAG, bean-shaped accessory gland; DAG, duct of accessory gland; ED, ejaculatory duct; EP, endophallus; G, gonopore; GJ, gland junction; MS median division of vesicles; OAG, opaque accessory gland; PZ, posterior zone; SV, seminal vesicle; T, testis; TAG, tubular accessory gland; TRAG, transparent accessory gland; VD, vas deferens; VE, vas efferens. (From Chen 1984, with permission, from the Annual Review of Entomology, Volume 29 © 1984 by Annual Reviews www.AnnualReviews.org)

try of male products and the physiological mechanisms of their proximate action on females. It was Parker (1970e) who first suggested that certain aspects of insect reproductive physiology may represent the products of sexual selection via sperm competition, a thesis that has gained considerable strength from recent attempts to incorporate evolutionary thinking into this area of research (Thornhill and Alcock 1983; Parker and Simmons 1989; Cordero 1995; Eberhard and Cordero 1995; Eberhard 1996). In this chapter I examine the functional significance of secretory products transferred by

Table 4.1

Male accessory gland products and their possible role in sperm competition avoidance.

Species	*Accessory Gland Product*	*Response of Female*	*Source* (Notes)
ORTHOPTERA			
Acrididae			
Locusta migratoria	Spermatophore/mating plug	↓R ↑O	Parker and Smith 1975; Lange and Loughton 1985
	Seminal fluid		
Schistocerca gregaria	Seminal fluid	↑O	Leahy 1973a,b
Melanoplus sanguinipes	Seminal fluid	↑O	Pickford et al. 1969; Friedel and Gillot 1976
Chorthippus brunneus	Seminal fluid	↑O	Butlin et al. 1987
Gomphocerus rufus	Seminal fluid	↓R ↑O	Hartmann and Loher 1996
Gryllidae			
Teleogryllus commodus	Prostaglandin synthetase + arachidonic acid	↓R ↑O	Loher 1979, 1981; Stanley-Samuelson and Loher 1986; Stanley- Samuelson et al. 1986 (2)
Acheta domesticus	Prostaglandin synthetase	↑O	Destephano and Brady 1977 (2)
Gryllus bimaculatus	Seminal fluid	↓R ↑O	Bentur et al. 1977; Simmons 1988a; Loher et al. 1993 (2)
Tettigoniidae			
Requena verticalis	Seminal fluid	↓R	Gwynne 1986 (2)
Kawanophila nartee	Seminal fluid	↓R	Simmons and Gwynne 1991 (2)
Decticus verrucivorus	Seminal fluid	↓R ↑O	Wedell and Arak 1989; Wedell 1993a (2, 3)
BLATTOIDEA			
Blatella germanica	Mating plug	↓R	Cochran 1979
HEMIPTERA			
Prokelisia dolus	Seminal fluid	↓R	Heady 1993 (2, 3)
Balclutha incisa	Seminal fluid	↓R	Nuhardiyata 1998 (2, 3)

Table 4.1 (*cont.*)

MECOPTERA			
Bittacus apicalis	Seminal fluid?	↓R ↑O	Thornhill 1976b
Harpobittacus nigriceps	Seminal fluid?	↓R ↑O	Thornhill 1983
DIPTERA			
11 families, 29 species	Various	↓R ↑O	See full review of Chapman et al. 1998 (1–5)
Anopheles spp.	Mating plug	↓R?	Giglioli and Mason 1966 (3)
	Seminal fluid	↓R	Craig 1967; Yuval and Fritz 1994 (2)
Aedes spp.	α matrone	↑O	Lum 1961; Liles 1965; Craig 1967; Fuchs et al. 1969;
	α & β matrone	↓R	Klowden and Chambers 1991 (3, 5)
Culex tarsalis	Seminal fluid	↓R ↑O	Young and Downe 1982, 1983, 1987
Drosophila melanogaster	Sex peptide, Acp26Aa	↓R ↑O	Gilbert 1981; Scott 1986a; Harshman and Prout 1994; Chap-
	Acp36DE, Acp62F, EST 6	↑sperm storage + displacement	man et al. 1995; Chen 1996; Wolfner 1997; Price et al. 1999 (1–3, 5)
Drosophila hibisci	Mating plug	↓R	Polak et al. 1998; Polak et al. 2000 (3)
Cyrtodiopsis whitei	Spermatophore/mating plug	↓R	Lorch et al. 1993
Lucilia cuprina	Seminal fluid	↓R ↑O	Smith et al. 1989; Barton Browne et al. 1990; Smith et al. 1990; Cook 1992 (1–3)
Ceratitis capitata	Seminal fluid	↓R	Farias et al. 1972; Blay and Yuval 1997; Chapman et al. 1998; Miyatake et al. 1999 (1, 2, 5)
Musca domestica	Seminal fluid	↓R	Riemann et al. 1967 (2)
COLEOPTERA			
Callosobruchus maculatus	Seminal fluid	↓R ↑O	Fox 1993a,b; Eady 1995; Fox et al. 1995; Savalli and Fox 1998a, 1999 (1–4)
Stator limbatus	Seminal fluid	↑O?	Savalli and Fox 1998b (1, 2, 4)
Acanthoscelides obtectus	Seminal fluid	↓R ↑O	Huignard 1983
Caryedon serratus	Seminal fluid	↑O	Boucher and Huignard 1987

Diabrotica virgifera	Seminal fluid	↑O	Sherwood and Levine 1993
Dytiscus alaskanus	Mating plug	↓R?	Aiken 1992
Leptinotarsa decemlineata	Peptide		Smid 1997, 1998
HYMENOPTERA			
Bombus terrestris	Mating plug	↓ sperm storage	Duvoisin et al. 1999
Carebara vidua	Mating plug		Robertson 1995
LEPIDOPTERA			
Noctuoidea			
Pseudaletia separata	Seminal fluid	↓R	He and Tsubaki 1991, 1992 (2, 3)
Trichoplusia ni	Seminal fluid	↑O	Karpenko and North 1973; Hagan and Brady 1981; Ward and Landolt 1995 (4)
Spodoptera litura	Seminal fluid	↑O	Sridevi et al. 1987
Bombycoidea			
Bombyx mori	Seminal fluid	↑O	Yamaoka and Hirao 1977 (2)
Pyraloidea			
Heliothis virescens	Seminal fluid Ecdysone	↓R ↑O	Mbata and Ramaswamy 1990; Raina and Stadelbacher 1990; Ramaswamy and Cohen 1992; Park et al. 1998 (2, 3)
Heliothis zea	Seminal fluid	↓R	Raina 1989
Helicoverpa zea	57-amino-acid peptide	↓R	Kingan et al. 1993, 1995 (2)
Helicoverpa armigera	Peptide	↓R ↑O	Fan et al. 1999
Plodia interpunctella	Seminal fluid	↓R ↑O	Lum and Flaherty 1970; Lum and Brady 1973; Cook and Gage 1995 (2)
Tortricoidea			
Platynota stultana	Juvenile hormone	↓R ↑O	Webster and Cardé 1984 (2)
Choristoneura fumiferana	Seminal fluid	↓R ↑O	Delisle and Hardy 1997 (1, 4)
Gelechioidea			
Opisina arenosella	Seminal fluid	↓R ↑O	Santhosh-Babu and Prabhu 1987 (2)

Table 4.1 (*cont.*)

Papilionoidea	Mating plugs	↓R	See table 4.2 (1, 3)
Danaeus plexippus	Seminal fluid	↓R ↑O	Oberhauser 1988, 1989, 1992, 1997; Oberhauser and Hampton 1995; Oberhauser 1997 (1, 3, 4)
Jalmenus evagoras	Seminal fluid	↑O	Hughes et al. 2000 (2–4)
Pieris rapae	Seminal fluid	↓R	Sugawara 1979; Obara 1982; Bissoondath and Wiklund 1996a; Cook and Wedell 1996 (1–3)
Pieris napi	Seminal fluid	↓R ↑O	Sugawara 1979; Obara 1982; Wiklund et al. 1993; Kaitala and Wiklund 1994, 1995; Bissoondath and Wiklund 1996a; Cook and Wedell 1996; Wiklund et al. 1998 (1, 3, 4)
Heliconius cydno	Seminal fluid	↓R	Boggs, unpublished, in Thornhill and Alcock 1983 (2, 3)
Heliconius charitonius	Seminal fluid	↓R	Boggs and Watt 1981 (3)
Dryas julia	Seminal fluid	↓R	Boggs and Watt 1981 (3)

Notes: 1. variation in male ability to transfer seminal products; 2. dose-dependent response in female; 3. male costs (males become depleted with successive copulations, have reduced lifespan, or both); 4. female benefits (e.g., nutrients for egg production and/or increased lifespan); 5. female costs (e.g. reduced lifespan). ↓R decreased female sexual receptivity and/or remating ability; ↑O increased oviposition and oogenesis.

males to females at copulation, and the extent to which sexual selection via sperm competition could have played a role in their evolution.

4.2 Mating Plugs

In a diversity of insect orders males transfer their ejaculates packaged in a spermatophore (Mann 1984), and it is clear that some male accessory glands are responsible for the production of structural proteins necessary for the formation of the spermatophore (Leopold 1976). Spermatophores show considerable morphological variation across taxa. In the Tettigoniidae spermatophores are generally attached to the external genitalia of the female and can be highly elaborated structures, consisting of a sperm-containing ampulla as well as a secondary sperm-free mass, known as a spermatophylax (Gwynne 1997; see section 9.3). In other taxa such as flies (Diptera) and grasshoppers (Acrididae) the spermatophore is placed internally, within the female's reproductive tract (Gregory 1965; Kotrba 1996). In some mosquitoes (Culicidae), the male first transfers free sperm followed by a solid mass of material, referred to as the "mating plug." The material is secreted by the accessory glands and persists within the reproductive tract of the female (Gillies 1956; Lum 1961; Giglioli and Mason 1966). The functional significance of the culicine mating plug is unclear, and it has been concluded that the plug represents an evolutionary remnant of the spermatophore (Lum 1961; Giglioli and Mason 1966). Mating plugs are perhaps most obvious in the butterflies (Lepidoptera). After copulation the butterfly spermatophore lies in the corpus bursa. In some species the spermatophore is associated with a secondary structure, the spermatophragma or sphragis, that fills and/or extrudes from the genital opening (Drummond 1984).

Although mating plugs have been recognized in a number of taxa (table 4.1), surprisingly little work has focused on the adaptive significance of these structures. Parker (1970e) and Boorman and Parker (1976) argued that mating plugs function to delay and/or reduce the probability of female remating, so that they are favored by sexual selection as structures for the avoidance of sperm competition. Moreover, Parker (1970e) suggested that, in general, spermatophores that persist within the reproductive tract of females could serve the dual function of delivering the ejaculate and plugging the reproductive tract, thereby avoiding sperm competition from future males. The sperm competition hypothesis yields predictions for male reproductive success at two levels. First, within species, the presence of a mating plug should significantly reduce the probability that rival males can achieve insemination. A corollary of reduced rival insemination success is that the observed patterns of sperm utilization in species with mating plugs should favor the plugging male. Second, across species, the females of those species in which males transfer mating plugs should have a lower lifetime mating frequency than species without mating plugs.

Plugs as Barriers to Remating

The efficacy of mating plugs in preventing female remating has been investigated in a number of taxa. Within-species studies have examined the success of male mating attempts with plugged and unplugged females. In general, the evidence suggests that mating plugs and/or internalized spermatophore remnants do represent short term barriers to reinsemination. Giglioli and Mason (1966) reported that only five of 3866 wild-caught *Aedes melas* had two plugs in their reproductive tracts, and that in staged matings males were never successful in copulating with females that had been plugged by a previous male. Polack et al. (2001) showed how in the fruit fly *Drosophila hibisci* the probability of a second copulation increased with an experimental decrease in plug size. Parker and Smith (1975) found that 26% of copulation attempts were aborted when a previous spermatophore was present in the reproductive tract of female locusts (*L. migratoria migratorioides*). Those males that appeared to be successful took some 4 h before they achieved copulation, compared with just 50 min when attempting to copulate with females that did not have a previous spermatophore in their reproductive tract. Bimodal copulation durations are characteristic of stalk-eyed flies (*Cyrtodiopsis whitei*) where copulations with virgin females are over four times longer than copulations with recently mated females (Lorch et al. 1993). Moreover, short copulations do not result in the transfer of sperm. Lorch et al. (1993) concluded that the presence of a spermatophore acted as a mating plug in preventing subsequent males from achieving insemination. The efficacy of the plug in preventing successful copulation by future rivals may be short lived, however, since short copulations were restricted to attempts made within 6 min of the preceding male's copulation (Lorch et al. 1993). Nevertheless, the spermatophore remains in the female's reproductive tract for 2–60 min and blocks access to the paired spermathecae (Kotrba 1990) so that it may represent a barrier to sperm storage, even if copulation is achieved (Lorch et al. 1993).

Mating plugs that block the reproductive tract of the female can serve only as short-lived barriers to future males. Long-term barriers would prevent the female from ovipositing eggs and thereby be detrimental to both male and female reproductive success; mating plugs are necessarily ejected by the female when she oviposits. Thus, in locusts, the spermatophore is a barrier only to future males that attempt to copulate before the female oviposits. Once the clutch is laid and the plug ejected, rival males attain copulation as readily as when copulating with virgin females (Parker and Smith 1975). Moreover, the potential costs to females of having their reproductive tracts blocked by copulating males may lead them to eject mating plugs

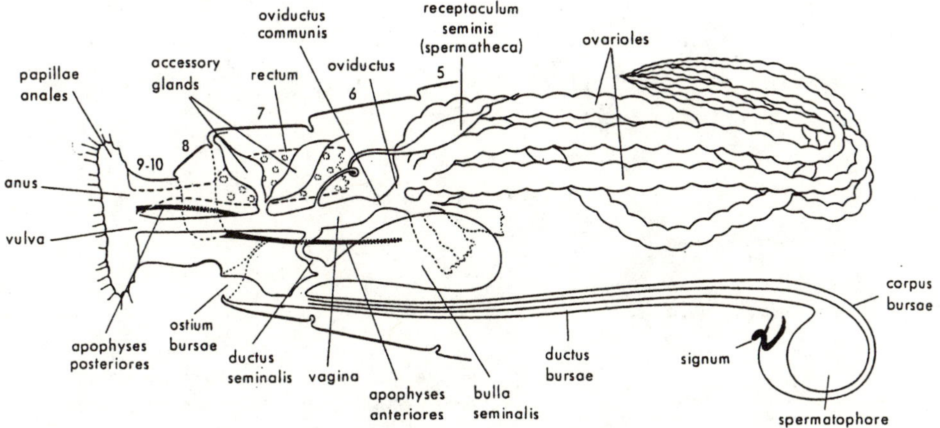

Figure 4.2 Diagrammatic illustration of the internal and external female genitalia of the ditrisian Lepidoptera. The reproductive tract has two openings, the ostium bursae, through which the male delivers the spermatophore to the corpus bursae. The spermatophore must be ruptured by the signum before sperm can be released. Sperm must then travel the length of the ductus bursae and cross to the spermatheca via the ductus seminalis. Eggs are layed via the second opening, the vulva, so that blocking of the ostium bursae does not interfere with oviposition (from Nielsen and Common 1991, © CSIRO Australia 1991).

before oviposition, thereby circumventing male attempts to prevent remating (Lorch et al. 1993; Eberhard 1996).

Blocking of the reproductive tract is less problematic for females in the ditrisian Lepidoptera, where there are two genital openings. The male transfers a spermatophore to the corpus bursa gaining access via a secondary reproductive opening, the ostium bursae (fig. 4.2). The sperm migrate from the spermatophore in the corpus bursa via the ductus seminalis to the spermatheca. Eggs are laid through the primary reproductive opening, the ostium oviductus. Thus, blocking of the copulatory opening has no impact on the female's ability to oviposit. Perhaps not surprisingly it is in the Lepidoptera that we see the greatest evidence of evolutionary divergence in mating plug morphology.

The leidopteran mating plug is an extension of the spermatophore that blocks the ostium bursae. Butterfly mating plugs exhibit highly variable and species-specific morphology (Drummond 1984; Orr 1995; fig. 4.3). In some, the plug is a simple internal extension of the spermatophore (often referred to as a spermatophragma; Ehrlich and Ehrlich 1978) that is attached to the stalk of the spermatophore and fills the length of the ductus bursae. In others, the male's accessory secretions exude from the ostium bursae to form a small external covering, or in their most elaborate form (the so-called "sphragids") they can be large, hollow, foliate structures that are moulded by the male during copula and attached to the female with a belt of material that

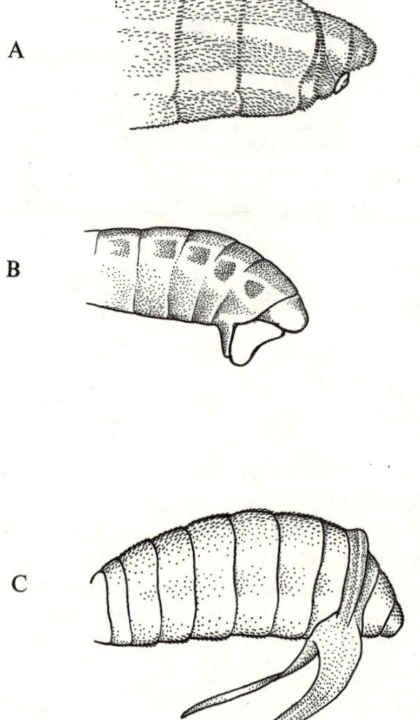

Figure 4.3 Variation in the morphology of Lepidopteran mating plugs. (A) The mating plug of *Graphium sarpedon* (ranked as plug type 1) is a small caplike structure covering the sinus vaginalis (drawn from Matsumoto and Suzuki 1995). (B) In *Atrophaneura alcinous*, the mating plug (ranked as plug type 3) is an amorphous mass that extrudes from the ductus bursae (drawn from Matsumoto and Suzuki 1995). (C) The most elaborate mating plug or sphragis (ranked as plug type 6) of *Euryades corethrus* is held in place by a collar that surrounds the female's abdomen (drawn from Orr 1995).

can surround the entire abdomen (Drummond 1984; Matsumoto and Suzuki 1995; Orr 1995). Production of the mating plug represents a significant cost of reproduction for males that is manifest both at the inter- and intraspecific levels. Orr (1988, 1995) and Matsumoto and Suzuki (1995) noted that across butterfly species there appears to be a trade-off between expenditure on the mating plug and expenditure on the spermatophore. Orr (1995) presented a formal analysis of the proposed interspecific trade-off from data on 73 species of butterflies from the families Papilionidae and Nymphalidae, identifying a dichotomy between species that invest more in the spermatophore (and to the extent that spermatophore size and ejaculate size are correlated, sperm transfer) than the mating plug, and those that invest predominantly in blocking the female's reproductive tract with a sphragis at the expense of producing a large spermatophore. Orr's (1995) analysis was not based on any phylogenetic hypothesis for the species concerned and unfortunately is confounded by phylogeny; 45 of 47 species with low investment in mating plugs were papilionids compared with 14 of 26 species with a sphragis ($\chi^2 = 18.96$, df 1, $P < 0.001$). Thus, the apparent dichotomy may to some extent reflect common ancestry within the two families of butterflies rather than an evolutionary trade-off between investment in the spermatophore and investment in the mating plug. Matsumoto and Suzuki (1995) found a similar trend for six species within the papilionids although, again, species with relatively high or low investment in the mating plug fall exactly into two different subfamilies. More convincing evidence for the costs of plug production come from intraspecific studies of male investment. Matsumoto and Suzuki (1992) found that male expenditure on the mating plug in *Atrophaneura alcinous* declined some 270% across four successive matings. Orr (1988) reported similar declines in male expenditure on mating plugs for *Cressida cressida, Toides richmondia, Princeps aegeus*, and *Acraea andromacha*, indicating that, in general, resources available to males for the production of mating plugs are limiting. Moreover, the rate of decline in plug expenditure for the sphragid-producing *A. andromacha* was greater than the decline in spermatophore expenditure, while the opposite was true for *T. richmondia* and *P. aegeus* which produce a small internal plug (Orr 1988). A similar phenomenon has been reported in anophaline mosquitoes, where males can produce no more than two mating plugs, yet can continue to inseminate in excess of five females (Giglioli and Mason 1966).

The efficacy of mating plugs in preventing females from remating has been examined for just four species of butterfly. In the nymphalid *Euphydryas chalcedona* and the papilionid *A. alcinous*, males transfer a large mating plug that exudes from the ostium bursae (Dickinson and Rutowski 1989; Matsumoto and Suzuki 1992). Dickinson and Rutowski (1989) examined the efficacy of mating plugs by transplanting plugs from mated females to virgins and removing them from mated females (fig. 4.4). To control for the experimental manipulation, they removed and reattached mating plugs to mated females. Males had a significantly lower probability of copulating with virgin females to which mating plugs had been experimentally attached than with normal virgins, demonstrating that the mating plug was an efficient barrier to remating. However, unplugged mated females did not copulate more frequently than controls because females exhibited rejection behavior that was equally efficient in deterring courting males. The presence of a mating plug had no influence on the frequency with which males approached females or on their courtship persistence, indicating that the plug was not a visual deterrent to male mating attempts. These experiments show that, although the mating plug can represent a barrier to reinsemination once females regain receptivity, in the short term male and female interests over female remating probably coincide. Moreover, they also show that, while mating plugs may reduce the probability of remating, some males nevertheless are able to mate with plugged females. Matsumoto and Susuki's (1992) study of *A. alcinous* suggests that variation in the size of the mating plug delivered by a male has a major impact on its effect on female remating. The probability that a pair could be successfully hand coupled decreased with increasing size of the plug transferred by a previous male, and consequently the probability that a second male could transfer a spermatophore declined

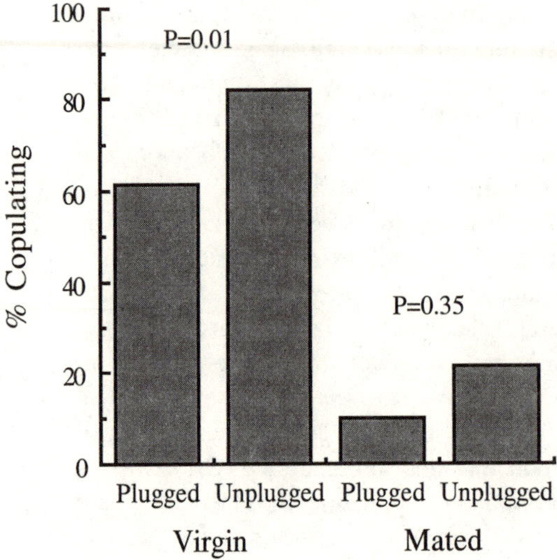

Figure 4.4 Experimental study of the efficacy of the mating plug in *Euphydryas chalcedona* (Lepidoptera: Nymphalidae). Experimental manipulations included attaching plugs to virgin females or removing the plugs of mated females. Data for a control group of mated females that had the plug removed and replaced are included in the mated and plugged treatment group. Plugs prevented males from copulating with virgin females. Mated females had a low probability of copulating because of female mate rejection behavior. (Data from Dickinson and Rutowski 1989)

with increasing size of the first male's mating plug. There was significant variation in plug size that was attributable to male body size, so that large males would be more successful in preventing a female from remating than would small males, thereby imposing directional sexual selection on both plug and male body size.

Female remating has similarly been examined in two species of sphragis producing butterflies, *C. cressida* and *A. andromacha* (Orr 1988; Orr and Rutowski 1991). In both of these species courtship is absent and males seize females in the air and attempt to copulate with them once brought to the ground. Orr and Rutowski (1991) examined the behavior of free-flying males presented with females that were either virgin, mated with an intact sphragis, or mated with the sphragis experimentally trimmed. Males were equally likely to approach females in all three groups, but the probability that a male would bring a female to the ground declined from 92% for virgin females to 50% for females with a trimmed sphragis and only 10% for females with an intact sphragis. Males never attempted to copulate with females bearing a sphragis. In a second experiment, Orr and Rutowski (1991) presented males with mated females that had an intact sphragis, a trimmed sphragis, or had had the sphragis completely removed. As in the first experiment, the sphragis had a major influence on the probability that the male would bring the female to the ground; 90% of females without a sphragis were taken down, compared with 60% for those with a trimmed sphragis and 23% for those with an intact sphragis. Moreover, 90% of males attempted to mate with females that had no sphragis, compared with 27% with a trimmed sphragis and only 3% with an intact sphragis. These results show quite clearly that the elaborated sphragis of *C. cressida* represents both a physical barrier and a visual deterrent to future males. Similar results were obtained for the sphragis-bearing nymphalid *A. andromacha* (Orr 1988). Interestingly, most sphragis-producing butterflies adopt a similar forced copulation, and it is tempting to speculate that a lack of female mate rejection behavior following copulation, typical in most other butterflies, may have played some role in favoring the evolution of highly elaborated and costly mating plugs characteristic of these species (Orr 1995).

A logical prediction that arises from these intraspecific studies is that, all else being equal, the females of species in which males transfer a mating plug after insemination should have a lower lifetime mating frequency than species in which there is no mating plug. Butterflies are extremely useful models for examining mating frequency in the field because spermatophores persist in the corpus bursae so that mating history can be directly determined by dissection. The validity of this method relies on a number of assumptions, however (Drummond 1984): males transfer only one spermatophore per copulation, insemination is not possible without spermatophore transfer, and spermatophores remain recognizable in the corpus bursae.

There is good evidence that males transfer just one spermatophore per copulation, but severely depleted males may be able to inseminate without use of a spermatophore and, in some groups at least, gradual absorption of spent spermatophores can occur (Drummond 1984). Insemination without spermatophores is likely to be a rare event, restricted to dense or highly female-biased populations where males have high mating frequencies (Drummond 1984). Spermatophore absorption, on the other hand, could be a significant problem in estimating mating frequency for some groups, leading to underestimates of the true mating frequency. Nevertheless, Drummond (1984) notes that even in species with spermatophore absorption the sclerotized column of the spermatophore often persists in the corpus bursae, facilitating estimates of female mating history.

A further problem with field estimates of female mating frequency is that spermatophore counts vary consistently with female age (Pliske 1973; Ehrlich and Ehrlich 1978; Drummond 1984) so that a true estimate of female lifetime mating frequency would require samples to be collected at the end of the flight season and/or estimates to be assessed on the basis of female age. Unfortunately, spermatophore counts rarely control for female age or the time during the flight season at which the sample is taken, so that they probably underestimate the true mating frequency of a species. The extent of this problem can be assessed from the data provided by Drummond (1984). Many of the studies cited quote both the mean spermatophore count for nonvirgin females in the sample, as well as the maximum number of spermatophores observed. The maximum number of spermatophores probably provides an overestimate of the typical mating frequency of a species, but, nonetheless, gives a good indication of the potential for remating by females. The data in Drummond (1984) show that across 110 samples covering 79 species of Lepidoptera, mean spermatophore count is in general a good predictor of the maximum degree of female remating observed (fig. 4.5). For samples in which female age class was noted, the data indeed show that females in the oldest age class have higher spermatophore counts than the sample as a whole (complete sample, 1.72 ± 0.11; oldest age class, 2.19 ± 0.22; t = 4.816, df 48, P = 0.0001). However, examination of the data in fig. 4.5 shows that it is only in highly polyandrous species that female mating frequency is underestimated by mean spermatophore counts. With these caveats in mind, in general the mean number of spermatophores counted in the ostium bursae can provide an accurate assessment of female mating frequency.

Ehrlich and Ehrlich (1978) tested the hypothesis that mating plugs reduce female mating frequency across 71 species of butterfly from 14 subfamilies by comparing mean spermatophore counts obtained from species with mating plugs with those without mating plugs. However, they failed to find support for the hypothesis; using species as independent data points, the mean spermatophore count for species that did not have plugs was 1.30 compared with 1.25 for species with plugs. A phylogenetically based test that compared the mean spermatophore counts of plugged and nonplugged subfamilies similarly failed to support the hypothesis. Ehrlich and Ehrlich (1978) concluded that the females of species with mating plugs were clearly able to remate, given the evidence of multiple spermatophores within the corpus bursae, and that the function of mating plugs remained unclear.

There are a number of reasons why species with mating plugs may, nonetheless, exhibit multiple mating. Remating is possible within a small time window between mating and plug hardening (Labine 1964; Matsumoto and Suzuki 1995). Variation in plug size within species, derived either from male body size differences, or from accessory gland depletion following multiple male mating, can have a major impact on plug efficacy (Matsumoto and Suzuki 1995). Moreover, in species with mating plugs, we should expect selection to favor any adaptation in males that allows them to circumvent the plugging attempts of their rivals (Parker 1984). Without antagonistic selection of this nature, selection to maintain the plug will be withdrawn (Boorman and Parker 1976). Thus, species of butterfly in which males transfer a mating plug have sharp, shovel-like aedeagi that appear to function in plug penetration and/or removal (Matsumoto and Suzuki 1992; Orr 1995). In reality, we should expect plugs to reduce the probability of female remating, rather than remove it completely.

Ehrlich and Ehrlich's (1978) analysis utilized a simple categorical approach; species were classified as plugged or unplugged, despite the considerable variation in plug size and morphology that exists across species (fig. 4.3). Given that within-species variation in plug size can influence female remating, across-species variation in plug morphology is likely to have a major impact on function. Furthermore, it is difficult to judge the precise

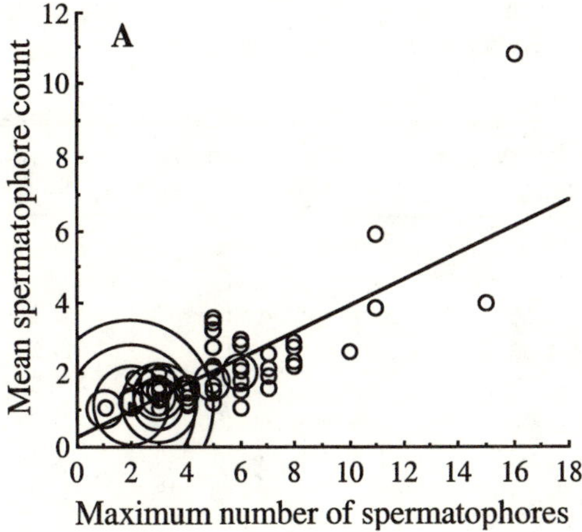

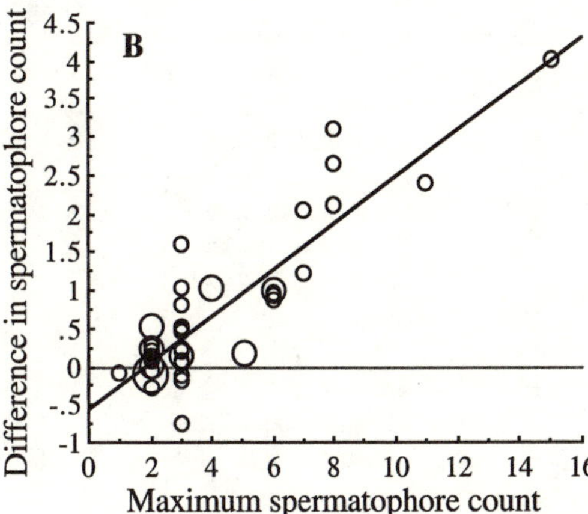

Figure 4.5 Field estimates of female mating frequency across Lepidoptera, based on spermatophore counts from the ostium bursae. (A) The sample mean number of spermatophores provides a good predictor of the maximum potential for female remating (r^2 = 0.66, $F_{(1,\ 108)}$ = 212.88, $P < 0.001$). (B) However, calculating the difference in spermatophore counts between the entire sample mean and the mean for females in the oldest age class shows that female remating rates are increasingly underestimated by the sample mean with increasing degree of female remating (r^2 = 0.76, $F_{(1,\ 47)}$ = 153.98, $P < 0.001$). Increasing symbol size represents increasing overlap of data. (Data from 110 samples of 79 species of Lepidoptera in Drummond 1984)

criteria used by Ehrlich and Ehrlich (1978) in categorizing species as having plugs. Often the neck of a spermatophore can project along the length of the ductus bursae, and appear to block the osteum bursae, although there is no plug (Drummond 1984; see fig. 4.2), and reports of plugs in species such as *Pieris rapae* and *P. protodice* by Ehrlich and Ehrlich (1978) are not supported by observations made over several hundreds of dissections of recently mated females (N. Wedell, personal communication; R. Rutowski, personal communication). I have therefore collected available data on plug morphology and on mean spermatophore counts from the literature, and from colleagues in this area of research (table 4.2) in order to reevaluate the reduced mating frequency hypothesis for plug evolution.

Species without plugs were selected for the analysis from Ehrlich and Ehrlich (1978) on the basis that more than ten specimens had been exam-

Table 4.2
Estimates of female mating frequency across butterflies and variation in mating plug morphology.

Species	Spermatophore count[a]	Plug Type (Rank, % of Spermatophore)[b]	Source
Papilionidae			
Parnassiinae			
Parnasius glacialis	1.05	Large sphragis (6, 96%)	1, 2
Parnasius clodius	1.00	Large sphragis (6)	2–5
Parnasius phoebus	1.00	Large sphragis (6)	2, 5
Parnasius stubbendorfi	1.06	Large sphragis (6)	1
Luehdorfia japonica	1.04	Large sphragis (6, 97%)	1, 2
Luehdorfia puziloi	1.04	Large sphragis (6, 94%)	1, 2
Papilioninae			
Graphium sarpedon	1.80	Small plug (1)	1
Battus philenor	1.73	No plug (0)	2, 6
Cressida cressida	1.00	Large sphragis (6, 99%)	2, 4
Parides proneus	1.00	Small sphragis (4, 36%)	2
Atrophaneura alcinous	1.20	Large plug (3, 21%)	1, 2, 7
Trogonoptera brookiana	1.67	Large plug (3, 39%)	2, 4
Papilio machaon	1.21	Small plug (1, 5%)	1, 2
Papilio polyxenes	1.33	Small plug (1, 7%)	2, 8
Papilio glaucus	1.73	Small plug (1, 8%)	2, 3, 6
Papilio aegeus	1.33	Large plug (3, 24%)	2, 4
Papilio protenor	1.87	Small plug (1)	1, 2
Papilio maackii	1.38	Small plug (1)	1
Papilio bianor	1.83	Small plug (1)	1
Papilio polytes	1.00	Medium plug (2, 12%)	1, 2
Papilio memnon	1.59	Medium plug (2, 15%)	1, 2
Pieridae			
Coliadinae			
Eurema lisa	1.35	No plug (0)	9
Colias philodice	1.33	No plug (0)	4
Pierinae			
Pieris napi	1.90	No plug (0)	10
Pieris rapae	2.28	No plug (0)	10
Riodinidae			
Riodininae			
Nymphidium cachrus	3.13	No plug (0)	4
Nymphalidae			
Danainae			
Danaeus plexippus	2.23	No plug (0)	2, 11
Satyrinae			
Heteronympha penelope	1.00	Medium sphragis (5, 74%)	2
Euptychia penelope	1.00	No plug (0)	4

Table 4.2 (*cont.*)

Species	*Spermatophore count*[a]	*Plug Type (Rank, % of Spermatophore)*[b]	*Source*
Heliconiinae			
Acraea egina	1.93	Medium sphragis (5, 84%)	2, 12
Acraea natalica	5.37[c]	No plug (0)	2, 13
Acraea andromacha	1.43	Medium sphragis (5, 82%)	2, 14
Acraea eponina	1.29	Medium sphragis (5, 72%)	2, 12
Acraea encedon	1.05	Large plug (3, 25%)	2, 12
Nymphalinae			
Euphydryas chalcedona	1.31	Large plug (3)	15, 16
Euphydryas editha	1.27	Large plug (3)	4, 17
Ithomiinae			
Ithomia drymo	2.91	No plug (0)	4

Sources: 1. Matsumoto and Suzuki 1995; 2. Orr 1988; 3. Shields 1967; 4. Ehrlich and Ehrlich 1978; 5. Orr 1995; 6. Burns 1968; 7. Matsumoto and Suzuki 1992; 8. Lederhouse 1981; 9. Rutowski 1978, pers. comm.; 10. Wedell, pers. comm.; 11. Pliske 1973; 12. Owen et al. 1973; 13. Pierre 1986; 14. Epstein 1987; 15. Rutowski and Gilchrist 1987; 16. Rutowski et al. 1989; 17. Labine 1964.

[a]Mean number of spermatophores per mated female.

[b]Ranks were based on descriptions of mating plugs and on quantitative estimates of the proportion of the spermatophore that is devoted to the plug (from Orr 1988). A plug *sensu stricto* (ranks 1–3) was defined as an amorphous mass of material placed internally and/or visible externally, exuding from the ostium bursae. Sphragids (ranks 4–6) have species-specific morphology and are molded by specialized structures associated with the male genitalia (Orr 1995).

[c]Orr (1988) provides only the maximum number of spermatophores in the ostium bursae (14). The mean value was estimated from the relationship shown in figure 4.5A.

ined. I have not included any of the lycaenids since they appear to routinely digest spermatophores rapidly in the ostium bursae, as indicated by mean spermatophore counts for nonvirgin females of less than 1.0 (Ehrlich and Ehrlich 1978). Neither have I included species in the genus *Heliconius*. Although plugs do not occur in this group, females are generally monandrous, avoiding remating by utilizing antiaphrodisiac pheromones transferred by the male during copulation (Gilbert 1976). I used a qualitative ranking system to describe mating plug morphology (table 4.2; see fig. 4.3). In many cases there were data available in the form of the proportional weight of the plug relative to the spermatophore (Orr 1988) on which this ranking was based. I have utilized these data to reexamine the hypothesis that mating plugs function in reducing the probability that females will remate. A reduced probability of remating should be reflected in a lower absolute mating frequency in species with mating plugs. More specifically, the females of those species with large external and elaborated mating plugs (sphragids) should have a lower mating frequency than females of species with small, simple, internal mating plugs.

The hypothesis was tested using comparative analysis by independent contrasts (Purvis 1991) to look for an evolutionary association between changes in mating plug size and changes in female mating frequency. I used the phylogeny for the Papilionoidea proposed by Janz and Nylin (1998). Within the Papilioninae, those species not included in Janz and Nylin (1998) were placed according to Ae (1979). The phylogenetic relationship of *A. andromacha* to its congeners could not be established so it was excluded from the analysis. There is no information regarding branch lengths for the phylogeny used. Moreover, branch lengths could not be estimated from the number of taxa at the end points because the analysis considered only a small sample of species. I therefore assumed a punctuated model of evolution by setting all branch lengths equal. Variances in raw contrasts were homogeneous, confirming that the model of evolution used was adequate for the analysis (Purvis 1991).

The data support the hypothesis that mating plugs decrease the probability of female remating; across species there was a significant decline in female mating frequency, as estimated from spermatophore counts, with increasing plug size and/or morphological exaggeration (fig. 4.6). There are two caveats that should be considered. First, the relationship observed in fig. 4.6 may be conservative because the mating frequency of species without mating plugs could be underestimated, given that the accuracy of mean spermatophore counts declines with increasing female remating (fig. 4.5). Nevertheless, the mating frequency of sphragid-forming parnassiinids may also be underestimated, given their tendency toward spermatophore digestion (Drummond 1984). Second, while mating plugs may function in reducing female mating frequency, not all species of butterfly with low female mating frequency have mating plugs. For example for many species males avoid the risks of sperm competition from future males by alternative physiological adaptations, such as the transfer of accessory gland products in the seminal fluid (see section 4.3). Moreover, female mate rejection behaviors are also important in determining remating (Dickinson and Rutowski 1989). Thus, low female mating frequency can occur when plugs are absent, weakening the analysis in fig. 4.6.

In conclusion, both inter- and intraspecific patterns of female remating support the notion that mating plugs in butterflies, and possibly other taxa (table 4.1), have arisen in the context of avoiding sperm competition from future rivals by reducing the probability of female remating.

Effect of Plugs on Sperm Utilization

While plugging the female's reproductive tract may reduce the probability that a rival male will mate some time in the future, it is clear that mating plugs are not totally effective in preventing the female from remating. Males may have adaptations for circumventing existing plugs, while females may

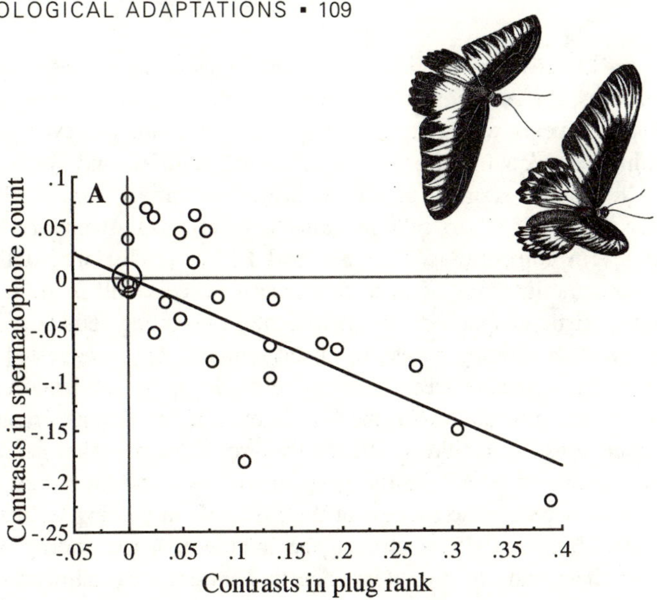

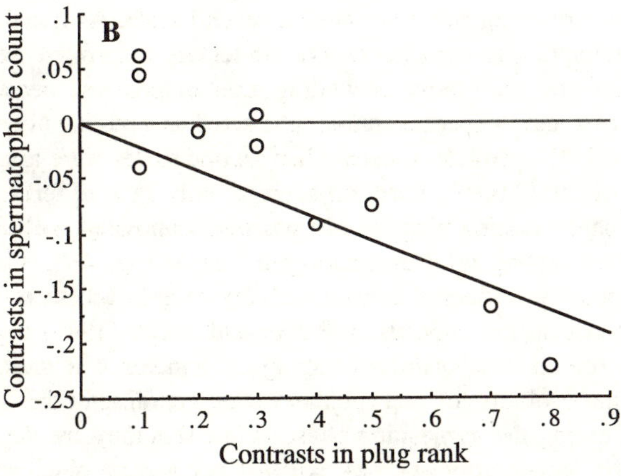

Figure 4.6 Comparative analysis by independent contrasts (CAIC) of the correlated evolution of mating plugs and reduced female mating frequency in butterflies. In (A), the plug rank described in table 4.2 was used as a continuous variable, and the "Crunch" algorithm in CAIC (Purvis 1991) was used to examine correlated changes in spermatophore counts and plug types across the phylogeny (Felsenstein 1985). There was a significant evolutionary association between the size of the mating plug and female remating frequency ($F_{(1,\ 28)}$ = 41.78, $P < 0.0001$). Regression forced through the origin (Grafen 1989) shows that increases in male expenditure on the mating plug are associated with reductions in female remating. Strictly speaking, plug rank is not a continuous variable. Thus, in (B) the "Brunch" algorithm was used to identify phylogenetically independent lineages within which a change in plug rank (a discrete character) has occurred (Burt 1989). There was a significant negative association across lineages, such that larger changes in plug rank were associated with larger reductions in the continuous variable, female mating frequency ($F_{(1,\ 9)}$ = 28.07, $P < 0.001$).

influence the effectiveness of male-positioned plugs in preventing them from remating. Thus, the review above shows quite clearly that remating is common even in species with mating plugs. It is not clear, however, whether such matings are equally effective in achieving fertilization. Mating plugs may also hinder the storage and/or utilization of sperm from rival males even when the physical barrier to remating is breached. In their early review of insect sperm competition, Boorman and Parker (1976) argued that the eventual outcome in terms of male fertilization success will to some extent reflect antagonistic adaptation; in one direction preventing female remating, while in the other gaining access to mated females. They suggested that if plugs are highly effective barriers to future rivals, sperm utilization should favor the first male to mate with the female even if future males manage to secure copulations; P_2 should be effectively zero. If, on the other hand, some males are capable of breaching the plugs placed by previous males, the effect may be to decrease the success of the first male in gaining fertilizations.

Parker and Smith (1975) examined the influence of the mating plug on paternity in their study of migratory locusts. Females were allowed to mate with a single male before being allocated to one of two treatments; females were presented with a second male immediately, or allowed to oviposit a clutch of eggs before being presented with a second male. As noted above, second males attempting to copulate before the female oviposited were less likely to be successful than those copulating after oviposition, because the presence of the first male's spermatophore tube acted as a mating plug (Parker and Smith 1975). Nevertheless, even when second males were apparently successful in copulating, they gained, on average, only 38% of fertilizations when the first male's mating plug was in position, compared with 87% of fertilizations when mating with an unplugged female (fig. 4.7). Thus, the mating plug appears to influence a male's ability to gain access to the female's sperm stores during copulation. Parker and Smith (1975) suggested that the greater reproductive value of unplugged females was most likely responsible for the evolution of precopulatory mate guarding in this species. Males guarding during the oviposition phase ensure that they are the first to copulate with the female after she has expelled her mating plug, and thus gain fertilizations in the batch of eggs laid at the subsequent oviposition. The spermatophore of *C. whitei* appears to function in a similar manner. Lorch et al. (1993) found that sperm utilization changed from a first-male advantage to sperm mixing, depending on whether the first male's spermatophore was in place during the female's second copulation, and in general low mean values of P_2 in Diptera are associated with insemination via spermatophores that may act as mating plugs (George 1967; Curtis 1968; Dame and Ford 1968; Linley 1975).

Ridley (1989) tested the hypothesis that mating plugs resulted in reduced fertilization success for second males in his comparative analysis of sperm displacement in insects. All else being equal, the efficacy of a species mating

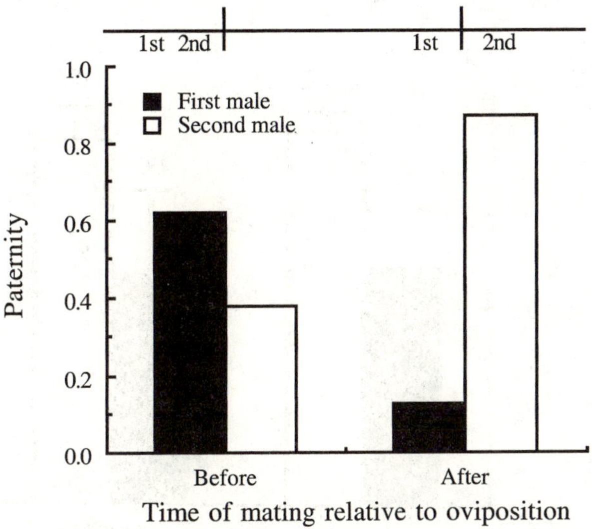

Figure 4.7 Outcome of paternity studies in the migratory locust, *Locusta migratoria migratorioides*, following double matings occurring either before or after oviposition. The first male to mate leaves the tube of his spermatophore within the reproductive tract of the female. Prior to oviposition, second males have difficulty in achieving copulation, and when they do they sire fewer offspring than the first male. The first male's spermatophore tube is ejected with the egg mass so that copulation after oviposition is easily achieved and second males gain the majority of fertilizations when the female next oviposits. The spermatophore tube thus functions as a mating plug. The scale bar above the figure represents the timing of copulations relative to oviposition, shown by the vertical bars. (Data summarized from Parker and Smith 1975)

plug might be reflected in the average value of P_2 for that species; across species the mean P_2 might increase with decreasing plug efficacy, because an increasing proportion of females will be successfully reinseminated by second males. However, all else is unlikely to be equal. The strength of the effect of plugs on P_2 will depend on the mechanisms of sperm competition involved. Where mechanisms of sperm precedence and/or displacement give the second male to successfully inseminate the female a strong fertilization advantage, then the effect of decreasing plug efficacy on species mean P_2 values may be apparent. On the other hand, with sperm mixing any effect is likely to be weak and difficult to detect (see section 2.8; Simmons and Siva-Jothy 1998). Nevertheless, in support of the hypothesis, Ridley (1989) found that across species, mean P_2 values were significantly lower for those with mating plugs compared with those without mating plugs. However, after controlling for phylogeny using his method of outgroup comparison, the significance of this relationship was lost, arguably due to a lack of statistical power resulting from the low number of comparisons involved (fig. 4.8). Given the naiveté of the assumptions underlying the hypothesis, it is remark-

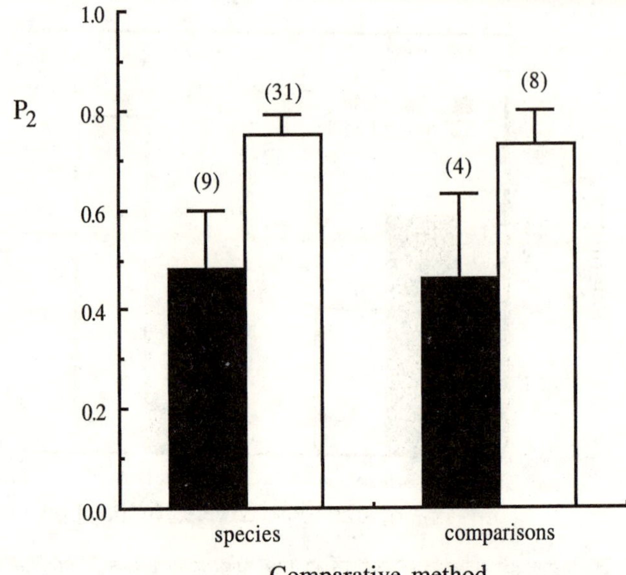

Figure 4.8 Comparative analysis of the hypothesis that mating plugs reduce the fertilization success of future males mating with a female. The data are the mean (± SE) proportion of offspring sired by the second male with a doubly mated female, plotted for species with (solid bars) and without mating plugs (open bars). Analysis was significant when conducted across species (t_{38} = 2.9, P = 0.006) but not so when performed using outgroup comparisons to control for phylogeny (t_{10} = 1.81, P = 0.100). (From Ridley 1989)

able that the trend toward a fertilization advantage for plugging males was so strong.

An examination of the distributions of P_2 values may provide greater insight into the functional significance of mating plugs in the context of sperm utilization. In their study of *Locusta migratoria*, Parker and Smith (1975) found that the variance in P_2 was considerably greater when double matings occurred in the presence of a mating plug; some matings resulted in complete first-male fertilization success while others resulted in complete second-male fertilization success. Low average values of P_2 are generally associated with high variance (fig. 2.4), and Simmons and Siva-Jothy (1998) suggested that this variance may result from the breakdown of mechanisms that prevent sperm from future males entering the sperm storage organ(s); P_2 values may be high when the plug's protection is breached, or very low in cases where the plug remains intact. Such an effect was noted by Lorch et al. (1993) in their study of *C. whitei*.

One striking consistency seen in the Lepidoptera is distributions in P_2 values that are bimodal, the two modes occurring at zero and 1 (table 2.3). Lepidopteran spermatophores are placed in the ostium bursae and the spermatophore tube must be aligned with the ductus seminalis to enable sperm to migrate to the spermatheca (fig. 4.2). Because spermatophores remain in the ostium bursae they have the potential to act as mating plugs in a manner analogous to the spermatophore tube of *L. migatoria* (fig. 4.2). Second males must dislodge the previous spermatophore and align their own for successful insemination. Failed insemination occurs in 5–30 % of matings with virgin females (Drummond 1984) because of failure to correctly align the spermatophore. In her study of *Euphydryas editha*, Labine (1966) noted that, in the cases where P_2 was zero, the first spermatophore had prevented proper orientation of the second so that second-male sperm remained in the ostium bursae and never reached the spermatheca. Similar observations were made by Brower (1975) in his study of *Plodia interpunctella*. Structural proteins involved in the manufacture of spermatophores in insects, including the lower Lepidoptera, are derived from accessory gland secretions that are readily digested. The incorporation of chitin into the walls of spermatophores is a feature of the higher Lepidoptera and appears to aid in their persistence in the female reproductive tract (Drummond 1984). Thus, indigestible spermatophores that act as mating plugs may have arisen as an adaptation for the avoidance of sperm competition from future males.

Alternative Functions for Mating Plugs

Although the presence of mating plugs can delay future males from mating with a female, it is clear that in many cases female mate rejection behavior is both necessary and sufficient to prevent sperm competition. Giglioli and Mason (1966) pointed out that in mosquitoes female mate rejection behavior, affected by male seminal products (table 4.1), seems to preempt the necessity for males to plug their mate's reproductive tract in order to prevent further mating. Indeed, Dickinson and Rutowski's (1989) experimental analysis of mating plug function in *Euphydryas chalcedona* shows that female mate rejection behaviors play the predominant role in preventing remating, at least in the short term (fig. 4.4). An alternative hypothesis for the evolution of mating plugs was put forward by Hinton (1964), who suggested that plugs may serve to prevent the loss of sperm from the female's reproductive tract. Although such a function cannot be considered as an adaptation for the avoidance of sperm competition from future males, it nevertheless may still represent an adaptation under selection via sperm competition (Parker 1970e). Males who are able to ensure that a greater quantity of their ejaculate is retained by the female should be able to gain a greater proportion of fertilizations than males whose sperm are lost from the female's reproductive tract. The tendency for females to eject sperm after copulation has been suggested as a possible mechanism whereby females can select among potential mates (Eberhard 1996). Thus, it is entirely plausible that mating plugs, in some instances, could represent a counteradaptation in males to avoid sperm ejection.

There has been little effort to test the sperm retention hypothesis. Certainly, in anophaline mosquitoes the presence of a mating plug is not necessary for successful insemination (Giglioli 1963). Polak et al. (1998) recently addressed this function in their study of *Drosophila hibisci*. They found that the sperm mass was localized about the ventral receptacle and spermathecal duct junctions by the mating plug. Males depleted by previous matings produced plugs that were greatly reduced in size, and sperm were found throughout the uterus. These data support the notion that the plug retains sperm in a position where they are more likely to be taken into storage. Like anophaline mosquitoes, female *D. hibisci* with plugs did not remate, although again this may represent female behavioral strategies since males never actually mounted or attempted to copulate with recently mated females despite persistent courtship (Polak et al. 1998). The mating plug of anophaline mosquitoes is also transferred after the sperm mass (Giglioli and Mason 1966) and may similarly serve to localize sperm at the site of storage. In contrast, the plugs of *Aedes* spp. and *Psorophora* spp. are found in the uterus surrounded by the sperm mass, rather like the plug produced by depleted male *D. hibisci*, so that they are clearly not involved in sperm localization (Lum 1961). In *Aedes* spp. and *Psorophora* spp. the plug is dissolved within 2–3 h and it may be that the granular secretion is not a mating plug at all, but rather accessory gland secretions that influence female receptivity (section 4.3).

Insemination reactions are common in the genus *Drosophila* (Markow and Ankney 1988; Pitnick et al. 1997). The insemination reaction is manifest as a swelling of the female vagina caused by the male's ejaculate and it has been interpreted as a mating plug (Patterson 1946; Parker 1970e). However, more recent ultrastructural examination of the insemination reaction of five species of *Drosophila* has revealed that the insemination reaction is not a single phenomenon (Alonso-Pimentel et al. 1994). It seems that currently two species are known to have true mating plugs (*D. hexastigma* and *D. hibisci*; Alonso-Pimentel et al. 1994; Polak et al. 1998). One species was identified as having a true insemination reaction (*D. mojavensis*) while three were identified as having a sperm sac; a gelatinelike substance that surrounds the sperm mass, analogous to the spermatophores identified in a number of other Diptera (Kotrba 1990). The functional significance of insemination reactions and sperm sacs remains unexplored, although some work suggests that the insemination reaction represents a female immune response (Asada and Kitagawa 1988) perhaps against the toxic peptides present in the male's seminal fluid (Chapman et al. 1995; Wolfner et al. 1997).

4.3 Seminal Products

Sperm are transferred within the seminal fluid which is itself comprised of a variety of secretory products originating from the male accessory glands.

Male seminal products have been recognized as having significant impacts on female reproductive physiology and behavior (Leopold 1976; Chen 1984; Eberhard 1996). Typically, seminal products decrease female receptivity to future matings and increase the probability of oviposition and/or the number of eggs oviposited (table 4.1). The impact of seminal products on female reproduction is widespread, being reported from 59 species of insect from six orders. The most commonly represented taxa in table 4.1 are the flies (Diptera) and moths (Lepidoptera), probably because of the significance of manipulating female reproductive physiology in the context of pest management. Nevertheless, the data in table 4.1 show that changes in female receptivity and/or oviposition are taxonomically widespread responses to insemination. The involvement of the male accessory glands in female responses to insemination is evident from experimental manipulations of female receptivity and oviposition by the surgical implantation of male accessory glands, or the injection of accessory gland homogenates (table 4.1 and references therein).

The evolutionary implications for male reproductive success of changes in female physiology and behavior following insemination seem clear. First, the longer the female remains unreceptive subsequent to mating, the longer will be the period over which a male will avoid sperm competition from future rivals. Second, the more offspring a female produces within her period of unreceptivity, the greater will be the copulating male's share of his mate's lifetime reproductive success. Below I review the evidence that seminal products of the accessory glands may have evolved, in part, as sexually selected adaptations arising from sperm competition. I begin by reviewing aspects of seminal fluid form and function in three of the most researched taxa (the Orthoptera, Lepidoptera, and Diptera) before considering the details of alternative routes for the general evolution of seminal products in insects.

Orthopteran Refractory Periods

Following insemination, female tettigoniids enter what has been called a refractory period during which they fail to respond to the sexual signals of males. Experimental manipulations of the duration of spermatophore attachment, and thus the volume of ejaculate transferred to the female, indicate that the duration of the female's nonreceptive period is strongly dose dependent (Gwynne 1986; Wedell and Arak 1989; Simmons and Gwynne 1991) (fig. 4.9). These data imply that some factor within the ejaculate is responsible for the temporary cessation of phonotaxis (female orientation to acoustically signaling males). Moreover, Wedell (1989) showed that the onset of egg deposition and the total number of eggs laid by female *Decticus verrucivorus* during the refractory period were also dependent on the amount of ejaculate received. Thus, males benefit from altered female sexual behavior and physiology, because there will be little risk of competition for paren-

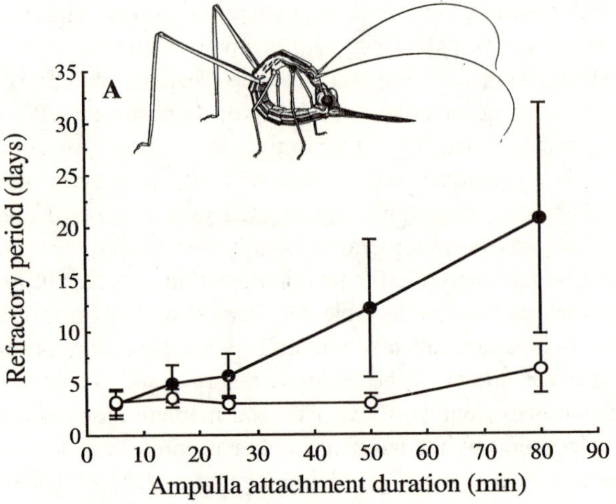

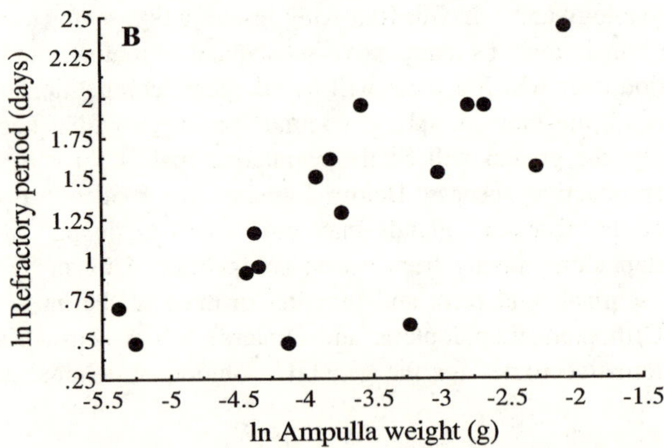

Figure 4.9 Effects of ejaculate volume on the period during which female tettigoniids fail to respond to sexually signaling males, the refractory period. (A) Female *Kawanaphila nartee* show an increase in the refractory period with increasing volume of ejaculate transferred to the spermatheca. Ejaculate volume was manipulated by varying the time that the sperm-containing ampulla of the spermatophore was attached to the gonopore. Data are presented for two groups of females: the first were females fed ad libitum with pollen (solid circles) and the second were females kept with limited access to pollen. The dose-dependent response was apparent only when females were well fed, indicating that the long-term response to male seminal products is dependent on female interests (Simmons and Gwynne 1991). (B) Comparative analysis across sixteen species of tettigoniid of the relationship between mean ejaculate (ampulla) weight and the mean duration of the female's refractory period. Species with large ejaculates have longer refractory periods, as might be expected if larger ejaculates contain greater quantities of receptivity-inhibiting seminal products (Wedell 1993b).

tage from future males during the period of elevated offspring production. Cessation of female phonotaxis and elevation of oviposition following insemination has also been demonstrated in gryllids (see review by Stanley-Samuelson and Loher 1986; table 4.1) but in this taxon more is known of the mechanisms involved. Males transfer quantities of an enzyme system, prostaglandin synthetase, as well as a precursor of prostaglandin, arachidonic acid, within their seminal fluid (Stanley-Samuelson and Loher 1983). Arachidonic acid is converted to prostaglandin within the female's spermatheca. This hormone enters the hemolymph, is taken up by the ovaries, and stimulates oviposition while suppressing phonotaxis (Destephano and Brady 1977; Loher 1981; Loher et al. 1981; Stanley-Samuelson and Loher 1985). Injections of prostaglandin into virgin females show that the response is dose dependent (Stanley-Samuelson et al. 1986). The incorporation of prostaglandins and arachidonic acid into ejaculates is not a unique feature of the gryllids (e.g., Brenner and Bernasconi 1989).

The duration of the refractory period of female tettigoniids has also been shown to depend on their own nutritional status; females on low-quality diets have shorter refractory periods than those on high-quality diets (Gwynne 1990; Simmons and Gwynne 1991; fig. 4.9A). The proximate mechanism behind this effect may lie in the availability of substrate on which seminal enzymes might act. For example, polyunsaturated fatty acids upon which prostaglandin synthetase can act may be derived in part from the female's diet (Dadd 1981) so that her nutritional status influences the amount of prostaglandin synthesized by the male's seminal enzymes. Nevertheless, it is not known if diet influences the response of female gryllids, or if prostaglandin synthetase is the active product in male tettigoniid seminal fluid.

Comparative evidence supports the dose-dependent role of the ejaculate of male tettigoniids in inducing female refractory periods. Wedell (1993b) examined the female refractory period of 28 species of tettigoniid from 16 genera. Using genera as independent data points in her analysis, Wedell (1993b) found a positive association between the length of the female period of unreceptivity to calling males and the weight of ejaculate provided by males (estimated from the weight of the ejaculate containing ampulla of the spermatophore) (fig. 4.9B). Although there was no control for phylogenetic relationships below the generic level, the data are at least suggestive of an evolutionary association between these variables. Induction of a longer refractory period in females should be favored by sexual selection because of the reduction of female mating frequency and thus sperm competition risk. There was no association between ejaculate weight and number of eggs laid during the refractory period in this comparative data set (Wedell 1993b), suggesting that the observation in *D. verucivorous* (Wedell and Arak 1989) may be species specific.

Lepidopteran Ejaculates

Mating-induced changes in female sexual receptivity are well documented in moths and butterflies (table 4.1). In pyraloid, tortricoid, and noctuid moths, virgin females emit pheromones to attract males for mating. Mating has the affect of reducing pheromone titer and thus reduces the probability that females will attract and mate with a second male (see reviews by Raina et al. 1994; Ramaswamy et al. 1994). Pheromone biosynthesis in the female is under the control of a neurosecretion from the brain–suboesophageal ganglion–corpa cardiaca complex (Raina and Klun 1984; Tang et al. 1989; Ramaswamy et al. 1994). The male seminal product responsible for the termination of pheromone production, often referred to as a pheromonostatic factor, has been shown to enter the hemolymph in *Heliothis zea* and probably acts directly on the female's brain (Raina 1989). Pheromone production can be restored in mated females by injections of pheromone biosynthesis activating neuropeptide, implying that the pheromonostatic factor transferred in the seminal fluid inhibits the production of this neurosecretion by mated females (Raina 1989).

Mechanical stimulation of the female's reproductive tract also seems to be important in affecting her response to male seminal products; cutting the ventral nerve cord of female *Helicoverpa zea* inhibits the postmating suppression of pheromone production (Kingan et al. 1995), implying that the presence of a spermatophore in the corpus bursae is necessary for the full female response. Raina (1989) showed that the presence of a spermatophore in the corpus bursae was essential for the response to pheromonostatic factor, although the presence of sperm was not (see also Raina and Stadelbacher 1990; Foster 1993). In fact, although seminal products are active in inhibiting pheromone titer in *Heliothis virescens*, they do not affect female calling behavior (Ramaswamy et al. 1994). Rather, the mechanical stimulation generated by the spermatophore appears to be responsible for the cessation of calling (Kingan et al. 1993). The significance of reduced pheromone titer for male *H. virescens* was demonstrated by Ramaswamy et al. (1994). They found that unmated females had a mating probability of 1.0 compared with just 0.6 for mated females of equivalent age. Thus, reducing the pheromone titer of females reduces the risk of sperm competition for males by 40%. This effect is undoubtedly underestimated due to the crowded conditions in which laboratory experiments were performed (Ramaswamy et al. 1994).

It is not clear what the active component of the seminal fluid is, although ecdysone has been implicated; injections of >1.0 μg of 20-OH-ecdysone into virgin female *H. virescens* reduced the pheromone titer to that of mated females (Ramaswamy and Cohen 1992). However, the source of ecdysone is unclear. Raina et al. (1994) were able to demonstrate reduced pheromone titer in females mated to castrated males, suggesting that the source of the ecdysone was probably not the testes, and that the female's response was not sperm dependent. Nevertheless, Ramaswamy et al. (1994) found an order of magnitude higher concentration of ecdysone in the testes than in the accessory glands and injections of testis or accessary gland extracts have equivalent effects on female behavior.

Juvenile hormone has been implicated as the seminal product responsible for increased oviposition in moths (Webster and Cardé 1984; Park et al. 1998). In Lepidoptera that do not have mature oocytes at the time of eclosion, juvenile hormone secreted by the corpora allata of the female initiates egg maturation and oviposition (e.g., Nijhout and Riddiford 1974, 1979). Adult male moths are unable to synthesize juvenile hormone but they do produce juvenile hormone acid that is methylated in the accessory glands (Bhaskaran et al. 1988). In the majority of moths examined, male abdomens contain significantly more juvenile hormone than do female abdomens (Gilbert and Schneiderman 1961; Williams 1963), and large quantities are transferred to the female in the seminal fluid (Shirk et al. 1980; Park et al. 1998). It appears that females may be limited in their ability to produce juvenile hormone (Ramaswamy et al. 1997) and its donation by males may enhance egg maturation and oviposition in *Platynota stultana* and *H. virescens* (Webster and Cardé 1984; Park et al. 1998). Juvenile hormone has no influence on pheromone titer or female sexual receptivity (Raina 1989; Ramaswamy and Cohen 1992). Thus, different seminal products appear to serve different functions.

The importance of mechanical stimulation from the spermatophore in the corpus bursae on female sexual receptivity is well documented in the butterflies (fig. 4.10). Sugawara (1979) identified a pair of nerves that innervate the corpus bursae of *Pieris rapae* and connect with the ventral nerve cord. Inflation of the corpus bursae, either by mating and spermatophore deposition or by experimental injection of silicone oil, increased the spontaneous rate of afferent impulses along the bursal nerves. Moreover, the number of impulses increased with the volume of oil injected, exhibiting a dose-dependent response (fig. 4.10). The frequency of mate refusal responses by female *P. rapae* followed a similar pattern, both with variation in natural spermatophore volume, and with variation in the volume of injected silicone oil. Thus, stimulation of the stretch receptors in the corpus bursae is responsible for the reduced sexual receptivity of female *P. rapae* to remating. Moreover, the period over which females remain unreceptive to further matings is also associated with spermatophore size in a number of moths and butterflies (e.g., Oberhauser 1989; Kaitala and Wiklund 1994; table 4.1 and references therein) so that males depleted by previous matings induce shorter periods of unreceptivity than do virgin males, or those allowed to recover from previous mating activity. The period over which females remain unreceptive depends largely on the time required for females to break down spermatophores; the larger the ejaculate, the longer it takes to digest (Oberhauser 1992). As with moths, a humeral factor has also been implicated in

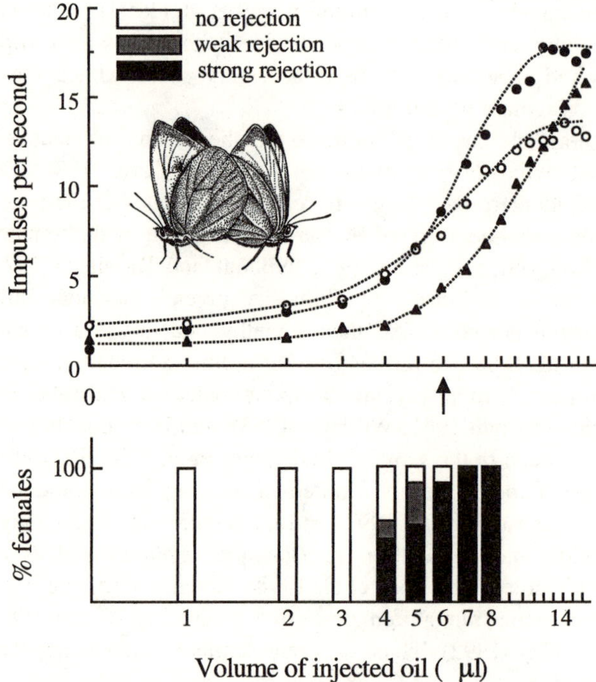

Figure 4.10 The response of stretch receptors in the corpus bursae for three female *Pieris rapae*, elicited by the experimental introduction of different volumes of mineral oil (*top*). The volume of a typical male's ejaculate is indicated by an arrow on the *x*-axis. Variation in the proportion (numbers are the number of females injected) of females rejecting copulatory attempts of males is also shown as a function of the volume of oil injected into the corpus bursae (*bottom*). The stretch receptors exhibit a dose-dependent response so that variation in the volume of ejaculate transferred by males results in variation in the probability that females will remate. (Redrawn from Sugawara 1979)

the induction of the mate refusal behavior of *P. rapae* (Obara 1982), although in general the physiological aspects of reduced sexual receptivity in butterflies are not well studied. Conversely, more detailed evolutionary studies of female receptivity are available for butterflies than for moths. Given (i) that reduced receptivity to further mating by females will provide a selective advantage to males from the avoidance of sperm competition, and (ii) that larger ejaculates induce longer periods of unreceptivity, selection should favor an increase in the size of ejaculates produced by males. Selection for increased ejaculate size should be greater in polyandrous species than in monandrous species because the risk of sperm competition, and so the benefits of reduced female receptivity, will be greater. Svärd and Wiklund (1989) examined the weight of ejaculates across eleven species of Pierid and nine species of Satyrid butterflies and found that the weight of male ejaculatory products increased with increasing degree of polyandry, as estimated by spermatophore counts in field-collected females (see fig. 4.5). Karlsson (1995) has since confirmed this evolutionary association across 21 species of butterfly from three families (Papilionidae, Pieridae, and Nymphalidae), after controlling for phylogenetic association (fig. 4.11). He also examined the amount of resources allocated to reproduction by male butterflies, by determining the nitrogen content of the male abdomen (Karlsson 1996). There was a positive association between the degree of polyandry and abdomen nitrogen content in males, as would be expected if males responded to increased risk of sperm competition by increasing their investment in receptivity-inhibiting seminal products (fig. 4.11). Increased ejaculate weight and reproductive expenditure could reflect increases in sperm production under sperm competition risk (Gage 1994). However, studies of *P. rapae* show that ejaculate weight and sperm numbers are actually negatively rather than positively associated (Cook and Wedell 1996; Watanabe et al. 1998) so that ejaculate weight reflects seminal fluid content rather than sperm. Further evidence of directional selection on male ejaculate weight comes from the observation that the males of polyandrous species are able to produce accessory secretions more rapidly than those of monandrous species, and thus maintain their expenditure on the ejaculate across successive matings (Svärd and Wiklund 1989).

The constituents of butterfly seminal fluids have not been studied in great detail. Marshall (1985) examined the protein and lipid composition of *Colias philodice* and *C. eurytheme* spermatophores and found that proteins represented 12% and lipids 10% of total male secretion. The lipid fraction was composed of equal quantities of hydrocarbons and nonhydrocarbons (cholesterol, diglycerides, triglycerides, and free fatty acids) and traces of phospholipids. Bissoondath and Wiklund (1995) found that, after controlling for ejaculate weight, the amount of ejaculate protein was positively associated with degree of polyandry across 11 species of butterfly, and although their analysis did not control for phylogeny the association is consistent with the notion that the protein fraction may be the focus of selection. Accordingly, Bissoondath and Wiklund (1996b) also found that the males of polyandrous species were able to maintain the protein content of their seminal fluids across multiple matings while those of monandrous species are rapidly exhausted. These phenomena are expected, given the association between polyandry and male resource allocation to reproduction observed by Karlsson (1996).

Drosophila Seminal Fluids

Our most detailed understanding of male seminal fluids and their function come from studies of the dipteran genus *Drosophila*, and in particular *D. melanogaster*. Manning (1962, 1967) first observed that female sexual recep-

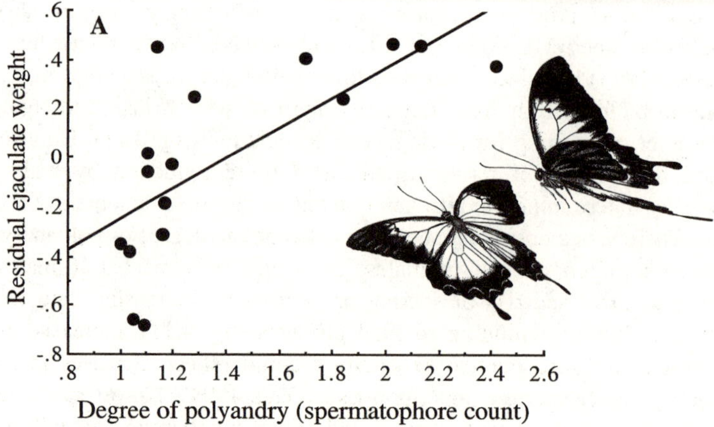

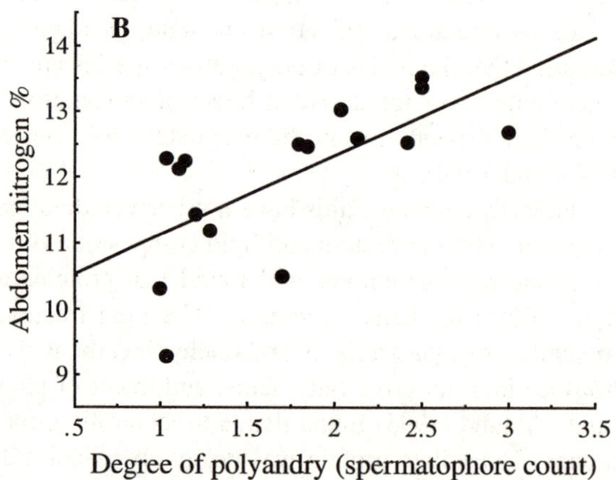

Figure 4.11 Across species of butterfly there is a positive association between the degree of polyandry and (A) the weight of ejaculate transferred by males at copulation, and (B) the amount of nutrient reserves devoted to reproduction, estimated as % total body nitrogen found in the abdomen. In butterflies, ejaculate size has a significant effect on the female's period of receptivity and thus on risk of sperm competition from future males. If selection has favored increased ejaculate size to avoid sperm competition, we would predict that polyandrous species should be subject to greater selection for increased ejaculate size than monandrous species due to the greater risk of sperm competition. Comparative analysis using independent contrasts shows that the cross-species relationships observed among ejaculate mass (A), male allocation to the ejaculate (B), and degree of polyandry are not dependent on phylogeny, and thus there is an evolutionary association between these variables as predicted by sperm competition theory (see Karlsson 1995, 1996).

tivity and egg production were altered by mating; in his experiments females become unreceptive to further mating for a minimum of 4 days after mating, but receptivity was fully restored after 9 days. Receptivity returned after 24 h if females did not receive sperm during copulation, and the timing of return to receptivity for those females receiving sperm was dependent on the time at which their sperm stores became depleted. Sperm dependence of long-term changes in female behavior following mating has since been verified many times (e.g., Pyle and Gromko 1978; Gromko et al. 1984b; Letsinger and Gromko 1985), and indeed, Gromko et al. (1984b) show that it is the storage of sperm in the ventral receptacle that is important. The ventral receptacle is innervated (Miller 1950) and the nervous center controlling receptivity is the brain (e.g., Tomkins and Hall 1983), suggesting that mechanical stimulation arising from sperm may influence long-term female responses to male seminal products in much the same way as is seen in the Lepidoptera. Clearly, short-term responses are induced by seminal products, given that the receptivity of females inseminated by mutant spermless (*XO*) males is the same as those inseminated by normal (*XY*) males for the first 8–10 h after insemination (Scott 1987).

Early attempts to elucidate the factors responsible for mating-induced changes in female receptivity and oviposition involved the transplantation of male accessory glands into the abdomens of virgin females (Garcia-Bellido 1964; Merle 1968). Such operations were successful in inducing the typical changes in behavior and physiology characteristic of mated females, thereby confirming the male accessory glands as the source of receptivity-inhibiting and oviposition-inducing seminal products. However, our detailed knowledge of *Drosophila* genetics coupled with advances in biochemical and molecular genetic techniques have allowed researchers to dissect the seminal fluid of *D. melanogaster* and identify precise molecules that are produced by the accessory glands and determine their mode of action on females. A vast body of literature has arisen from research in this area, and readers are directed to the detailed reviews of Kubli (1996), Chen (1996), and Wolfner (1997).

Ejaculations of seminal fluids and sperm appear to be temporally separated events in *Drosophila*; the delivery of seminal fluid begins almost immediately after the onset of copulation, although the first sperm do not arrive in the female's reproductive tract until 9–10 min later (Nonidez 1920; Fowler 1973; Gromko et al. 1984a). Stumm-Zollinger and Chen (1985, 1988) estimate that the *D. melanogaster* seminal fluid contains a minimum of 85 different proteins, but so far just a handful of these have been characterized and their function examined. Richmond et al. (1980) identified an enzyme, esterase 6 (EST 6), that is secreted by the anterior ejaculatory duct and is active in influencing female reproductive processes during and after insemination. EST 6 is transferred to the female within the first minute of copulation and rapidly enters the female hemolymph, where it persists for as long as 4 days after copulation, although there is a rapid decline in activity within the first 24 h (Meikle et al. 1990). Females mated with mutant males having a null allele (EST 6 ^{0}) at the esterase 6 locus fail to exhibit the decrease in receptivity seen in females mated with males having active (EST 6 $^+$) alleles (Scott 1986a). Esterase 6 may thus be involved in the suppression of female receptivity. The timing of the female's short-term receptivity response and the loss of esterase 6 activity are certainly congruent with such an interpretation, although females mated to null males do not always remate, suggesting the involvement of some other seminal product (Scott 1986a). The effect of esterase 6 on oviposition is not clear; females mated with males having the null allele had a higher initial egg production although long-term productivity of null and active mated females was identical (Gilbert et al. 1981). Thus, the activity of esterase 6 is not sufficient to explain the elevation in oviposition typical in mated females. Esterase 6 does appear to have a significant involvement in sperm storage and use. Females mated to EST 6$^+$ males release sperm from the ventral receptacle and spermathecae more rapidly than do females mated with EST 6 ^{0} males (Gilbert 1981). These results suggest that esterase 6 stimulates females to use the copulating male's sperm to fertilize eggs during the period immediately following copulation. Gilbert et al. (1981) argued that esterase 6 may also be involved in sperm displacement. If esterase 6 or the products of its enzymatic activities induced rapid loss of sperm from the female's sperm stores to the uterus at the onset of copulation, these sperm could be diluted with far greater quantities of the male's own sperm that arrive some 10 min later, so that the sperm restored by the female after copulation would be predominantly those of the copulating male. This model is certainly consistent with the observation that nonvirgin females mated with spermless second males have a significantly reduced productivity (Scott and Richmond 1990), although it is not known if esterase 6 contributes to the loss of previously stored sperm during copulation.

The details of accessory gland function have been elucidated by use of transgenic flies. Male *Drosophila* have two accessory glands that each consist of a muscular sheath surrounding a single layer of secretory cells that empty into the lumen of the gland (Bairati 1968). There are two types of secretory cells, the vast majority of which are the so-called main cells (Bertram et al. 1992). Kalb et al. (1993) produced male flies in which the main cells were destroyed during development by a toxin (diphtheria toxin subunit *A*, DTA) that was produced intracellularly in genetically engineered flies. Moreover, they were able to produce flies that differed in the degree of main cell ablation, but which produced near-normal quantities of sperm. Females mated to transgenic flies failed to elicit the reduction in receptivity and increased egg production seen in females mated with normal control males on the day following mating, even when they received normal quantities of sperm (fig. 4.12). These experiments show that the changes in fe-

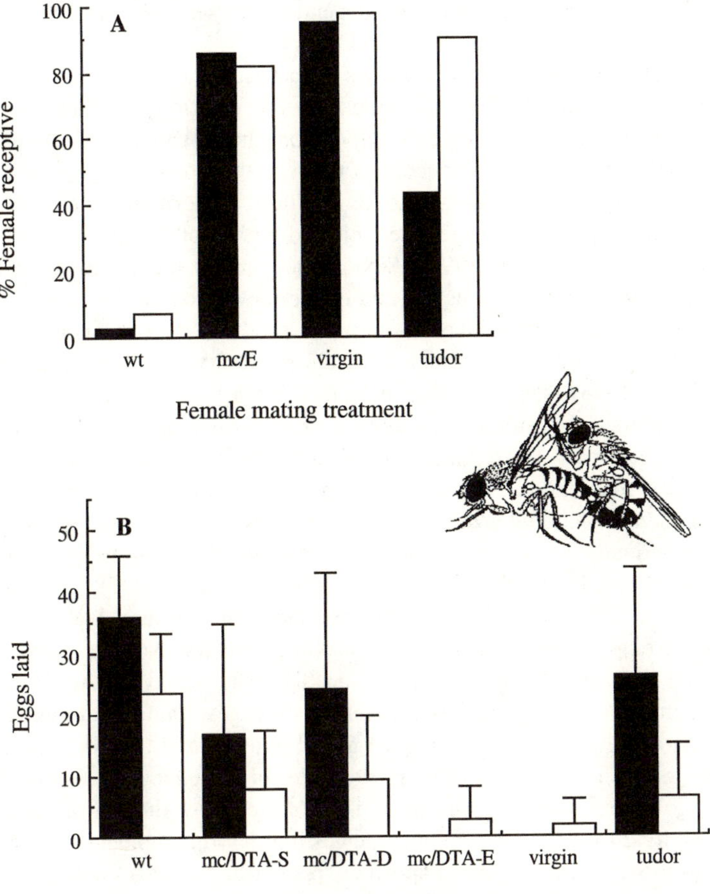

Figure 4.12 Dissecting the causes of postmating changes in (A) female receptivity and (B) oviposition behavior in the fruit fly *Drosophila melanogaster.* When mated with normal wild-type (*wt*) males, females have a reduction in sexual receptivity and an increase in oviposition relative to unmated (virgin) females. Males of differing genetic background were used to unravel the effects of sperm and accessory gland products. When mated with males with normal accessory gland products but unable to produce sperm (*tudor*), females responded to mating with a moderate decrease in receptivity and normal increase in oviposition on the first day after mating (solid bars). However, receptivity and oviposition returned to the virgin state on the second day (open bars), showing that sperm are required for the long-term response to mating. Males genetically engineered so that they did not produce seminal products derived from the main cells of the accessory glands (mc/DTA-E) were unable to induce postmating responses in females, showing that accessory gland products are involved in the short-term switch in female behavior and physiology. Males with greatly reduced (mc/DTA-D) or reduced (mc/DTA-S) main cell products produced intermediate responses, illustrating dose dependency. (Data from Kalb et al. 1993)

male behavior and physiology are dependent on seminal products secreted by the main cells.

Chen et al. (1988) injected fractions of methanol-extracted male accessory glands into virgin females and found only one fraction that had significant effects on female receptivity and oviposition, the so-called sex peptide. Virgin females genetically engineered to produce their own sex peptide exhibit the same lack of receptivity and oviposition as normally mated females (Aigaki et al. 1991), as do females injected with synthetically produced sex peptide (Schmidt et al. 1993). Sex peptide enters the hemolymph, from where it has access to the neural and endocrine centers of the female. It has been shown to activate juvenile hormone production in the corpora allata (Moshitzky et al. 1996), which is itself necessary for oocyte maturation (Riddiford 1993), suggesting a mechanism remarkably parallel to that suggested for the oviposition stimulation properties of moth ejaculates. Further characterization of *Drosophila* seminal products has come from sequencing the genes that encode accessory gland specific mRNA's (Wolfner et al. 1997). Genes encoding accessory gland proteins are referred to as Acp's and are individually identified on the bases of their chromosome position (sex peptide is encoded by the gene Acp70A; Wolfner et al. 1997). The first Acp to be identified was Acp26Aa, a prohormone with structural features resembling the egg-laying hormone of *Aplysia* (Monsma and Wolfner 1988). Herndon and Wolfner (1995) isolated mutant flies that lacked the Acp26Aa protein. Females mated with flies deficient in Acp26Aa laid fewer eggs on the day following mating than did those mated with normal control males, indicating that Acp26Aa is partially responsible for the change in female physiology following mating. Lack of transfer of Acp26Aa did not affect female receptivity 24 h following mating, suggesting that this molecule's sole function is to contribute to the oviposition response. Acp26Aa-deficient males were equally competent in avoiding sperm displacement and had equal fertilities to normal control males (Herndon and Wolfner 1995). Once transferred to the female's reproductive tract, Acp26Aa undergoes sequential proteolytic cleavages (Park and Wolfner 1995). Acp26Aa is not processed in unmated females genetically engineered to produce their own Acp26Aa. Neither is it processed in such females mated to males deficient in main cell products, esterase 6, or sperm. However, increasing the amount of main cell products increases the degree of Acp26Aa processing, strongly suggesting that another main cell product transferred along with Acp26Aa is responsible. It is not yet clear what molecules are involved, or why Acp26Aa requires processing. However, the fact that Acp26Aa remains unprocessed within the accessory glands of the male suggests that some form of protease inhibitor, possibly the serpin Acp76A (Coleman et al. 1995), prevents processing until it is itself deactivated in the female's reproductive tract (Park and Wolfner 1995; Wolfner et al. 1997).

A second Acp to have been investigated in some detail is Acp36DE, a large glycoprotein that remains within the female's reproductive tract (Bertram et al. 1996). Acp36DE is transferred to the female during the first 10 min of copulation. It is highly localized in the female's genital tract, binding to the ventral oviduct wall anterior to the openings to the sperm storage organs, and at the leading edge of the sperm mass. Posterior to the sperm mass is the mating plug, or sperm sac (sensu Alonso-Pimentel et al. 1994). Thus, the sperm mass is localized about the entrance to the sperm storage organs in much the same manner as is accomplished by the mating plug of *D. hibisci* (Polak et al. 1998), which is essential for sperm storage and retention (Neubaum and Wolfne 1999; Tram and Wolfner 1999). Acp36DE may act on the female nervous system to induce muscular contractions that move sperm into storage, since females with masculinized nervous systems fail to store sperm (Arthur et al. 1998). Thus, the molecule may play an important role in sperm competition. Indeed, Clark et al. (1995) show that allele variation at the Acp36DE loci is significantly associated with the avoidance of sperm displacement from future rivals. They also demonstrated an association between allelic variation at Acp26Aa, Acp29B, and Acp53E loci that influenced sperm displacement avoidance, although Herndon and Wolfner (1995) have since shown Acp26Aa not to be involved in sperm competition. Sperm storage may represent a second role for the serpin Acp76A (Wolfner et al. 1997). In mammals serpins are involved in the clotting of blood, and Acp76A could facilitate the clotting of semen to form the sperm sac (Alonso-Pimentel et al. 1994) that lies posterior to the sperm mass. The majority of Acp76A is found in the sperm sac when it is expelled from the female 2 to 4 h after mating (Coleman et al. 1995).

Scott and Richmond (1990) demonstrated that displacement of sperm from the mated female by subsequent males involved seminal fluid. They used *tudor* males that produce normal accessory secretions but no sperm to show a decline in the productivity of females previously mated to normal males and then remated to spermless males. Harshman and Prout (1994) extended these observations to include experiments with the genetically engineered males that lack main cell products. Decreased productivity of females was not apparent after mating with males deficient in main cells, supporting a role of Acp's in sperm displacement. Clark et al. (1995) examined what they termed "offensive" and "defensive" processes of sperm competition. Males successful in the offensive role are able to gain high levels of paternity when mating with previously mated females (in double matings sperm utilization shows high P_2 and low P_1). In contrast, males successful in the defensive role are able to resist future sperm displacement such that second males have a low success in fertilization (low P_2 and high P_1). Interestingly, Clark et al. (1995) found no correlation across males in their offensive and defensive abilities, suggesting that different Acp's are involved in each of these processes. Variation in loci coding for Acp's on the second chromosome (Acp's 26Aa, 29B, 36DE, and 53E) was associated with variation in sperm defence, while a single Acp on the third chromosome, Acp76A, had only a marginal association with variation in sperm offence. Allelic variation at the esterase 6 locus was not associated with variation in sperm offence or defence, suggesting that it is in fact not involved in sperm competition (see above). These data implicate male accessory gland proteins in mediating the avoidance of sperm competition from future rivals. However, there are problems associated with the experimental protocol adopted by Clark et al. (1995) that confound the interpretation of their data. They house virgin females for 2 h en masse with first males of known genotype, and then transfer them to vials containing second males with an alternative genotype, where they remain for 12 h. Following these exposures to males, Clark et al. (1995) assume that females have mated once with a male of each genotype. However, females are known to become unreceptive to further matings following copulation so that with their protocol there was no way of knowing whether females had in fact mated with the second type of male. Indeed, Service and Vossbrink (1996) found significant genetic variation in the ability of males to induce sexual unreceptivity to further mating by females. The Acp responsible for decreased female receptivity is Acp26Aa, one of the same Acp's that was associated with genetic variation in sperm offence in Clark et al.'s (1995) study. Thus, variation in P_1 may arise because of variation in the receptivity of females to second males, rather than any process associated with postcopulatory sperm competition.

The mode of action of Acp's in sperm competition is gradually being uncovered. The involvement of Acp36DE in the movement of sperm into and out of the sperm storage organ(s) (Neubaum and Wolfner 1999), coupled with the fact that the numbers of sperm stored generally has an impact on fertilization success in insects, including *D. melanogaster* (table 2.1), suggests that variation in the ability of males to introduce their sperm into the ventral receptacle may play an important role. Indeed, Chapman et al. (2000) have recently examined the role of Acp36DE in sperm displacement. Males with a null allele for Acp36DE have a lower P_2 than control males, because sperm from these males are not stored. Using spermless Acp36DE null males showed that seminal fluid itself was not sufficient to displace first-male sperm. Moreover, the recent discovery of Acp62F, which has a sequence similarity with a neurotoxin contained in the venom of a spider, *Phoneutria nigriventer*, that itself inhibits neuromuscular activity, points to a role for Acp62F in affecting sperm storage and removal, perhaps via inhibiting the muscular activity of the female's reproductive tract during insemination (Wolfner et al. 1997). The work of Chapman (2000) and others (Civetta 1999; Price et al. 1999; Gilchrist and Partridge 2000) has shown that both sperm and Acp's are involved in the displacement of rival sperm from the ventral receptacle. Females mated to spermless males lose fewer sperm from their sperm stores than do females mated to normal males (Chapman et al. 2000; Gilchrist and Partridge 2000). However, rival sperm are not displaced from the paired spermathecae (Civetta 1999; Price et al. 1999). Nevertheless, these rival sperm are not capable of fertilizing eggs after the second male copulates, strongly suggesting that they are incapacitated by accessory gland products delivered by the copulating male (Price et al. 1999). Further support for such an effect comes from short-term reductions in female fertility when females remate with the second of two wild-type males (Prout and Clark 2000). Gilchrist and Partridge (1995) found no differences in egg hatch when males remated with their previous mate compared with a female that had sperm stored from a rival male. These data show that the seminal fluid effects on sperm are not specific to rival sperm. The experiments of Price et al. (1999) suggest that sperm may become more susceptible to incapacitation the longer they have been in storage, perhaps explaining how males can transfer spermicidal compounds without compromising their own fertilization success.

Wolfner (1997) suggests that just 10–50 % of the total *D. melanogaster* Acp's have thus far been identified. Acp's clearly have a major impact on female behavior and physiology that is beneficial to the copulating male in terms of his ability to avoid sperm competition from future males (fig. 4.12). The *D. melanogaster* model has allowed us great insights into this phenomenon because it is known so well genetically. However, there is increasing evidence that Acp's play biologically active roles in other taxa. Acp's are being increasingly reported in other *Drosophila* species (e.g., Baumann 1974; Chen and Balmer 1989; Imamura et al. 1998). Chapman et al.'s (1998) recent review of the dipteran literature reported 29 species from 19 families in which male accessory gland secretions decrease female receptivity and increase oviposition. Moreover, the data in table 4.1 suggest that this phenomenon is taxonomically widespread.

Alternative Routes for Seminal Product Evolution

The natural selective advantage to females in refraining from reproduction until they have received appropriate stimuli from copulation are obvious; females thus responding will avoid producing infertile eggs. One potential signal available to females would be the mechanical stimulation arising from copulation and/or the presence of an ejaculate in the reproductive tract. Indeed, loss of receptivity in cockroaches is frequently the result of mechanical stimulation arising from the spermatophore (Roth 1962; Engelmann 1970; Smith and Schal 1990), and female response to male seminal products has frequently been shown to depend on appropriate mechanical stimulation associated with copulation and/or the presence of sperm (table 4.1 and references therein). Mechanical stimuli may be short lived, however, particularly for species in which there is no spermatophore, or where the spermatophore is lost shortly following its evacuation. Seminal fluid contains a variety of chemicals that function in the activation and nutrition of sperm, and/or in their transfer to the female (Mann and Lutwak-Mann 1981). Such chemical constituents of semen are naturally selected products of male reproductive physiology. However, females that are able to recognize and respond to such chemicals would be at a naturally selective advantage since they could avoid ovipositing infertile eggs before copulation, or when their sperm stores become depleted. That female insects do become depleted of sperm is well documented (Ridley 1988) and the duration of female postmating responses in *Drosophila* is clearly dependent on the number of sperm remaining (Gromko et al. 1984a). Thus, changes in female behavior and physiology following mating undoubtably have their evolutionary origin in a naturally selected context. Nevertheless, the advantages to males in changes in female behavior and physiology are also obvious; a male will avoid sperm competition from future rivals as long his mate refrains from remating and engages in egg deposition. Therefore, once females use cues from males to trigger their own reproduction, males able to invoke a longer period of unreceptivity and/or greater oviposition response will be favored by sexual selection. Males will be selected to produce greater quantities of seminal fluid than are required for their own physiological function, or to produce greater quantities of the particular element of seminal fluid that females use as a cue to their own fertility. Moreover, they could also be selected to incorporate female hormones into their ejaculate, even though such hormones may play no role in male reproductive physiology, if they can obtain increased fitness by "overdosing" females with such hormones (an example might be juvenile hormone that regulates female oogenesis, Nijhout and Riddiford 1974, 1979).

Eberhard (1996) has provided a thorough review of the evidence that sexual selection has played an important role in the evolution of male seminal products. He recognised five lines of evidence that favor the sexual selection hypothesis, and I will provide a brief outline of each here. For more details the reader is directed to Eberhard (1996). First, male products often have an invasive action, in that they enter the hemolymph and affect the central nervous and/or endocrine system of the female. One might expect natural selection on females to have favored the evolution of receptor sites in the reproductive tract itself. Second, females often require multiple cues from males. Thus, a response to seminal products often requires backup stimuli, such as mechanical stimulation from the reproductive tract, which guards the female against responding in a manner that may be disadvantageous for her. Moreover, there is often redundancy in male seminal products. For example, in *Drosophila* Acp26Aa and sex peptide both impact on female oviposition although the relative influence of each is currently unknown (Wolfner 1997), while in *Aedes* both α matrone, and to a lesser extent β matrone, impact on female receptivity (Fuchs and Hiss 1970; Young and Downe 1987). Redundancy in signals has been associated with sexual selection on secondary sexual traits (e.g., West-Eberhard 1983; Møller and Pomiankowski 1993) and can arise due to a conflict of interest between the sexes. Whenever a male seminal product evolves to manipulate female behavior or physiology to an extent that the exaggerated response becomes detrimental to female fitness, natural selection should favor females that cease responding to the seminal product in question and use an alternative product to assess the amount of sperm in storage (Eberhard 1996). Third, female responses to male seminal products are typically dose dependent (table 4.1). If females required simple information regarding the storage of a sperm supply sufficient to produce fertile eggs, one might expect them to exhibit an all or nothing response (Rice 1984). Such responses are rare, however. The oviposition response of the field cricket *Gryllus bimaculatus* may be one example since females do not elevate oviposition in response to a single mating, but require a level of insemination equivalent to three copulations per day of life to trigger increased oviposition (Simmons 1988a). More normally, females show graded responses in receptivity and oviposition, related to variation in the quantity and/or quality of seminal products transferred by individual males (table 4.1). Fourth, traits subject to sexual selection typically exhibit rapid and divergent evolution (West-Eberhard 1983, 1984) leading to species specificity in both morphology and function. Evidence from *Drosophila* suggests considerable evolutionary divergence in the protein constitution of male accessory gland products (Chen 1984; Bownes and Partridge 1987; Coulhart and Singh 1988; Chen and Balmer 1989). Heterospecific matings demonstrate that these products show species specificity. Although sex peptide from *D. melanogaster* is effective in altering female physiology and behavior in crosses with females of its sister species *D. simulans, D. mauritiana*, and *D. sechellia*, the responses are much reduced compared with conspecific crosses (Stumm-Zollinger and Chen 1988). Furthermore, *D. melanogaster* sex peptide is totally ineffective in the more distantly related *D. funebris* (see Eberhard 1996 for a complete review of this literature). In a recent study, Price (1997) used spermless male *Drosophila* to show that heterospecific seminal fluid is ineffective in sperm displacement across members of the *simulans* species group, providing a postcopulatory mechanism of species isolation. Finally, seminal products active in influencing female behavior and physiology often mimic the female's own hormones. As discussed above, male moths frequently incorporate juvenile hormone into their seminal fluids, although it appears to serve little physiological function in the reproductive male. Rather, juvenile hormone is released by the corpora allata of the female and regulates oogenesis (Nijhout and Riddiford 1974; Riddiford 1993). Incorporation of prostaglandin and ecdysone into the seminal fluid by males of some Orthoptera and Lepidoptera provides interesting parallels (table 4.1).

If sexual selection is involved in the evolution of male seminal products, which seems reasonable given the above, there are a number of mechanisms through which selection can operate. The selective advantage due to sperm competition avoidance has been stressed throughout this chapter and is, in its purest sense, a model of intrasexual selection. However, it is clear that female processes are directly involved, implying a process of postcopulatory cryptic female choice (sensu Eberhard 1996). As with the evolution of male genitalia discussed in chapter 3, to discuss postcopulatory female choice and sperm competition as independent processes is to obfuscate reality; the two models are not mutually exclusive since selection imposed by females will act on variation in male ability to avoid and/or engage in sperm competition. However, there are different classes of models of female choice which have different evolutionary implications (Kirkpatrick and Ryan 1991). One model implicit in the discussion above is the direct model of sensory bias or sensory exploitation (Ryan et al. 1990; Ryan and Rand 1993); male seminal products may evolve because (i) females have a predisposition to respond to these products in a naturally selected context, and (ii) males benefit from female responses in the context of sperm competition or its avoidance. In the early stages of evolution, females could obtain immediate benefits if seminal products are associated with the number of sperm present in the ejaculate, which in turn affects immediate female fertility. After elaboration of the male seminal product, sexual conflict might arise if male products become detrimental to female reproduction, so that antagonistic coevolution between seminal products and female response to such products ensues, generating rapid and divergent evolution (Holland and Rice 1998). However, Cordero (1995, 1998), Eberhard and Cordero (1995), and Eberhard (1996) have proposed that male seminal products could evolve under an indirect model of preference evolution. Under Fisherian selection, it is argued that females gain indirect benefits associated with their responses, because they will produce sons better able to elicit responses in females. At the same time, females produce daughters who respond to males with seminal products, so that the process becomes self-reinforcing until checked by natural selection. The seminal products are arbitrary traits in that they signal nothing regarding the quantity and/or quality of the ejaculate provided by the male. Alternatively, seminal products could arise as honest signals of ejaculate quality and/or quantity. Seminal products could be honest signals if they were costly for males to produce. The benefits obtained by females from their preferences could be direct or indirect. If the information contained in the signal pertains to the quantity and/or the quality of the ejaculate, females responding to honest male signals may ensure their fertility by continuing to mate until they have an ejaculate that is sufficient to maintain high fertilization success. Alternatively, the benefits may be indirect insofar as the signal contains information regarding sperm and/or male quality which itself may be genetically correlated with offspring viability (Cordero 1995; Eberhard and Cordero 1995; Cordero 1996; Eberhard 1996; Cordero 1998).

There are empirical data available that can be used to address some of the predictions generated by the models proposed for seminal product evolution.

All models predict that there should be variation in male ability to produce seminal products and that some of that variation should be genetic. A number of studies have found phenotype variation in the quantity and/or quality of male seminal products that can influence the response of females in a quantitative manner (table 4.1). For example, male body size has been shown to influence ejaculate size and the magnitude of female postmating responses in the sheep blowfly *Luculia cuprina* (Cook 1992) and the seed beetles *Callosobruchus maculatus* and *Stator limbatus* (Fox et al. 1995; Savalli and Fox 1998b). Variation in male diet and/or larval environment appears to generate variation in ejaculate quality and female postmating responses in moths (Lum and Flaherty 1970; Lum and Brady 1973; He and Tsubaki 1991, 1992; Delisle and Hardy 1997) and butterflies (Wedell 1996). Genetic variation in ejaculate size has been demonstrated in *C. maculatus* (Savalli and Fox 1998a) and there is a genetic basis to the influence of male accessory gland products on sperm competition in *Drosophila* (Service and Fales 1993; Clarket al. 1995; Service and Vossbrink 1996; Hughes 1997). Thus, the raw material for sexual selection appears to be present.

Seminal products have generally been found to be costly for males to produce. For the majority of species studied, experiments have shown that males rapidly become depleted of accessory gland products with successive matings. Females mated with depleted males have postmating responses that are reduced in magnitude compared to those exhibited by females mated with virgin males, or males allowed time to recover from mating activities (see table 4.1 and references therein). In *L. cuprina*, females mated with depleted males also have reduced fertility, indicating that the size of the ejaculate may well be an honest signal of the immediate fertility benefits associated with ejaculate size (Smith et al. 1990). Moreover, Cook (1992) found that the costs of seminal products were size dependent; small males became depleted of seminal products sooner than did large males, as indicated by the refractory periods they induced in sequentially mated females. If this size-dependent cost of accessory gland function had a genetic basis, there would be evidence that females could produce sons of superior quality by responding to the variation in seminal products their fathers transfer at copulation. However, given that ejaculate size and sperm number appear to be highly correlated in this species (Smith et al. 1990) manipulations that control the numbers of sperm are required to establish the relative importance of male seminal products for female postmating behavior.

Delisle and Hardy (1997) found that male moths, *Choristoneura fumiferana*, reared on high-quality diets (high-quality males) induced longer refractory periods in females than did males reared on low-quality diets (low-quality males). Ejaculate size was probably not responsible for this effect, because males reared on artificial diets produced the largest spermatophores but induced refractory periods similar to those induced by low-quality males. Thus, the quality of the ejaculate appears to influence female receptivity. Delisle and Hardy (1997) found that females mated with low-quality males had a lower fertility and total fecundity than females mated to high-quality males. These data strongly suggest that ejaculate quality may represent an honest signal of direct benefits for female reproduction in terms of the number of sperm available to fertilize eggs, and possibly the amount of nutrients available for the production of eggs (see below).

The evolutionary costs of seminal products have been studied in some detail using *Drosophila*. Pitnick et al. (1997) found a positive evolutionary association between male expenditure on the ejaculate (estimated from counts of radioactive isotopes recovered from mated females) and male age at first reproduction across 34 species of *Drosophila*; species that transfer large ejaculates require long periods of prereproductive feeding, indicating that seminal products represent a significant evolutionary cost for males. Within-species studies of male *D. melanogaster* show that males exposed to females have a reduced survival compared with those not exposed to females, or exposed only periodically (Prowse and Partridge 1997). However, by far the greatest cost was the reduction in fertility of exposed males; males with constant exposure to females showed complete sterility when 80% of their cohort were still alive. This reduced fertility appeared to be associated with declines in both accessory gland volume and sperm number, although, contrary to the claim of Lefevre and Jonsson (1962), reduced fertility appeared to be more closely associated with decline in sperm number (Prowse and Partridge 1997). Given that female responses to male seminal products are dependent on sperm numbers in *Drosophila* (Gromko et al. 1984a), the evidence suggests that females assess their fertility directly, rather than relying on seminal products to honestly signal the quantity of sperm available. The same is true for many species of moth (table 4.1), and in butterflies, reductions in ejaculate volume are associated with increases rather than decreases in sperm numbers (Cook and Wedell 1996; Watanabe et al. 1998) so that seminal fluid products are not honest signals of the quantity of sperm transferred.

Signaling costs per se are not specific to handicap models of preference evolution so that the observation that seminal products are costly to produce does not in itself allow us to distinguish between competing hypotheses of sexual selection (Johnstone 1995). Longevity and/or long-term fecundity costs of seminal products could be balanced by fitness benefits obtained during early matings, conforming with a Fisherian model. The evidence pertaining to the direct benefit and handicap hypotheses discussed above is mixed and inconclusive. Evidence that female preference increases general offspring performance, not just the performance of sons, would be required as support for the indirect handicap hypothesis. Currently no study has attempted to fully explore these models in relation to seminal products. Thus, based on the available data it is not yet possible to draw general conclusions regarding the mechanism(s) by which female preferences favor the evolution of seminal products in males.

For selection to operate on male seminal products via sexual conflict there must be a cost to females derived from the products they receive from males. Mating has been shown to decrease female survival in ten species of Diptera from five families (Chapman et al. 1998) as well as a number of other insect taxa, including Orthoptera (Dean 1981; Burpee and Sakaluk 1993) and Coleoptera (Murdoch 1966). There is very good evidence from *D. melanogaster* that this cost of mating arises from the transfer of seminal products by the male. Early work showed that exposure to males decreased female longevity (Fowler and Partridge 1989). Part of this cost of exposure to males arises from differences in egg production for mated and unmated females (Partridge et al. 1987) and part from nonmating effects of courtship (Partridge and Fowler 1990). Decreased longevity is not associated with the receipt of sperm (Chapman et al. 1993). Chapman et al. (1995) utilized the main-cell-deficient transgenic flies produced by Kalb et al. (1993) to show that decreased female longevity was a direct consequence of receiving the accessory gland products from males that function in sperm competition (fig. 4.13). The recent discovery of a toxin in male seminal fluid (Acp62F, Wolfner et al. 1997) may well contribute to this female cost of reproduction. That this cost generates antagonistic coevolution between males and females was elegantly demonstrated by Rice (1996). Rice (1996) established two populations of flies. In the experimental population, females were discarded at each generation so that male evolution was able to proceed without being countered by changes in the female line of their population. At each generation, these males were bred using females from a second population of flies that were not coevolving with the experimental population. After 30 generations, Rice (1996) examined the fitness of the evolving male population when mated with noncoevolving females. He found that the evolving males had increased fitness relative to controls, in terms of their ability to induce remating by previously mated females, to induce unreceptivity to remating in those females, and to resist sperm displacement from future males when those females later regained receptivity. Thus, the net fitness (offspring production) of males was increased by 24% over the 30 generations of evolution. More importantly, when the cost of reproduction for females was assessed, females that were mated to males from experimental lines had a decreased survival relative to controls, an indication that increased male fitness came at the expense of decreased female fitness (see also Rice 1998a). These results demonstrate that male evolution is generally constrained by antagonistic coadaptation by females in *D. melanogaster*. It would be instructive to further dissect the process of sexual conflict in this system. For example, sexual conflict may be specific to particular male seminal products. Chapman et al. (1996) have shown that mutant females (*dunce*) who fail to

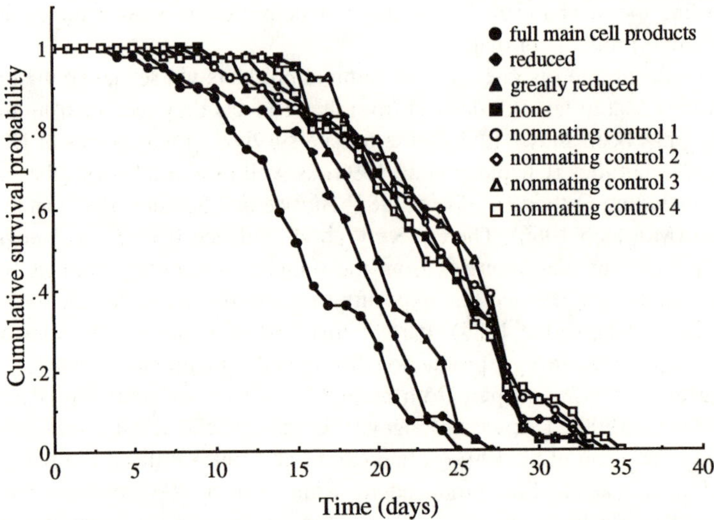

Figure 4.13 The cost of male seminal product action for female *Drosophila melanogaster.* The data (redrawn from Chapman et al. 1995) show survival curves for females mating with normal males having full main cell products or males genetically engineered to have reduced, greatly reduced, or no main cell products (see also fig. 4.12). When males fail to transfer main cell products, females survived as long as nonmating virgin controls. However, with increasing expression of main cell products in males, females showed a dose-dependent reduction in survival relative to controls.

respond to sex peptide suffer an increased cost of mating relative to normal females. Females with the *dunce* mutation have a higher mating frequency than normal females because they fail to become unreceptive. As a consequence their exposure to the toxic effects of other male accessory gland products will be greater. Thus, female responses to some constituents of male seminal fluid may be in their best interests so that sexual conflict over their action will be reduced or absent.

Other evidence for sexual conflict over male seminal product action comes from studies of gryllid and tettigoniid Orthoptera. Simmons and Gwynne (1991) found that female responses to seminal products, in terms of reduced receptivity to further mating by female *Kawanaphila nartee*, were dependent on the female's own nutrition. Since females can benefit from mating via the acquisition of food gifts provided by males (Simmons 1990a), there appears to be a conflict of interest over the number of times females mate. When resource limited, females ignore male seminal products (figure 4.9). Female *Teleogryllus commodus* metabolize and excrete more prostaglandin than reaches the ovaries, suggesting that the level of activity that best suits the male may not be in the female's own interests (Stanley-Samuelson and Loher 1985). However, currently too few taxa have been studied to claim a general cost of male seminal products on female reproduction, or to support a general role for sexual conflict in seminal product evolution (table 4.1). It should also be noted that a demonstration of female costs per se is not in itself evidence for sexual conflict. If there are direct and/or indirect fitness benefits for females associated with their responses to male seminal products, these may outweigh the costs of their action on lifespan so that there will be no antagonistic coevolution in females (Cordero 1998).

Seminal products that increase female egg production are often interpreted as nutrient contributions made by the male to female reproduction (e.g., Simmons and Parker 1989; Boggs 1990; Wiklund et al. 1993; Kaitala and Wiklund 1994; Fox et al. 1995; Karlsson 1995; Savalli and Fox 1998b). Implicit in this interpretation is that male seminal nutrients can evolve under natural selection for male parental investment (Trivers 1972). The attractiveness of the paternal investment interpretation is understandable, given that females do appear to have increased fitness as a direct consequence of receiving male-derived seminal products. However, as Eberhard (1996) has recently pointed out, distinguishing between nutritive and non-nutritive functions of male seminal products is not easy. Eberhard (1996) provides a number of lines of evidence that can provide distinctions. For example, when the target organ(s) of male products are in the female's nervous or humeral system and the products are not incorporated into eggs or female nutrient reserves, a nutritive function can be ruled out. In this case, removal of the target organ(s) function should remove the effect of male products. Likewise, if male products fail to have an effect on closely related species, or if artificial stimulation of females using mechanical or chemical mimics induces the same responses in females, then the male products are unlikely to be nutritive. In some cases, female responses are too rapid to be dependent on male nutrient donations (Simmons and Parker 1989) or last longer than the male product does within the female, while in others the active male substance is known to be a mimic of the female's own hormones rather than a nutrient source (see Eberhard 1996). These types of evidence clearly preclude a nutritive role for male products. However, often the evidence is equivocal. For example, if the target organ(s) are the ovaries so that male products are found in developing eggs it is difficult to rule out a nutritive role. Moreover, male products can have the incidental effect of providing a nutrient source to females even though their primary mechanism of action is via an influence on female physiology and behavior. Nevertheless, I concur with Eberhard (1996) that the evidence for evolution of male seminal products via natural selection for male parental investment is often equivocal or negative.

An elevation of female egg production immediately following mating cannot be taken as evidence for a nutritive function of male seminal products. The burst of egg production following mating may have a negative impact on egg production in later life, so that lifetime female fitness is unchanged by male seminal products. Male fitness is obviously increased, because the male will obtain a greater proportion of a female's lifetime fecundity due to the burst in oviposition that he is able to elicit. Thus, the majority of species studied in table 4.1 are noted to have no benefit to females derived from male seminal products. That is not to say females do not obtain benefits, simply that in many cases the appropriate experiments have not been performed.

The typical method for examining male nutritive contributions has been to inject males with amino acid mixtures labeled with radioactive isotopes, and to examine female reproductive tissue and/or eggs for traces of male-derived radioactivity after copulation (Friedel and Gillot 1977; Boggs and Watt 1981; Huignard 1983; Boucher and Huignard 1987; Bownes and Partridge 1987; Wedell 1993a). Again, these types of experiments alone cannot be used as evidence for a nutritive role of male seminal products. First, injected amino acids may not be incorporated into the male seminal products that influence female oviposition. If they are not, and are transferred as free amino acids, they could be used by females in producing eggs although this is not evidence that the male seminal product is similarly used by females. Second, if they are incorporated into male seminal products, often these enter the hemolymph and therefore come into direct contact with the ovaries and developing eggs. Moreover, if the ovaries are the target of male seminal product action we should expect to see radioactivity in the ovaries and eggs even when there is no nutritive role. For example, in Bownes and Partridge's (1987) study of *D. melanogaster* and *D. pseudoobscura*, male-derived radioactivity was found in the ovaries and developing eggs within 24 h of copulation with labeled males. Male seminal products have the effect of increasing egg production. This type of evidence is often taken as indicative of a nutritive role for male seminal products. However, as Bownes and Partridge (1987) point out, the protein content of the entire male accessory gland complex is equivalent to just five eggs, and only a small proportion of accessory gland content is transferred at copulation. The radioactively labeled amino acid used by Bownes and Partridge (1987) was methionine, an essential amino acid in oogenesis (Sang and King 1961). Thus, if transferred as free amino acid in the ejaculate it would be rapidly incorporated into eggs. Chapman et al. (1994) fully investigated the nutritive function of male seminal products using both genetic and phenotypic comparisons. If females obtain nutritive benefits from seminal products, those maintained on a low diet should show greater lifetime fecundity and/or longevity as a result of multiple mating compared with females reared on a high-quality diet. However, although diet had a significant impact on lifetime fecundity, Chapman et al. (1994) found no significant differences between females continually exposed to males, and those provided with a male one day in three. In an artificial selection experiment, Chapman et al. (1994) utilized populations of *D. melanogaster* that had been allowed to evolve for five years on either a low-quality or a high-quality diet. If females obtain significant nutritional benefits from male seminal products, selection should favor the evolution of increased mating frequency in females in the low-quality-diet populations. However, Chapman et al. (1994) found no significant difference in the remating rates of females from high- and low-diet populations. Thus, the evidence overwhelmingly rejects a nutritive function for male seminal products, despite the incorporation of male-derived radioactively labeled amino acids, and the short-term elevation in fecundity that results from mating. A second example is Wedell's (1993a) study of *D. verrucivorus*. She noted that male-derived radioactivity accumulated in the ovaries 3–4 days after mating, and in mature eggs 10–12 days after mating. However, increased oviposition of mature eggs by females occurred during the refractory period of females, 1–5 days after mating, so that the temporal delay precludes a nutritive function for male seminal products. Moreover, females do not have an increase in their lifetime fecundity as a consequence of mating (Wedell and Arak 1989).

Markow and Ankney (1988) found an association between the strength of the insemination reaction (swelling of the vaginal wall following mating; Patterson 1946) and the incorporation of radioactively labeled amino acids into female reproductive tissues across 19 species of *Drosophila*. They concluded that the positive relationship was suggestive of a nutritive function of male seminal products, based on the assumption that the insemination reaction was effective in reducing the risk of sperm competition from future rivals while the male's nutrients were being utilized by the female. However, a detailed phylogenetic examination using 34 species of *Drosophila* (Pitnick et al. 1997) failed to find support for this association. It is also now known that the insemination reaction differs across species and a simple "mating plug" interpretation is unjustified (Alonso-Pimentel et al. 1994; section 4.2; Pitnick et al. 1997). Pitnick et al. (1997) did find phylogenetic variation in the magnitude of incorporation of male-derived radioactive isotopes. However, they also conclude, for the reasons outlined above, that without further study an interpretation based on nutritive effects of male seminal products is equivocal.

Some studies do provide evidence that seminal products can increase female fitness, as well as that of the copulating male. Fox's (1993b) study of *Callosobruchus maculatus*, for example, demonstrated that multiply mated females had a significantly greater lifetime fecundity than singly mated females. Similarly, lifetime fecundity and/or longevity has been shown to be elevated by multiple mating in the moths *Pseudaletia unipuncta* (Svärd and McNeil 1994), *Trichoplusia ni* (Ward and Landolt 1995), and *C. fumiferana* (Delisle and Hardy 1997), as well as the butterflies *Danaeus plexippus* (Oberhauser 1989), *Pieris napi* (Wiklund et al. 1993), and *Colias erytheme* (Rutowski et al. 1987). In general, the females of polyandrous butterflies appear to devote fewer of their own resources to reproduction than do monandrous species, perhaps because the males of polyandrous species devote more resources to their ejaculates, which thus represent a significant nutrient pool with which to produce eggs (Karlsson 1995; Karlsson 1996). Thus, in the Lepidoptera there is good evidence for a general beneficial effect of male seminal products on female fitness. Nevertheless, Morrow and Gage's (2000) recent comparative study of moths provides evidence that male seminal products evolve in response to sperm competition. They used testis mass as an index of sperm competition risk; numerous comparative analyses have shown how sperm competition risk favors an increase in sperm production and thus testis mass (see chapter 7). If seminal products evolved in the context of parental investment, an increase in sperm competition risk should favor a reduction in male investment because of reduced paternity assurance (Wright 1998). In contrast, Morrow and Gage (2000) found that accessory gland mass was positively associated with testis mass across 130 species of moth. It could be argued that increased testis mass should be favored to assure confidence of paternity in species with high male investment. However, Morrow and Gage (2000) also found that accessory gland mass was positively associated with the size of the female's corpus bursae. Since stretch receptors in the bursa contribute to loss of female receptivity (Sugawara 1979), an increase in the size of the bursa should favor increased volumes of accessory secretions, and thus increase accessory gland mass, in order to prevent females from remating. There is no reason to expect coevolutionary change in accessory gland size and female reproductive tract morphology where the seminal products serve a purely nutritive function.

Parker and Simmons' (1989) theoretical analysis of the evolution of nuptial feeding in insects suggests that the evolutionary origin of male-derived nutrients is unlikely to be via natural selection for male parental investment. In the early stages of their evolution, gifts would be small. Parker and Simmons' (1989) models suggest that, when a female receives a gift that contributes only a moderate supplement toward reproduction, it will be optimal for her to *reduce* gametic output. By doing so she increases her reproductive rate and gains a higher lifetime fecundity (Parker and Simmons 1989). From the male perspective, however, he will have reduced fitness due to the immediate reduction in oviposition at the next clutch, unless of course he has confidence of paternity for the female's entire lifespan. Sperm competition studies clearly demonstrate that genetic monogamy, if it exists at all, is an extremely rare phenomenon in nonsocial insects (table 2.3). Thus, natural selection is predicted to act against male nutrient investment and cannot explain the origin of seminal products. In contrast, even small increments in the amount of seminal products that inhibit female remating or otherwise improve a male's success in avoiding sperm competition will be immediately favored by sexual selection. As seminal products increase in magnitude, they could begin to impact on female fecundity in a beneficial manner, as a side effect of their evolved function in sperm competition. Once established, further elaboration could occur due to sexual selection by female choice for males providing ejaculates of higher quality, or via natural selection on males for increased parental investment (Simmons and Parker 1989).

In conclusion, although male seminal products may prove beneficial in some species, the bulk of evidence suggests that their principal function is to influence female behavior and physiology in a manner that increases male fitness in the context of sperm competition avoidance (see also Eberhard 1996).

4.4 Pheromones

Mature virgin females are generally highly attractive to sexually active males, and female receptivity is often signaled via the release of pheromones (Thornhill and Alcock 1983; Lewis 1984; Eisner and Meinwald 1995). There will be a selective advantage for males that are able to reduce the attractiveness of their mates, since this will reduce the risk of mating by, and sperm competition from, future males. During mating, male *D. melanogaster* transfer to the female cuticle a hydrocarbon, 7-tricosene, that is virtually absent from the virgin female. Deposition of this pheromone has the effect of reducing female attractiveness to future males for the first 3–4 h after mating (Tomkins and Hall 1981; Scott and Richmond 1985; Scott 1986b). Females begin to generate and emit their own 7-tricosene 6 h after mating and the deposition of this so-called antiaphrodisiac by males is thought to reduce the female's attractiveness until she is able to synthesis her own pheromone. 7-tricosine is the major cuticular hydrocarbon in male *D. melanogaster* so that reduced attractiveness of females arises due to sexual mimicry (Scott 1986b; Scott and Jackson 1988). There was some suggestion that female attractiveness may be regulated by male seminal products. Work by Mane et al. (1983) suggested that the seminal fluid enzyme esterase 6 metabolized the seminal fluid component *cis*-vaccenyl acetate to produce the antiaphrodisiac *cis*-vaccenyl alcohol. However, subsequent work showed that this product was not produced in vivo (Vander Meer et al. 1986) and that it is not necessary for decreased postmating attractiveness (Scott and Richmond 1987). Tram and Wolfner (1998) have used males lacking main cell function to show that male accessory gland products have no influence on the sexual attractiveness of females.

Male-donated antiaphrodisiac pheromones have also been reported to be transferred to females during mating in mealworms, *Tenebrio molitor* (Happ 1969), solitary bees, *Centris adani* (Frankie et al. 1980), and *Heliconius* and *Pieris* butterflies (Gilbert 1976; Andersson et al. 2000). Female *H. erato* store the male-derived antiaphrodisiac pheromone in a special gland at the tip of the abdomen, referred to as the "stink club." After copulation, females release male-derived pheromone to deter future males from attempting copulation. Gilbert (1976) found that the antiaphrodisiac odors produced by males were race specific. This observation is paralleled by studies of *Drosophila* strains in which the effectiveness of male-derived pheromones in deterring rivals was also found to be strain specific (Scott and Jackson 1988). In *Drosophila*, the antiaphrodisiac action is only short lived, and may serve to prevent rival matings until seminal products that inhibit receptivity (section 4.3) have their full effect in the mated female. In *Heliconius* the antiaphrodisiac can last the female's entire lifespan so that they mate only once. The observation that female *Heliconius* actively utilize male-derived pheromones to deter future males suggests that there is no conflict between males and females over monandry. In *P. napi*, however, females benefit from remating because of the nutrients contained within the seminal fluids. There is a decrease in pheromone titer with time since mating, which results in females becoming more attractive to males. It is not yet known if it is females themselves who regulate the decline in pheromone efficacy (Andersson et al. 2000), although females clearly regulate their receptivity in relation to their nutrient requirements (Kaitala and Wiklund 1994).

4.5 Summary

The products of male accessory glands play a variety of roles in delaying and/or preventing females from mating with future males, which in turn reduces the risk of future sperm competition. Physical barriers to future males result from the presence of spermatophore remnants left in the female's reproductive tract. These barriers may be short lived and female controlled, because the spermatophore must be ejected in order for females to oviposit. In some groups of butterflies mating plugs have become sufficiently elaborated in size and structure to ensure that females remain monogamous throughout life. This is possible because females have separate copulatory and oviposition openings to their genital tracts. Chemical products such as antiaphrodisiac pheromones that repel future males can also reduce the risk of future rival matings. Chemical products of the accessory glands that are secreted into seminal fluid have been shown to induce a variety of behavioral and physiological changes in females that result in decreased sexual receptivity and egg production immediately following mating. The precise molecules involved in postmating responses have been identified in some cases and the mechanisms of action established. Female responses to seminal products are not guaranteed, however, and often depend on the storage of sufficient sperm to maintain their fertility. Thus, there may be conflict between males and females in the action of seminal products on female reproduction. While it is clear that the avoidance of sperm competition is an important selective agent in the evolution of male seminal products, the necessary interaction with female-imposed selection has not been well studied.

A number of competing hypotheses can explain the evolution of seminal products via female preferences. Perhaps the most parsimonious routes are via direct sensory exploitation of female responses to insemination that arise under natural selection on females to avoid infertility or direct selection operating via honest signals that convey information regarding the quantity and/or quality of ejaculate received. Alternatively, females could obtain indirect benefits for their offspring if, by responding to males providing greater ejaculatory stimulation, they produced sons better able to stimulate females to avoid sperm competition. If the quality of seminal products relates to male quality, females could even gain direct benefits by responding to male seminal products through the production of offspring of superior viability. Currently there is little or no empirical evidence to distinguish between these competing hypotheses of preference evolution, and more evolutionary studies are required in this area of insect physiology. Sexual conflict has undoubtedly played a role in seminal product evolution in *Drosophila*, but an understanding of the generality of sexual conflict requires detailed quantitative genetic and phenotypic examinations of the costs of seminal products on female fitness in a greater diversity of taxa. In some cases, notably the Lepidoptera, male seminal products can prove beneficial to both males and females, because females can digest the proteinaceous products of male seminal fluids and utilize these nutrients in producing eggs. However, distinguishing between nutritive and non-nutritive functions of male seminal products is not easy, and requires a demonstration that female lifetime fecundity is elevated as a direct consequence of receiving male seminal products. In general, it seems that female benefits, where they occur, are unlikely to have been important in the early origin of male seminal products.

5 Avoidance of Sperm Competition III: Behavioral Adaptations

5.1 Introduction

The ability to temporarily or permanently inhibit female receptivity to further mating represents a highly adaptive strategy for reducing the risks of sperm competition from future males (chapter 4). Although the manufacture of mating plugs and/or seminal products may be costly, males are at least free to continue mate searching and thus potentially increase their reproductive success through the acquisition of further matings. However, reduced sexual receptivity may not always be in the female's best interests and thus considerable conflict between males and females may occur over remating (Simmons and Gwynne 1991). Even with slight levels of continued female receptivity, males who leave their mates prior to oviposition will be at a selective disadvantage if there is a chance that females will encounter and mate with another male. Moreover, because of the antagonistic nature of selection pressures that arise under sperm competition (Parker 1970e, 1984), there will be an immediate selective advantage to males that are able to dissolve or penetrate a rival's mating plug, or to persuade and/or coerce females with reduced receptivity to remate (Parker 1974, 1984). The only truly effective way to ensure paternity at the next oviposition would be to remain with and defend the female from the mating attempts of rivals until the clutch is fertilized and laid.

In a variety of insect taxa, reproductive behavior includes a prolonged period of male-female association that is apparently not required for insemination. Parker (1970d, 1970e) defined these mating associations as "passive phases" in order to distinguish them from the phase of active linkage of male and female genitalia that constitutes copulation, and argued that postcopulatory passive phases were important in the avoidance of sperm competition. Parker (1970e) also noted that in many species copulation can be prolonged beyond the time necessary for insemination, and suggested that the male may act as a living mating plug after insemination, thereby preventing the female from mating with rival males. Postcopulatory associations that are thought to function in the avoidance of sperm competition have become collectively known as strategies of mate guarding.

Table 5.1
Alternative hypotheses for postcopulatory mating associations.

Hypothesis	Prediction
Paternal investment	Females gain immediate benefits via enhanced foraging, reduced harassment from mate-searching males, or reduced predation, which increases the number and/or quality of offspring.
Ejaculate transfer	Enhanced transfer of sperm and/or accessory gland products.
Multiple mating	Males increase their fertilization success via multiple inseminations with female.
Cryptic female choice	Males increase probability of paternity via postinsemination courtship of female.
Mate guarding	Males reduce probability that female will accept sperm from rival males.

Although it is often assumed that prolonged mating associations constitute a form of mate guarding, as pointed out by Thornhill (1984b) it is vital to assess alternative hypotheses when evaluating the adaptive significance of traits believed to have arisen under sperm competition. In his review, Alcock (1994) identified a number of alternative hypotheses for the evolution of prolonged associations between males and females following copulation (table 5.1). Females might gain immediate benefits so that prolonged association by males with their partners might represent a form of paternal investment which evolves under natural selection. This hypothesis predicts that females and males will benefit directly from the association via enhanced offspring production. Eberhard (1991) suggested that females may increase their own fitness by assessing males during postmating courtship displays and cryptically choosing, by selective sperm storage and/or utilization, those males that are better able to provide adequate stimulation. The cryptic female choice hypothesis thus predicts that male fitness will be positively associated with some aspect of their behavioral performance during the postcopulatory association. Alternatively, it may be only males that obtain benefits from prolonged mating associations. For example, although copulation may serve to transfer the ejaculate to the female's reproductive tract, time may be required for sperm to be transferred to the female's sperm storage organs, or for nonsperm products of the ejaculate to be transferred. Mating associations should then be subject to natural selection for efficient ejaculate transfer. Males may also need time to manipulate their own and/or rival sperm within the female's reproductive tract in order to increase their fertilization success. Moreover, males may prolong the association in order to mate repeatedly with the same female and thereby enhance their fertilization success by increasing the representation of their sperm in the female's sperm storage organs. While sperm manipulation and multiple matings are likely to be subject to selection under sperm competition, they are not relevant in the context of preventing rival males access to females, a function specific to the mate guarding hypothesis.

In this chapter I review studies of mating associations in insects and, by consideration of alternative hypotheses, examine the general impact of sperm competition avoidance on their evolution. Throughout the chapter the term "mate guarding" is used specifically for associations between males and females that function in the avoidance of sperm competition. I first review theoretical studies that have looked for general conditions that are likely to favor the evolution of mate guarding, before reviewing the empirical literature on postcopulatory associations for evidence of support for theoretical predictions of the mate guarding hypothesis, and for the alternative hypotheses outlined in table 5.1. Finally, I examine morphological and behavioral adaptations that function in mate guarding and end with a discussion of male mate choice in the context of sperm competition.

5.2 Theoretical Models of Mate Guarding

Parker (1974) adopted an optimality approach to analyze the evolution under sperm competition of male time investment in postcopulatory mate guarding. He considered that prerequisites for the evolution of postcopulatory guarding behaviour were a high enough encounter rate between mate-searching males and receptive females, and a long enough time frame from insemination to oviposition to allow sperm from different males to overlap and compete for fertilizations (if oviposition was instantaneous with the completion of insemination, there would be no eggs available for future rivals to compete over). Parker's (1974) approach was to balance the benefits in terms of fertilization gains obtained from guarding a given female with the costs associated with searching for and copulating with additional females. His formulations suggest that the maximum advantage to a male adopting a strategy of postcopulatory mate guarding within a population of nonguarding males occur when (i) the encounter rate with receptive females is low relative to the time from insemination to fertilization so that, for the same fertilization gain, guarding males will spend less time guarding than nonguarding males will spend finding additional females; (ii) the ratio of males to females is high so that intramale competition for receptive females is intense during the period from insemination to fertilization; (iii) the proportionate fertilization gain to the last male to mate is high, so that nonguarders run the risk of losing fertilizations to rival males; and (iv) at the time when guarding begins, a large proportion of the female's eggs remain to be fertilized.

Yamamura (1986) adopted a game theory approach in his analysis of the evolution of mate guarding strategies. He envisaged receptive females entering an area for oviposition that contained a constant number of males searching for mates. Incoming females arrived at a constant rate and encountered searching males at random. Copulation was instantaneous and when males guarded females, other males were unable to mate with them. If females were not encountered and mated within a constant time period, they oviposited and left the area (this parameter is analogous to Parker's 1974 period between insemination and fertilization). The analyzed scenario was thus analogous to a resource-based mating system. Males were able to adopt a strategy of guarding for some time less than or equal to the critical interval between mating and oviposition, and male fitness was evaluated as the number of eggs fertilized by that male per unit time. The evolutionarily stable strategy set derived can be represented in two dimensional space using the parameters *s*, the ratio of males to incoming females during the time period of oviposition, or the operational sex ratio of Emlen and Oring 1977, and *d*, a measure of the searching efficiency and thus encounter rate of males with receptive females (fig. 5.1). In general, a strategy of postcopulatory mate guarding for the entire period from copulation to egg deposition becomes evolutionarily stable when the operational sex ratio is male biased and the efficiency of searching for receptive females is moderate to high. Increasing the period between arrival and oviposition has the effect of changing the stable strategy from guarding to nonguarding (fig. 5.1). Increasing the proportion of eggs fertilized by the last male to mate with the female has the effect of increasing the parameter space in which guarding is the stable strategy (Yamamura 1986). Interestingly, in all cases examined there is a parameter zone in which both nonguarding and guarding strategies can be maintained in the population.

Yamamura and Tsuji (1989) also analyzed a situation in which males and females co-occurred within the same area, and where males searched for females at random through the habitat and copulated with them once located. The availability of receptive females thus declined with time. Yamamura and Tsuji (1989) analyzed two forms of this non-resource-based mating system that differed in the patterns of female reproduction; females either oviposited continuously, laying eggs at a constant rate throughout the reproductive period, or discretely, laying batches of eggs that were matured and then laid together. The results of the model were essentially the same as those obtained from Yamamura's (1986) original analysis, with the exceptions that in the resource-based mating system polymorphisms in guarding strategy were more likely to occur with high last-male fertilization success, while in the non-resource-based system polymorphisms were more likely to occur with low last-male fertilization success. Oviposition patterns had little impact on the ESS guarding strategy. Yamamura and Tsuji (1989) concluded that the product of the fertilization probability and the operational sex ratio was the most critical aspect in predicting whether a postcopulatory guarding strategy promotes male fitness.

More recently Fryer et al. (1999b) used an ESS approach to analyze what

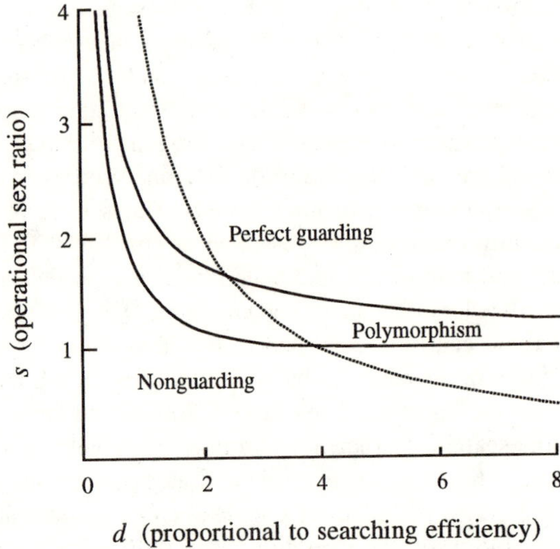

Figure 5.1 Results of a game theory model in which conditions are sought for the evolution of postcopulatory mate guarding as an evolutionarily stable strategy. The ESS can be classified by two parameters, the operational sex ratio (*s*) equal to the number of males divided by the number of incoming females during a period of time (*t*), and a measure of searching efficiency (*d*) that will depend on an individual male's ability and environmental conditions at the mating site, such as visibility. The parameter space below the lower solid line is where nonguarding is the ESS, while the parameter space above the upper solid line is where a guarding strategy is the ESS. Between these boundaries lies a parameter space where a mixed strategy is possible, with both guarding and nonguarding strategies being stable. Postcopulatory mate guarding becomes more advantageous as the number of searching males increases. Increasing time (*t*) will have the effect of decreasing *s* and increasing *d* so that a point *s,d* will move through the parameter space for guarding to the parameter space for nonguarding (dotted line). Thus, for example, if the period between a female arriving at the mating site and ovipositing increases, the advantages associated with postcopulatory guarding decrease. (Redrawn from Yamamura 1986)

they called a two-round sperm competition model. The model was based on the somewhat unrealistic constraint that males could mate just twice, and they could either allocate their entire sperm supply in the first round of mating and guard their mate, or they could allocate some variable proportion of their sperm to the first female and then search for and allocate the remaining sperm to a second female. They included in their analysis a parameter that defined the weight (r) given to the sperm of the second male to mate with the female, or the second-male advantage in fertilization, and a parameter that described the probability that a guard would successfully retain his mate when challenged (ϕ). Their analysis showed that as ϕ increased so the region in which a pure guarding strategy was stable also increased, and that this effect was magnified by increases in the proportion of sexually active males. The probability of successful defense, ϕ, required for guarding to be stable decreased with increasing fertilization advantage, r, for the second male. The model of Fryer et al. (1999b) thus behaves rather similarly to those of Yamamura (1986) and Yamamura and Tsuji (1989).

Both optimality (Parker 1974) and ESS (Yamamura 1986; Yamamura and Tsuji 1989; Fryer et al. 1999b) approaches to modeling the evolution of mate guarding converge on a number of evolutionary predictions: sperm competition should favor postcopulatory mate guarding when (1) females show continued receptivity after copulation; (2) the intensity of competition among males for females is high (measured as either the operational sex ratio or the density of males at the site of mating); (3) male mate-searching efficiency is high; (4) the period between insemination and oviposition is short; (5) the probability of retaining a female when challenged is high; and (6) the last male to mate gains a high proportion of the fertilizations at oviposition. Note that, although postcopulatory mate guarding may still be adaptive at low or moderate levels of last-male fertilization success (see below), the trait simply becomes increasingly likely to evolve with higher levels.

5.3 Evidence for Mate Guarding in Insects

Overview

Extended mating associations appear widespread in the insects, with information now available on 125 species from ten different orders (table 5.2 at the end of this chapter). The quality of the data varies from simple behavioral observations to experimental examinations of the relative costs and benefits of mating associations for both males and females. The three general forms of mating association, contact, noncontact, and copulatory, appear taxonomically widespread, suggesting that they are not phylogenetically constrained. Moreover, there is considerable variation both within and between species that facilitates the testing of alternative hypotheses for the evolution of these associations in a comparative context.

The majority of associations in table 5.2 are postcopulatory. Nevertheless, some species show periods of association that also occur prior to insemination or both prior to and after insemination. In general, precopulatory associations are thought not to have arisen in the context of the avoidance of sperm competition, but rather as a means of monopolizing access to females until such time as they become receptive to mating (Parker 1970e, 1974). Grafen and Ridley (1983) adopted a game theory approach to show how precopulatory mate guarding would arise whenever opportunities for mating are restricted in time by limited female receptivity. In fact, the guarding of females prior to copulation is not uncommon in the insects, particularly in species where females become unreceptive to further matings so that there is no sperm competition. Pupal mating by male heliconid butterflies is a classic example (Gilbert 1976; Deinert et al. 1994). Males compete for access to and defend pupae from which females are about to emerge. Copulation occurs immediately the pupal case is ruptured and females do not mate again (see Thornhill and Alcock 1983 for a review of this phenomenon in the insects).

Male dung flies, *Sepsis cynipsea*, mount and remain in precopulatory association with females throughout oviposition. Females are receptive to copulation only after oviposition, so that males successful in defending females from the takeover attempts of rivals and gaining copulation when females finally become receptive have their paternity assured in the female's next oviposition (Parker 1972a,b). A similar behavior occurs in locusts, *Locusta migratoria*, and experiments have shown that male fertilization success is greatest when copulation follows oviposition, explaining why precopulatory guarding of females during oviposition is favored in this species (Parker and Smith 1975; see fig. 4.7).

Thus, in general, precopulatory guarding is unlikely to have been selected in the context of the avoidance of sperm competition. Nevertheless there may be exceptions. Temporal variation in sperm competition risk can result in adaptations to the timing of insemination. In the monarch butterfly *Danaeus plexippus*, copulation is extended up to 14 h depending on the time of day at which the pair initiate copulation (Svärd and Wiklund 1988b). Dissection of individuals forced apart showed that sperm were not transferred until late in the copulation, and generally not until nightfall. Thus, this association is rightly seen as a preinsemination association. Oviposition occurs on the day following copulation and extending copulation until after nightfall when mate searching has ceased appears to prevent the female from being mated by rival males prior to oviposition. Similar tactics of preinsemination mate guarding are prevalent in the ischnuran damselflies where pairs remain in copula for several hours after males have removed sperm from the female's sperm storage organs (Sherman 1983; Convey 1989; Moore 1989; Cordero 1990; McMillan 1991; Alcock 1992; Sawada 1995; table 5.2). Sperm transfer takes just a few minutes and occurs at the end of the mating association. Again, the duration of the copulatory association is dependent on the time of day that it is initiated, with sperm transfer and uncoupling occurring at the end of the mate-searching period when the risk of remating by females with rival males is low.

Variation with Sperm Competition Risk

One of the major theoretical predictions for the mate guarding hypothesis is that changes in sperm competition risk (the probability that a female will mate with a rival male) should favor changes in male investment in postcopulatory mate guarding (section 5.2). Evidence for a sensitivity to increased risk of sperm competition is widespread in studies of postcopulatory associations. Of the studies in table 5.2, twenty-four reported increased duration of mating association with increased intensity of competition among males for access to females, measured indirectly as the degree of male bias in the operational sex ratio and/or increases in male density, or directly by observations of interference by rival males.

Patterns of postcopulatory associations are well documented in the odonates, and studies of anisopteran mating systems provide strong evidence for the mate guarding hypothesis (table 5.2). Males of many dragonflies typically establish territories around the banks of streams or water bodies where females come to oviposit. During copulation males remove some of the sperm stored from previous rivals and reposition remaining sperm before delivering their own ejaculate so that they will have close to complete paternity when the female oviposits, if she does not mate again (see section 3.3). There is considerable variation in postcopulatory guarding behavior both within and between species (table 5.2). In some species males release their mates after insemination and hover close above them until they begin to oviposit. Males may then begin to move away from their partners for increasing periods, to patrol their territories or perch on nearby vegetation. If a rival male encounters and attempts to grasp the ovipositing female, the guarding male will chase the intruder away before resuming a hovering position above his mate. Alternatively, males may remain in the tandem position throughout oviposition, dropping periodically to the water surface to allow their mate to oviposit. Within the libellulids males of some species adopt both contact and noncontact guarding. On completion of oviposition, the pair flies up from the pond surface before separating. Females leave the pond while males remain and search for additional mates.

The intensity of mate guarding in anisopterans appears to be dependent on the risk of sperm competition from rival males (table 5.2). For example, in her study of *Pachydiplax longipennis*, Sherman (1983) found a qualitative change in male guarding behavior with increasing density of searching males at the pond (fig. 5.2). The proportion of males that did not guard at all, or that hovered briefly before perching nearby, declined with increasing male density, while the proportion of males that hovered continually during oviposition and chased intruding males increased. McMillan (1991) developed a guarding intensity index for *Plathemis lydia* based on a ranked scale with hovering less than 0.5 m from the female being recognized as intense guarding and territory patrolling or perching away from the ovipositing female being recognized as weak guarding. As with *P. longipennis*, McMillan (1991) found that the intensity of postcopulatory mate guarding increased with both increasing number of males at the pond and increasing frequency of aggressive interactions between males (fig. 5.2). Moreover, McMillan (1991) was able to experimentally manipulate the intensity of mate guarding by focal males by removing searching males from the pond. Under experi-

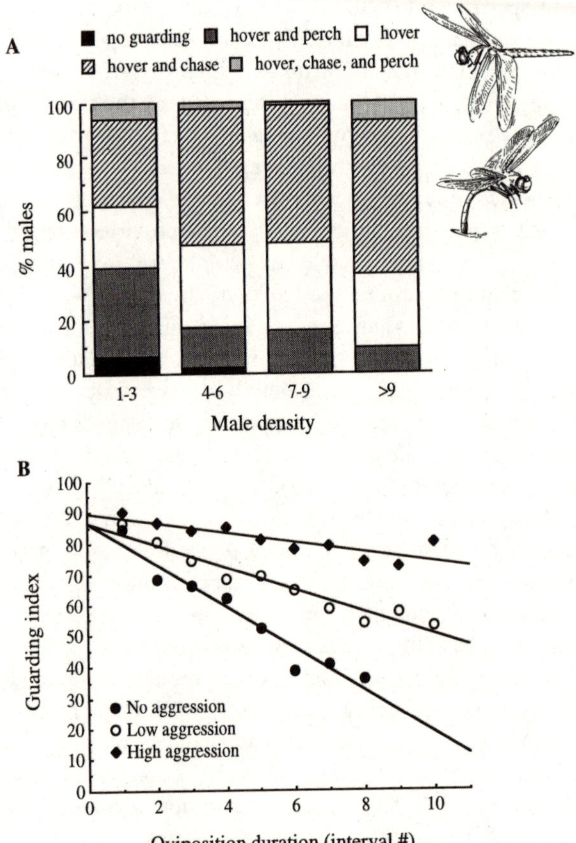

Figure 5.2 Plasticity in the postcopulatory mate guarding behavior of dragonflies in response to increased risk of sperm competition. (A) The proportion of male *Pachydiplax longipennis* adopting no or weak mate guarding (brief hovering followed by perching) decreases as the density of searching males at the pond increases, while the proportion of males adopting strong mate guarding (close hovering and chasing of rival males) increases (data from Sherman 1983). (B) Changes in the intensity of mate guarding by male *Plathemis lydia* throughout oviposition under three levels of rival male harassment (low aggression: 1–2 aggressive interactions/30 s; high aggression: 3 or more aggressive interactions/30 s). Guarding index was computed from the proportion of time males spent performing behaviors that were ranked for intensity. Perching and territory patrolling were ranked 0, far hovering ranked 1, and near hovering (<0.5 m from female) ranked 2. The intensity of male guarding declined through oviposition, but the rate of decline was greater under low aggression than under high aggression. (Data from McMillan 1991)

mentally reduced male density, focal males had a reduced intensity of mate guarding compared with males guarding under natural male densities either before or after the experimental manipulations.

The genus *Sympetrum* adopts both contact and noncontact guarding dependent on the risk of sperm competition. Uéda (1979), working with *S. parvulum*, found that males adopt alternative mate-searching strategies, either defending territories at the water body or wandering throughout the area in search of arriving females. Territorial males always adopted noncontact guarding of the form seen in *P. longipennis* and *P. lydia*. Likewise, wandering males adopted noncontact guarding when the density of territories was low, but an increasing proportion of wanderers remained in tandem, adopting contact mate guarding, when the density of territorial males at ponds increased. Alternative tactics appeared to be due to male dominance status and could change from day to day. Males unsuccessful in competitive interactions would presumably be unable to defend females during noncontact guarding, so that remaining in contact with them yields the lowest risk that their female will remate with a rival. For example, Tsubaki and Ono (1985) found that wandering males were less successful in noncontact mate guarding than territorial male *Nannophya pygmaea*. In contrast, male *S. sanguineum* are never territorial and all males exhibit both contact and noncontact guarding. Convey (1989) found that males remained in tandem during the entire oviposition period when interference rates from searching males during the early stages of oviposition were high, but adopted noncontact guarding for the later 42% of the oviposition period when interference rates were low. Similar results were reported for *Zygonyx natalensis* by Martens (1991).

Risk of sperm competition usually declines through the course of the oviposition period. As the cumulative proportion of the female's clutch laid increases, so the fitness costs for a male associated with female remating will decline. Accordingly, we see that the intensity of mate guarding by males declines over the course of oviposition; males hover at the beginning of oviposition but become increasingly likely to patrol and/or perch later in the oviposition period. In *S. sanguineum*, males switch from tandem contact guarding at the beginning of oviposition to noncontact hovering and/or perching later in oviposition (Convey 1989). The rate of decline in mate guarding intensity was found to depend on the population risk of sperm competition in *P. lydia* (fig. 5.2). Alcock (1992) similarly examined changes in the intensity of mate guarding during oviposition in *Paltothemis lineatipes*. He concluded that the duration of strong mate guarding (hovering) was dependent on visual confirmation that the female was ovipositing safely. By experimentally interrupting oviposition during the strong guarding phase, he was able to manipulate the duration of strong guarding by males, which was positively associated with the time required for females to find a new oviposition site. In contrast, interrupting females later in the oviposition period, when males had begun weak guarding (perching and/or patrolling), did not cause them to switch to strong mate guarding, presumably because the females involved had already laid the majority of their eggs.

There are two assumptions underlying the conclusion that the observed plasticity in anisopteran mate guarding arises because of variation in the risk of sperm competition: first, that male density and/or male-biased sex ratios result in a greater chance that females will remate with a rival male before they have deposited their clutch of eggs, and, second, that the different forms of mate guarding behavior are indeed associated with variation in the success of mating attempts by rival males. There is good empirical evidence for both of these assumptions. First, in Sherman's (1983) study of *P. longipennis*, the proportion of females disturbed by mate-searching males, the proportion of females leaving the pond because of harassment by searching males, and the proportion of females mated per visit to the pond all increased with increasing male density. Moreover, guarding by the male resulted in a significant reduction in these three measures of female disturbance and remating. The degree of sexual interference between territorial male *S. parvulum* and wandering males increased with the density of territorial males at the pond (Uéda 1979), and the probability of remating by rivals increased with male density in Wolf et al.'s (1989) study of *Leucorrhinia intacta*. Moore's (1989) study of *Libellula luctuosa* also showed that the proportion of male copulation attempts that were successful increased with the number of males present at ponds. Second, McMillan (1991) found that the probability of a successful copulation by a rival male of *P. lydia* was dependent on the type of guarding adopted by a female's current mate. When males adopted hovering 87–91 % were successful in retaining their mate after a bout of harassment, compared with just 67% of males who were perching or patrolling their territory. Experimental manipulations of guarding male *S. sanguineum* revealed that 78% of females separated from their guarding male oviposited without remating, compared with 99% of females that were left in tandem with their mate. In *S. obtrusum*, tandem guarding was 100% successful in avoiding rival remating compared with a success of 88% for noncontact guarding (Singer 1987). These figures show that for the avoidance of sperm competition, contact guarding is the most effective form of mate guarding and perching the least effective.

There is also good experimental evidence for the impact of sperm competition risk on mate guarding from a number of other insect groups. For example, the duration of mate guarding can be increased by the addition of rival males or by experimentally increasing the degree of male bias in the operational sex ratio in water striders (Arnqvist 1992a), stick insects (Sivinski 1983), tiger beetles (Shivashankar and Pearson 1994), grasshoppers (Muse and Ono 1996), and hemipteran bugs (McLain 1980; Sillen-Tullberg 1981). However, perhaps the best evidence for variation in sperm competition risk favoring the evolution of male mate guarding strategies comes from studies of the soapberry bug, *Jadera haematoloma* (Carroll 1988, 1991, 1993; Carroll and Corneli 1995). This species occurs in the south central United States where it breeds in dense aggregations on its host plants. Carroll (1988) reported differences in the reproductive ecology of *J. haematoloma* between populations in Oklahoma and Florida, including differences in the duration of copulation. Copulations in Florida last on average for 10 h while in Oklahoma they last on average for 26 h but up to 128 h. Carroll (1988, 1991) concluded that the prolonged copulation is a form of copulatory mate guarding because (i) insemination is achieved in just 10 min, (ii) the probability of a female remating with a rival is greater when the male is removed than when he is allowed to remain in copula, and (iii) the male's fertilization success is reduced by 60% following a rival copulation. The observed difference in the duration of copulatory mate guarding between populations is consistent with selection for the avoidance of sperm competition; in Florida the sex ratio in aggregations is 1:1 while in Oklahoma it varies due to sex-biased mortality of females, and can be skewed as high as five males per female. In a "common garden" experiment Carroll and Corneli (1995) exposed F2 male offspring from each population reared under identical laboratory conditions to varying sex ratio treatments and found that only males bred from the Oklahoma population showed phenotypic plasticity in their mate guarding behavior; with increasing male bias, Oklahoma males had a decreased probability of switching mates between ovipositions and a corresponding increase in the duration of copulatory mate guarding (fig. 5.3). The principal factors influencing a male's mate guarding behavior appeared to be an increase in search time for single females and a decline in the probability of finding a single female with increasing male bias in the sex ratio, both key parameters in theoretical models of the evolution of mate guarding (Parker 1974; Yamamura 1986; Yamamura and Tsuji 1989). These results show that phenotypic plasticity in mate guarding strategies has a genetic basis. Moreover, the evolutionary history of variable sex ratios in Oklahoma, coupled with population variation in the fitness benefits to males from mate guarding (fig. 5.3) and a lack of gene flow between populations (Carroll and Boyd 1992), has resulted in an evolutionary divergence in mate guarding strategies across the soapberry bug's geographic range.

Comparative evidence from five species of the tiger beetle genus *Cicindela*, also provides evidence of divergent evolution of phenotypic plasticity in guarding behavior. Shivashankar and Pearson (1994) found that across *Cicindela* spp. the mean duration of contact guarding varied from 6 to 70 min (see table 5.2 at end of chapter) and the operational sex ratios of natural populations varied from close to unity to 2:1. Although the mean duration of mate guarding was not correlated with the degree of male bias in the operational sex ratio across species, those species with male-biased sex ratios

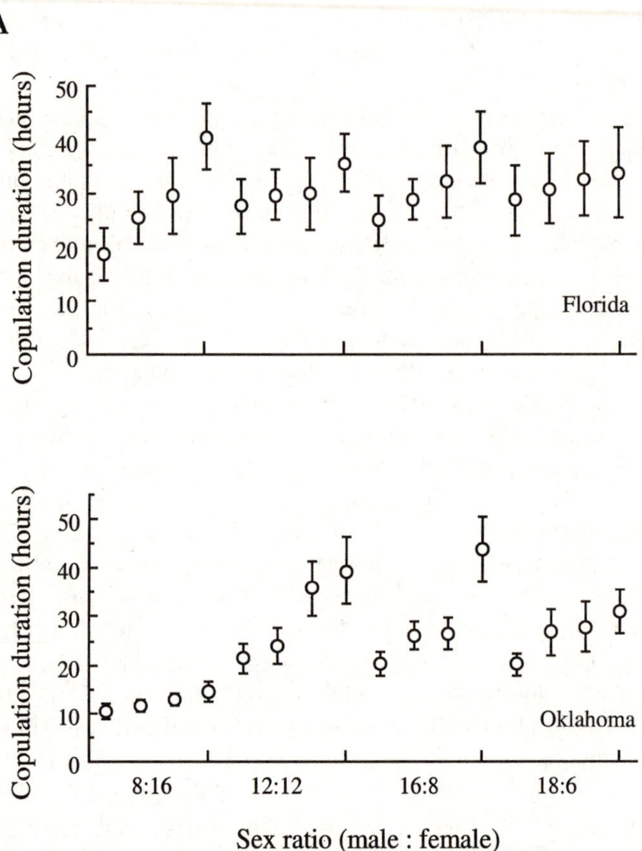

Figure 5.3 Plasticity in postcopulatory mate guarding in the soapberry bug *Jadera haematoloma*. (A) Results from a common garden experiment in which populations obtained from Florida and Oklahoma were reared under laboratory conditions for two generations before examining variation in their guarding behavior. In natural populations in Oklahoma, male-biased sex ratios appear to favor a greater frequency and duration of contact mate guarding than occurs in Florida, where sex ratios do not deviate from unity. The common garden experiment suggests that plasticity in guarding behavior has a genetic basis and that the two populations have undergone evolutionary divergence. F2 males from Oklahoma increased the duration of guarding in response to an experimental increase in the ratio of males to females. F2 males from Florida were not sensitive to the experimental manipulations (from Carroll and Corneli 1995). (B) Estimated fitness payoffs for guarders (diamonds) and nonguarders (circles) in the Oklahoma (solid symbols) and Florida (open symbols) populations. Fitness was calculated from empirical estimates of the fertilization success of the last male to mate (72% for Florida and 58% for Oklahoma), the time to guard for one oviposition (19.2 h for Florida and 21.6 h for Oklahoma), the search time per copulation (9.6 h for Florida and 26.2 h for Oklahoma), and the probability that an unguarded female remates ($\approx$ 1 for both populations). Guarding is not profitable for males in Florida unless they remain with females for four ovipositions or more. In Oklahoma, the guarding strategy rapidly becomes the more profitable one primarily because of the higher search costs for finding additional mates (redrawn from Carroll 1993).

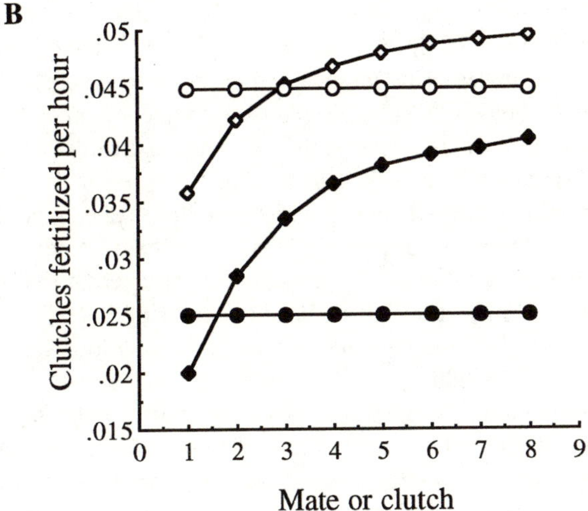

Figure 5.3 (*cont.*)

showed significant changes in the duration of guarding when a rival male was experimentally introduced. Species with sex ratios close to unity did not exhibit phenotypic plasticity in guarding behavior.

Polymorphisms

The models of Yamamura (1986) and Yamamura and Tsuji (1989) predict that polymorphisms in guarding behavior can be maintained under conditions of low to moderate levels of male competition (fig. 5.1) and there is some evidence to suggest that such polymorphisms do exist. The proportion of males adopting mate guarding behavior shows considerable variation in populations of odonates. Even in high population densities of the dragonfly *L. luctuosa*, only 50% of males guard their mates during oviposition (Moore 1989). In damselflies, the proportion of males guarding can vary from 90% in *Calopteryx maculata* to 85% in *C. aequabilis* and just 49% in *C. amata* (Meek and Herman 1990), and in *Mnais pruinosa pruinosa* only territorial males guard mated females (Nomakuchi et al. 1984). Moreover, variation in the frequency of guarding behavior within species appears sensitive to the parameters of Yamamura's (1986) theoretical models. In the case of *Calopteryx*, variation in male searching efficiency is associated with variation in guarding frequency; both *C. maculata* and *C. aequabilis* females oviposit above water where they are conspicuous to mate-searching rivals. In contrast, *C. amata* exhibits submerged oviposition, so that the reduced probability of finding ovipositing females for rival males may favor an increase in the frequency of males that do not exhibit mate guarding behavior.

In Moore's (1989) study of *L. luctuosa*, a decrease in male density resulted in an increase in the frequency of guarding behavior from 50% of males to 72% of males. This might seem contrary to theoretical prediction; increased male density results in reduced frequency of guarders, while Yamamura's (1986) models predict that the ratio of guarders to nonguarders should increase as male density is increased. However, in *L. luctuosa*, territory owners obtain virtually all matings and time spent guarding females may be costly in terms of a male's ability to maintain his territory and thus gain access to additional mates. With increasing competitor density, the costs of mate guarding may outweigh its benefits, favoring a decline in guarding. In Yamamura's (1986) models, fitness payoffs to guarding and nonguarding males were set to be equal, thereby assuming that alternative tactics represent mixed evolutionarily stable strategies, for which there is currently very little evidence (Gross 1996). This problem is illustrated by Mangan's (1979) study of the cactus fly *Odontoloxozus longicornia*. In this species, males adopt alternative mate-securing tactics, either defending territories at sites of cactus necrosis where females oviposit, or searching randomly on the cactus for arriving females. Territorial males guard females during oviposition while searching males do not. Mangan (1979) estimated the fitness payoffs for alternative strategies from empirical measures of relative encounter rates with females, their egg deposition rates, the probabilities of mating, and time investments in courtship and copulation. He concluded that territorial mate guarding males had a fitness payoff some five times that of nonguarding males. Nevertheless, Mangan (1979) noted that when the costs of territoriality increased, for example under conditions of high male density or when declining female mobility reduced encounter rates, males stopped maintaining territories and began searching.

Costs and Benefits

An understanding of the evolution and maintenance of mate guarding strategies clearly requires insight into both costs and benefits of the behavior. The principal benefit of mate guarding to males, by definition, is that guarding should reduce the probability that the female will mate with a rival male before she has laid her clutch of eggs, so that mate guarding males should have a higher fertilization success, on average, than nonguarding males. There is good empirical evidence that postmating associations do indeed decrease the probability of female remating (table 5.2 at the end of the chapter). Observations based on natural variations in male guarding tendencies, as well as male removal experiments, clearly demonstrate that females remate sooner when unguarded (e.g., Sherman 1983; Fincke 1986; Singer 1987; Carroll and Corneli 1995; Dickinson 1995). In species where the last male to mate fertilizes the majority or all of the eggs laid, as is generally the case for odonates (table 2.3), decreased female remating is equivalent to increased male fertilization success. However, fertilization benefits will on average be much lower in species where sperm from rival males mix in the female's sperm stores, reducing the potential benefits to males for mate guarding. With strong first-male sperm precedence, males should abandon their mates immediately. Theoretical models suggest that mate guarding is more likely to evolve with strong last-male fertilization success (Parker 1974; Yamamura 1986; Yamamura and Tsuji 1989). However, whether mate guarding evolves or not will depend strongly on the costs associated with the behavior, as well as the potential benefits. Thus, even with low last-male fertilization success, relatively low costs could still favor a mate guarding strategy. Conversely, with high last-male fertilization success, high costs could favor a nonguarding strategy. The pattern of sperm utilization per se will therefore have little predictive power for patterns of mate guarding, and vice versa (Simmons and Siva-Jothy 1998). Rather, it is the balance between fitness cost and benefit for males that will determine the nature of mate guarding strategies (Alcock 1994).

Parker's (1970d) study of the yellow dung fly *S. stercoraria*, was the first cost-benefit analysis to consider the evolutionary maintenance of postcopulatory associations in a quantitative manner. After copulation males disengage their genitalia but remain mounted above the female for around 16 min while she oviposits. This time represents a cost to males because, by increasing the time investment required per female, it decreases the overall rate at which males can encounter additional females and thus their rate of egg gain. On the other hand, since the last male to mate with a female fertilizes some 80% of the female's clutch, a male that abandons his mate before oviposition will suffer a considerable loss of eggs if the female remates. From observations of male and female interactions around dung pads, Parker (1970d) was able to calculate the probability that recopulation by a rival male would occur either during copula or during oviposition for different densities of competing males. Combined with information on the average percentage of a female's clutch sired by the last male to mate (around 80%), number of eggs laid by a female (around 43), and time required to find (140 min), copulate with (32 min), and guard (16 min) an incoming female, Parker (1970d) calculated the number of eggs gained per unit time for a variant male lacking the guarding phase in a population of normal males that guarded their mates. When there were no males searching at the dung, a nonguarding variant was at a selective advantage. However, the variant's fertilization rate declined extremely rapidly as male density increased from zero to densities above 0.1 males per 500 cm^2 of dung. The fitness advantage to the guarding males reached some 400% at the average density of males found around dung pats in nature (15 males per 500 cm^2). Even at their lowest densities of 1.3 males per 500 cm^2, guarding would be favored over a nonguarding variant (Parker 1970d).

Carroll (1993) examined several cost and benefits believed to be important for male mate guarding decisions in his Oklahoma and Florida populations of *J. haematoloma*, and used these parameters to estimate fitness payoffs for males that guard their mates compared with those that fail to guard across varying numbers of oviposition cycles (fig. 5.3). His analysis showed that for a single oviposition cycle guarders had lower fitness than nonguarders, although the fitness differential was greatest in Florida, primarily due to differences in male search times, which were nearly three times higher there than in the male-biased sex ratios of Oklahoma. For the Oklahoma population, the fitness of guarding males rapidly increased and surpassed that of nonguarders with increasing numbers of ovipositions. Nonguarders in Florida had superior and/or equal fitness across the first three ovipositions. In other words, males who guarded in Florida had lower fitness because they paid the cost of lost mating opportunity with the greater availability of females that occur there (fig. 5.3). These models provide a qualitative fit to the observed differences in mate guarding behavior between populations; in Florida, mate guarding is less frequently observed and guarding durations are often shorter than in Oklahoma (Carroll 1988, 1993; Carroll and Corneli 1995). Carroll's (1993) analysis suggests that in both populations males would gain the maximum fitness if they remained with their mates indefinitely, a prediction that deviates from observation for both populations. Nevertheless, other parameters that would affect this prediction that were not incorporated into Carroll and Corneli's (1995) analysis would be the impact that guarding has on male feeding ability, and thus long-term survival, and/or the probability that females would survive to produce multiple clutches under natural conditions. Both parameters should act to decrease the benefits associated with long-term guarding. For example, Sih et al. (1990) found that food deprivation decreased the duration of contact mate guarding in the water strider *Gerris remigis*. Since females have increased foraging success when paired with males (Wilcox 1984; Clark 1988), while males are unable to feed, reduced guarding duration most likely reflects the interests of hungry males.

Dickinson (1995) conducted both male removal and female removal experiments in a population of blue milkweed beetles, *Chrysochus cobaltinus*. With male removal, females remated sooner than when males were allowed to guard, confirming the benefits of mate guarding for males. With female removal, males remated sooner than when they were left to guard, illustrating the cost of mate guarding for males. Dickinson (1995) used her observed remating rates for males and females to model the cumulative fitness payoffs to males over a 2 h period that were associated with guarding or dismounting and searching for a new mate. Since she had no information on the fertilization success of last males in this species, she estimated fitness payoffs across the full range of possibilities from zero to complete paternity. Dickinson (1995) found that, with last-male paternity below 40%, males that leave immediately after insemination have a higher fitness than those who remain with their mates (fig. 5.4). For levels of paternity above 40%, guard-

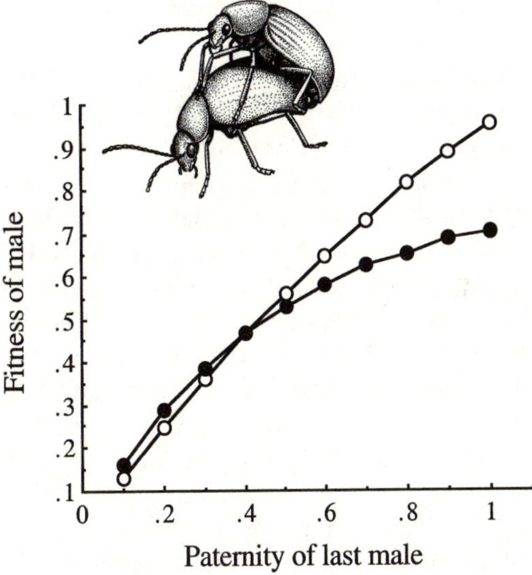

Figure 5.4 Simulated fitness at 2 h of male blue milkweed beetles *Chrysochus cobaltinus* that leave their mates after 20 min of copulation (solid circles) and those that remain mounted for postcopulatory mate guarding (open circles) as a function of the proportion of offspring sired by the last male to copulate. Fitness was calculated as the number of females with which a male mated devalued by loss of paternity due to remating by the female. For levels of paternity above 40%, guarding males have an increasing fitness advantage. (Redrawn from Dickinson 1995)

ing males have an increasing fitness advantage. In general, last male paternity in beetles is in excess of 0.5 (table 2.3), explaining why male *C. cobaltinus* guard their mates for an average of 1.7 h after the 20 min copulation. Dickinson (1995) also showed that, although mate guarding decreased a male's ability to feed, this cost did not impact on male survival. This study is an excellent illustration of how fertilization success interacts with male remating costs and the benefits of preventing females from remating in determining whether or not a male should exercise postcopulatory guarding.

Male remating costs have been proposed as a major factor in the evolution of noncontact as opposed to contact guarding in odonates (Alcock 1979b; Uéda 1979; Alcock 1983, 1994). However, the evidence for multiple mating by noncontact guarding males is mixed. In his studies of *C. maculata*, Alcock (1979b) showed that males would mate with and successfully guard additional females, while Waage (1979a) found that although males guarded multiple females they rarely mated with more than one. The populations studied by Alcock (1979b) and Waage (1979a) differed greatly in density, and the failure of males to copulate with additional females in Waage's (1979a) study may have arisen from an inability of males to capitalize on additional females under conditions of high intrusion rates at their territories (Alcock 1979b; Waage 1979a). Allowing nonmates to oviposit on territories may still hold adaptive significance for males, however, as it would decrease the probability that an intruding male would capture and copulate with the guarding male's own mate. Although the acquisition of multiple mates by males has been reported in a handful of noncontact guarding species, in general the males of noncontact guarding species do not routinely engage in multiple matings (Waage 1984; Tsubaki and Ono 1985).

An alternative cost of contact mate guarding proposed by Waage (1984) and Sherman (1983) is that it prevents the males of territorial species from maintaining their territories during the time it takes females to oviposit. In support of this hypothesis, many territorial odonates do appear to exhibit noncontact guarding (Waage 1984; table 5.2 at end of chapter), and within species territorial males exhibit noncontact guarding while their wandering conspecifics exhibit contact guarding (Uéda 1979). Nevertheless, noncontact guarding has also been reported from nonterritorial species (e.g., Sherman 1983; Convey 1989), suggesting that the cost of territory loss for contact guarders is also inadequate as a general explanation for the evolution of contact versus noncontact guarding.

The potential association between territoriality and the form of mate guarding can be further examined using studies of water striders (Hemiptera: Gerridae). Water striders inhabit the water surface of ponds, streams, and even the ocean, where they feed on other arthropods that become trapped on the water surface. Those species that have been studied can be broadly classified as exhibiting one of two types of mating system (Andersen 1997; Arnqvist 1997b). Either they exhibit scramble competition polygamy, where males sit and wait for females on the water surface or actively search the surface for foraging females, or they exhibit resource defense polygamy, where males establish territories at preferred oviposition sites and signal to females via vibrations transmitted on the water surface. Most species exhibit postcopulatory mate guarding for periods of seconds to weeks (table 5.2). Consistent with Sherman's (1983) and Waage's (1984) hypothesis, species with resource defense polygamy, where males must defend and maintain their territories in the face of rival intrusion, have been shown to adopt noncontact mate guarding. In contrast, in species with scramble competition polygamy the male often remains mounted upon the female and adopts contact mate guarding (Arnqvist 1997b). Despite this apparent association, phylogenetic examination of gerrid mating systems (Andersen 1997) fails to support the notion that noncontact guarding arises because of the costs of potential territory loss associated with contact guarding. Complete information on mating behavior is available for little more than thirty species mostly belonging to the genera *Aquarius*, *Gerris*, and *Limnoporus* (Andersen 1994; Arnqvist 1997b). Optimization of mating behavior onto the phylogeny of these genera reveals that scramble competition polygamy is probably the ancestral state in these taxa (fig. 5.5). Evolutionary transitions to resource defense polygamy (territoriality) occur once in each of the genera. In *Limnoporus* there is no concomitant change from contact to noncontact mate guarding. In *Aquarius* only one of four species for which there is information shows the expected change from contact to noncontact guarding, and the transition from scramble competition to resource defense in *G. swakopensis*

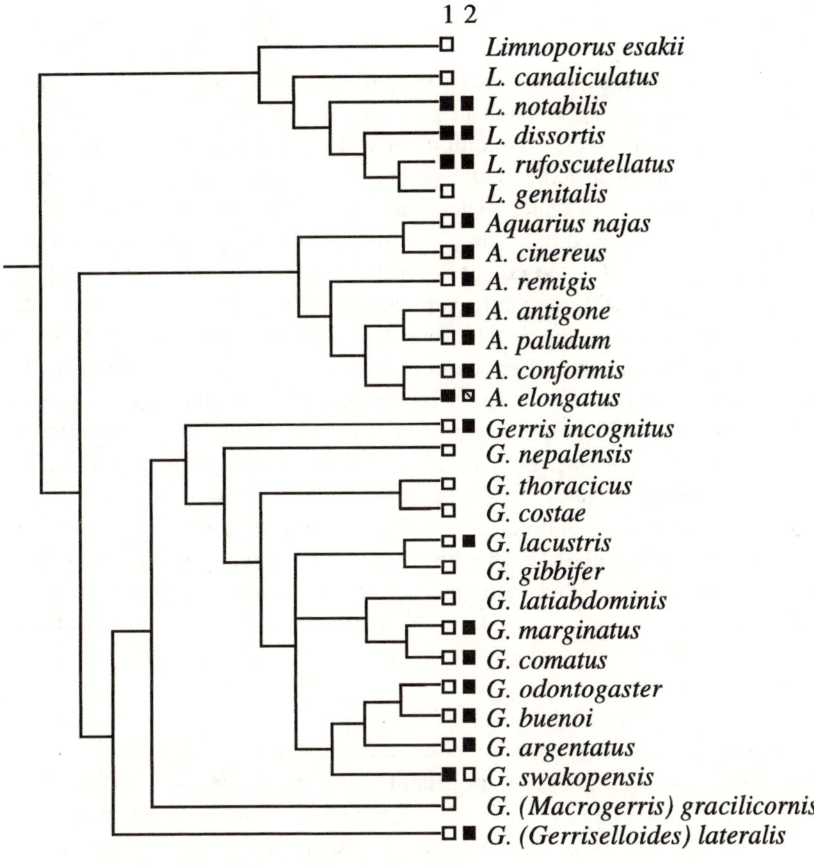

Figure 5.5 Optimization of mating system onto the reconstructed phylogeny of water striders (Gerridae). The ancestral mating system is one of scramble competition polygamy (open squares). Territoriality (solid squares) has arisen once in each of the three lineages of *Limnoporus*, *Aquarius*, and *Gerris*. The form of mate guarding adopted by each species is shown where information is available. The predominant pattern is one of contact mate guarding. Contrary to the hypothesis that contact guarding is costly for males who must maintain territories, transitions to territoriality have not generally been associated with changes from contact to noncontact guarding. (After Andersen 1997)

appears to be associated with a complete loss of postcopulatory mate guarding. These data provide little support for the notion that noncontact guarding arises as an adaptive alternative to contact guarding in territorial species.

While most discussions have concentrated on the social costs of alternative mate guarding behaviors, Singer's (1987) study of the libellulid *Sympetrum obtrusum* revealed an unappreciated yet important physiological cost.

Contact guarding males spend nearly three times longer in sustained flight than do males adopting noncontact guarding, making contact guarding an energetically more expensive tactic despite its greater efficacy. The proportion of males adopting noncontact guarding as opposed to contact guarding was dependent on ambient conditions of wind speed and temperature. Measurements of thoracic temperature revealed that males switch tactics based on their physiological state; males with thoracic temperature below 30°C almost exclusively used noncontact guarding while males with thoracic temperatures above 33°C used contact guarding. Thoracic temperature was positively associated with ambient temperature and negatively associated with wind speed, so that ambient conditions of low temperature and/or high wind speed made the sustained flight required for contact guarding physiologically impossible for males. Thus, the energetic costs of mate guarding, as well as the social costs, will play an important role in determining the occurrence and nature of mate guarding. In general there is compelling evidence to suggest that changes in the nature of mate guarding (from maintaining visual contact to close association and sustained contact) represent a continuum in intensity that is associated with both elevated cost and elevated efficacy. The type of behavior adopted, and thus the costs incurred, appear to depend predominantly on the perceived risk of sperm competition from rival males.

Sexual Conflict

The foregoing discussion has not considered female interests. While it is true that females can sometimes gain benefits from postmating associations (see section 5.4), prolonged periods of association, essentially under male control, may not always be in the female's best interests. The conflict that is expected to ensue when female interests are not served (Parker 1979) provides good evidence that prolonged mating associations have evolved primarily to serve male interests in avoiding sperm competition. Female costs have rarely been examined. In her study of *Tramea carolina*, Sherman (1983) found that single females oviposited faster than did those held in tandem by contact guarding males, and thereby spent less time at the pond surface where they risk predation (Waage 1984). However, the best evidence for costs to females arising from postcopulatory mate guarding by males comes from studies of water striders (table 5.2). In this group of insects there is strong conflict over the mating decision, which is manifest as precopulatory struggles, and over the duration of the mating association, which is manifest as postcopulatory struggles (Rowe et al. 1994; Arnqvist 1997b).

During contact mate guarding, female water striders must carry males on their backs for extended periods while they continue to forage on the water surface. Several studies have revealed that females suffer reduced mobility as a direct consequence of mate guarding males. Studies of female movements on the water surface show that the stride length and stride rate during locomotion are both depressed when a female moves with a guarding male on her back (Arnqvist 1989a; Fairbairn 1993; Amano and Hayashi 1998). Reduced mobility appears to have serious implications for female survival (Fairbairn 1993; Rowe 1994). Arnqvist (1989a) found that paired female *Gerris odontogaster* were three times more likely to be taken by predatory backswimmers (*Notonecta* spp.) than were solitary females. Similar results were obtained from studies of backswimmers preying on *G. buenoi* (Rowe 1994), and Fairbairn (1993) found that mating female *Aquarius remigis* were twice as likely to suffer predation by frogs than were solitary females. Part of the increased risk of predation arises because backswimmers are attracted to disturbances at the water surface that arise during pre- and postmating struggles (Rowe 1994). During struggles, predation is female biased by 59%. However, mating and postcopulatory guarding also increases predation risk, and because males are mounted above females it is generally the female (89% of cases) that is taken (see Arnqvist 1989a; Rowe 1994). Finally, contact mate guarding has also been found to be energetically expensive for females. Watson et al. (1998) found that female *A. remigis* with guarding males or solder weights on their backs had a 24% and 28% increase in energy expenditure, respectively, compared with females skating unencumbered on the water surface. Moreover, energy expenditure increased to 43% over unencumbered females when guarded females were escaping from potential predators. Thus the evidence suggests that prolonged mating associations may indeed represent a significant fitness cost for females at least in this group of insects.

The evidence suggests that female water striders adopt what has become known as convenience polyandry (Thornhill and Alcock 1983) in that they will resist the mating attempts of males when the costs of mating are high, but will accept superfluous matings when the costs of resistance exceed those of mating (Arnqvist 1992b; Rowe et al. 1994; Arnqvist 1997b; Watson et al. 1998). Moreover, it seems that the duration of postcopulatory mate guarding is similarly dependent on female interests (Rowe 1992). Male interests are best served by prolonging the mating association by contact mate guarding. Studies have shown that guarded females are less subject to male harassment and/or spend less time resisting additional males than do solitary females (Wilcox 1984; Krupa and Sih 1993; Vepsäläinen and Savolainen 1995; Amano and Hayashi 1998). Male harassment rates also increase with male density or male bias in the operational sex ratio (Arnqvist 1992a; Rowe 1992; Krupa and Sih 1993). Guarding increases a male's fertilization success (Jabłoński and Kaczanowski 1994) and, as predicted by the mate guarding hypothesis, the duration of postcopulatory guarding is longer in male-biased than in female-biased operational sex ratios (see table 5.2). Nevertheless, female interests predict the same changes in duration of mate guarding as do male interests. Under high male densities, increased harassment rates will increase the costs of resistance for females, both in terms of increased energy expenditure associated with struggles, and because of the increased risks of predation that premating struggles attract. Thus, at high male densities or male-biased operational sex ratios, female interests should be best served by accepting guarding in order to reduce harassment rates from unpaired males (Rowe 1992).

Recent studies have examined the relative importance of male and female interests in determining the duration of mate guarding in water striders. Jabłoński and Vepsäläinen (1995) attempted to disentangle male and female influences by examining the guarding behavior of *G. lacustris* males when pairs copulated either in the presence of other males or in the presence of other females. They found that the duration of guarding was greater in the presence of rival males than in the presence of females. Jabłoński and Vepsäläinen (1995) estimated the degree of female acceptance from measures of the time from the end of copulation to the occurrence of three consecutive bouts of postcopulatory struggling (long periods prior to resistance being indicative of female acceptance). They estimated male persistence in mate guarding as the proportion of the last 10 min of guarding that females spent struggling, with the assumption that nonpersistent males would release females after only small amounts of struggling compared with persistent males. In support of the mate guarding hypothesis, males were more persistent when copulations occurred in the presence of rival males than when they occurred in the presence of females or in isolation from conspecifics. In support of the convenience polyandry hypothesis, females showed greater acceptance of guarding in the presence of conspecific males than of females, or when copulating in isolation. Thus, both male and female processes appear important in extending the guarding period.

In a related study, Vepsäläinen and Savolainen (1995) experimentally disentangled male and female effects on the duration of reproductive behaviors by manipulating both the ambient operational sex ratio within which experimental subjects copulated, and the operational sex ratios experienced by subjects prior to experimental pairing. Experimental male and female subjects were allowed to interact within populations that contained four other males or four other females. Subject males had been held for 60 h prior to experimental tests either with five other males or with five females. Subject females had similarly been held either with five other females or with five males. Thus, within pairs, males and females experienced the same ambient operational sex ratio but had different prior experiences of operational sex ratio. Vepsäläinen and Savolainen (1995) found that only the male's previous experience had an impact on the duration of reproductive behaviors. Neither the operational sex ratio experienced by females prior to copulation nor the operational sex ratio experienced by pairs during copulation affected copulation duration or the duration of postcopulatory mate guarding (fig. 5.6). However, when males had previously experienced a female-biased operational sex ratio both the duration of copulation and the duration of mate

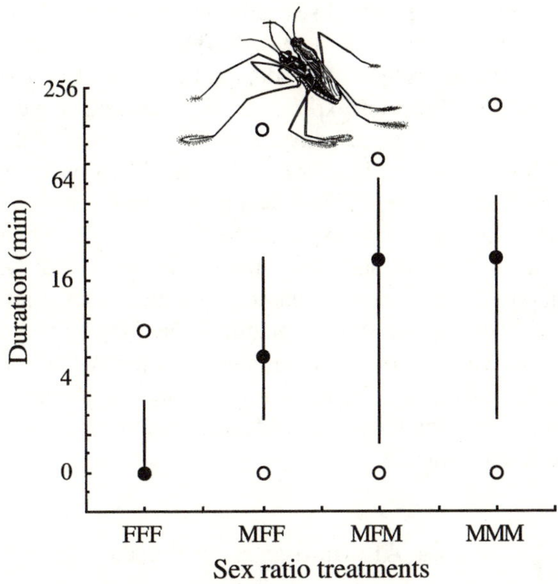

Figure 5.6 The duration of postcopulatory contact guarding in the water strider *Gerris lacustris,* following experimental manipulation of the operational sex ratio experienced by males and females prior to and during copulation. The data are the $\log_2$ transformed median (solid circle), upper and lower quartile range, and maximum and minimum values (open circles). Sex ratio treatments read from left to right are the sex ratio experienced by the female and by the male prior to copulation, and the ambient sex ratio during copulation (e.g., FFF, female experienced a female-biased sex ratio, male experienced a female-biased sex ratio, copulation occurred in the presence of excess females). Only the sex ratio experienced by males prior to copulation influenced the duration of postcopulatory guarding. Males reduced the duration of guarding when they had experienced a female-biased sex ratio. (Data from Vepsäläinen and Savolainen 1995)

guarding were short, but when males had experienced a male-biased operational sex ratio they extended the duration of copulation and postcopulatory mate guarding (Vepsäläinen and Savolainen 1995). The data thereby support the notion that the duration of mate guarding represents a manifestation of male interests. The results also shed light on the resolution of conflicts of interest between male and female. During experimental trials males were observed to be the sex responsible for the termination of guarding in 86% of cases in which they had previously experienced a female-biased operational sex ratio. Thus, when males perceived the risk of sperm competition from future rivals to be low, they dismounted and left females before the onset of conflict over mating duration. In contrast, when both males and females had experienced a male-biased operational sex ratio, 55% of matings were terminated by females after the onset of struggling, but when females had experienced a female-biased operational sex ratio the proportion of matings terminated by struggling females was increased to 95%.

A somewhat similar protocol was adopted by Lauer et al. (1996) in their study of *A. remigis*. The prior experience of males and females was again manipulated but in this case the probability of females accepting a mating was the variable of interest. Females were more likely to accept a mating and males were more insistent if they had experienced a male-biased operational sex ratio than if they had experienced a female-biased operational sex ratio. The data on water striders thus support the notion that female avoidance of sexual harassment in male-biased populations may alleviate conflict over mating and postcopulatory mate guarding, favoring a resolution in favor of male interests. The data on water striders also illustrate how female interests can play an important role, both in the evolution of mate guarding by males, and in determining the success of male attempts to avoid sperm competition from future rivals.

5.4 Alternative Hypotheses

As Thornhill (1984) has pointed out, sperm competition is often the only adaptive context considered in studies of insect reproductive biology, and this has been particularly true for studies of postcopulatory associations. Nevertheless, there are a number of alternative explanations for prolonged mating associations that should be considered (table 5.1).

Paternal Investment

In one of the first reports of a postcopulatory association, Foster (1967b) suggested that the male dung fly *S. stercoraria* cooperates with the female by remaining mounted during oviposition and in this way provides her protection from interference by searching males so that she can deposit her eggs. Immediate benefits accruing to females and/or their offspring are here considered as a form of paternal investment (sensu Trivers 1972). Of course such fitness benefits would also be accrued by males, but in this case the behavior should not be subject to sexual selection under sperm competition, but rather natural selection for efficiency of offspring production. Parker's (1970d) quantitative study of the postcopulatory association in *S. stercoraria* considered a number of alternative hypotheses for its evolution, including the paternal investment hypothesis. He first considered the fitness benefits that might accrue to females from an increased survivorship; when disturbed by cattle, males fly from the dung surface carrying their mates. Parker (1970d) estimated the maximum possible advantage that could accrue to females from being carried from the dung on disturbance to be just 3.3%. He also considered the possibility that postcopulatory associations may increase oviposition efficiency. He found that attended and unattended females laid eggs at an equivalent rate of around 14 seconds per egg. Moreover, Parker (1970d) calculated that by engaging in mating and a postcopulatory association females can save around 60 min in depositing a batch of eggs by avoiding harassment from searching males so that females who are protected during oviposition could produce their next batch of eggs sooner. However, given that the production of a clutch takes around 8 days, the selective value of the postcopulatory association is just 0.52%. Thus, the overall selective benefit to females arising from the postcopulatory association is fairly small, compared with the selective advantage of 400% bestowed on males through the avoidance of sperm competition (Parker 1970d).

Nevertheless, there is good evidence to suggest that the females of some species can gain immediate benefits from postcopulatory associations (table 5.2). In some odonates, the female submerges in order to oviposit on subaquatic vegetation where she is inaccessible to rival males. Nevertheless, males continue to guard their mates, sometimes submerging with them (Robert 1958) but more usually shifting from contact to noncontact guarding as the female submerges (Miller and Miller 1981; Fincke 1986). Fincke (1986) examined the costs and benefits of mate guarding for males and females in the damselfly *Enallagma hageni*. The longer the female remained submerged, the more of her clutch she oviposited, and the more likely males were to abandon their submerged mate. When females resurface, they usually do so with enough force to break the surface tension of the water and fly away. Typically they are seized by males as they surface. Some 96% of guarding males are successful in seizing their mates if they return to the surface within 10 min. They will then fly them to a new site to continue oviposition. However, after 40 min of submerged oviposition 50% of females were seized by nonmates because their mates had abandoned them. Given the risk that females would be seized by nonmates when they resurfaced without having oviposited their entire clutch, Fincke (1986) was able to calculate that a guarding male fertilized 30% more eggs than he would have if he abandoned his mate immediately after she had submerged. Females also benefit from guarding, and from the tactic of rival males of attempting to seize resurfacing females. Some 20% of females become trapped at the water surface, and of these a minimum of 6% will drown. Fincke (1986) calculated that the probability of drowning for females was 0.09 per clutch and 0.16 over the course of an average female lifespan. Male guarding reduced this probability by 29% and the combination of guarding and the search for resurfacing females by other males reduced it by 71%, a considerable fitness benefit to females.

Observational studies suggest that harassment by searching male dragonflies often interrupts oviposition so that guarded females are able to deposit a greater number of eggs than unguarded females (e.g., Waage 1978; Sherman 1983; McVey and Smittle 1984; Spence and Wilcox 1986; Moore 1989). For example, in their study of the dragonfly *Nannophya pygmaea*, Tsubaki et al. (1994) examined experimentally the effects of mating and mate guarding on oviposition. They compared oviposition behavior of females that had mated and were guarded with those that had mated and had their guarding male removed and with those that oviposited alone without first mating. By comparing the two groups of females that oviposited alone, Tsubaki et al. (1994) found that mating increased the total frequency within the oviposition episode that females left their perch and flew out over the water to oviposit. Moreover, comparing mated females who were guarded with those whose guard had been removed revealed that guarding increased the time that females actually spent ovipositing, even in the absence of interference from rival males. This may be important for females because it decreases the period during which they are exposed to aquatic predators and increases the amount of time for foraging. Tsubaki et al. (1994) also found that female reproductive success was increased by ovipositing with a guarding male because males tended to hold territories in areas where the probability of egg hatching was higher than average.

Finally, some studies suggest that mate guarding can enhance a female's foraging ability. McLain (1981) has shown that singleton soldier beetles *Chauliognathus pennsylvanicus* foraging on flowers are more likely to be knocked off by intraspecific competitors, in this case wasps, than those engaged in mating and postcopulatory guarding. In this way, mating females are able to forage more efficiently than solitary females. Likewise, solitary female water striders experience considerable sexual harassment from searching males while they forage on the water surface. Wilcox (1984) found that single female *A. remigis* had a sixteenfold increase in time required to repel harassing males compared with females subject to contact guarding. As a result, guarded females spent more time foraging on the water surface and had a higher prey capture rate than did single females. Given the sexual conflict that exists over mating and postcopulatory mate guarding in water striders, the proposed foraging benefits for females that arise from guarding predict that hungry females should be more willing to engage in mating and its associated mate guarding than satiated females. To test this prediction, Clark (1988) manipulated both ambient operational sex ratio and female nutritional status in his study of mate guarding in *A. remigis*. In accord with both the female foraging benefits and mate guarding hypotheses, the duration of mating associations was increased in male-biased operational sex ratios. However, when operational sex ratio was held constant, Clark (1988) found no difference in the duration of the mating association for starved or fed females. On the other hand, male nutritional status does influence the duration of the association, with starved males terminating mating associations sooner than fed males (Sih et al. 1990). Studies of *G. buenoi* suggest that mating associations decrease, rather than increase, female foraging efficiency (Rowe 1992). The data for water striders are therefore inconsistent with the hypothesis that males guard their mates to enhance foraging efficiency, and more consistent with the avoidance of sperm competition favoring mate guarding.

It is clear that the evolution of postcopulatory associations will be strongly dependent on female behavior. If females are unreceptive to further mating, either due to the efficacy of male seminal products (chapter 4) or because remating incurs costs to females, males will gain no selective advantage from postcopulatory associations because their paternity will already be assured. Of course, any adaptive benefits accruing to females from mating associations can also accrue to males; if females are able to oviposit more efficiently, or reduce the risk of death between oviposition bouts, males will have increased fitness through the greater number of progeny produced by their mate. Nevertheless, a postcopulatory association would not spread via natural selection for male parental investment unless the benefits accrued by the male's own offspring were sufficient to outweigh the costs associated with lost reproductive opportunities (Parker 1984). However, if females derived significant benefits from mate guarding, such as the increased survival noted for *E. hageni* (Fincke 1986) and for the stick insect *Diapheromera veliei* (Sivinski 1983), and there was even slight variation in female receptivity following mating that favored guarding by males, full female receptivity could evolve rapidly. Female benefits could thereby alleviate any conflict of interests between males and females over mate guarding and favor the coevolution of full female receptivity to further mating due to natural selection on females as postcopulatory guarding evolved via sexual selection (Parker 1970d, 1984). It is clear from these considerations that the alternative hypotheses for the evolution of postcopulatory associations in table 5.1 may not be mutually exclusive.

Ejaculate Transfer

In a number of insect taxa insemination is indirect in that sperm are packaged into a spermatophore that is either attached to the female's genital opening or inserted into her bursa copulatrix. The majority of these taxa exhibit postcopulatory associations, the adaptive significance of which has been extensively studied, particularly in the crickets (Orthoptera: Gryllidae) (table 5.2). After attaching the spermatophore, a male cricket maintains antennal contact with his mate, stridulating aggressively and blocking the female's path should she attempt to move. In katydids (Tettigoniidae) and fishflies (Megaloptera), males remain attached to the female for several hours after spermatophore transfer (table 5.2). In addition to the mate guarding hypothesis, these taxa have been used to examine the alternative ejaculate transfer and multiple mating hypotheses (table 5.1).

The ejaculate transfer hypothesis suggests that male postcopulatory associations function in ensuring that the spermatophore remains attached long enough to facilitate complete sperm transfer (Gerhardt 1913). In the crickets *Gryllus bimaculatus* (Simmons 1986), *Acheta domesticus* (Sakaluk and Cade 1983), and *Gryllodes sigillatus* (Sakaluk 1984), transfer of sperm from the spermatophore takes around 50–60 min after which females remove and sometimes consume the empty spermatophore. Early removal of the spermatophore would be detrimental to male fitness, and indeed in some cases has been shown to reduce fertilization success (Simmons 1987b; Sakaluk and Eggert 1996). In accord with the ejaculate transfer hypothesis, spermatophores often remain attached for longer when a male remains with the female than when he is experimentally removed (fig. 5.7). Within some species the duration of guarding is positively associated with the duration of spermatophore attachment (Simmons 1986; Sakaluk 1991; Simmons 1991b; Hockman and Vahed 1997) and the intensity of male guarding behavior declines after 20 min of spermatophore attachment, coincident with a reduction in the rate of sperm transfer (Khalifa 1949; Loher and Rence 1978; Simmons 1986, 1987b). However, contrary to the ejaculate transfer hypothesis, across those species of gryllid that have been studied so far, there does not appear to be a general association between the duration of guarding and the duration of spermatophore attachment (fig. 5.7), suggesting that the ejaculate transfer hypothesis alone is insufficient as a general explanation for the evolution of postcopulatory associations in this taxon.

According to the phylogeny presented by Gwynne (1997), postcopulatory association had a single origin within the Gryllidae. However, there also appear to have been a number of independent evolutionary origins of a second mechanism for ensuring complete ejaculate transfer in the gryllids. In *G. sigillatus*, the spermatophore consists of the sperm-containing ampulla as well as the spermatophylax, a gelatinous mass that is the product of the male's accessory glands. When the female attempts to remove the spermatophore after copulation, the spermatophylax detaches from the ampulla and the female first consumes this mass before attempting again to remove the ampulla. The time it takes the female to consume the spermatophylax is just sufficient to allow sperm transfer from the ampulla to the female's sperm storage organ (Sakaluk 1984). Thus, postcopulatory association for ejaculate protection seems redundant in this species. Indeed, Sakaluk (1991) found that in the case of *G. sigillatus*, male presence had no impact on the duration of spermatophore attachment (see fig. 5.7). The fact that the evolution of a spermatophylax was not associated with the secondary loss of postcopulatory association in *G. sigillatus* supports the conclusion that ejaculate transfer per se is not a sufficient explanation for this behavior. Studies by Sakaluk (1991) and by Frankino and Sakaluk (1994) have demonstrated that

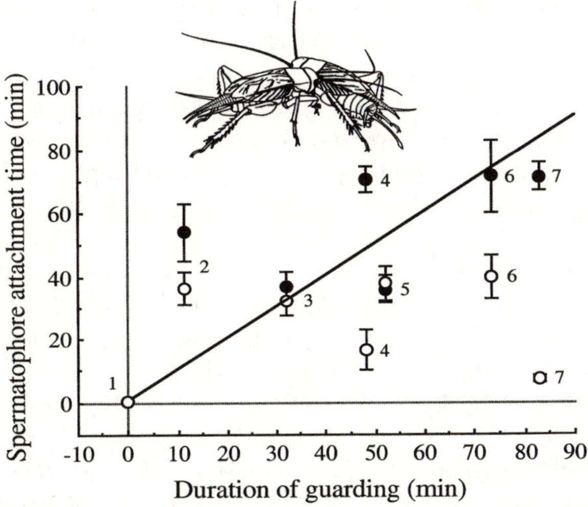

Figure 5.7 After attaching a spermatophore during copulation, male gryllids typically guard their mate, responding aggressively if she attempts to remove the spermatophore or leave the area. Studies of a number of species have demonstrated that the mean (± SE) duration of spermatophore attachment is longer when the male is allowed to guard his mate (solid circles) than when he is removed after spermatophore attachment (open circles). However, across species there does not appear to be a general relation between the duration of guarding and the duration of spermatophore attachment. The line equating perfect concordance between guarding and spermatophore attachment duration is indicated. While guarding may contribute to the successful transfer of sperm from the spermatophore before it is removed, mate guarding may also serve other adaptive functions. 1, *Cycloptiloides canariensis*—note that in this species males do not guard (Dambach and Beck 1990); 2, *Balamara gidya* (Evans 1988); 3, *Acheta domesticus* (Khalifa 1950; Sakaluk and Cade 1983); 4, *Gryllus bimaculatus* (Simmons 1986); 5, *Gryllodes sigillatus* (Sakaluk 1991); 6, *Teleogryllus natalensis* (Hockman and Vahed 1997); 7, *T. commodus* (Loher and Rence 1978; Evans 1988). (Vignette from Loher and Rence 1978 with permission of Blackwell Wissenschafts-Verlag)

postcopulatory association functions to reduce the probability that females will mate with rival males, a finding congruent with the sperm competition hypothesis for the evolution of mate guarding.

The association between postcopulatory associations and the occurrence of a spermatophylax can also be examined in the tettigoniids. In contrast to the gryllids, spermatophylax feeding is the ancestral character state and there has been a secondary origin of postcopulatory association in a number of taxa (Gwynne 1997). For example, three species of the genus *Coptaspis* (Tettigoniidae: Conocephalinae) have been shown to have extended postcopulatory associations lasting up to 7 h (Wedell 1998). In support of the ejaculate transfer hypothesis, across species the duration of the association increases with the weight of the spermatophore and thus the amount of ejaculate to be transferred. Associated with the origin of postcopulatory association in *Coptaspis* spp. is a secondary loss of the spermatophylax (the spermatophylax is rudimentary and does not protect the ejaculate during insemination). A second independent origin of postcopulatory association occurs in the Meconematinae. As with *Coptaspis*, for the two *Meconema* spp. that have been studied, a longer period of postcopulatory association is related to the transfer of a larger ejaculate. For both species postcopulatory association coincides with a secondary loss of the spermatophylax (Vahed 1996; see also Vahed 1998a).

A remarkable evolutionary convergence occurs in the Megaloptera, which show patterns of postcopulatory association and spermatophylax production identical to those seen in the tettigoniids. Male dobsonflies, *Protohermes* spp., produce a spermatophylax on which females feed while ejaculate is transferred from the ampulla of the spermatophore (Hayashi 1992). In contrast, male fishflies do not produce a spermatophylax, but remain in contact with their mates for up to 6 h after spermatophore transfer (Hayashi 1996). Across species, the duration of postcopulatory association appears to be sufficient to facilitate the complete transfer of sperm, so that species that transfer smaller ejaculates have a shorter period of postcopulatory association. Finally, in the alderflies, *Sialis* spp., males neither guard nor provide a spermatophylax (Hayashi 1999b). Instead, sperm transfer is very rapid, taking just 6–7 min. In alderflies it seems that adaptations to ensure ejaculate transfer are unnecessary. A parallel occurs in the gryllid *Cycloptiloides canariensis*, where sperm are transferred within 30 s of spermatophore transfer and males do not exhibit postcopulatory associations (Dambach and Beck 1990). These comparative patterns provide strong support for the ejaculate transfer hypothesis for postcopulatory associations in taxa where sperm transfer is indirect.

It is interesting to note that in the tettigoniids females have a postmating refractory period that is induced by male seminal products (section 4.3). This may account for the lack of co-occurrence of spermatophylax feeding and postcopulatory association. In contrast, female *Gryllodes* retain sexual receptivity and will mount courting males even when they have a previous male's spermatophore attached. This difference may explain why postcopulatory association in this lineage has been maintained even though males produce a spermatophylax. It is also interesting that some tettigoniid lineages have switched from spermatophylax feeding to postcopulatory association. The answer may lie in the relative costs of spermatophylax production and prolonged mating associations. Production of a spermatophylax is costly for males, particularly so in nutrient-limited habitats, and can decrease male potential reproductive rate (Simmons 1995c). Guarding may represent a cost-effective method of ejaculate transfer when the costs of spermatophylax production become too high.

A number of taxa in table 5.2 have been described as exhibiting prolonged copulatory associations because they are thought to extend copulation beyond the time necessary for sperm transfer. In some instances these conclusions are well founded, based on an examination of sperm transfer events. For example, in the ischnuran damselflies males remove rival sperm and then remain in copula for several hours, delivering their own ejaculate at the end of the association, presumably when the risk of takeover by rival males is reduced (e.g., Robertson 1985; Cordero 1990). However, in other studies where copulation duration has been found to depend on the operational sex ratio, it has been concluded that prolonged copulatory associations constitute copulatory mate guarding on the grounds that short copulations facilitate full fertility (e.g., Rubenstein 1989; Carroll 1991). However, the alternative hypothesis, that prolonged copulation facilitates the transfer of larger ejaculates, has not been considered.

In general, the numbers of sperm required to achieve complete fertility for females is small, relative to the numbers of sperm actually transferred by males. For example in *S. stercoraria*, full fertility is achieved after just 5 min of the average 36 min copulation (Parker et al. 1993). In general, female insects store only a proportion of the total sperm transferred by males (see review in Eberhard 1996). As we shall see in chapter 7, there is now an abundance of evidence to suggest that males adjust the number of sperm transferred when the risk of sperm competition is high. Increases in sperm numbers serve to increase a male's fertilization success when his sperm must compete with those of rival males for fertilizations. It is therefore difficult to distinguish between the mate guarding hypothesis and the ejaculate transfer hypothesis for prolonged copulations in insects without a complete understanding of sperm transfer and utilization patterns.

The first study to examine prolonged copulation was Sillén-Tullberg's (1981) study of the lygaeid bug *Lygaeus equestris*. Copulation duration was increased under male-biased operational sex ratios but Sillén-Tullberg (1981) found that a male's fertilization success was not increased with extended copulations, supporting the notion that prolonged copulation served to prevent females from remating with rival males rather than to increase the numbers of sperm transferred. However, McLain's (1980) study of the green stink bug *Nezara viridula* found that prolonged copulation actually reduced the fertilization success of rival males that copulated subsequently (fig. 5.8). These data are consistent with an interpretation of sperm precedence in which longer copulations result in the transfer of larger ejaculates that rival males are unable to compete with. Similarly, in his study of the water strider *A. remigis*, Rubenstein (1989) found that the last male's fertilization success was dependent on his copulation duration. Such a relationship is inconsistent with the notion that sperm transfer is completed early in the mating association; if this were the case fertilization success should be independent of copula duration as it is in *L. equestris* (Sillén-Tullberg 1981) and *Neacoryphus bicrucis* (McLain 1989). Rather, prolonged copulation may be asso-

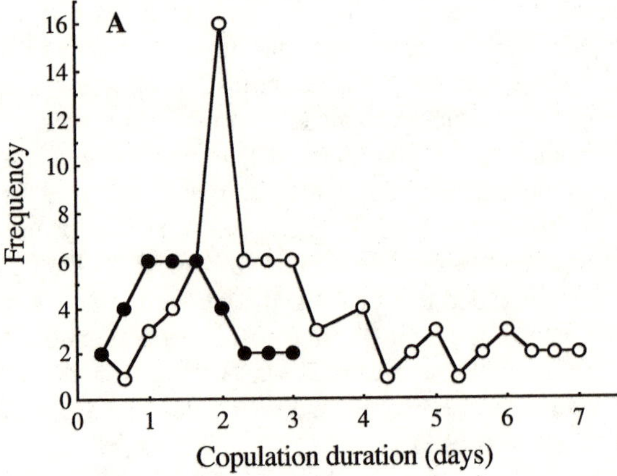

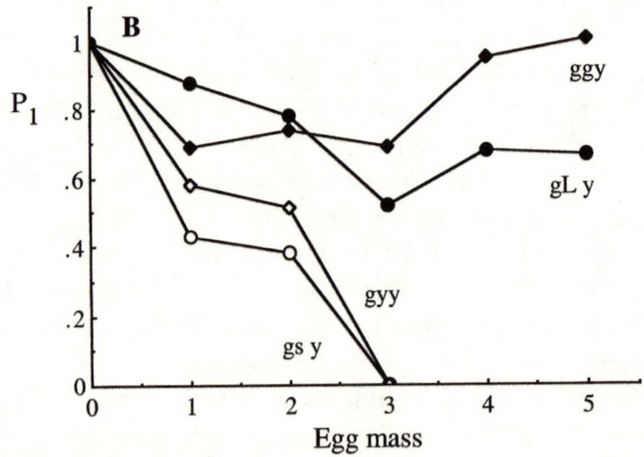

Figure 5.8 Adaptive significance of prolonged copulation in the green stink bug *Nezara viridula.* (A) Copulation duration was significantly longer when pairs were held in a male-biased operational sex ratio (4 males : 2 females) (open circles) compared with pairs held in a female-biased operational sex ratio (2 males : 4 females) (solid circles), consistent with the hypothesis that males exhibit copulatory mate guarding. In (B), the fertilization success of subject males subsequent to a rival copulation was examined using genetic markers, green (g) and yellow (y) bodies. Green males were mated first, followed by yellow males. Eggs fertilized by green males produce black nymphs so that the proportion of offspring sired by the first male (P_1) could be estimated from the percentage of black nymphs. When the first male had a long copulation (gL y), he had a higher fertilization success following a rival copulation than when he had a short copulation (gs y). The fertilization bias toward first males could be reproduced by allowing first males two short copulations instead of a single long copulation, indicating that long copulations involve the transfer of more sperm than do short copulations. The data are therefore consistent with the hypothesis that prolonged copulation functions in transferring a larger ejaculate when the risks of sperm competition are high. (Data from McLain 1980)

ciated with increasing numbers of sperm transferred and thus increasing fertilization success. Dickinson's (1986, 1988) study of the milkweed beetle *Labidomera clivicollis clivicollis* shows quite clearly that prolonged copulatory associations facilitate the transfer of greater numbers of sperm in this species, both within copulations and across successive copulations with the same female, and that prolonged copulatory associations function to increase the male's fertilization success. Of course, prolonged copulation may serve the dual function of preventing rival males from copulating and facilitating the transfer of large ejaculates for competition with rivals who may have copulated previously or may copulate later in the female's lifetime. Both mate guarding and ejaculate transfer hypotheses predict increased copulation duration with increasing male bias in the operational sex ratio. Thus, detailed examinations of ejaculate transfer and sperm utilization are required to distinguish between these hypotheses.

Multiple Mating

When examining the postcopulatory behavior of house crickets *A. domesticus*, Khalifa (1950) found that spermatophore attachment duration was independent of male presence (fig. 5.7) and this led him to suggest that the male's postcopulatory association functioned in maintaining contact with the female to facilitate repeated inseminations, a suggestion echoed by Alexander and Otte (1967) in their review of gryllid reproductive morphology and behavior. Postcopulatory associations might in general function to increase male mating frequency with the same female if there is a selective advantage to such behavior. Simmons (1987b) found that male fertilization success was associated with the number of matings with individual females in *G. bimaculatus*, and in general, because gryllids typically exhibit sperm mixing, the more matings a male obtains with a female the greater will be his fertilization success. The multiple mating hypothesis for postcopulatory associations in gryllids predicts that the time required for spermatophore replenishment for the male should be equal to or less than the period of postcopulatory association, so that males remain in contact with females until they are ready to remate and then begin to court a second time. In general this is the case for gryllids (Zuk and Simmons 1997), although *G. sigillatus* is an exception. In this species, the necessity for males to produce a spermatophylax means that the remating interval is some eight times longer than the duration of the postcopulatory association so that remating is unlikely to represent a selective pressure favoring the behavior in this species.

Sivinski (1983) found that male stick insects *Diapheromera veliei* periodically reinserted the penis up to nine times during the 3–156 h of mating association. It is not known if insemination occurred on each insertion but it is entirely possible that the male remains mounted upon the female to acquire multiple matings with her. A recent study of pine engraver beetles *Ips pini* (Lissemore 1997) has shown how a male's fertilization success increases over the course of the 5 to 6 weeks that the pair remain within his gallery. Similarly, male fertilization success increases with the duration of pair associations in carrion beetles *Necrophorus vespilloides* (Müller and Eggert 1989). It appears that multiple mating by the male with his partner facilitates the dilution and/or displacement of rival sperm over the course of the mating association. Males of the brooding aquatic bug *Abedus herberti* also insist on repeated copulations throughout the mating association, thereby ensuring paternity of the eggs they will eventually brood (Smith 1976, 1979). As with prolonged copulation, multiple inseminations with the same female may be favored when the risks of sperm competition are high so that multiple mating can play a significant role in favoring the evolution of prolonged periods of association between males and females. Nevertheless, as with the other hypotheses outlined above, the multiple mating hypothesis and the mate guarding hypothesis need not be mutually exclusive since mate guarding may serve the dual function of increasing the male's mating frequency and preventing rival males from copulating.

Cryptic Female Choice

The final and most recent hypothesis for the evolution of postcopulatory associations is the proposal of Eberhard (1991, 1996) that males may prolong the mating association in order to deliver courtship that increases the probability that the female will use his sperm in fertilizing their eggs. Females are proposed to bias sperm storage and/or fertilization toward males better able to stimulate them during postcopulatory associations, thereby exerting cryptic female choice. Few studies have examined this hypothesis in detail and those that have yield mixed evidence.

Male carrion flies *Dryomyza anilis* search for females around carcasses or droppings where females come to oviposit. With mature females copulation lasts for less than 1 min, following which the male remains mounted and begins to tap the female's abdomen with his genitalia. On average the male performs 22 bouts of tapping interspersed with periods of quiescence that last around 2 min. The pair will then move to the carcass to oviposit. After a period of oviposition, males will recopulate and perform a further series of tapping bouts, followed by another bout of oviposition. The male's fertilization success is dependent on the number of tapping sequences performed and the number of matings (Otronen 1990, 1994). Thus, postcopulatory associations in this fly do appear to increase a male's fertilization success which is dependent on his behavior during the postcopulatory association. Nevertheless, multiple matings are also a significant component of fitness for males that will favor prolonged association (cryptic female choice in this system and others is discussed in more detail in section 9.4).

The possible significance of male behavior directed at females during postcopulatory associations has also been examined in the water striders *G. lateralis* and *G. lacustris* (Arnqvist and Danielsson 1999a; Danielsson and Askenmo 1999). During associations, male *G. lateralis* hold the female by folding their midlegs around those of the female and vibrate their bodies, causing the couple to oscillate vertically for 1–3 seconds. Arnqvist and Danielsson (1999a) showed that, although there were consistent differences between males in the frequency of this copulatory courtship, variation in the behavior did not influence a male's fertilization success, counter to the predictions of the cryptic female choice hypothesis. Similarly, Danielsson and Askenmo (1999) found no correlation between the duration of postcopulatory mate guarding and fertilization success in their study of *G. lacustris*. In the latter study, takeovers by rival males were not possible since pairs were isolated from rival males. However, when rival males are present guarding males do have a higher fertilization success than males experimentally prevented from guarding (Jabłoński and Kaczanowski 1994) because rivals are able to seize and copulate with unguarded females. Thus, in the case of *G. lacustris* postcopulatory guarding increases male fertilization success in the context of rival exclusion, but not in the context of cryptic female choice. With so few studies available it is too early to assess the general significance of cryptic female choice in the evolution of prolonged postcopulatory associations. Moreover, as with other hypotheses, cryptic female choice and mate guarding need not be mutually exclusive hypotheses for the evolution of postcopulatory associations.

5.5 Adaptations for Efficient Guarding

Whenever there is competition for a limited supply of reproductive females, selection should favor traits in males that are associated with superior competitive ability. In many taxa it has been established that an increase in the intensity of competition for females is associated with an increased rate of interference with copulating pairs by searching males (table 5.2). Interference takes the form of attempts by searching males to displace or "takeover" females from copulating or guarding males. Such competition is predicted to favor adaptations in males for increased efficiency in retaining females when challenged.

Morphological Adaptations

Chapter 3 of this volume considered the importance of sexual selection for the evolution of male genitalia. One hypothesis for the evolution of complex male genitalia is that the various clasping devices, processes that are inflated within the female's reproductive tract, or heavy spination on the aedeagus, may function as holdfast devices that aid the male in retaining his mate during takeover attempts. It was Richards (1927) who first proposed that sexual selection via competition during takeover would favor the evolution of complex genital morphology. The possibility has since been discussed by Parker (1970e) and Thornhill and Alcock (1983), and devices that apparently hold females during copulation have been assumed to function in the context of avoiding takeover, often without consideration of alternative hypotheses (Thornhill 1984b). Yet, surprisingly, there have been no empirical studies of the potential influence of variation in male genital morphology on the ability of males to resist takeover. Studies of the functional significance of the dorsal clamps in scorpion flies (Thornhill 1984b) and the grasping apparatus of male water striders (Arnqvist 1989b; Arnqvist et al. 1997) suggest that these structures function in increasing a male's success in forcing copulations upon females, rather than in preventing rival males from copulating. Nevertheless, clasping genital cerci are used by the males of some species, such as stick insects (Sivinski 1979), in order to remain attached during protracted mating associations and competition with rival males.

In yellow dung flies, 36% of males are subject to takeover either during copula (4%) or during the guarding period when females are ovipositing (32%) (Parker 1970b). During struggles between guarding and searching males, the guarding male will attempt to reestablish genital contact with his mate. The male's external genitalia consist of two enlarged hooklike paralobes that are inserted into the female's genital opening (Parker 1969). These genital hooks make it extremely difficult to separate the male from the female, and indeed Parker (1970b) found that the probability of successful takeover was significantly lower for pairs in genital contact. Following a successful defense, males will immediately withdraw their genitalia so that the female can continue to oviposit. In contrast, following a successful takeover, the newly mounted male will copulate for an average of 20 min, depending on the number of eggs the female has left to lay (Parker et al. 1999). These observations suggest that male genital hooks could function in decreasing the probability of takeover. It is not unreasonable to assume that larger males will have larger hooks (they certainly have larger aedeagi; Parker and Simmons 2000) and larger males are more successful in resisting takeovers than are small males (Sigurjónsdóttir and Parker 1981). Of course, large males are also likely to have greater physical strength than small males, but if physical strength per se was the relevant factor we would expect larger males also to be more successful in takeover attempts, which is not the case (Sigurjónsdóttir and Parker 1981).

It has been suggested that prolonged postcopulatory associations involving contact between male and female may favor the evolution of sexual size dimorphism. The costs to females that are associated with carrying males for extended periods will also represent fitness costs for males. For example, high energetic costs of carrying males will result in females diverting resources away from egg production, thereby reducing the number of offspring fathered by the male. Alternatively, increased predation risk for females will result in reduced probability of offspring production for males. All else being equal, selection should thereby favor adaptations in males that minimize these costs, and male dwarfism is one avenue by which the burden on females could be reduced. It is often the case that species with extended periods of contact association following insemination exhibit sexual dimorphism. In his survey of 155 species of stick insect, Sivinski (1979) found that males were on average 30% smaller than females. Moreover, species with below average male to female ratio were those with the longest mating associations recorded in the Phasmatoidea (see also Carlberg 1983). Nevertheless, all things are unlikely to be equal. Sexual selection may act to increase male size in the context of competitive interactions between guarding and searching males. Thus, Sivinski (1983) found that the ratio of male to female body size for species that occur in high-density situations is higher than the average.

Although Sivinski's (1979) study revealed some interesting patterns, it was qualitative in that it did not control for phylogenetic association. Sexual size dimorphism has been studied extensively in the water striders (review in Arnqvist 1997b). Fairbairn (1990) also considered the possible impact of loading constraints during postcopulatory associations on the evolution of sexual dimorphism in this group. She examined the duration of mating associations across seven species of water strider and found that, although the duration of associations differed significantly across species, there was no correlation between the duration of associations and the ratio of female to male body size, suggesting that loading constraints during pairing have not been generally responsible for the overall sexual dimorphism in body size across the gerrids considered. Nevertheless, the duration of associations was significantly related to the residuals from a regression of female body size on male body size; species with larger females than expected for the size of males had longer mating associations, suggesting that loading constraints may favor deviations from the underlying sexual dimorphism. Fairbairn's (1990) analysis showed that the differences in size ratios among taxa could not be due to phylogenetic association (see also Andersen 1994, 1997). Because larger males are more successful in premating struggles, sexual selection for increased male size occurs in low-density populations or those with female-biased operational sex ratios (Rowe et al. 1994; Arnqvist 1997b). Moreover, Rowe and Arnqvist (1996) found that for three species of *Gerris* large males actually guard for longer than do small males. These factors may counter any selection for reduced male size during postcopulatory associations. Thus, a complete explanation for observed patterns of sexual dimorphism in water striders is likely to involve a complex interaction of selection pressures acting on males and females (Arnqvist 1997b).

BEHAVIORAL TACTICS

Although Parker (1970e, 1970d) described postcopulatory associations as passive phases, he noted that males were far from passive during guarding, adopting often ritualized rejection behaviors in response to the takeover attempts of rival males. Guarding male *Scatophaga stercoraria* adopt a series of rejection responses when contacted by searching males. Contacts with searching males result in the guarding male raising his midleg on the side of contact so that the searching male is fended away from the pair. More commonly, searching males pounce onto the back of a guarding male, upon which the guard will raise both midlegs into a horizontal position and elevate his body above the female by pushing against her thorax with his straightened forelegs (Parker 1970b). This has the effect of doubling the distance between the searching male and the ovipositing female. In this way the guarding male prevents the searcher from making contact with his mate and as a result interactions rarely continue for longer than a second or two. Attacks are only escalated if the searching male makes physical contact with the female, suggesting that some contact sex recognition operates (Parker 1970b). As discussed above, when struggles ensue, the guarding male will attempt to regain genital contact. Thus, the stylized rejection responses of *S. stercoraria* appear to effectively reduce the probability of escalated takeover attempts, which occur in just 7% of all encounters (Parker 1970b).

A similar strategy of increasing the distance between rival and female by standing upright on straightened legs occurs in the crane flies (Adler and Adler 1991), while in libellulid dragonflies males hover above their ovipositing female and will fly rapidly backward and upward on the approach of a rival, thereby knocking it away from the female. If intruders persist, guarding males leave their mate to drive the rival away from the area (Waage 1984). An apparently unique postcopulatory behavioral response has been reported by Alcock and Forsyth (1988) in their study of the rove beetle *Leistotrophus versicolor*. Males search for females at dung and/or carrion where they come to capture flies. Following copulation males become highly aggressive toward their mates, and drive them away from the area in which they are likely to be contacted by other searching males. The probability of postcopulatory aggression toward females increases when rival males are present. Alcock and Forsyth (1988) considered a number of alternative hypotheses for postcopulatory aggression, including the notion that it reduced competition for resources or represented a signal of sperm transfer. However, the data were most consistent with the hypothesis that the behavior reduced the probability that a female would remate with a rival male.

Removing the female from areas of high sperm competition risk is a frequently observed behavioral response both during and after copulation. In many butterflies there is a postnuptial flight during which the male flies with the female hanging beneath him to an area away from the site of mate location (Pliske 1975; Rutowski 1979; Thornhill and Alcock 1983). In many Diptera that form mating swarms, the male will fly with his mate to an area away from the swarm before copulating with her (Downes 1970; Neems et al. 1998; Sadowski et al. 1999), and those that meet at oviposition sites will leave and finish copulation in the surrounding vegetation (Alonso-Pimentel and Papaj 1999). Parker (1971) noted that in *S. stercoraria* the proportion of pairs emigrating from the dung to the surrounding grass to copulate was dependent on the density of males searching on the dung. Copulations away from the dung surface are costly for males; they increase the cycle time (the time taken to find, copulate with, and guard a female during oviposition) because the temperature in the surrounding grass is lower, resulting in copula durations some 30–35 % longer. This led Parker (1971) to examine the adaptive significance of emigration. From field observations of the spatial distribution and mating activity of males and females on and around the dung, he calculated the encounter rates with searching males sustained by copulating pairs at a range of male densities (fig. 5.9). Not surprisingly, the number of encounters sustained by pairs per minute decreased rapidly with distance from the dung in the downwind direction. This is because males typically search for females on, or upwind of, the dung surface (Parker 1970a). The number of encounters sustained also increased with the density of searching males. Patterns of emigration behavior were predicted by the patterns of encounter sustained; over 80% of males emigrated to a position downwind of the dung and moved on average 90 cm, where encounter rates were close to zero (fig. 5.9). Taking into account the probability of takeover and the proportion of eggs fertilized by the last male to copulate with the female prior to oviposition, Parker (1971) was able to estimate the fitness, in terms of fertilization gain, of males that emigrate or remain on the dung surface to copulate. Below densities of 5 males per 500 cm^2 of dung, it is advantageous for males to remain on the dung surface and reduce their cycle time by copulating at higher temperatures. However, as male density increases above 5 males per 500 cm^2 of dung, it becomes increasingly advantageous to emigrate to the surrounding grass, despite the costs of increased cycle time (fig. 5.9). Field observations show that the threshold density for emigration (the density at which >50% of pairs emigrate from the dung) is around 5 males per 500 cm^2 of dung. These data show that removal of females from areas of high sperm competition risk can result in significant fitness benefits for males.

5.6 Male Mate Choice

For species in which selection has favored adaptations in males that result in the copulating male gaining the majority of fertilizations, previously mated females are of equal reproductive value to males as unmated females be-

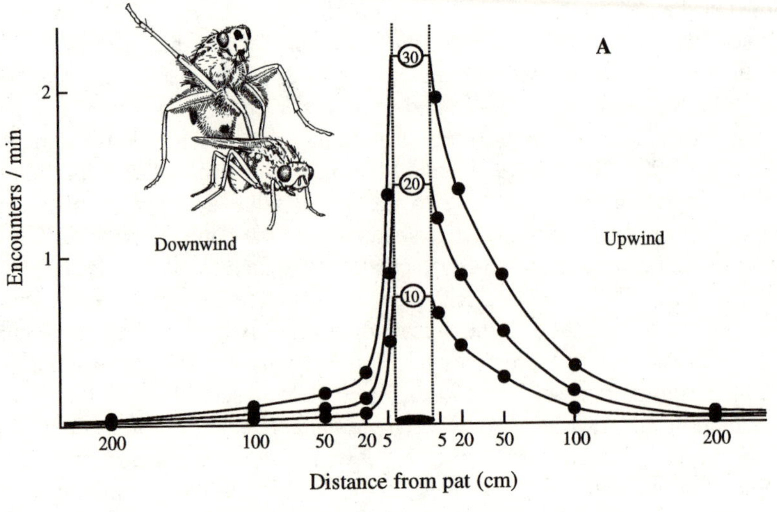

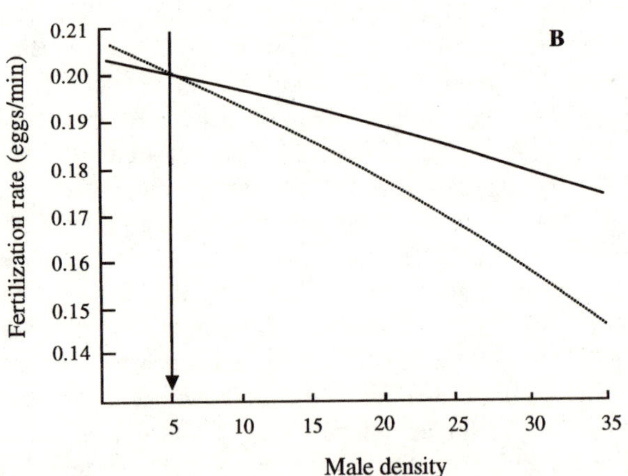

Figure 5.9 Adaptive significance of emigration from the site of mate searching in the yellow dung fly *Scatophaga stercoraria*. (A) The encounter rate of copulating pairs with searching males on and around the dung when the density of searching males is 10, 20, or 30 males per 500 cm^2 of dung. Males search for females on the dung surface and on the upwind side of the dropping, where the rate of arriving females is greatest, so the encounter rates of copulating pairs with searching males is greatest in these areas. The encounter rates also increase with the density of searching males. To reduce the encounter rate with searching males, and thus the probability of takeover, paired males emigrate to the downwind side of the dung and over a distance of around 90 cm. (B) The fertilization gain for males that emigrate (solid line) and those that fail to emigrate (dotted line) for various densities of searching males. Below a density of 5 males per 500 cm^2 of dung, males remaining on the dung have a fitness advantage because of the shorter cycle time that results from faster copula durations at warmer temperatures. However, above densities of 5 males per 500 cm^2 of dung, males that emigrate have the higher fitness gain due to the avoidance of takeover. The threshold point (vertical line) beyond which more than 50% of males emigrate is predicted to occur at 5 males per 500 cm^2 of dung. This is close to the value observed in the field. (Redrawn from Parker 1971)

cause sperm competition is effectively avoided. However, in species where males are unable to avoid sperm competition from previous males, the reproductive value of mated females will be reduced compared with unmated females. If mating were cost-free, males would maximize their fitness by accepting all females encountered, irrespective of the number of eggs they fertilized. However, mating is rarely cost-free. Males are limited in their ability to produce ejaculates (Dewsbury 1982; chapter 7) and in the time available for searching for and copulating with females (Parker 1978). As such, selection is expected to favor male behavior that maximizes not just the number of matings achieved, but the fertilizations gained per unit investment in mating. Thus, we should expect to see the evolution of male mate choice whereby males avoid sperm competition by rejecting previously mated females (Schwagmeyer and Parker 1990).

Male mate choice in the context of sperm competition has been examined in *Drosophila*. Markow (1982) found that single male *D. mojavensis* offered one unmated female and one female that had mated 24 h previously were significantly more likely to court the unmated female. However, male discrimination was lost when mated females had mated 48 to 96 h previously. Interestingly, the reproductive value of mated females increases with time since mating, because the proportion of offspring fertilized by a second male to copulate increases from just 66% at 24 h to 97% after 48 h. Similarly, male *D. hibisci* preferentially court and mate with previously unmated females, and amongst unmated females they prefer young females. Polak et al. (1998) ascribed this pattern of male mate choice to the existence of a mating plug that should favor the first male to mate at the time of the female's first oviposition. Unfortunately, data on sperm utilization in this species are unavailable. Nevertheless, strong first-male paternity does appear to be responsible for male mating preferences in the tettigoniid *Requena verticalis*, where males were shown to prefer young unmated females in both the field and laboratory (Simmons et al. 1994).

Lewis and Iannini (1995) examined the fitness consequences for male mate choice in their study of the flour beetle *T. castaneum*. In this species there is a slight second male advantage at the time of fertilization that is lost as sperm mix in the spermatheca, so that males should obtain fewer offspring when mating with a previously mated female compared with an unmated female. Lewis and Iannini (1995) placed a single male with five unmated and five previously mated females and observed their mating behavior. Males approached and contacted unmated females significantly more often than previously mated females and the copulation rate with unmated females was higher. Lewis and Iannini (1995) incorporated into their experiments phenotypic markers so that they could assign offspring to focal males. They found that the observed differences in male behavior directed toward unmated and previously mated females resulted in a fitness advantage for male preference, with males obtaining on average 52% more offspring when mating with unmated females. It is not clear how males can distinguish between mated and unmated females. In the case of *T. castaneum*, pheromones may be involved, because in their study Arnaud and Haubruge (1999) found that males not only discriminated against mated females when offered both mated and unmated females, but when given the choice between two mated females, one of which was the male's own previous mate, males chose the female with whom they had not mated. Males do produce sex pheromones (Lewis and Austad 1994) and these may be transferred to the female at the time of mating. The observation that males avoid females with whom they have already mated suggests some form of self-recognition.

Males are often assumed to incur trivial costs in mating so that the possibility that selection should favor male mate choice has been largely neglected (Dewsbury 1982). In general the costs of ejaculate production (see table 7.1 below), coupled with the fact that sperm competition will generate variance in the reproductive value of females to males, are predicted to generate selection for male mate choice. Indeed, studies of insects suggest that male choice for unmated females, and thus the avoidance of sperm competition, may be widespread (Pickford and Gillot 1972; Barrows and Gordh 1978; Rutowski 1982; Johnson and Hubbell 1984; Wiklund and Forsberg 1985; Brown and Stanford 1992; Wang and Millar 1997).

5.7 Summary

Postcopulatory mating associations are amongst the best-studied effects of sperm competition in insects. Theoretical considerations predict that selection should favor adaptation for mate guarding by males when females retain even slight levels of sexual receptivity following copulation, the interval between copulation and oviposition is short, mate-searching efficiency is high, and the last male to copulate with the female gains a significant proportion of fertilizations. Empirical studies from a variety of insect taxa support the mate guarding hypothesis. The incidence, nature, and duration of postcopulatory associations are sensitive to variation in the operational sex ratio, and thus the risk of sperm competition from future rivals. Nevertheless, there may be considerable sexual conflict over mate guarding which may not always be in the female's best interests. Conflicts of interest may be resolved in favor of males if the costs to females of sexual harassment from searching males exceed those incurred as a result of mate guarding.

Alternative hypotheses for the evolution of prolonged mating associations are often neglected in studies of insect mating systems. Females can obtain significant fitness benefits from mating associations, in the form of protection from sexual harassment, increased rates of oviposition, and even increased survival probability. Such benefits are likely to favor the coevolution of complete female sexual receptivity and postcopulatory mate guarding by males. It is unlikely that mating associations will evolve entirely under natural selection for male parental investment because of the costs to males associated with reduced mating opportunities. In species where insemination is indirect via an externally attached spermatophore, postcopulation associations appear to function in the efficient transfer of the ejaculate. Prolonged copulation is often viewed as copulatory mate guarding and there is some evidence that this is the case. However, it may also serve to increase the amount of ejaculate transferred when males face increased risk of sperm competition. Postcopulatory associations can function in providing the male with access to his mate for repeated inseminations and for copulatory courtship that persuades the female to utilize his sperm for fertilizations. Nevertheless, few of the hypotheses for prolonged mating associations offer mutually exclusive explanations for its evolution, and sperm competition is likely to play an important role in them all.

Whenever postcopulatory associations occur, selection will favor adaptations that improve the efficiency of the association. Morphological traits such as complex male genitalia may function as holdfast devices during takeover attempts from rival males, and male dwarfism may function in reducing the costs incurred by females in carrying males for protracted periods. However, empirical evidence for morphological adaptations to mate guarding is limited. Stylized rejection responses in males appear to function in defending females from takeover, and males may often remove females from areas of high sperm competition risk in order to avoid takeover attempts by searching rivals.

Sperm competition can induce variance in female reproductive value. Where mating incurs nontrivial costs for males, we should expect selection to shape male behavior so as to maximize the reproductive returns per unit investment in mating. Males should not mate indiscriminately, but rather should mate with females offering the greatest reproductive return. Evidence suggests that, in the absence of mechanisms for gaining sperm precedence, male mate choice for unmated females, and thus the avoidance of sperm competition from previous males, can be favored.

Table 5.2

The incidence, nature, and fitness consequences of mating associations in insects.

Species	Association	Duration	Cause of Variation	Effect on Male and/or Female Fitness	Source
ODONATA					
Zygoptera					
Argia vivida	Post contact		Location		Conrad and Pritchard 1990
Calopteryx aequabilis	Post noncontact				Meek and Herman 1990
Calopteryx amata	Post noncontact				Meek and Herman 1990
Calopteryx maculata	Post noncontact	10–15 min		Harassment(−)[b]; oviposition(+)	Waage 1973, 1978; Alcock 1979b, 1983; Meek and Herman 1990
Calopteryx maculatum	Post noncontact				Johnson 1962
Calopteryx splendens	Post noncontact				Siva-Jothy and Hooper 1995
Coenagrion puella	Post contact	86 min			Banks and Thompson 1985
Coenagrion scitulum	Pre contact	2 sec–81 min	Time of day, male density		Cordero et al. 1995
Enallagma cyathigerum	Post contact & noncontact		Submerged oviposition		Miller and Miller 1981
Enallagma hageni	Post noncontact	Up to 117 min		Female remating(−)[a]; survival(+)[b]	Fincke 1986
Hetaerina vulnerata	Post contact & noncontact	30–60 min	Submerged oviposition		Alcock 1982
Ischnura elegans	Copulatory	5 h			Miller 1987a
Ischnura gemina	Post contact	36 min			Hafernik and Garrison 1986

Ischnura graellsii	Copulatory	1–5 h	Male density, time of day, female mating history		Cordero 1990
Ischnura ramburi	Copulatory	1.5–6.7 h	Time of day		Robertson 1985
Ischnura senegalensis	Copulatory	7 h			Sawada 1995
Lestes sponsa	Post contact				Stoks et al. 1997
Mnais pruinosa pruinosa	Post noncontact		Location, alternative strategy		Nomakuchi et al. 1984; Siva-Jothy and Tsubaki 1989a,b
Nososticta kalumberu	Post contact		Interference		Thompson 1990
Anisoptera					
Acisoma panorpoides inflatum	Post noncontact	21–160 s			Hassan 1978
Aethriamantha rezia	Post noncontact				Hassan 1981
Erythemis simplicollis	Post noncontact			Oviposition(+)[b]	McVey and Smittle 1984
Leucorrhinia intacta	Post noncontact				Wolf et al. 1989
Leucorrhinia rubicunda	Post noncontact				Pajunen 1963
Libellula depressa	Post noncontact				Utzeri and Dell'Anna 1989
Libellula luctuosa	Post noncontact		Male density	Oviposition(+)[b]	Moore 1989
Libellula pulchella	Post noncontact				Pezalla 1979
Nannophya pygmaea	Post noncontact			Oviposition(+)[b]	Tsubaki and Ono 1985; Tsubaki et al. 1994
Orthemis ferruginea	Post noncontact				Harvey and Hubbard 1987
Orthetrum chrysostigma	Post noncontact	0 or 85 s	Alternative male tactics		Miller 1983

Table 5.2 (*cont.*)

Species	*Association*	*Duration*	*Cause of Variation*	*Effect on Male and/or Female Fitness*	*Source*
Pachydiplax longipennis	Post noncontact		Male density	Female remating(−)[a]; Oviposition(+)[b]; Harassment(−)	Sherman 1983
Paltothemis lineatipes	Post noncontact	30–130 s	Time for female to start oviposition		Alcock 1987
Plathemis lydia	Post noncontact		Male density		Jacobs 1955; Campanella and Wolf 1974; McMillan 1991
Sympetrum danae	Post contact				Michiels 1992
Sympetrum depressiusculum	Pre & post contact	≈ 7 h			Miller et al. 1984
Sympetrum obtrusum	Post contact & noncontact		Physiological state; temperature; wind speed	Female remating(−)[a]	Singer 1987
Sympetrum parvulum	Post contact & noncontact		Alternative male strategies; male density		Ueda 1979

Sympetrum sanguineum	Post contact &/or noncontact	222 s contact, 164 s noncontact	Male density	Oviposition(−)[b]	Convey 1989
Tramea carolina	Post contact & noncontact		Density	Oviposition(−)[b]	Sherman 1983
Urothemis assignata	Post contact & noncontact				Hassan 1981
Zygonyx natalensis	Post contact & noncontact		Interference	Oviposition time(−)[b]	Martens 1991
ORTHOPTERA					
Grylloidea					
Acheta domesticus	Post noncontact	52 min		Insemination(+)[a]	Khalifa 1950; Sakaluk and Cade 1980
Balamara gidya	Post noncontact	9–16 min		Insemination(+)[a]	Evans 1988
Gryllodes sigillatus	Post noncontact	32 min		Insemination(+)[a]; female remating(−)	Sakaluk 1991; Frankino and Sakaluk 1994
Gryllus bimaculatus	Post noncontact	48 min	Parasite load; male size; pair relatedness	Insemination(+)[a]; male remating(+)	Simmons 1986, 1990c, 1991b
Gryllus pennsylvanicus	Post noncontact		OSR		Souroukis and Murray 1995
Teleogryllus commodus	Post noncontact	83 min		Insemination(+)[a]	Loher and Rence 1978; Evans 1983; Evans 1988
Teleogryllus natalensis	Post noncontact	20–120 min		Insemination(+)[a]; female remating(−)	Hockman and Vahed 1997

Table 5.2 (*cont.*)

Species	*Association*	*Duration*	*Cause of Variation*	*Effect on Male and/or Female Fitness*	*Source*
Tettigonioidea					
Coptaspis spp.	Copulatory	Up to 7 h	Ejaculate size	Insemination(+)[a]	Wedell 1998
Meconema meridionale	Copulatory	81 min		Insemination(+)[a]	Vahed 1996
Meconema thalassinum	Copulatory	17 min		Insemination(+)[a]	Vahed 1996
Uromenus rugosicollis	Copulatory	100 min		Insemination(+)[a]	Vahed 1998a
Acridoidea					
Atractomorpha lata	Post contact	9–16 h	OSR		Muse and Ono 1996
Dichromorpha viridis	Post contact	17–55 h			Johnson and Niedzlek-Feaver 1998
Schistocerca gregaria	Pre and/or post contact	up to 48 h			Popov 1958
PHASMATODEA					
Baculum sp 1.	Copulatory	11–12 h			Carlberg 1987a
Diapheromera covilleae	Copulatory				Sivinski 1979
Diapheromera veliei	Copulatory	3–156 h	Sex ratio	Predation(−)[b]	Sivinski 1983
Extatosoma tiaratum	Copulatory	15–19 h			Carlberg 1987b
Necroscia sparaxes	Copulatory	79 days			Gangrade 1963
HEMIPTERA					
Dysdercus bimaculatus	Copulatory	1–6 days		Female remating(−)[b]	Carroll and Loye 1990
Jadera haematoloma	Copulatory	10 min–11 days	Sex ratio, genetic	Fertilization success(+)[a]; female remating(−)	Carroll 1988, 1991, 1993; Carroll and Corneli 1995

Lygaeus equestris	Copulatory		Sex ratio, egg load		Sillen-Tullberg 1981
Neacoryphus bicrucis	Copulatory		OSR		McLain 1989
Nezara viridula	Copulatory	6 h–7 days	OSR	Fertilization success(+)[a]	McLain 1980
Oncopeltus fasciatus	Copulatory		Time of day		Walker 1979
Parastrachia japonensis	Copulatory		OSR		Tsukamoto et al. 1994
Phyllomorpha laciniata	Copulatory				Kaitala and Miettinen 1997
Rhinocoris albopilosus	Post contact				Odiambo 1959
Gerridae					
Aquarius najas	Post contact	Weeks–months			Vepsäläinen and Nummelin 1985; Murray and Giller 1990
Aquarius paludum	Post contact	18.2 h		Mobility(−)[b]; harassment(−)	Amano and Hayashi 1998
Aquarius remigis	Copulatory	50–405 min	Hunger, predation risk, OSR	Fertilization success(+)?[a]; harassment(−)[b]; foraging(+)[b]; energetic expenditure(+); predation(+)	Rubenstein 1984; Wilcox 1984; Clark 1988; Rubenstein 1989; Sih et al. 1990; Wilcox and Di Stefano 1991; Fairbairn 1993; Krupa and Sih 1993; Weigensberg and Fairbairn 1994; Lauer et al. 1996; Watson et al. 1998
Eurymetra natalensis	Post noncontact				Nummelin 1988
Gerris buenoi	Post contact	22.5–83 min	Sex ratio; male and female size	Predation risk(+)[b]; foraging(−)	Rowe 1992, 1994; Rowe and Arnqvist 1996
Gerris elongatus	Post noncontact		Season; male size		Hayashi 1985

Table 5.2 (*cont.*)

Species	*Association*	*Duration*	*Cause of Variation*	*Effect on Male and/or Female Fitness*	*Source*
Gerris lacustris	Post contact	30 s–170 min	OSR; female size	Fertilization success(+)[a]	Jabłoński and Kaczanowski 1994; Jabłoński and Vepsäläinen 1995; Vepsäläinen and Savolainen 1995; Rowe and Arnqvist 1996
Gerris lateralis	Post contact	11 min to >48 h	Male and female size		Arnqvist 1988; Rowe and Arnqvist 1996
Gerris odontogaster	Post contact	40–90 min	Sex ratio; density		Arnqvist 1992a,b, 1997b
Limnoporus dissortis	Post contact	5–20 min		Oviposition(+)[b]	Spence and Wilcox 1986; Wilcox and Spence 1986
Limnoporus notabilis	Post contact	5–20 min		Oviposition(+)[b]	Spence and Wilcox 1986; Wilcox and Spence 1986
Limnoporus rufoscutellatus	Post contact	>1 h			Vepsäläinen and Nummelin 1985; Nummelin 1987; Arnqvist 1997b
Metrocoris histrio	Post contact				Koga and Hayashi 1993; Arnqvist 1997b
Microvelia austrina	Post contact	2–35 h	Hunger, predation risk	Fertilization success(+)[a]; female remating(−)	Travers and Sih 1991

Potamobates tridentatus	Post contact	Hours			Wheelwright and Wilkinson 1985
Ragadotarsus anomalus	Post noncontact				Wilcox 1972
Rhagadotarsus hutchinsoni	Post noncontact	42 s			Nummelin 1988
Tenagogonus albovittatus	Post contact				Nummelin 1988
COLEOPTERA					
Brentus anchorago	Pre and post noncontact				Johnson 1982
Chauliognathus pennsylvanicus	Post contact			Foraging(+)[b]	Mason 1980; McLain 1981
Chrysochus cobaltinus	Post contact			Female remating(−)[a]	Dickinson 1995
Cicindela aurofasciata	Post contact	17–50 min	Presence of rival		Shivashankar and Pearson 1994
Cicindela bicolor	Post contact	7–37 min	Presence of rival		Shivashankar and Pearson 1994
Cicindela calligramma	Post contact	62 min			Shivashankar and Pearson 1994
Cicindela catena	Post contact	22 min			Shivashankar and Pearson 1994
Cicindela fastidiosa	Post contact	28–36 min	Presence of rival		Shivashankar and Pearson 1994
Cicindela marutha	Post contact	683 min			Kraus and Lederhouse 1983
Dermestes maculatus	Post contact	Up to 60 min	OSR		Archer and Elgar 1999
Dytiscus alaskanus	Post contact				Aiken 1992
Ips pini	Post noncontact	4–6 weeks			Lissemore 1997
Labidomera clivicollis	Post contact	Up to 2.5 days		Fertilization success(+)[a]	Dickinson 1986, 1988
Leptinotarsa decemlineata	Post contact				Boiteau 1988

Table 5.2 (*cont.*)

Species	Association	Duration	Cause of Variation	Effect on Male and/or Female Fitness	Source
Monochamus scutellatus	Post contact	130–215 min	Male size		Hughes 1981; Hughes and Hughes 1982, 1985
Nemognatha nitidula	Post contact	0–8.4 min	Mating location		Brown and Stanford 1992
Odontota dorsalis	Post contact	5 min–64 h			Kirkendall 1984
Ontholestes cingulatus	Post noncontact	Up to 16 min		Female remating(−)[a]	Alcock 1991
Pachyta collaris	Post contact	Up to 20 h			Michelsen 1963
Rhytirrhinus surcoufi	Post contact	Up to 30 days			Dumont 1920
MEGALOPTERA					
Neochauliodes sinensis	Post contact	2.2 h		Insemination(+)[a]	Hayashi 1996
Parachauliodes continentalis	Post contact	6 h		Insemination(+)[a]	Hayashi 1996
Parachauliodes japonicus	Post contact	4.8 h		Insemination(+)[a]	Hayashi 1996
PLECOPTERA					
Pteronarcella badia	Copulatory	16–170 min	Sex ratio		Zeigler 1991
DIPTERA					
Antocha saxicola	Post noncontact	5.8 min			Adler and Adler 1991
Dactylolabis montana	Post noncontact	2.3 min			Adler and Adler 1991
Dryomyza anilis	Post contact	Up to 150 min	Female maturity	Male remating(+)[a]; fertilization success(+)	Otronen 1990

Limnonia simulans	Post noncontact	0.9 min			Adler and Adler 1991
Odontoloxozus longicornis	Post noncontact		Alternative strategies		Mangan 1979
Plecia nearctica	Post contact	3 days			Thornhill 1976a, 1980b
Scatophaga anilis	Post contact				Parker 1970d
Scatophaga stercoraria	Post contact	16.5 min		Female remating(−)[a]; fertilization success(+); harassment(−)[b]	Foster 1967b; Parker 1970d; Borgia 1981
HYMENOPTERA					
Aphytis melinus	Post contact			Fertilization success(+)[a]	Allen et al. 1994
Cotesia rubecula	Post noncontact			Female remating(−)[a]; fertilization success(+)	Field and Keller 1993
Nomadopsis puellae	Copulatory	1–30 min	Time of day		Rutowski and Alcock 1980
LEPIDOPTERA					
Danaeus plexippus	Copulatory	1–14 h up to 27 h in constant light	Time of day		Svärd and Wiklund 1988b
Leucinodes orbonalis	Copulatory	15–30 min			Srivastava and Srivastava 1957

Notes: Pre, post refer to the timing of mate guarding relative to copulation. Three forms of mating association are recognized: copulatory—males remain in genital contact with the female; contact—males break genital contact but remain mounted; noncontact—males dismount but remain in close proximity to the female. Causes of variation in the form of mate guarding or in the duration of guarding are noted.

[a]Affects male fitness.

[b]Affects female fitness.

+/− direction in which fitness is affected.

6 Copula Duration

6.1 Introduction

The primary function of copulation is the transfer of sperm and other seminal fluid products. All else being equal, copulation might be viewed as an act of mutual benefit to male and female. However, Parker (1970e) suggested that internal fertilization, and thus the act of copulation, may have arisen partly via sexual selection through sperm competition that favored males able to place their sperm closer to the site of egg maturation and release. The duration of copulation is often highly variable, both within and between species. In some insects, copulation appears to be extended beyond the time required for ejaculate transfer, and this behavior has been interpreted as a mechanism of mate guarding whereby males reduce the risk that their mates will copulate with rival males before the current clutch of eggs is laid (see section 5.3). In accord with the mate guarding hypothesis, variation in copula duration is sometimes associated with risk of rival copulation, for example, copulations are longer when the density of males is high or the operational sex ratio is biased toward an excess of males. However, as was noted in section 5.4, extended copulation may also function to increase the number of sperm and/or amount of seminal fluid transferred. As we shall see in chapter 7, increased risk of sperm competition can favor an increase in the number of sperm transferred by the copulating male (Parker et al. 1997) if this increases his probability of fertilization when in competition with the sperm of rival males. Thus, both the mate guarding and ejaculate transfer hypotheses predict variation in copula duration with male density and/or bias in the OSR. However, only the ejaculate transfer hypothesis predicts an association between copula duration and the proportion of offspring sired by the copulating male, a relationship for which there is some evidence (table 6.1). It is also clear that for some species much of the time spent in copula is spent manipulating rival sperm within the female's reproductive tract (chapter 3). Thus, variation in copula duration may result from variation in the extent to which males mechanically influence the degree to which their own sperm will be subject to competition with those of rivals, rather than variation in ejaculate transfer. For example, Waage (1984) noted that in general sperm removers tend to have rather longer copula durations (1–20 min) than do sperm repositioners (7–20 s). Moreover, for sperm removers longer copula durations are associated with the removal of greater numbers of sperm (e.g., Siva-Jothy 1987b; Siva-Jothy and Tsubaki 1989a). Whenever success

Table 6.1

Within-species variation in copula duration and its consequences for male fitness.

Species	Copula Duration	Cause of Variation	Male Fitness Consequences	Source
Odonata				
Ischnura elegans	25–191 min	Female mating history		Cooper et al. 1996a
Mnais pruinosa pruinosa	Trimodal, 55 s, 77 s, and 147 s	Alternative male tactics, location	Degree of sperm removal	Siva-Jothy and Tsubaki 1989a,b
Mnais pruinosa costalis	Trimodal, 59 s, 70 s and 139 s	Alternative male tactics, location		Watanabe and Taguchi 1990, 1997
Leucorrhinia intacta	Bimodal, 1–3 min or 3 to >11 min	Alternative male tactics, location	Fertilization success ↑	Wolf et al. 1989
Orthetrum cancellatum	Bimodal, 21 s or 894 s	Location	Degree of sperm removal	Siva-Jothy 1987b
Orthetrum chrysostigma	Bimodal, 1–2 min or 5 min–>1 h	Alternative male tactics		Miller 1983
Orthetrum coerulescens	1.5–13 min		Fertilization success ↑	Hadrys et al. 1993
Sympetrum danae	6 min–>1h	Time of day, male mating history	Fertilization success ↑	Michiels 1992
Sympetrum parvulum	1 min–>20 min	Alternative male tactic, male density		Ueda 1979
Orthoptera				
Decticus verrucivorus	2–9.5 min	Female weight, female mating history		Wedell 1992
Hemiptera				
Gerris lacustris	5–20 min	OSR		Vepsäläinen and Savolainen 1995

Table 6.1 (*cont.*)

Species	*Copula Duration*	*Cause of Variation*	*Male Fitness Consequences*	*Source*
Gerris lateralis	10–180 min	Male mating history	Sperm transfer and fertilization success ↑	Arnqvist and Danielsson 1999b
Gerris odontogaster	40–90 min	OSR, male density		Arnqvist 1992a, 1997b
Gerris remigis	Bimodal, 30 min or >4.5 h	OSR, predation risk, hunger	Fertilization success ↑	Clark 1988; Sih et al. 1990
Neacoryphus bicrucis	Bimodal, 1–8 h or 8–24h	OSR	Fertilization success ↑	McLain 1989
Thyanta pallidovirens	Bimodal, 7–8 h or >17 h	Female mating history		Wang and Millar 1997
Coleoptera				
Adala bipunctata	70–350 min		Sperm transfer and fertilization success ↑	Ransford 1997; de Jong et al. 1998
Harmonia axyridis	35–135 min		Fertilization success ↑	Ueno 1994
Mecoptera				
Bittacus apicalis	1–31 min	Nuptial prey size	Sperm transfer ↑	Thornhill 1976b
Harpobittacus nigriceps		Nuptial prey size	Sperm transfer ↑	Thornhill 1983
Panorpa penicillata	40 min–3 h	Nuptial prey quality	Sperm transfer ↑	Thornhill 1979
Panorpa vulgaris	c.5 min–>230 min	Number of nuptial salivary masses; functional notal organ	Sperm transfer and fertilization success ↑	Thornhill and Sauer 1991
Diptera				
Ceratitis capitata	90 min–5h	Male size and diet	Fertilization success ↑, female remating ↓	Saul et al. 1988; Taylor and Yuval 1999
Cyrtodiopsis whitei	Bimodal, 10 s or 50 s	Female mating history		Lorch et al. 1993

Drosophila mojavensis	60 s–480 s	Male size, female size, genetic	Early oviposition	Krebs 1991
Drosophila pseudoobscura	Average 6.2 min	Male and female mating histories	Unknown	Snook 1998
Drosophila persimilis	Average 7.9 min	Male and female mating histories	Unknown	Snook 1998
Drosophila affinis	Average 1.4 min	Male and female mating histories	Unknown	Snook 1998
Empis borealis	6–40 min	Nuptial prey size	Sperm transfer ↑	Svensson et al. 1990
Rhagoletis juglandis	Bimodal, <30 s or >10 min–1 h	OSR, location	Unknown	Alonso-Pimentel and Papaj 1996, 1999
Scatophaga stercoraria	2–179 min	Male size, female size, genetic, location, temperature, diet, male mating history, egg load	Fertilization success ↑	Parker 1970f; Ward and Simmons 1991; Parker and Simmons 1994; Mühlhäuser et al. 1996; Parker et al. 1999

in sperm competition is a positive function of copula duration we should expect selection to act on males to increase the amount of time they spend copulating. Nevertheless, selection pressures that oppose increased copula duration may also arise, for example, where increased copula duration reduces net male fitness via a reduction in the number of females inseminated. The observed copula duration should then represent a balance of selection pressures acting on males. Moreover, the copula duration that maximizes male fitness may not be in the female's best interests. Females are often able to realize their maximum fertilization capacity after a brief period of copulation during which only a small proportion of the total ejaculate is received; because of sperm competition males often transfer many more sperm than are required for fertilization alone (chapter 7; Eberhard 1996). Extended copulations may be costly for females, for example if copulation increases the female's risk of predation (Fairbairn 1993; Rowe 1994) or reduces the time available for oviposition (Sherman 1983). Alternatively, females could use copula duration as a mechanism for choosing among potential mates, either because longer copulations facilitate the receipt of larger ejaculates that outcompete sperm stored from previous males, or because longer copulations allow the assessment of males during copulatory courtship (Eberhard 1996). Selection on females to control copulation is likely to result in sexual conflict over copula duration so that the observed duration is a manifestation of both male and female interests. Here I review studies of variation in insect copula duration, the extent to which males and females control copula duration, and the role of sperm competition in its evolution.

6.2 Sperm Displacement: Optimal Copula Duration in Dung Flies

The most detailed analysis of copula duration in any insect must be that of the yellow dung fly *Scatophaga stercoraria*. Male flies gather at fresh droppings where they await the arrival of gravid females who come to lay their egg batch in the dropping. Males seize arriving females and fly with them to the surrounding grass to copulate (section 5.5). Virtually all arriving females have mated previously so that the sperm stores are filled to their maximum capacity (Parker et al. 1990). Studies of sperm utilization following multiple matings show that the proportion of the female's egg batch that is sired by the last male to copulate increases with exponentially diminishing returns as copula proceeds (Parker 1970f; fig. 2.6). The data conform to a model of sperm competition in which the male delivers his ejaculate directly into the female's sperm stores and there is a constant rate of displacement of sperm from the stores, coupled with instantaneous random mixing (see box 2.1). Hence, the longer the male copulates, the more likely he is to displace his own sperm, thereby slowing the rate of increase in his fertilization gain.

Mathematically, under direct displacement the fertilization gain per minute of copulation for the second male to mate can be expressed as

$$P_2(t) = 1 - \exp(-ct), \qquad [6.1]$$

where t is copula duration and c the constant rate of sperm displacement expressed as a proportion of the total sperm store displaced (Parker and Stuart 1976; Parker and Simmons 1991). We should expect selection to shape male copulation behavior in such a way as to maximize lifetime reproductive success. Parker and Stuart (1976) and Parker (1970f) adopted the marginal value theorem approach to predict the copula duration that would maximize male lifetime fitness; a male should stop copulating when the marginal gain from mating with the current female falls below that expected from finding and copulating with another female. The average time required to find a gravid female was estimated from field observations as 140 min (Parker 1970a). Because the density of searching males around the dropping is high, a male must guard the female while she lays her batch of eggs if he is to prevent her from copulating again with the subsequent loss of fertilizations (Parker 1970d). Males must guard for an average of 16.5 min so that the total time cost associated with finding and copulating with a female is 156.5 min. The optimal copula duration predicted by marginal value theorem can be obtained by drawing a tangent from the total time cost to the curve describing the relation between time spent copulating and fertilization gain under sperm competition (fig. 6.1). Mathematically, the optimal copula duration t^* can be approximated as

$$t^* \approx \left(\frac{1}{c}\right)\left(1 + \frac{1}{c\mu + 1}\right) \ln(c\mu + 1) \qquad [6.2]$$

(Stephens and Dunbar 1993). t^* thus depends only on the rate of sperm displacement c and the total time cost μ. The optimal copula duration predicted by marginal value theorem is approximately 42 min, close to the observed average copula duration of 35.5 min (Parker and Stuart 1976).

Since Parker and Stuart's (1976) original analysis, a number of additional factors have been examined that can potentially affect the observed copula duration. For example, 32% of pairs suffer takeovers from males searching on the dropping. Males that acquire females by takeover have a considerably shorter search time so that including this effect will reduce the average total time cost and hence the predicted optimal copula duration (Charnov and Parker 1995). The incorporation of time costs associated with feeding and the replenishment of sperm supplies also has the effect of decreasing the optimal copula duration (Ward and Simmons 1991; Parker 1992a). The gain rate with virgin females is considerably greater than with mated females because males do not need to displace sperm. Thus, theoretically the optimal copula duration with virgin females should be shorter (Parker et al. 1993). Any of these factors could in principle account for the fact that the original

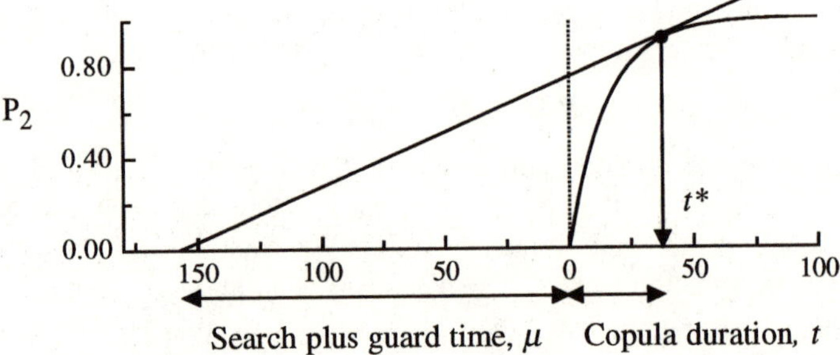

Figure 6.1 Graphical solution for the optimal copula duration in dungflies (modified from Parker and Stuart 1976). Sperm displacement conforms to a process in which there is random displacement of sperm from the female's sperm stores coupled with instantaneous random mixing. Fertilization gain (the proportion of eggs fertilized by the copulating male, P_2) rises with exponentially diminishing returns as copula proceeds. The empirically derived curve is described by the function $P_2 = 1 - \exp(-0.065t)$. The maximum rate of fitness gain for the copulating male is found by constructing a tangent to the gain curve from the total time cost (μ) of 156.5 min, associated with searching for and guarding a female. The copula duration at the maximum rate of fitness gain is equal to the optimal copula duration t^* and occurs at 35.5 min.

predicted optimal copula duration was higher than that observed, although the effect of virgins has been shown to be small because virtually all females arriving at the dropping are mated (Parker et al. 1993). However, the most significant refinements of the analysis have come from an incorporation of phenotypic differences between males.

Simmons and Parker (1992) examined variation in the rate of fitness gain across males of different sizes and found that large males had a higher constant rate of sperm displacement c than did small males. Previous analyses of optimal copula duration assumed all males to have the same rate of fitness gain, estimated from the sample of males in Parker's (1970f) original analysis. However, the fact that the curve of fertilization gain with time spent copulating rises more steeply for large males will have the effect of reducing their optimal copula duration (equation [6.2]). Because c shows continuous variation across the natural range of male sizes (fig. 6.2) we should expect copula duration to be negatively dependent on male size. However, there is a second reason why male size should influence copula duration. The probability of takeover decreases with increasing male size (Sigurjónsdóttir and Parker 1981) so that the interval between successive matings will be size dependent; the biggest males should experience about half the time costs, μ, of the smallest males who gain virtually no takeovers (Parker and Simmons 1994). Reducing μ should also reduce t^* (equation [6.2]). Indeed, observations have confirmed that copula duration decreases with male size (Ward and Simmons 1991; Simmons and Parker 1992; Parker and Simmons 1994). Considering these two effects, Parker and Simmons (1994) calculated phe-

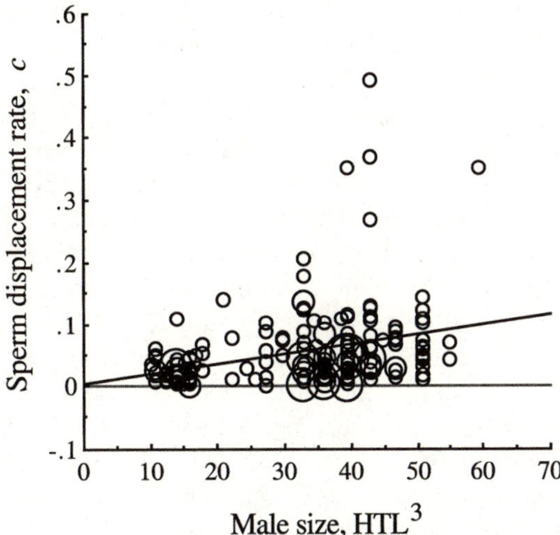

Figure 6.2 Relationship between the rate of sperm displacement and the size of male dung flies, measured as the length of the hind tibia cubed (HTL^3). Data were obtained from empirical studies of fertilization gain for males using the irradiated-male technique. Each male's displacement rate was calculated from his observed fertilization gain in a double mating, P_2, using a rearrangement of equation [6.1]: $c = [-\ln(1 - P_2)]/t$ (see Parker and Simmons 1994, 2000). Larger males were found to have a higher constant rate of sperm displacement than smaller males (see also Simmons and Parker 1992).

notypic profiles of optimal copula duration across the natural male size range. Of the two effects, the size-dependent rate of sperm displacement had the principal impact on the predicted optimal copula duration. Together, the two effects gave a reasonable fit to the observed copula duration for average sized males but the model predictions deteriorated at the small end of the phenotypic size distribution due to an overestimation of the predicted copula duration (see fig. 6.4 below).

The foregoing analyses assume that females have no influence on the optimal copula duration. Females arriving at the dung are able to deposit a fully fertile batch of eggs because of sperm stored from previous matings. Thus, copulation incurs an average time cost of around 35 min for each additional copulation the female allows. Therefore, selection might be expected to favor reduced copula duration for females so that the observed copula duration may also reflect female interests. Females have been shown to adopt violent rejection responses when mounted, wherein they sway from side to side at high frequency (Borgia 1981). However, this behavior appears to be more common among females who have completed oviposition and for whom copulation provides no benefit (Parker 1970c). Parker (1970c) examined the possibility that female behavior could influence copula duration by observing copulations involving living and dead females. He found that female behavior was important in curtailing copulation when females had already deposited their batch of eggs. However, for newly arriving females or those separated from males during copula but before oviposition, there were no significant differences in copula durations for living or dead females so that female behavior does not appear to be responsible for variation in copula duration with fully gravid females. A cost-benefit analysis taking into account the time required to successfully repel a male and the encounter rate with searching males at the dropping suggested that gravid females actually gain time by allowing copulation because the male's postcopulatory guarding phase effectively protects them from further male harassment (Parker 1970c,d). Thus, gravid females that accept copulation are at a selective advantage. The same is not true for females who have deposited their batch of eggs, because they leave the dropping to forage in areas away from male harassment and have nothing to gain from the male's guarding phase.

Although females may exert no direct influence on copula duration, aspects of female morphology may exert indirect selective influences. The model of constant random sperm displacement (equation [6.1]) assumes that sperm are transferred at a constant rate directly into the site of storage and utilization. However, knowledge of the morphology of the female's reproductive tract precludes such a mechanism. The female has three (sometimes four) chitinous spermathecae, each connected to the bursa copulatrix by a narrow spermathecal duct (fig. 1.2). The aedeagus appears to have no direct access to the spermathecae but rather, as is typical for Diptera (Lachmann 1997), aligns with the openings of the spermathecal ducts during ejaculation (Hosken et al. 1999; Hosken and Ward 2000). Sperm are thus transferred to the bursa copulatrix and appear to be exchanged between the bursa and the spermathecae via muscular contractions of the spermathecal ducts (Simmons et al. 1999b; Hosken and Ward 2000). Simmons et al. (1999b) developed an indirect model of sperm displacement that took into account the rate of ejaculate exchange which, unlike the rate of sperm displacement in the direct model of equation [6.1], is a property of the female.

Simmons et al. (1999b) envisaged a system in which, during copulation, sperm are ejaculated into the bursa at a constant rate, z sperm per minute, and exchanged between the spermathecae and the bursa by the female at a constant rate, a sperm per minute (fig. 6.3). At time t, z sperm from the copulating male are delivered into the bursa and a random sample of $+a$ sperm are shunted from the bursa to the spermathecae. Following random instantaneous mixing $-a$ sperm are then transported back to the bursa. Thus, during each time unit of a male's copulation, a proportion $c' = a/S_{max}$ (where S_{max} is the maximum number of sperm stored in the female's sperm stores) of the sperm store is lost and replaced by input of sperm from the bursa. This cycle of exchange of sperm between spermathecae and bursa is repeated throughout copulation. Sperm entering the sperm store are always a mixture of the copulating male's sperm and rival sperm, unlike the direct

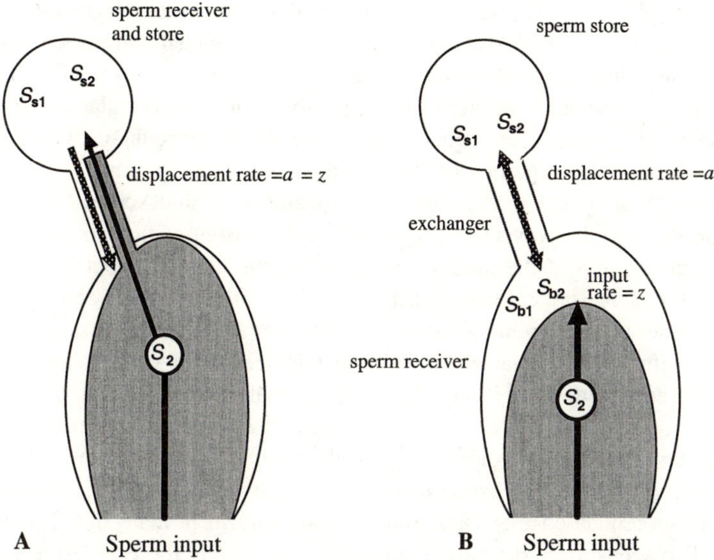

Figure 6.3 Schematic representation of direct (A) and indirect (B) sperm displacement in yellow dung flies *Scatophaga stercoraria*. With direct displacement (A) the male delivers his ejaculate directly into the spermathecae or sperm stores (females generally have three [fig. 1.3] although only one is depicted here) so that the rate of sperm transfer by males, z, equals the rate of sperm entry to the stores, a. At any time the sperm stores contain a proportion S_{s1} of male 1 sperm and a proportion S_{s2} of male 2 sperm. As male 2 sperm enter the sperm stores, there is instantaneous mixing and an equal proportion of the sperm stores are displaced to the bursa copulatrix. With indirect displacement (B) the male ejaculates into the bursa copulatrix or sperm receiver at constant rate z. The total bursal sperm consists of a proportion S_{b1} and S_{b2} from male 1 and male 2, respectively. Sperm are exchanged between the sperm stores and the receiver by the spermathecal ducts at a rate a. Again there is instantaneous mixing of sperm in the sperm stores, and an equal proportion of sperm then return to the sperm receiver. See text for more details. (After Simmons et al. 1999b)

model of equation [6.1], which assumes that sperm entering the spermathecae are pure sperm from the copulating male. The predicted relationship between fertilizations gained and time in copula under the indirect model of Simmons et al. (1999b) concurs very closely with that of the direct model because the number of sperm ejaculated into the bursa is very high relative to the number of rival sperm exchanged from the spermathecae; essentially sperm returning to the spermathecae are predominantly those of the copulating male.

Although incorporating female processes into the sperm displacement model has little effect on the average predicted fertilization gain for males, it does have a significant effect on the fit between observed and predicted copula durations. Parker and Simmons (2000) used their indirect model of sperm displacement to examine the mechanism by which larger males obtained a higher rate of fertilization gain. They recognized two possible limits for the way that sperm flow rates might scale with male body size. Sperm are typically propelled through the ejaculatory duct by muscular contraction of the male's reproductive tract (Davey 1985b). If musculature is independent of male size, sperm outflow should be dictated purely by the cross sectional area of the ejaculatory duct apparatus so that sperm flow rate z should scale to the power 2 with some linear measure of male body size. At the other extreme, if the mass of ejaculatory musculature and corresponding ejaculatory force increases in direct proportion with male body size, z should scale to the power 3 with some linear measure of male body size. As in the model with direct displacement, P_2 under indirect displacement should rise more steeply with time in copula for large males. In the direct model, the rate of sperm displacement $c = z/S_{max}$, so that P_2 rises more steeply simply because z is higher (equation [6.1]). In the indirect model, however, the rate of sperm displacement, distinguished as c', is now equal to a/S_{max}, the product of two female characteristics and so independent of male size. Nevertheless, displacement of rival spermathecal sperm S_{s1} by the copulating male's sperm S_{s2} is achieved by the bursal sperm S_{b2} component of the exchanged sperm a (fig. 6.3), which is male size dependent; when z is higher the number of S_{b2} sperm in the bursa relative to S_{b1} sperm is higher, thereby making the S_{b1} component of exchanged sperm a more dilute. As z becomes very large compared with a, S_{b1} relative to S_{b2} becomes negligible, so that virtually pure S_{b2} is exchanged and displacement will no longer increase with male size but will be limited only by the female's exchange mechanism. From marginal value theorem the optimal copula duration occurs when $P_2'(t)$ (where the prime denotes the differential of P_2 with respect to t) equals the maximum rate of gain obtainable from the habitat (Charnov 1976; Parker and Stuart, 1976). There is no explicit formulation for the maximum rate of gain under indirect displacement so Parker and Simmons (2000) determined $P_2'(t)$ by computer simulation using a value of the exchange rate (77 sperm min^{-1}) estimated from the observed c' of 0.062 and S_{max} of 851 sperm. For both scaling relationships, the predicted optimal copula duration declined with male size, as observed, with the smallest males having a higher predicted optimal copula duration if z scales with exponent 3 than if it scales with exponent 2 (fig. 6.4). An examination of the residual values of observed minus expected copula durations in fig. 6.4 revealed that the phenotypic profiles of optimal copula duration based on the indirect model of sperm displacement provided much better fits to the observed data than did the profile obtained from the direct displacement model. Although both slightly underestimate the true copula duration, the indirect model with scaling relation 3 provided the best fit (Parker and Simmons 2000). Thus, including a female-mediated process into the mechanism of sperm displacement greatly

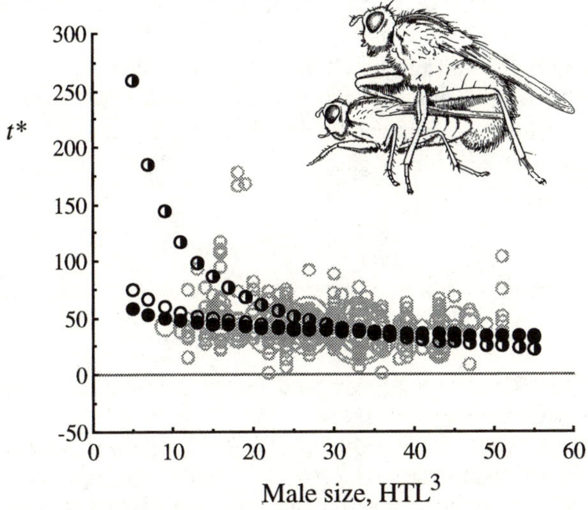

Figure 6.4 The relationship between male body size and copula duration in yellow dung flies. The observed relationship is negative and highly significant, and the data are depicted by the shaded symbols. Superimposed upon the observed data are the phenotypic profiles of optimal copula duration using two models of sperm displacement. The direct model assumes that males transfer sperm directly to and displace sperm directly from the sperm storage organs of the female, in the case of dung flies the spermathecae. The predicted optimal copula duration (shown as half-filled circles) is close to that observed for average-sized males, and the profile fits reasonably well across most of the normal male size range. However the model's prediction deteriorates toward the small end of the male size distribution. On average, the direct model overestimates the copula duration by some 6 min. The indirect model provides a much closer fit to the observed data. Here sperm are transferred to the bursa, and exchange occurs between the spermathecae and the bursa by a female exchange mechanism. Two forms of the model are shown, one in which male sperm delivery rate scales with the power 2 (solid circles) and one in which it scales to the power 3 (open circles). The latter provides the best fit to the observed data. Both indirect displacement profiles fit the observations for small males much better than does the direct displacement profile and remove the obvious disparity between observed and predicted optimal copula durations in dungflies. (From Parker and Simmons 2000)

improved the power of the marginal value theorem approach in predicting male copula behavior.

Charnov and Parker (1995) identified two further female parameters that could potentially influence the optimal copula duration. Larger females generally contain more eggs (Parker 1970f; Borgia 1981) so that the absolute fitness gain in terms of eggs fertilized, G, should be greater with larger females. Moreover, the rate of fertilization gain through sperm displacement under both direct and indirect models depends on the maximum sperm storage capacity of females, S_{max}. An increase in sperm storage volume with female size would mean that males should experience a slower rate of sperm displacement with larger females. Indeed, Parker et al. (1999) have since found a positive relationship between female size and spermathecal volume and a negative relationship between female size and rate of sperm displacement. Given that the optimal copula duration with a female of size H_i occurs at the maximum rate of fitness gain per male per minute during mate searching, R_{max}, where

$$R_{max} = G_i P'_{2i}(t^*) \quad [6.3]$$

(e.g., see Parker et al. 1999), combining equations [6.1] and [6.3] yields

$$R_{max} = G_i c_i \exp(-c_i t^*) \quad [6.4]$$

so that if either G_i or c_i increases with female size the result should be an increase in t^* with larger females.

Parker et al. (1999) used both the direct and indirect models of sperm displacement and their empirically derived relationships between sperm storage capacity, egg number, and female size to predict the optimal copula duration for males of fixed size. Both direct and indirect models yielded the predicted increase in t^* with female size. Parker et al. (1999) explored two forms of the indirect displacement model, one in which the female's exchange rate a was allowed to scale in a manner identical to the scaling of sperm storage volume and one in which a was fixed at the average value estimated by Simmons et al. (1999b). Interestingly, observations of copulating flies showed that copula duration did increase with female size, and the scaled-a version of the indirect model provided an almost perfect fit to the observations (fig. 6.5). Again, taking female involvement into account provided better predictive power in the analysis of copula duration. Parker et al. (1999) also examined the relative impact of size dependent variation in egg content and sperm storage capacity in determining copula duration by holding constant each parameter in the scaled-a version of the indirect displacement model. When egg content was fixed at its average value, the dependence of copula duration on female size virtually disappeared (fig. 6.6). In contrast, when sperm storage was fixed at its average value and only egg content allowed to vary, the relation between t^* and female size was much closer to the full model prediction (fig. 6.6). Thus a male's copula duration is more strongly influenced by the effect of female reproductive value, in terms of potential eggs gained, than by the change in spermathecal volume. This may explain why males adjust their copula duration to egg content alone when copulating with females after takeover (Parker et al. 1999).

One major assumption behind the marginal value theorem approach to modeling adaptation in male behavior is that there is genetic variation in copula duration upon which natural selection can act. Using a full sib design, Mühlhäuser et al. (1996) found significant variation in copula duration due to male body size, and significant variation both within males (from which

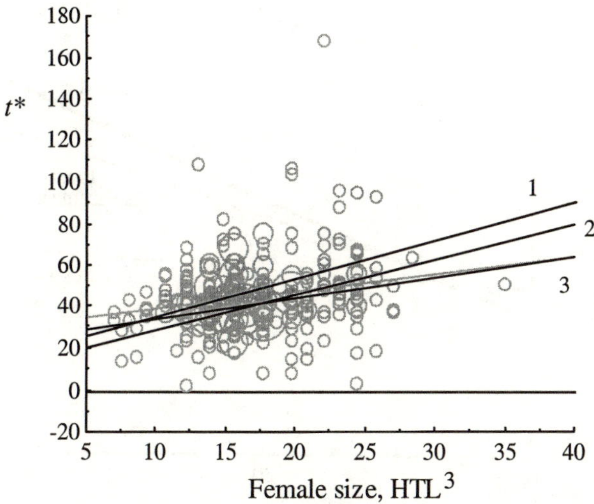

Figure 6.5 The relationship between female body size and copula duration in yellow dung flies. The observed copula durations are significantly positively related to female body size and illustrated as shaded circles. Also shown are the predicted relationships between optimal copula duration and female body size based on two models of sperm displacement. All models predict a relationship between copula duration and female size. With direct displacement (1) the optimal copula duration is overestimated across the entire female size range. Indirect models (2 and 3) are both closer to the observed relationship. When female exchange rate *a* between bursa and spermathecae is fixed at the observed average of 77 sperm min^{-1}, the predicted copula duration is also overestimated. However, when *a* is scaled with female body size in a manner identical to the scaling of spermathecal volume (3), the fit to the observed relationship is almost exact. (From Parker et al. 1999)

they calculated an index of flexibility in copula duration) and between families. Their analysis revealed that between 33% and 39% of the phenotypic variation in copula duration was due to additive genetic variance. This level of heritable variation is in general agreement with quantitative genetic studies of behavior (Mousseau and Roff 1987; Falconer and Mackay 1996) and similar in magnitude to the heritability of copula duration in *Drosophila melanogaster* (Gromko et al. 1991). Mühlhäuser et al. (1996) also examined the phenotypic and genetic correlations between copula duration and body size. Although the genetic correlation between body size and copula duration was itself not significant, there were significant phenotypic and genotypic correlations between the within-male flexibility in copula duration and male body size. The quantitative genetic data for dung flies thus confirm the potential for natural selection to shape phenotype-dependent copula duration in an adaptive manner.

The analysis of copula duration in yellow dung flies shows how selection can shape male behavior in an adaptive manner. While female influences are

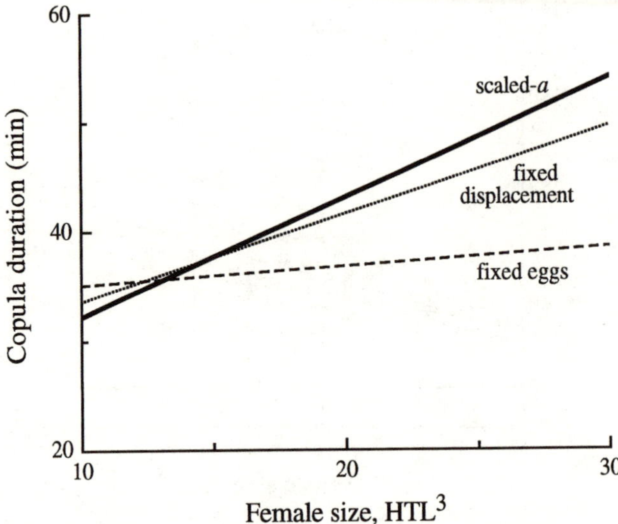

Figure 6.6 The relative effects of variation in female size-dependent fecundity and sperm displacement rate on the predicted relationship between optimal copula duration and female size under the scaled-*a* version of indirect sperm displacement in yellow dung flies. When sperm displacement rate is fixed to the population average, the relationship changes very little from the one including both egg number and displacement parameters. However, when egg number is fixed and only sperm displacement rate is allowed to vary with female size, the relationship between copula duration and female size virtually disappears. This suggests that male fitness gain in terms of absolute eggs fertilized is the more important selective agent in shaping male copula duration to female size. (From Parker et al. 1999)

clearly important selective pressures in shaping male behavior, the ultimate factor responsible for observed male behavior appears to be sperm displacement, and the maximization of male fitness gain in the face of sperm competition.

6.3 Copula Duration with Sperm Mixing

The optimality models used in the analysis of dung fly copula duration were designed specifically for an insect in which there is constant random sperm displacement, either direct or indirect, with instantaneous mixing. Nuyts and Michiels (1993) presented a general optimization approach in which they used game theoretical modeling to examined the impact of long-term sperm mixing and subsequent changes in sperm precedence patterns on insect copula duration. For example it is often the case that sperm do not mix in the female's sperm stores until some time after copulation has terminated, so that reductions in fertilization success due to sperm competition are not realized by males in ovipositions occurring immediately following copulation, but later in the female's reproductive life (e.g., Schlager 1960; Siva-Jothy and Tsubaki 1989a, 1994; see review in table 2.3). The two approaches differ in that the displacement models discussed in section 6.2 optimize male behavior for a single mating or egg batch, while the sperm mixing models optimize male behavior over a number of successive clutches. Thus, the male's gain in the mixing models is divided into immediate fertilization (at the current batch) $P_2(t)$ and a factor $P_2(t,h)$ (in future batches) representing the change in P_2 over a period of time h since copulation of duration t. Longer copula durations are assumed to result in higher $P_2(t,h)$. If eggs are not laid at the time of copulation, then only the factor $P_2(t,h)$ should be relevant. Nuyts and Michiels' (1993) models generated a number of predictions that were consistent with those generated from the sperm displacement models used to analyze dung fly copula duration. In general, long-term mixing of sperm within the female's sperm storage organs should favor increased copula duration if males optimize over more than one bout of oviposition. This is because males with longer copulations will transfer more sperm and thus have a higher fitness gain in future egg batches than males with shorter copulations. Thus, assuming that males respond to the population average for the parameter considered, with long-term sperm mixing and optimization over multiple batches, (i) a greater probability of takeover either during or after oviposition (risk of future sperm competition) should have the effect of increasing optimal copula duration; (ii) increasing the interval between copulation and oviposition should increase optimal copula duration; (iii) increasing search time should increase optimal copula duration; and (iv) the time and energy costs associated with reestablishment of territories should act to decrease the optimal copula duration because cost should increase with time in copula. When individuals can adjust their copula duration to local conditions, they are predicted to (i) decrease the optimal copula duration in response to increased risk of takeover; and (ii) increase the optimal copula duration with increased clutch size and thus fitness gain (Nuyts and Michiels 1993).

6.4 Evidence Consistent with an Optimization of Copula Duration

Nuyts and Michiels (1994) searched for qualitative fits to the predictions of their optimization models in empirical data from odonates. The odonates seem particularly relevant to models based on long-term sperm mixing. In damselflies males first remove rival sperm from the female's sperm storage organs and then deliver their own ejaculate (e.g., Waage 1979b; Siva-Jothy and Tsubaki 1989a; see section 3.3). The majority of time spent in copula represents the sperm removal phase. Cooper et al.'s (1996a) study of *Isch-*

nura elegans showed that copula duration was shorter with virgin females than with mated females, probably due to the fact that males did not need to remove sperm from virgins before delivering their own ejaculate. For previously mated females, longer copulations have been shown to result in greater proportions of rival sperm being removed (Siva-Jothy and Tsubaki 1989a; Siva-Jothy and Hooper 1995). Moreover, although males are able to fertilize virtually all eggs deposited immediately following copulation, increasing the interval between copulation and oviposition allows sperm to mix in storage and reduces the copulating male's fertilization success to a level that is dependent on the amount of sperm removed from storage during copulation (e.g. Siva-Jothy and Tsubaki 1989a; Cordero and Miller 1992). In dragonflies, males also reposition rival sperm so that immediately following copulation their own sperm obtain the majority of fertilizations. As with damselflies, fertilization success declines with the interval between copulation and oviposition due to long-term sperm mixing (McVey and Smittle 1984; Hadrys et al. 1993; Siva-Jothy and Tsubaki 1994). Thus the odonates fulfill many of the assumptions of Nuyts and Michiels' (1993) theoretical models.

There is considerable variation in copula duration within species of damselflies and dragonflies that appears to depend largely on the location of mating and/or the mate-securing tactic adopted by males (table 6.1). Territorial males typically have higher encounter rates with females and are less subject to takeover attempts than so-called wandering males who are not territorial and/or search for females away from the site of oviposition (Ueda 1979; Miller 1983; Siva-Jothy 1987b). Moreover, because territories are generally established at the site of oviposition, females encountered on territories are more likely to be ready to deposit a batch of eggs than those encountered at feeding sites, so that the period between copulation and oviposition will be shorter (Siva-Jothy 1987b). Despite the obvious benefits associated with territorial defense, establishing and maintaining territories is costly for males (Marden and Waage 1990; Plaistow and Siva-Jothy 1996). Each of these parameters is predicted to reduce optimal copula duration (Nuyts and Michiels 1993) and, in general, territorial male odonates have been shown to copulate for less time than males adopting alternative mate-securing tactics (table 6.1).

In their detailed study of copula duration in *Mnais pruinosa pruinosa*, Siva-Jothy and Tsubaki (1989a,b) identified three discrete mating tactics each characterized by a different duration of copula. Territorial males defend small pieces of dead wood that females utilize as oviposition sites, and copulate with females as they enter their territory boundaries. Sneak males perch outside territories and copulate with ovipositing females while owners are occupied with other females or in disputes with other males. Opportunistic or wandering males encounter females at random, often at feeding sites away from suitable areas for oviposition. As generally observed, territorial males had a higher encounter rate with females than did sneaks, while opportunistic males had the lowest encounter rate. Moreover, the probability of immediate oviposition was close to 1 for territorial males and sneaks that copulate at the oviposition site, but just 0.16 for opportunistic males. As expected sneak males had the shortest copulations while those of territorial and wanderer males were some 40% and 63% longer, respectively (Siva-Jothy and Tsubaki 1989a,b). These data are concordant with the predictions of Nuyts and Michiels' (1993) models in that the higher encounter rates of territorial males and the shorter period between copulation and oviposition are both predicted to reduce the optimal copula duration. Moreover, in the closely related species *M. p. costalis*, there is within-tactic variation in copula duration; sneaks copulating on the territory have 33% shorter copulations than do sneaks copulating outside the territory. Wanderer males have the longest copula durations and territorial males the shortest (Watanabe and Taguchi 1990, 1997). Sneaks who copulate within a male's territory are likely to have an increased probability of being detected and thus suffer a greater risk of takeover. The shorter copula duration of sneaks on territories is congruent with the within-male prediction of Nuyts and Michiels (1993, 1994). However, that sneaks in general have the shortest copula duration is counter to Nuyts and Michiels (1993, 1994) between-male prediction. Siva-Jothy and Tsubaki (1989b) performed a series of experimental manipulations which allow us to establish the extent to which each parameter is associated with copula duration in *M. p. pruinosa*. Experimentally varying the encounter rate with females or the risk of takeover by rival males had no influence on copula duration. Neither did variation in female fecundity. The factor most important appeared to be the site of female capture; males capturing females on a territory had shorter copula durations than males capturing females in nearby vegetation (fig. 6.7). Two aspects of site of capture could theoretically reduce optimal copula duration; the shorter time from copulation to oviposition on territories and the costs associated with territory defense. Unfortunately it is not yet possible to judge the relative importance of these parameters in *M. p. pruinosa*. Nevertheless, variation in copula duration appears to represent an adaptive male response and one not influenced by females because decapitation of females in copula has no impact on copula duration, but decapitation of males does (see also Miller 1987a for experimental data on *Ischnura elegans*; Siva-Jothy and Tsubaki 1989b).

Location of copulation, or more specifically the presence of oviposition substrate, has also been shown to influence copula duration in dung flies (Ward and Simmons 1991) and fruit flies *Rhagoletis juglandis* (Alonso-Pimentel and Papaj 1999). Parker (1992a) extended his optimality analysis to examine the possible factors that might contribute to the extended copula duration observed in matings away from cattle droppings. So-called extra-dung matings must be common in dung flies because less than 3% of females arriving at the dung are virgins (Parker et al. 1993). Nuyts and Michiels' (1993) sperm mixing models would predict that the delay between

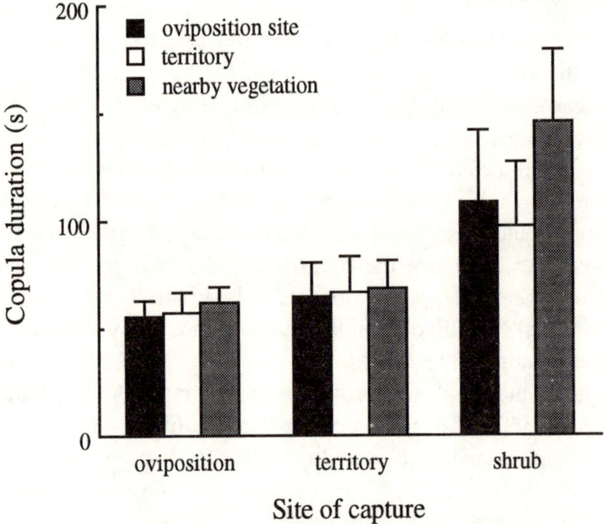

Figure 6.7 Variation in copula duration in the damselfly *Mnais pruinosa pruinosa*. Tethered females were presented to territorial males either at the oviposition site within their territories, perched away from the oviposition site but within the territory boundaries, or in nearby vegetation outside the male's territory. After capture, pairs were coaxed via the female's tether to a different site for copulation. The experiment thus examined the effects of both site of capture and site of copulation on resultant copula duration. Only site of capture influenced male behavior, with copula duration being longer when females were captured outside a male's territory. The effect of site of copulation was marginally short of significance, and there was no interaction. The data suggest that males may assess female intention to oviposit immediately after copulation from the location in which they are captured and modify their copula duration accordingly. (From Siva-Jothy and Tsubaki 1989b)

copulation and oviposition resulting from extradung matings should act to increase copula duration. For dung flies, however, virtually all females copulate on arrival at fresh dung so that males copulating away from the dung should expect a greatly reduced fertilization success because of sperm displacement when the female goes to the dropping to lay her next batch of eggs. Parker (1992a) adopted an evolutionarily stable strategy approach, offering two scenarios for the optimization of copula duration in extradung matings. In the first he assumed that males adopted discrete alternative tactics of either searching for extradung matings or searching for females on the dung. The proportion of an extradung mater's sperm expected to be displaced before the next oviposition was actually found to be irrelevant, so that any differences in the optimal copula duration for males adopting alternative tactics related entirely to differences in the cycle time (time spent searching for and copulating with a female). Since the search time away from droppings is likely to be very much greater than the search time on the dropping, the expectation was that extradung matings should be longer than matings on the dung, as observed (Ward and Simmons 1991). Using stochastic dynamic programming, Nuyts (1994) independently modeled the copula duration of dung flies, optimizing over a single day. He reached the same conclusion, that males remaining away from the dung for the entire day should copulate for longer than males searching on the dung. In his second scenario, Parker (1992a) assumed that males typically searched in both habitats and performed both extradung and dung matings. Here he found that males should spend less time in each extradung copulation because the gain is less, due to the inevitable displacement of sperm, than for their matings at the dropping. Although the observed data best fit the model in which males concentrate either on extradung or dung matings, it is not yet known if males indeed adopt such discrete mate-searching tactics.

In their study of fruit flies, Alonso-Pimentel and Papaj (1999) similarly found that copulations on fruit (the oviposition site) were shorter than copulations away from fruit, evidence that could be viewed as supporting sperm competition theory. However, Alonso-Pimentel and Papaj (1999) quite rightly erred on the cautious side in drawing conclusions from their observation; although casual field observations suggest that encounter rates are higher for males on fruit there are no quantitative data available, and the mechanism of sperm storage and utilization is not known for this species. Assuming that the sperm utilization patterns of *R. juglandis* are similar to those of its congener *R. pomonella*, the last male to copulate fertilizes more than 80% of eggs deposited (Myers et al. 1976). However, fertilization gain is not related to copula duration (Opp et al. 1990) making the sperm displacement models of Parker and his coworkers inappropriate. Neither does fertilization gain change with long-term sperm mixing (Opp et al. 1990) so that the models of Nuyts and Michiels (1993) are probably also inappropriate for predicting copula duration. Alonso-Pimentel and Papaj (1999) suggest that females may be more receptive on fruit than away from fruit, which they argued could result in shorter copulations. The alternative argument, longer copulations with more receptive females, seems equally plausible. Operational sex ratio was also found to have a major, additive effect on copula duration; copulations in a male-biased operational sex ratio were longer. In a previous study Alonso-Pimentel and Papaj (1996) examined the effect of operational sex ratio in detail, demonstrating that increased density of males had a greater influence on copula duration than did increased density of females. The data supported the notion that risk of sperm competition favored increased copula duration by males. In natural populations operational sex ratios in the vicinity of fruit tend to be male biased, so that the presence of fruit may be used by males as a proximate cue for assessing the potential risk of sperm competition. Moreover, if fertilization gain is not related to copula duration, the extended copulations under conditions of sperm competition risk may represent copulatory mate guarding (section 5.3), rather than increased ejaculate transfer.

The problem of distinguishing between male and female control over copulation duration is a difficult one. An elegant approach to the problem was that of Vepsäläinen and Savolainen (1995) discussed in section 5.3. Through experimental manipulations of the sex ratios experienced by males and females prior to and during mating activity, they were able to show that increased postcopulatory mate guarding in male-biased populations of the water strider *G. lacustris* was a male-controlled phenomenon rather than a female-controlled one. Likewise, copula duration in water striders often varies with operational sex ratio, with copulations lasting longer in male-biased than in female-biased populations (table 6.1; Arnqvist 1997b). Vepsäläinen and Savolainen's (1995) study demonstrated that, as with changes in postcopulatory mate guarding, increased copula duration in male-biased sex ratios was due primarily to changes in male rather than in female behavior. Sperm mixing appears to be the rule in those water striders studied so far (Arnqvist 1988; Rubenstein 1989; Arnqvist and Danielsson 1999a; Danielsson and Askenmo 1999), with increased copula duration resulting in increased fertilization success (Rubenstein 1989). Given the increased risk of takeover associated with male-biased sex ratios (Arnqvist 1992a; Rowe 1992; Krupa and Sih 1993) the increased copula duration is consistent with Nuyts and Michiels' (1993) models.

6.5 Female Influences

Despite the limited evidence for female influence over copula duration in the examples discussed above, the interests of females in mating cannot be ignored. There are a number of studies in which females have been shown to exert considerable influence over copula duration. In his study of the cactophilic fruit fly *Drosophila mojavensis*, Krebs (1991) partitioned both phenotypic and genetic variation in copula duration between males and females by examining interactions between individuals of different strains obtained from different geographic regions. There was significant between-strain variation in copula duration. Crosses between males and females from different strains revealed that males had the predominant influence on copula duration since the copula duration observed was more similar to that of the strain from which the male was derived. Nevertheless, female origin also had a significant influence on copula duration. There was no interaction between male and female origin, suggesting that behavioral interactions between the sexes did not account for the sex of origin result. Variation in copula duration due to males was some five times greater than variation due to females. Observations of copula duration for hybrid males and females indicated that copula duration was inherited additively for both sexes, and that for males there was probably a large autosomal influence. Krebs (1991) also examined intrastrain variation in copula duration, finding that large males had longer copula durations than did small males, and that copula duration increased with female size, a pattern reminiscent of that found in *Scatophaga stercoraria* (see section 6.2). Longer copulations were found to result in an increased probability that females would oviposit before remating, although the time to remating was unaffected. Thus, from the male's perspective, selection should favor increased copula duration since it would increase the probability of siring the female's clutch, or 66% of it where females have mated previously (Markow 1988). Moreover, since long-term sperm mixing decreases fertilization success (Markow 1988) we might expect optimization by males to favor long copula durations (Nuyts and Michiels 1993). Sexual conflict over male adaptations to sperm competition are known from the congener *D. melanogaster* (Chapman et al. 1995; Rice 1996) and a number of other Diptera (Chapman et al. 1998). The seminal products responsible for changes in female reproductive physiology reduce female survival (see section 4.3), so that selection on females might be expected to favor short copulations, perhaps accounting for the observed male and female influences over copula duration in *D. mojavensis* (Krebs 1991).

In some species females can have greater control over copula duration than do males and can influence both the number of sperm transferred and male fertilization success (Linley 1975; Linley and Hinds 1975b). This is particularly clear in species where males feed their mates during copulation (table 6.1). For example, in his now classic studies of hanging flies, Mecoptera, Thornhill (1976b, 1979, 1983, 1984a) showed that copula duration increases with the size of prey item offered by the male (fig. 6.8). Females terminate copulations sooner with males offering small prey items than with those offering large prey items. Short copulations result in the transfer of fewer sperm. Moreover, females retain sexual receptivity and deposit fewer eggs when copulations are short. Because females are likely to gain significant nutritional and survival benefits from male donations, Thornhill suggested that nuptial feeding behavior has evolved via female choice for male investment; males offering large prey items gain more fertilizations than those offering small prey items. A consequence of selection on males via female choice is that males have evolved a mechanism of prey assessment in which they will consume small prey items themselves, but retain large prey items for mating. Nevertheless, despite female control there does appear to be a copula duration that is optimal for the male. Males with large prey items will terminate copulations after 20 min, seize back the item of prey from the female, and use it to copulate with a second female (Thornhill 1976b). Thus, although copula duration is strongly influenced by females, male interests are also involved. In general, there appears to be considerable sexual conflict over copula duration in this group of insects (for further discussion see chapter 9).

One final example of conflict over copula duration comes from the bean weevil *Callosobruchus maculatus*. Here sclerotized spines on the tip of the male aedeagus puncture the female's internal genital tract, and the damage

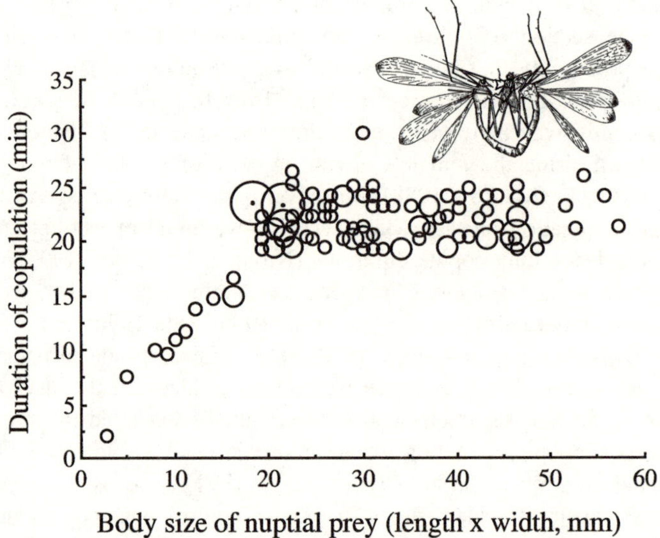

Figure 6.8 Female control over copula duration in the hanging fly *Hylobittacus apicalis*. Males use prey items of large size for attracting and mating with females. Females feed on male donations during copulation and will terminate copulations earlier with males that offer small prey items than with those that offer large prey items. Variation in copula duration directly influences the number of sperm transferred by males and the probability that females will deposit eggs before remating. Males with prey items larger than 20 mm^2 are able to copulate for 20 min. Moreover, males will seize back the prey item and use it to copulate with additional females, suggesting that a copulation of 20 min serves male interests. (From Thornhill 1976b)

can decrease female lifespan (Crudgington and Siva-Jothy 2000). Although the adaptive significance of male morphology and female damage is as yet unclear, Crudgington and Siva-Jothy (2000) were able to show that female kicking behavior toward the end of copulation was effective in reducing the length of copula by some 32%, and that females who terminated copula sooner through kicking sustained less genital tract damage than those prevented from kicking. This example shows how costs of mating for females are important in shaping the behavior we observe.

6.6 Alternative Explanations

Some of the studies in table 6.1 reiterate the need to consider alternative explanations for observed male behavior (Thornhill 1984b). Copula duration in stalk-eyed flies *Cyrtodiopsis whitei* shows bimodal variation with peaks at 10 s and 50 s (Lorch et al. 1993). This variation is also associated with variation in the numbers of sperm transferred and male fertilization success;

copulations shorter than 40 s do not result in the transfer of sperm. It is unlikely that males optimize their fitness gain with a female by failing to inseminate her. Rather, it seems that short copulations represent failed copulation attempts due to the presence of a mating plug, in this case the previous males' spermatophore, in the female's reproductive tract (Lorch et al. 1993).

Variation in copula duration can also be a manifestation of cryptic male mate choice. Copula durations in stink bugs *Thyanta pallidovirens* are either short or long depending on the female's previous mating history (Wang and Millar 1997). When males copulate with virgin females they do so for longer than 17 h and transfer some 17% of their body weight in ejaculate. However, males terminate copulations with nonvirgin females after 8 h, apparently without transferring an ejaculate. Moreover, males also terminated copulations with small females. It would appear that male seminal products contribute to female fecundity in this species so that males may preferentially inseminate larger virgin females because they will provide a greater reproductive return for their investment (see section 5.6). A similar preference is seen in the tettigoniid *Decticus verrucivorus*, where males spend longer in copula with nonvirgin females, and females of low body weight and fecundity, before providing them with a smaller spermatophore (Wedell 1992). These examples show that, while males appear to behave in a manner that would maximize their reproductive success, the copula durations observed are not a simple product of optimization in relation to fertilization gain under sperm competition.

6.7 Summary

The duration of copulation is often positively associated with the number of sperm transferred and/or with the degree of sperm removal and/or repositioning. Where sperm transfer and/or manipulation of rival ejaculates influences fertilization success, all else being equal selection via sperm competition should favor an increase in copula duration. Detailed theoretical and empirical analyses of copula duration in yellow dung flies suggest that males adopt an optimal copula duration under sperm displacement, that maximizes fitness gain given the costs associated with finding females. Female characteristics such as sperm storage capacity and fecundity also impact on the optimal copula duration. Optimality analyses have been extended to other taxa, most notably the odonates, in which territorial males generally copulate for less time than males adopting alternative mate-securing tactics. Theoretically, the factors responsible for variation in copula duration are the interval between copulation and oviposition and thus time available for sperm mixing, the risk of takeover from rival males, the costs of territory defense, and female fecundity. Studies of both damselflies and dragonflies lend some support for theoretical predictions. Where females gain direct benefits from copulation their interests do not appear to be in conflict with those of males so that male interests are manifest in observed behavior. Nevertheless, female interests can conflict with those of males so that the observed copula duration may represent a resolution of sexual conflict. Moreover, females may use copula duration as a mechanism of mate choice, or they may reduce copula duration to avoid the costs of mating. Unfortunately, female interests have not been widely studied and their general impact is difficult to assess.

7 Sperm in Competition I: Strategic Ejaculation

7.1 Introduction

A recurring expectation in studies of the evolution of animal mating systems is that males should be limited in their reproductive success only by the numbers of females they can encounter. The underlying rationale is that sperm are relatively cheap to produce compared with eggs (Bateman 1948; Trivers 1972; Arnold and Duvall 1994) so that males have the higher potential reproductive rate (Clutton-Brock and Parker 1992; see section 1.2). However, as Dewsbury (1982) pointed out, sperm are rarely transferred as individual functional units. Rather, they are packaged in ejaculates, often several million at a time, along with other reproductively important substances that make up the seminal fluid. In reality, there is good evidence that ejaculates represent a significant cost of reproduction for males (table 7.1).

The males of many insects require a period of time after adult emergence during which they accumulate resources for ejaculate production (Foster 1967a; Woodhead 1984; Oberhauser 1988). Moreover, once mated, males must replenish their resources before they can mate again and/or produce an ejaculate of comparable size and quality (Simmons 1986; Svärd and Wiklund 1989; Simmons et al. 1992; Arnqvist and Danielsson 1999b). Thus, in a number of butterflies and tettigoniids, spermatophore mass has been shown to increase with male age and with the interval between successive matings (table 7.1). These data suggest that ejaculate production is costly for males in terms of both time investment and nutrient allocation. Indeed, qualitative and quantitative variation in diet can have a significant influence on the rate at which males can produce ejaculates (Simmons 1995c), the number of sperm contained within those ejaculates (Gage and Cook 1994), the quality of seminal fluid proteins (Bissoondath and Wiklund 1995, 1996a,b; Wedell 1996; Blay and Yuval 1997), and the effectiveness of ejaculates in gaining fertilizations (Simmons and Parker 1992). Both interspecific (Wedell 1994) and intraspecific (Simmons 1995b; Simmons et al. 1999a) studies of tettigoniids point to a trade-off between the amount of resources allocated to ejaculates and the time it takes males to recover from mating. Moreover, Simmons et al.'s (1992) study of the energetics of mate attraction and ejaculate production in *Requena verticalis* demonstrated that energy expenditure on the ejaculate was maintained under resource limitation at the expense of a

Table 7.1
Some examples of constraints on ejaculate production for male insects.

Species	*Male Constraint*	*Evidence*	*Source*
ORTHOPTERA			
Gryllus veletis, Gryllus pennsylvanicus	Gut parasites	↓ spermatophore production	Zuk 1987
Gryllodes sigillatus	Developmental instability	↓ sperm numbers	Farmer and Barnard 2000
Kawanaphila nartee	Diet	↑ remating interval	Simmons 1995c
Poecilimon veluchianus, Poecilimon affinis	Remating interval	↑ spermatophore size	Heller and von Helverson 1991
Poecilimon mariannae	Parasitoid	↓ spermatophore production, size, and sperm number	Lehmann and Lehmann 2000
Requena verticalis	Male size, gut parasites, and diet	↓ spermatophore production	Simmons et al. 1992; Simmons 1993, 1995b
	Multiple mating	↑ remating interval	
	Remating interval	↑ spermatophore size	
BLATTODEA			
Diploptera punctata	Male age, duration of larval development	↑ sperm numbers	Woodhead 1984
HEMIPTERA			
Gerris lateralis	Remating interval	↑ sperm numbers, ↓ oviposition stimulation	Arnqvist and Danielsson 1999b
MEGALOPTERA			
Protohermes grandis	Remating interval	↑ spermatophore size	Hayashi 1993
DIPTERA			
Lucilia cuprina	Multiple mating	↓ sperm numbers, ↓ inhibition of female remating	Smith et al. 1990
Drosophila melanogaster	Multiple mating	↓ fertility, ↓ lifespan	Prowse and Partridge 1997
	Parasitoid	↓ oviposition stimulation	Fellowes et al. 1999
Drosophila nigrospiracula	Ectoparasites	↓ testis mass	Polak 1998
Drosophila grimshawi	Diet	↓ testis mass	Droney 1998

Table 7.1 (*cont.*)

Species	*Male Constraint*	*Evidence*	*Source*
Drosophila pachea, Drosophila acanthoptera, Drosophila wassermani	Sperm length, multiple mating	↓ sperm numbers	Pitnick and Markow 1994b
Scatophaga stercoraria	Diet Multiple mating	↓ fertilization success ↓ testis size	Ward and Simmons 1991; Simmons and Parker 1992
LEPIDOPTERA			
Plodia interpunctella	Diet Viral infection Rearing conditions	↓ sperm numbers ↓ fertility ↓ fertility	Lum and Flaherty 1970; Gage and Cook 1994; Sait et al. 1994, 1998
Pseudaletia separata	Male age Remating interval	↑ spermatophore size ↑ spermatophore size	He and Tsubaki 1992
Heliothis virescens	Male age Multiple mating	↑ spermatophore size ↓ spermatophore size	LaMunyon 2000
Colias erytheme	Multiple mating	↓ spermatophore mass	Rutowski et al. 1987
Danaeus plexippus	Male age Remating interval	↑ spermatophore size ↑ spermatophore size	Oberhauser 1988
Jalmenus evagoras	Mating Remating interval	↓ spermatophore mass ↑ spermatophore size	Hughes et al. 2000
Pieris rapae	Mating	↓ spermatophore mass	Bissoondath and Wiklund 1996a; Cook and Wedell 1996
Pieris napi	Mating	↓ spermatophore mass	Kaitala and Wiklund 1995; Bissoondath and Wiklund 1996a
Polygonia c-album	Diet	↓ ejaculate quality	Wedell 1996

reduction in energy expended on mate attraction. A trade-off between mate searching and ejaculate production was also evident in Gage's (1995) study of *Plodia interpunctella*; increased investment in testes was associated with decreased investment in the thorax and head (flight muscle and visual system) and a reduced lifespan. Studies of insect parasitism also support the notion of a trade-off between somatic maintenance and ejaculate production. Direct evidence comes from Fellowes et al.'s (1999) study of encapsulation responses of *Drosophila melanogaster* infected with the hymenopteran parasitoid *Asobata tabida*. Infested males were unable to induce the egg laying response in females, presumably because they were compromised in their ability to synthesize the necessary accessory gland product responsible for this effect (section 4.3). Indirect evidence comes from the observation that males infected with nutrient-depleting gut parasites, ectoparasitic mites, or parasitoids have reduced expenditure on ejaculate production (Zuk 1987; Simmons 1993; Polak 1998; Lehmann and Lehmann 2000). Thus, male expenditure on the ejaculate is constrained by immediate resource availability. There is also evidence that males are limited in their reproductive potential in the long term. Smith et al.'s (1990) study of *Lucilia cuprina* showed that the number of sperm contained in the ejaculate declined with repeated mating, and that the effectiveness of the seminal fluid in inducing unreceptivity in females also declined. In water striders *Gerris lateralis*, the number of sperm transferred, paternity, and the effectiveness of oviposition stimulants are all decreased by recent mating (Arnqvist and Danielsson 1999b). Prowse and Partridge (1997) found that mating reduced the lifespan of male *D. melanogaster*. Moreover, mating had a much more pronounced effect on male fertility. Males with a history of mating were completely sterile when more than 80% of their unmated cohort were still alive. Sterility was attributable to a reduction in sperm numbers, and importantly the effect of mating depletion was irreversible. It is clear from these studies that ejaculates are costly for males to produce and can severely limit their reproductive rate.

Because ejaculates are costly for males to produce, Dewsbury (1982) predicted that males should be strongly selected to partition their ejaculates optimally with regard to the number of females they inseminate and the expected fitness gain from those females. In his prospective analysis he suggested that the operational sex ratio might have a significant impact on a male's allocation strategy. When the operational sex ratio is biased toward an excess of females, males should allocate the minimum amount of sperm necessary to fertilize as many of those females as possible. When there is an excess of males, Dewsbury (1982) argued that males might be expected to allocate more to an individual female, given the increased costs of finding additional females. Moreover, he suggested that male mate choice should be favored if the number of females a male is capable of fertilizing is limited; males should allocate more sperm to females who offer a greater reproductive return (section 5.6).

In chapter 5 we saw how selection via sperm competition could favor variation in the duration of postcopulatory mate guarding. Because mate guarding reduces the male's ability to search for and copulate with additional females, selection should favor plasticity in male behavior, so that males guard females only when the risk of female remating (and lost paternity) is high, relative to the benefits of finding and copulating with additional females. Given the limitations imposed on ejaculate expenditure, selection should similarly favor plasticity in ejaculation strategies. Specifically, variation in the risk of sperm competition should favor variation in the amount of ejaculate allocated to each female; males might be predicted to invest minimally in females when the risk of sperm competition is low in order to fertilize more females, but to increase their ejaculate expenditure in fewer females when the risk of sperm competition is high. Recently, a theoretical basis for the analysis of ejaculation strategies has been developed (see Parker 1998), and in this chapter I present a review of the theory and the extent to which variation in insect ejaculate expenditure conforms to its predictions.

7.2 Sperm Competition Games

A fundamental assumption in the theoretical analyses of Parker and his colleagues (Parker 1990a,b; Parker et al. 1996, 1997) is that sperm from two or more males are present within the reproductive tract of females (or in the spawning environment for external fertilizers) so that they compete for access to fertilizations. Sperm are assumed to be utilized at random from the female sperm storage organ(s), what Parker called a fair raffle (or lottery), so that the payoffs to competing males depend on the ejaculation strategies (relative numbers of sperm transferred) adopted by other males in the population. The analyses require an evolutionarily stable strategy (ESS) approach (Maynard Smith 1982), hence the term sperm competition games, and predict the ESS level of ejaculate expenditure. The predictions generated are specific to systems in which sperm competition conforms to the raffle principle. They may not be relevant to species in which the copulating male displaces sperm from previous males, because with sperm displacement the payoffs in terms of fertilization gain become increasingly independent of the ejaculate of the first male. As sperm displacement approaches 100%, a male's expenditure on his ejaculate will be subject to a simple optimization (see the example of *S. stercoraria* in section 6.2) rather than competitive games between males.

Sperm competition games are also based on the assumption that expenditure at one mating comes at a cost in terms of a male's ability to expend resources on additional matings, an assumption for which there is good evidence (table 7.1). A male's fitness is then equal to the product of his gain per mating and the number of matings obtained (fig. 7.1). In any given population, fitness gain is assumed to vary continually so that individuals using different strategies (ejaculate expenditure per mating) will have different fitness gains. The ESS expenditure occurs where the fitness gain is maximized. At the ESS, individuals that use a strategy in which expenditure differs from the ESS will be unable to invade the population (fig 7.1B). However, if the population is not at the ESS, the ESS strategy may not be the strategy that invades fastest, though the population will generally move toward the ESS (fig. 7.1B).

Sperm competition becomes important in the models because the gain per

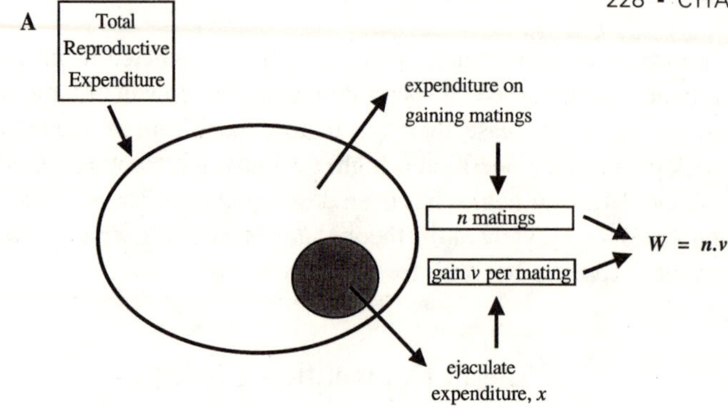

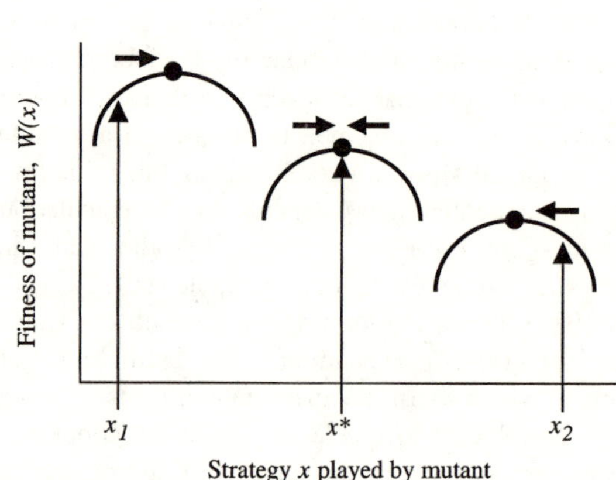

Figure 7.1 The logic of sperm competition games. (A) In sperm competition games, reproductive expenditure is partitioned between expenditure on gaining matings (e.g., searching, signaling to attract females, or competition with other males for access to females) and expenditure on the ejaculate. Thus, expenditure on the ejaculate reduces the number of females a male can inseminate. Expenditure, x, on the ejaculate increases the gain in terms of fertilization success with a given female, v, but reduces the number of matings possible, n. Male fitness, W, is therefore the product $n \cdot v$. (B) In a population playing the ESS strategy, x^*, all mutants playing $x \neq x^*$ will have a lower fitness, so that the fitness function rises to a peak at x^* and then declines. Thus, in a hypothetical population playing strategy x_1, a range of values of x higher than x_1 could spread. In a population playing x_2, a range of values of x lower than x_2 could spread. In a population at the ESS, x^*, no mutant can spread because all values of x have a lower fitness. The ESS can be solved analytically by differentiating the fitness function $W(x, x^*)$ with respect to the ejaculation strategy x and evaluating at $x = x^*$ so that

$$\left[\frac{W(x,x^*)}{\partial x}\right]_{x = x^*} = 0; \text{ subject to } \left[\frac{\partial^2 W(x,x^*)}{\partial x^2}\right]_{x = x^*} < 0$$

to ensure that x^* is a maximum. Since fitness is the product of $n \cdot v$, the fitness of a male playing strategy x in a population playing strategy x^*, $W(x, x^*) = n(x, x^*) \cdot v(x, x^*)$ and from the above at the ESS

$$-\left[\frac{n(x,x^*)}{n'(x,x^*)}\right]_{x = x^*} = \left[\frac{v(x,x^*)}{v'(x,x^*)}\right]_{x = x^*},$$

where the prime denotes the differential coefficient of n or v with respect to x. Note that n' is negative and v' positive, i.e., the number of matings decreases with expenditure, and the gain per mating increases with expenditure. (Modified from Parker 1998) mating will be reduced compared with a situation in which there is no sperm competition. Consider a species in which the probability of a female mating with two males (i.e., risk of sperm competition) equals q. Then on $1-q$ occasions there will be no sperm competition and a male will fertilize all of the female's clutch, while on q occasions a male will mate with a female who has either mated already or will mate again in the future. With q females, a male will obtain a proportion of the fertilizations dependent on his expenditure on the ejaculate relative to the other male. Thus a mutant expending x in a population playing the ESS ejaculation strategy x^* will have a gain per mating of

$$v(x,x^*) = \varepsilon(1 - q)\frac{x}{x} + 2q\left(\frac{\varepsilon x}{x + x^*}\right). \qquad [7.1]$$

The first part of the right hand side of the equation refers to the eggs, ε, fertilized with that proportion of females in the population that mate just once ($x/x = 1$ so that the male gains all of the eggs) and the second part refers to the number of eggs fertilized when playing against a competitor playing by ejaculation strategy x^* (equal to the relative amount of ejaculate from both males in the sperm stores) for those females that have already mated and those females that will mate in the future, hence the multiple of 2 (Parker 1998). We should expect selection to act on ejaculate expenditure x in relation to q so that fitness is maximized. Intuitively we might expect ejaculate expenditure to increase with q.

Interspecific Predictions

Let us first consider the general impact of sperm competition risk on a species' ejaculation strategy. Parker's (1982) original analysis predicted that ejaculate expenditure should increase with sperm competition risk by a factor of one-quarter of the probability that at a given mating an individual male will be subject to competition with a previous male. This is different from sperm competition risk q, which is the population average probability of double matings by females. For analysis across species it is easiest to examine the effect of sperm competition based on this species average risk and assuming that males have no information regarding current risk. The qualitative prediction remains the same, but now ejaculate expenditure is predicted to increase by a factor of one-half the risk of sperm competition (fig. 7.2) (Parker et al. 1997). The above predictions are based on the assumption of a fair raffle. In some cases sperm competition may conform to a loaded raffle, in which one sperm of male 1 is worth more or less than one sperm of male 2. Such a situation may arise, for example, if females show some preference for using sperm from one of the males, or if sperm from one male are less viable than those of another male. Loading the raffle has the effect of reducing the slope of ejaculate expenditure versus risk of sperm

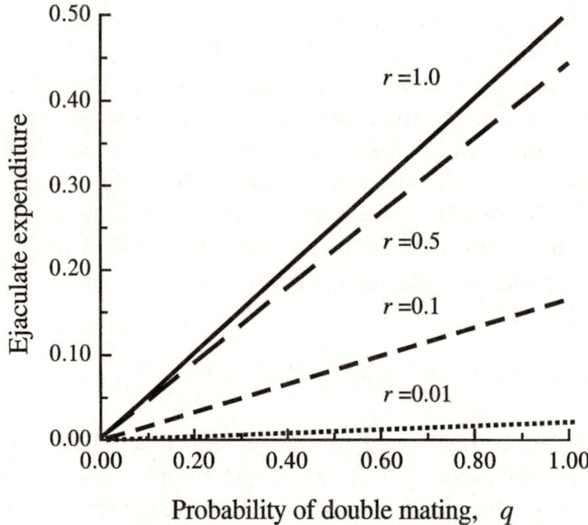

Figure 7.2 Interspecific predictions from sperm competition game models. The ESS ejaculate expenditure for varying degrees of unfairness, *r*, in the fertilization raffle. In general, ejaculate expenditure is predicted to increase with the population mean risk of double mating, *q*. As the raffle becomes loaded in favor of one of the competing males (*r*'s become increasingly < 1.0), the ESS ejaculate expenditure decreases. Severe loading makes ejaculate expenditure relatively insensitive to sperm competition risk. (After Parker et al. 1997)

competition (fig. 7.2). This illustrates the dependence of predictions on the raffle mechanism of sperm competition; as the loading becomes exceedingly high, as would be the case where the first male's sperm are removed from storage, ejaculate expenditure will become minimal and insensitive to q.

Parker et al. (1996) also examined the impact of sperm competition intensity on ejaculate expenditure. Sperm competition intensity refers to the number of males N in competition for a given batch of eggs, where $N \geq 2$. Although Parker et al.'s (1996) models initially considered external fertilizers, they are equally relevant to internal fertilizers where sperm competition conforms with the raffle principle (Parker 1998). Again, across species, as the average number of matings performed by females increases, the expenditure on the ejaculate by males should increase.

Selection for increased ejaculate expenditure should be reflected in an increase in spermatogenic tissue, and comparative studies across a variety of vertebrate taxa have now shown that increases in the degree of multiple mating by females are associated with increased testis size in males (Short 1979; Harcourt et al. 1981; Ginsberg and Rubenstein 1990; Møller 1991; Hosken 1997; Stockley et al. 1997; Hosken 1998). In general, vertebrate systems appear to conform with the raffle mechanism of sperm competition implicit in Parker's models (Dewsbury 1984; Birkhead and Møller 1992,

1998; Gomendio and Roldan 1993). However, in insects, a simple association between female mating frequency and testis size might not be expected because of the diversity of sperm competition mechanisms. Where sperm displacement is achieved by mechanical means, sperm numbers will be unimportant in determining the fertilization success of second males so that selection via female mating frequency is unlikely to favor increased testis size. On the other hand, where males use their own ejaculate to flush rival sperm from the sperm storage organs of females, sperm numbers will be unimportant at the time of fertilization yet selection could favor increased testis size because of the necessity for males to use their ejaculates for reasons beyond fertilization per se (Hosken and Ward 2001). Of course, selection may act on accessory glands rather than testes if it is seminal secretions rather than sperm that are involved in sperm displacement. These types of confounding factors need to be considered in any comparative analysis of insects, and when drawing conclusions from such analyses (section 2.8).

There are data from two groups of insects that can be used to assess the impact of sperm competition on ejaculate expenditure across species. In Lepidoptera sperm are transferred packaged in a spermatophore that is placed within the female's reproductive tract. There is no obvious process of the males genitalia, or of the spermatophore, that could mechanically displace sperm stored from rival males. Rather sperm must migrate from the ostium bursae to the site of storage and utilization (see fig. 4.2), either under their own propulsive force, and/or by movements of the female's internal reproductive tract (Drummond 1984; Tschudi-Rein and Benz 1990). Studies of sperm utilization show that when females mate with just two males, sperm utilization favors either the first or the second male (table 2.3); within species, cases of mixed sperm utilization following double matings are rare. Such a pattern would be consistent with a raffle loaded heavily in favor of one male. However, as the number of matings (intensity of sperm competition) increases, mixed paternity becomes more common (see section 2.7 and fig. 2.8). Thus, with increased intensity of sperm competition, sperm utilization may approach a fair raffle. Parker et al.'s (1996, 1997) models therefore predict that across species of Lepidoptera increased risk and intensity of sperm competition should favor increased expenditure on the ejaculate.

Svärd and Wiklund (1989) examined the influence of polyandry on the ejaculation features of two families of butterflies, the pierids and satyrids. Because empty spermatophores remain in the female's bursa copulatrix, it is possible to assess the degree of polyandry, and thus intensity of sperm competition, by performing spermatophore counts on wild-caught females (see fig. 4.5). Svärd and Wiklund (1989) found that the pierids were highly polyandrous (high intensity of sperm competition) while the satyrids were relatively monandrous (low risk of sperm competition). Accordingly, they found that the pierids had significantly larger ejaculate weights and, further, within the pierids there was a significant positive association between the degree of polyandry and both ejaculate weight and the rate at which males could produce sperm and accessory secretions. Svärd and Wiklund's (1989) study did not control for the effects of common ancestry on the variables they considered. In a more detailed study involving 74 species of butterflies from five families, Gage (1994) found that after controlling for body size and phylogeny there was a positive association between testis size and degree of polyandry, again estimated from spermatophore counts of wild-caught females (fig. 7.3). These data support the hypothesis that increased risk and intensity of sperm competition is associated with increased male expenditure on ejaculate production.

Nevertheless, because of the often complex mechanisms of sperm competition in insects, there are reasons why sperm competition could favor increased ejaculate size without direct competition between sperm. In his study of the moth *Choristoneura fumiferana*, Retnakaran (1974) noted the all-or-nothing pattern of parentage for the second of two males to mate, typical for Lepidoptera (table 2.3). He suggested that when the female sperm storage organs were filled by the first ejaculate there would be no room for second-male sperm, resulting in first-male parentage. Because the migration of sperm from the corpus bursae to the spermatheca occurs over an extended period of time, a second mating occurring shortly after the first could displace the first spermatophore before sperm have been transferred, thereby allowing the second male to fill the female's spermatheca and gain complete parentage. In support of his hypothesis, increasing the interval between matings had the effect of changing the pattern of parentage from complete second male to complete first male (see also Suzuki et al. 1996). Thus, spermathecal filling may be an important mechanism for the avoidance of sperm competition (Simmons and Siva-Jothy 1998). Moreover, in Lepidoptera, ejaculate size, and in particular the volume of seminal fluid and number of sperm transferred, are important determinants of the female's remating interval (section 4.3), so that increased ejaculate size will reduce the risks of future sperm competition. Both of these factors should favor increased investment in the ejaculate in order to prevent sperm competition, rather than to engage in it. In their comparative study of 178 species of moth, Morrow and Gage (2000) found positive associations between testis size and the volume of the spermatheca and between testis size and the volume of the corpus bursae. Stimulation of stretch receptors in the corpus bursae and the movement of sperm within the sperm stores are both factors that influence female remating (Thibout 1975; Sugawara 1979). Thus, both of these evolutionary associations are predicted if males are under selection to avoid sperm competition from future rivals.

Pitnick and Markow (1994b) found a positive association between testis size and female remating frequency in the *nannoptera* species group of the Drosophilidae. Again, this relationship could be taken as evidence for selection via sperm competition. The Drosophilidae are an interesting group in

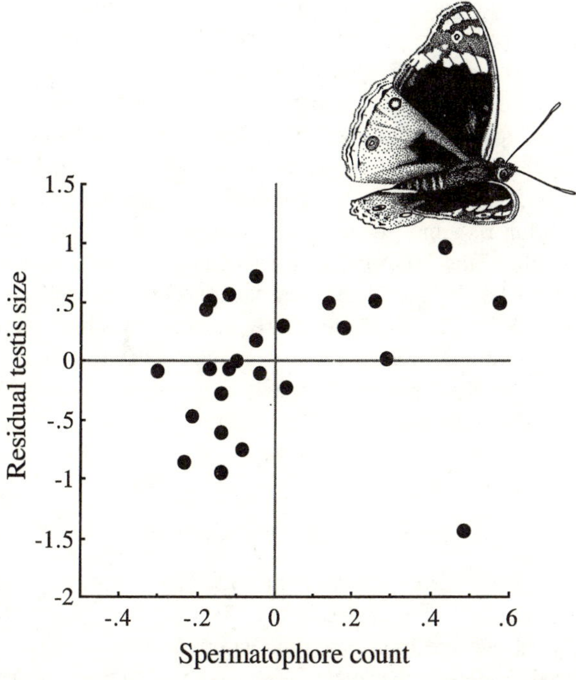

Figure 7.3 Sperm competition risk and intensity increase male expenditure on the ejaculate, estimated by testis mass, across twenty-eight species of butterfly. Residual values from a regression of testis mass on body mass correlate positively with mean spermatophore count (degree of polyandry). The comparative analysis controlled for phylogenetic inertia using the method of phylogenetic subtraction, so mean spermatophore counts have either positive or negative values, depending on the average expected for their family clades. (Redrawn from Gage 1994)

that sperm gigantism is common (see chapter 8). Across species, there is a positive association between sperm length and testis size (Pitnick 1996). Increased testis size appears to be associated with the production of giant sperm but not with increases in the number of sperm produced (Pitnick 1996). Further, there is a trade-off between sperm size and number such that sperm gigantism results in sperm limitation; males partition their limited supply of sperm among multiple females so that after a single mating females are submaximally inseminated (Pitnick and Markow 1994b; Pitnick et al. 1995a; Pitnick 1996). The role sperm competition plays in the evolution of sperm gigantism is unclear (see chapter 8). However, since females of giant-sperm-producing species are submaximally inseminated, the tendency to remate is increased and hence so is the risk of sperm competition. In *D. melanogaster* and *D. pseudoobscura*, sperm are relatively small and males maximally inseminate females (Gilbert 1981; Snook et al. 1994). In these species there is sperm displacement and incapacitation so that P_2 values are high (table 2.3) and the direct competition between the sperm from two males explicit in sperm competition game models is greatly reduced or absent. In contrast, as in *D. pachea* (Pitnick 1993), males of the giant-sperm-producing *D. hydei* may submaximally inseminate so that sperm from successive males are stored by the female (Markow 1985). Here sperm utilization conforms to a fair raffle so that selection at the level of true sperm competition is greater than for *D. melanogaster* or *D. pseudoobscura* (table 2.3). Thus, while sperm competition risk and intensity are associated with increased testis size across drosophilids, the relationship is not causal. Rather, increased sperm competition may be the inevitable consequence of increased sperm length (see also section 8.2). This example clearly demonstrates how the often complex mechanisms of sperm transfer, storage, and utilization found in insects can confound predictions based on models that assume a simple mechanism of sperm competition by numerical superiority.

Intraspecific Predictions

Within-species predictions become somewhat more complex because they can depend critically on what Parker (1990a) referred to as roles. Parker (1990a) analyzed two sperm competition games that differed from the previously discussed models in that males were assumed to have perfect information regarding their role, which was either favored (this might be, for example, the first male to copulate) or disfavored (correspondingly, the second male to copulate). The raffle was loaded in favor of the first male to mate with the female, hence his favored role. The models assume that females always mate with two males so that there is always sperm competition. In the first model Parker (1990a) assumed that roles were assigned at random so that males were equally likely to occupy the favored or disfavored role. In this case, the ESS ejaculation strategy was for males to invest equally in their ejaculates, despite knowledge of their current role. The result arises because across all matings the average fitness gain per female remains the same. If a male were to respond by increasing his expenditure when in the disfavored role, he would pay the cost of reduced numbers of matings, and therefore reduced frequency of being in the favored role (fig. 7.4). In contrast, if roles are assigned nonrandomly, so that some males are always in the disfavored role, it should pay those males to increase their expenditure on the ejaculate in order to compensate for their role (fig. 7.4). The greater the loading of the raffle, the greater should be the disparity in expenditure between favored and disfavored males.

Fryer et al. (1999a) have independently developed ESS models of ejaculation strategies. Unlike Parker's (1990a) models, Fryer et al. (1999a) allow females to mate with one or no other male so that males can face sperm competition with some females but not others. More importantly, Fryer et al.

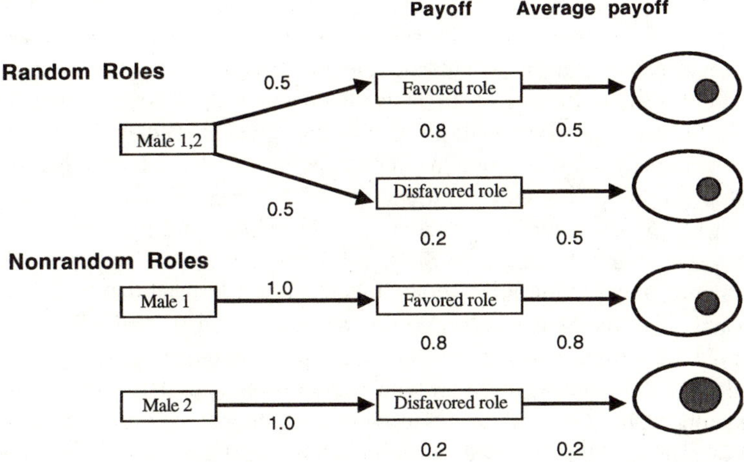

Figure 7.4 The loaded raffle model, in which each sperm of male 2 counts as *r* (where 0 < r < 1) in the raffle relative to each sperm of male 1. Males always face sperm competition. When roles are random each male has an equal probability of being in the favored role, which has the higher fitness gain. On average across all matings, both males obtain the same fitness gain, and the ESS expenditure on the ejaculate should be equal for both males despite knowledge of their role. When roles are nonrandom, one male is consistently in the disfavored role with the lower fitness gain. Here the ESS is for males in the disfavored role to increase their expenditure on the ejaculate to partially compensate for their low fitness gain (numbers refer to the probability of being in the favored or disfavored role, and the fitness payoffs in terms of the probability of fertilizing eggs). (Modified from Parker 1998)

(1999a) restrict their analysis to a two-round game, in which males can mate just twice. In general this may be biologically unrealistic, but it may be true for some species of butterflies in which the average mating frequency is less than 2 (Drummond 1984; with a balanced sex ratio the average numbers of matings by males and females must be the same). Fryer et al. (1999a) find that if all males and females mate twice then equal amounts of sperm should be inseminated by males in each round (i.e., males should allocate half of their ejaculate to each of the two females they will mate with). This result is reminiscent of Parker's (1990a) result for random roles. However, three factors serve to increase expenditure in round one. First, in a male-biased population where some males fail to mate in both rounds, males should inseminate more in the first round if they get the opportunity to mate. Second, the capacity to replenish sperm stores between rounds should increase the ESS expenditure in round one. Finally, depletion of sperm from the female's sperm stores due to oviposition between matings should favor increased expenditure in round one so that males can compete for fertilizations when females remate.

Recently, Mesterton-Gibbons (1999) reanalyzed Parker's (1990a) random roles model after modifying some of the assumptions. Unlike Parker (1990a), he assumed that in the absence of sperm competition there would be a risk that ejaculates would not contain enough sperm to achieve complete fertilization. In general sperm depletion can result in long-term declines in fertilization, favoring multiple mating by females (Ridley 1988). However, it is not clear if a single mating can be insufficient for complete fertilization. In some species even a fraction of a normal ejaculate can render females completely fertile (Parker et al. 1993) while in others single matings can result in inferior fertility (Simmons 1988a). With the slightest risk of incomplete fertilization, Mesterton-Gibbons (1999) showed that males in the favored role should expend more on the ejaculate than males in the disfavored role, irrespective of whether roles were assigned randomly or nonrandomly. Importantly, differences in these results appear to stem from the modeling of the trade-off between number of matings and ejaculate expenditure; Parker (1990a) used a multiplicative approach while Mesterton-Gibbons (1999) used an additive approach (the former multiplies and the latter subtracts a suitable discount or cost factor from the number of matings). Which approach is appropriate is a matter of conjecture. Ball and Parker (2000) have compared the two approaches. They show how the predictions of each model depend on complex interactions between the role typically occupied by a male (his phenotype) and whether he occupies the favored or disfavored role at mating. They conclude that the multiplicative model is the more accurate model and that it provides more detailed information. However, it must be remembered that theoretical models represent a means by which hypotheses are generated. Only empirical studies that address both the assumptions and predictions of various models can confirm or falsify the hypotheses they generate.

It is difficult to envisage a general situation in which males should consistently find themselves in a disfavored role. One situation in which roles will be nonrandom occurs in species where males adopt alternative mating strategies (Gross 1996). Often those males unsuccessful in competition for females will sneak copulations, for example, when guarding males are occupied in disputes with other males. By the nature of their alternative tactic, sneaks are always subject to sperm competition. Guards, on the other hand, will be subject to sperm competition with low probability p, dependent on the relative frequency of sneaks and guards in the population. Parker (1990b) analyzed a model specific to alternative mating strategies in which there is an asymmetry in information between males; sneaks always face sperm competition but guards only know p. The results were qualitatively similar to those of the nonrandom roles model; the ESS ejaculation strategy was for sneaks to expend more on the ejaculate than guards. However, as the frequency of sneak males increases in the population, the probability of sperm competition for males increases, removing the asymmetry in risk so that all males should then invest equally in their ejaculate (fig. 7.5).

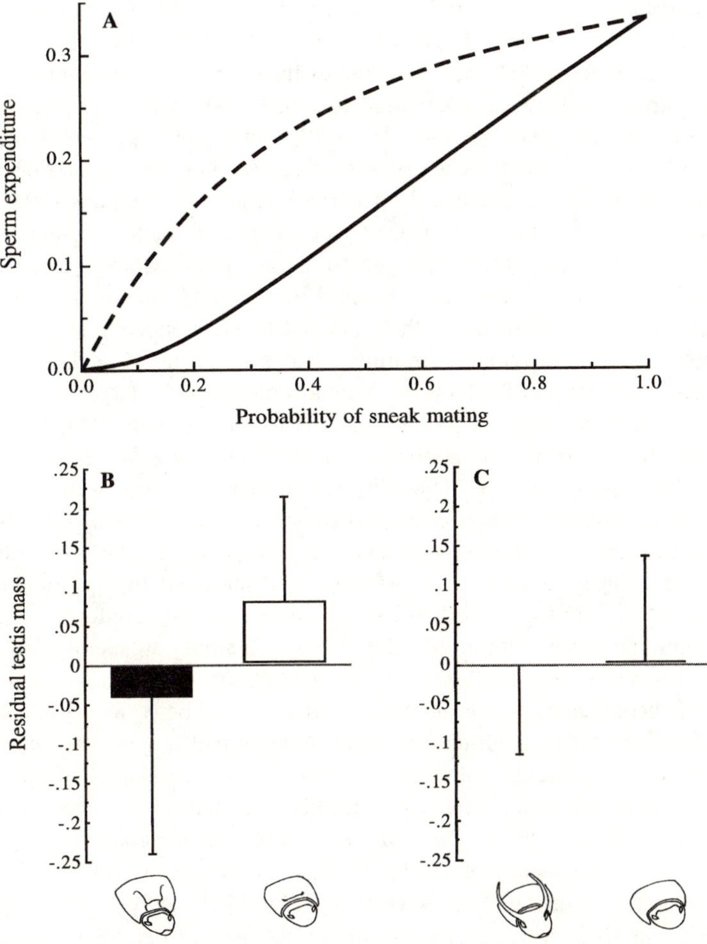

Figure 7.5 Ejaculation strategies adopted by males with alternative mating strategies. (A) The predicted ESS ejaculate expenditure for sneaks (the dashed curve) and guards (the solid curve) in relation to p, the probability that a sneak mating will occur for the case where the cost of sperm production for sneaks and guards is equal. There is asymmetry of information in that sneaks always "know" that there will be sperm competition with the guarding male. The ejaculation strategy of guards will be set by the population average probability of a sneak mating occurring, p. Sneaks should expend more on the ejaculate than guards. At low p, the ESS expenditure for sneaks and guards should be minimal, since there is very low risk of sperm competition. The disparity in expenditure increases with p initially, but as the risk of sneak matings increases, guards should increase their expenditure on the ejaculate, so that the expenditure of sneaks and guards will converge because sperm competition risk becomes symmetrical (see Parker 1990b). (B) Testis size, controlling for body size, across alternative mating tactics of the dung beetle, *Onthophagus binodis*. Hornless males (open bar) represent just 30% of the male population and sneak copulations with females guarded by horned males (solid bar). There is disparity in ejaculate expenditure, implicated by differences in testis size and confirmed by differences in ejaculate volume and sperm length. (C) In the congeneric species *O. taurus*, sneaks represent up to 60% of the male population, so that the probability of sneak matings is greatly increased. There is no evidence for a disparity in ejaculate expenditure (from Simmons et al. 1999c).

Dimorphism in male phenotype and behavior is not an uncommon feature of insects (Eberhard 1982; Eberhard and Gutierrez 1991; Tomkins and Simmons 1996; Emlen 1997), and two studies have examined variation in ejaculation strategies between alternative mating tactics. Male dung beetles in the genus *Onthophagus* are often dimorphic: some males produce horns on the head and/or pronotum whereas others remain hornless, appearing morphologically similar to females (Emlen 1997; Hunt and Simmons 1997). Dimorphism in body plan is associated with alternative mate-securing tactics (Cook 1990; Emlen 1997; Hunt and Simmons 1998; Moczek and Emlen 2000). Horned males monopolize females by guarding the entrances to tunnels beneath the dung in which the male and female cooperate in provisioning brood masses. In contrast, hornless males sneak into breeding tunnels and copulate with guarded females. A male's phenotype is largely dependent on the amount of resources provided by his parents (Emlen 1994; Hunt and Simmons 1997) so that the alternative tactics represent a conditional reproductive strategy (Gross 1996). By the nature of their tactics there is asymmetry in information on sperm competition risk; hornless males are always subject to sperm competition from guarding males, while guards are subject to sperm competition with low probability, determined by the number of sneaks in the population. Furthermore, sperm competition conforms to a raffle in onthophagines, with both males to mate obtaining on average an equal share of fertilizations (Tomkins and Simmons 2000). In accord with Parker's (1990b) theoretical expectation, after controlling for body size Simmons et al. (1999c) found that hornless male *O. binodis* had larger testes than did their horned conspecifics (fig. 7.5). Moreover, an examination of ejaculate characteristics revealed a similar pattern; hornless males had larger ejaculate volumes and longer sperm. The same was not true for the congeneric species *O. taurus*. The frequency of sneaks in *O. taurus* is close to 60% so that the risk of sperm competition for horned males is likely to be high compared with *O. binodis* in which sneaks constitute just 30% of the population. Theory predicts that the disparity in ejaculate expenditure should decrease with increasing risk for guards. Thus, the differences between these species in the relative ejaculate expenditure across morphs are also consistent with expectation. A second study of a dimorphic bee *Amegilla dawsoni* failed to provide evidence for differences in testis size or ejaculate size between alternative mating tactics, not surprisingly since it was also found that females were monandrous (Simmons et al. 2000).

The within-species games considered by Parker (1990a) assumed that females always mated with two males, so that there is always sperm competition. Parker et al. (1997) analyzed a series of models in which there was variation in the probability that females would mate twice and males had information about this risk of double mating (see also Ball and Parker 1998). The models ask how a male should adjust ejaculate expenditure given information that their current mate has not mated previously and is unlikely to mate again (which occurs with probability $1 - q$), or has already mated once or will mate a second time (both of which occur with probability q). That is to say, males have information that a particular female has a greater or lesser probability than the average probability q of mating twice. As established earlier, across-species ejaculate expenditure is predicted to rise with the probability of double mating. With perfect information, males should expend minimally on females who mate only once but increase their expenditure with females who will mate twice, to a level dependent on the average probability of double mating. However, if males can distinguish between mated and unmated females, but cannot assess whether previously unmated females will mate again in the future, ejaculate expenditures converge at the extreme values of q; when females rarely remate males should have minimal expenditure irrespective of the female's mating status. Conversely, if females always remate, males should expend equally on the ejaculate with unmated and mated females since they will always experience sperm competition (fig. 7.6). Thus, both levels of information predict that within species males should adjust their ejaculate expenditure in response to variation in the perceived risk of sperm competition.

Within-species variation in ejaculation strategies can arise at two levels. First, we have already discussed how the average probability of double matings by females can vary across species, leading to interspecific variation in ejaculate expenditure. Within species, populations may also differ in the average probability of double mating. Interspecific theoretical predictions regarding ejaculate expenditure should hold equally well across populations within species. Indeed, studies of two species of moth, *Plodia interpunctella* and *Pseudaletia separata*, provide strong support for this prediction (He and Tsubaki 1991, 1992; Gage 1995; He and Miyata 1997). These species are semelparous so that all resources for reproduction are accrued and distributed within the body during larval development. Female *P. interpunctella* emerging from high-density populations have a higher mating frequency than do females emerging from low-density populations, thereby imposing a higher average risk and intensity of sperm competition on males. Gage (1995) found that the males in these populations had an ejaculate expenditure that matched this increase in selection via sperm competition; males emerging from high-density populations had relatively larger testes and transferred more sperm than males emerging from low-density populations. Likewise, male *P. separata* emerging from crowded conditions produced heavier spermatophores that contained more apyrene sperm (He and Tsubaki 1992; He and Miyata 1997). In the case of *P. separata*, females from crowded conditions did not remate sooner, which may be due to the increased numbers of apyrene sperm transferred by males (section 8.3).

Second, within populations males may respond to current information on sperm competition risk (Parker et al. 1997). A growing number of empirical studies are lending support to the prediction that males strategically adjust

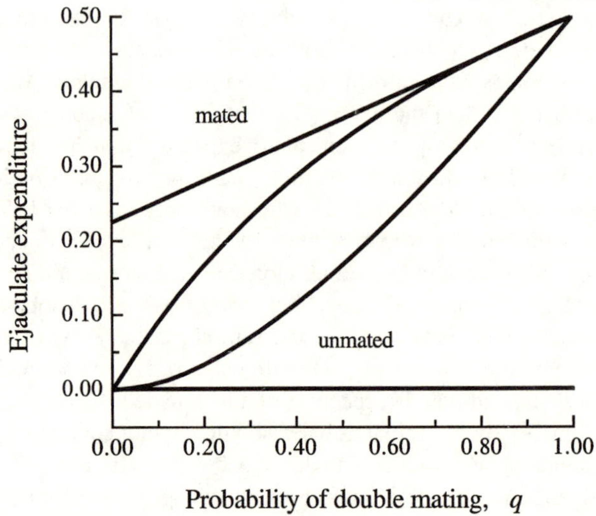

Figure 7.6 Intraspecific variation in ejaculate expenditure when males mate with previously unmated females (lower curve and line) or with mated females (upper curve and line). When males have complete information, the ESS ejaculate expenditure is to invest minimally in previously unmated females who they know will not mate again in the future, and to expend more with mated females, dependent on the population average probability of double mating. When males have incomplete information, they "know" that females have not mated previously but do not know if they will mate in the future. The ESS is to expend less on the ejaculate than with previously mated females, but to expend an amount dependent on the population average risk of double mating. (Modified from Parker 1998)

their expenditure on the ejaculate in response to perceived risk of sperm competition (table 7.2). Male crickets, Mediterranean fruit flies, and beetles all show an increase in the number of sperm transferred to females when copulation occurs in the presence of rival males (Gage 1991; Gage and Baker 1991; Gage and Barnard 1996), perhaps indicating that the risk of future sperm competition is high. In most of these cases, females will remate immediately after copulation so that the presence of rivals may signal immediate risk of sperm competition. Ransford (1997) failed to find an effect of male presence on ejaculate expenditure in two-spot ladybirds, *Adalia bipunctata*. This is not surprising since sperm competition appears to conform to a mechanism of sperm displacement (table 2.5) and a male's fertilization success is independent of the first males ejaculate (Ransford 1997). Thus, *A. bipunctata* should not conform with predictions based on raffle models. Studies of tettigoniids and butterflies have also failed to find an effect of male presence on the numbers of sperm transferred (Cook and Gage 1995; Wedell 1998). Interestingly, in these species females become refractory to further matings for some time after copulation, so the presence of rivals need

Table 7.2
Ejaculation responses to correlates of risk or intensity of sperm competition and correlates of female quality in insects.

Species	*Correlate*	*Response*	*Source*
ORTHOPTERA			
Acheta domesticus	OSR	↑ sperm numbers	Gage and Barnard 1996
	Female size	↑ sperm numbers	
Gryllodes sigillatus	OSR	↑ sperm numbers	Gage and Barnard 1996; Farmer and Barnard 2000
	Female symmetry	↑ sperm numbers	
Kawanaphila nartee	Female size and OSR	↓ spermatophore size	Simmons and Kvarnemo 1997
Requena verticalis	Female age	↑ sperm numbers	Simmons et al. 1993; Simmons 1995b
	Male investment	↑ sperm numbers	
Decticus verrucivorus	Mating status	↓ spermatophore size	Wedell 1992
Coptaspis sp. 2	Male presence	None	Wedell 1998
	Female weight	↑ sperm numbers	
	Mating status	↑ sperm numbers	
COLEOPTERA			
Onthophagus binodis	Alternative tactics	Sneaks ↑ sperm numbers	Simmons et al. 1999c
Onthophagus taurus	Alternative tactics	None	Simmons et al. 1999c
Tenebrio molitor	Male presence	↑ sperm numbers	Gage and Baker 1991
Adalia bipunctata	Male presence	None	Ransford 1997
DIPTERA			
Ceratitis capitata	Male presence	↑ sperm numbers	Gage 1991; Blay and Yuval 1997
	Male diet	↑ sperm numbers	
HYMENOPTERA			
Amegilla dawsoni	Alternative tactics	None	Simmons et al. 2000
LEPIDOPTERA			
Pseudaletia separata	Larval density	↑ sperm numbers	He and Tsubaki 1992; He and Miyata 1997
Plodia interpunctella	Mating status	↑ sperm numbers	Cook and Gage 1995; Gage 1995, 1998
	Male presence	None	
	Female age	↓ sperm numbers	
	Female weight	↑ sperm numbers	
	Larval density	↑ sperm numbers	
Pieris rapae	Mating status	↑ sperm numbers	Cook and Wedell 1996; Watanabe et al. 1998; Wedell and Cook 1999a,b
	Female weight	↑ sperm numbers	

not signal immediate sperm competition risk. Nevertheless, Cook and Gage's (1995) study of the moth *Plodia interpunctella* provided good evidence that males respond to cues relating to past and future risk. They allowed males to mate with females that had mated 6 days previously to males experimentally manipulated to deliver ejaculates of varying size. Males delivered more sperm to females that had already received a large ejaculate than to females that had received a small ejaculate or to unmated females (fig. 7.7). Cook and Gage (1995) argued that males could assess the size of past ejaculates since the emptied spermatophores of previous males remain in the ostium bursae, to which the male has direct access during copulation. They also found that males delivered more sperm to young virgin females than they did to old virgin females (fig. 7.7). The risk of future matings is reduced in old females because they are unlikely to live longer than the refractory period that results from mating (Cook and Gage 1995).

The induction of a refractory period facilitated by seminal fluid products is a mechanism by which males avoid sperm competition from future males (section 4.3), and the ability of males to provide sufficient seminal fluid products may be a direct cue for their assessment of future risk. Male butterflies become severely depleted of ejaculate material following a single mating (table 7.1) and females receiving smaller ejaculates will remate sooner than those receiving large ejaculates (Oberhauser 1989; Kaitala and Wiklund 1994; Bissoondath and Wiklund 1997; Wedell and Cook 1999a). Thus, the risk of future sperm competition is greater for males who have mated previously because of their inability to delay female remating. Accordingly, male *P. rapae* withhold sperm in the duplex, the male sperm storage organ, when mating for the first time, in order to increase the number of sperm transferred to their second mates with whom the future risk of remating is greater (Cook and Wedell 1996; Watanabe et al. 1998; Wedell and Cook 1999a,b). Moreover, the duration of the female's refractory period decreases with female body mass, and, as would be predicted, males transfer more sperm to larger females (Wedell and Cook 1999a). They also appear capable of making finer discriminations since, like *P. interpunctella*, they will transfer more sperm to females whose first mate was a seminal-fluid-depleted male (one that had transferred many sperm) than to females whose first mate had been previously unmated (one that had transferred relatively few sperm). The relative number of sperm transferred is a major determinant of fertilization success (Wedell and Cook 1998) and male *P. rapae* thereby appear to ejaculate strategically to maximize their reproductive success in the face of sperm competition. The data for *P. rapae* are in general agreement with the predictions of Parker et al.'s (1997) risk assessment models. They do not support the models of Fryer et al. (1999a), which predict greater expenditure of sperm on the first spermatophore than the second.

A further example comes from the work of Blay and Yuval (1997). They found that female Mediterranean fruit flies *Ceratitis capitata* mated to males

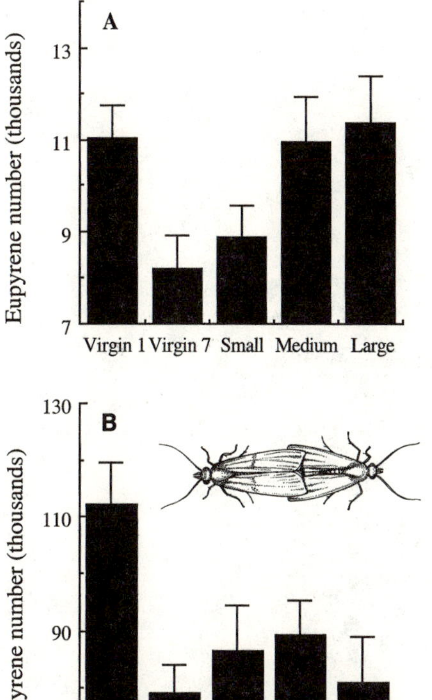

Figure 7.7 Male *Plodia interpunctella* increase their expenditure on sperm in relation to the risk of sperm competition. The figures show the numbers of (A) eupyrene and (B) apyrene sperm ejaculated by 1-day-old previously unmated males when copulating with females that varied in their insemination status. Two age groups of females were examined. The first were previously unmated (virgin) 1-day-old females and the second were 7-day-old females that were either unmated (virgin) or had been mated with males delivering quantitatively different ejaculates—small, medium, or large. Females were mated at 7 days to ensure they had recovered receptivity following their first mating. Variation in first ejaculate size was achieved by using males that had never mated before (and therefore produced a large ejaculate), had mated once before (produced a medium-sized ejaculate), or had mated twice before (produced a small ejaculate). Second males increased their expenditure on eupyrene sperm when in competition with large ejaculates. They were also sensitive to the age of virgin females, ejaculating more eupyrene and apyrene sperm when mating with young females that were more likely to mate again in the future. (From Cook and Gage 1995)

reared on a low-protein diet were more likely to remate than females mated to males reared on a high-protein diet. Seminal products are similarly responsible for loss of female receptivity following mating (Chapman et al. 1998; Miyatake et al. 1999) so that protein-deprived males may have been unable to produce effective seminal fluids. These males responded to the increased probability of their mates remating by increasing the number of sperm transferred.

Parker et al. (1996) analyzed a series of models in which they varied the intensity of sperm competition (the number of ejaculates competing for a given batch of eggs). These models were originally developed for externally fertilizing species but the predictions apply equally to internal fertilizers (Parker 1998). Parker et al. (1996) found that in general ejaculate expenditure should increase as the species or population average number of males (ejaculates) in competition increases (fig. 7.8). However, ejaculate expenditure for a given male at a given mating should decrease the greater the number of competitors. This might at first seem counterintuitive. However,

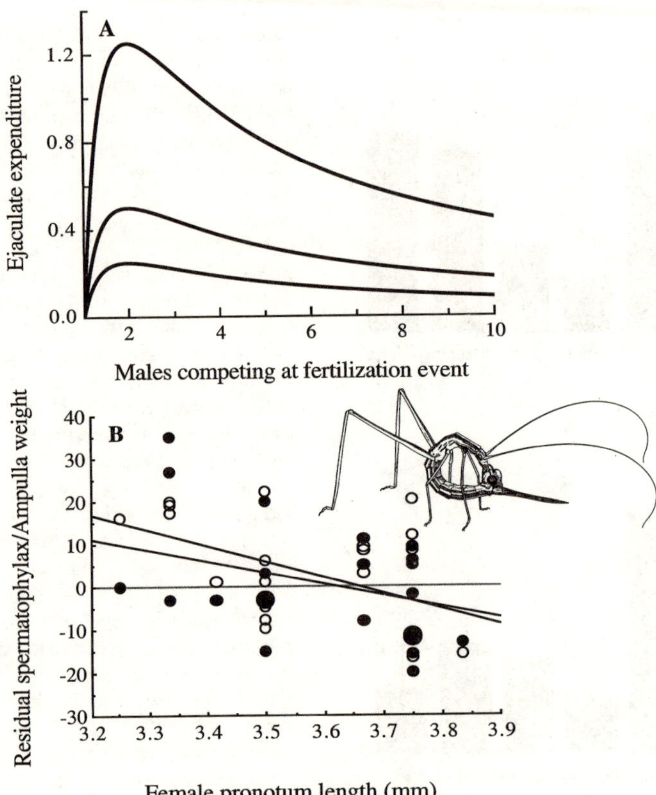

Figure 7.8 Intraspecific variation in ejaculate expenditure in relation to the intensity of sperm competition. (A) The predicted ESS ejaculate expenditure increases as the number of males in competition for fertilizations increases from 1 to 2, but decreases as the number of competitors increases beyond 2. Three curves are shown for populations that differ in the average number of males (N = 1, 2, or 5) that compete for one female's eggs. Across species, ejaculate expenditure increases with the intensity of sperm competition; but within species, males are predicted to conserve ejaculates when the value of females declines due to increased intensity of sperm competition (from Parker et al. 1996). (B) Male tettigoniids *Kawanaphila nartee* may be sensitive to increased sperm competition intensity. In this species, females compete for access to a limited supply of males, and larger females obtain more matings. Sperm mix in storage, so that the intensity of sperm competition is greater with larger females. After controlling for male body size, males transfer spermatophores with smaller sperm containing ampullae (solid circles) and smaller spermatophylaxes (open circles) to larger females (from Simmons and Kvarnemo 1997).

with a fair raffle, a male's fertilization gain will be inversely proportional to the number of ejaculates present. Therefore the value in terms of fertilization gains from a given mating falls rapidly as the number of males competing rises above 2, so that it should pay males to conserve their ejaculates for more profitable matings where sperm competition is less intense (fig. 7.8).

Currently there is only one study that provides evidence for an effect of sperm competition intensity on ejaculation strategies. In the courtship-role-reversed tettigoniid *Kawanaphila nartee*, females compete for access to a limited supply of males (Gwynne and Simmons 1990; Simmons and Bailey 1990). Larger females have a competitive advantage over smaller females and obtain a greater number of matings (Gwynne and Simmons 1990; Gwynne and Bailey 1999). Although larger females are more fecund, because of female competition the intensity of sperm competition should be greater within large females, reducing their reproductive value to males. Importantly, courtship role reversal is plastic in this species, dependent on resource availability; female remating is greatly reduced in populations with conventional courtship roles (Gwynne and Simmons 1990; Simmons and Bailey 1990). Simmons and Kvarnemo (1997) found that males transferred larger spermatophores when mating under conditions of courtship role reversal, as expected given the greater risk of sperm competition. Moreover, males transferred smaller spermatophores to large females with whom the intensity of sperm competition would be greater (fig. 7.8). Female size had no influence on spermatophore size in populations with conventional courtship roles (Simmons and Kvarnemo 1997).

7.3 Cryptic Male Choice

The central thesis of Dewsbury's (1982) contribution was that, given the costs of ejaculate production, males should be selected to partition their limited resources among females with whom they expect the highest reproductive returns; in short, they should exhibit mate choice. There are two possible levels of male mate choice. Males may either reject females of low quality as mating partners (section 5.6) or they may mate with all females and allocate their ejaculates in relation to female quality. I refer here to differential allocation of sperm by males as cryptic male choice because, unlike mate rejection, it occurs at a level that is not readily appreciated from behavioral observations of mating activity and as such is akin to the differential acceptance and/or use of sperm envisaged as cryptic female choice (Lloyd 1979; Walker 1980; Thornhill 1983; Eberhard 1996). There are several examples in the literature of cases in which males reject previously mated females as mates, presumably because sperm competition reduces their reproductive value (section 5.6). Here I examine how variation in ejaculation strategies reflect cryptic male choice.

Cryptic male choice is implicit in Parker et al.'s (1996) models of sperm competition intensity; males were predicted to reduce their expenditure on females as the intensity of sperm competition increased, because sperm competition reduced the reproductive value of females. Variation in ejaculation strategies with female quality were examined explicitly in the theoretical models of Galvarni and Johnstone (1998). They constructed models of optimal sperm allocation using stochastic dynamic programming techniques. They asked how a male's ejaculate expenditure should vary over the course of a single breeding season given a finite supply of sperm and random encounters with females that varied in quality. Using a diminishing returns function, the fitness payoff to a male was allowed to rise with ejaculate allocation to a level dependent on female quality. Galvarni and Johnstone's (1998) models predicted that males should expend an increasing amount on their ejaculate as female quality increased, irrespective of the amount of sperm they have remaining. Optimal ejaculate expenditure also increased through the season, because the probability of encountering further females declined with time. Finally, ejaculate expenditure was predicted to increase with amount of sperm remaining. Allowing replenishment of sperm increased expenditure in all cases.

Unlike sperm competition games, Galvarni and Johnstone's (1998) optimality models did not take into account what other males in the population were doing so they did not take into account sperm competition. To do that they used an alternative sigmoid function for fitness gain with increasing ejaculate expenditure. This function allowed fitness gain to rise to an asymptote dependent on female quality, but at low ejaculate expenditures females of high quality yielded lower gain. Essentially this might be the case if sperm competition were more intense in females of high quality. This is the case, for example, for *K. nartee*, where large females are more fecund but also mate with more males. With the sigmoid function males were still predicted to allocate more ejaculate to females of higher quality because of the greater potential gain and competition for that gain. However, as males become sperm depleted Galvarni and Johnstone's (1998) models predict that males should allocate sperm only to females of average quality, presumably because they are unable to compete in sperm competition for high-quality females.

Wedell (1992) found that male tettigoniids *Decticus verrucivorus* transferred larger spermatophores to previously unmated females than they did to mated females. Thus they allocated less sperm when in competition with a previous male. These data do not support the predictions of Parker et al.'s (1997) models. They are consistent with Fryer et al.'s (1999a) expectation that males should expend more on the ejaculate in the first round, particularly when females oviposit between rounds, so that the first male's sperm become depleted in the sperm stores. Wedell (1992) found that females laid more eggs during the refractory period following their first mating than they did in the second refractory period following their second mating. Indeed, females can lay up to 30% of their total lifetime egg production in the refractory period that follows their first mating, so that greater ejaculate expenditure with previously unmated females is also consistent with a greater expenditure by males in females of greater reproductive value (Galvarni and Johnstone 1998).

Body size is generally a predictor of fecundity in insects so that larger females should offer greater reproductive returns to males. A number of studies have found that males ejaculate more sperm when mating with larger females (Gage and Barnard 1996; Gage 1998; Wedell 1998; Wedell and Cook 1999a). Moreover, a recent study of the cricket *Gryllodes sigillatus* has shown that males deliver more sperm to females having lower levels of fluctuating asymmetry (Farmer and Barnard 2000), a trait reputed to convey reliable information on individual genetic quality (Møller and Swaddle 1997). These results lend some support to Galvarni and Johnstone's (1998) prediction that males allocate more of their limited reserves to females of greater reproductive value, and to the notion of cryptic male choice. However, at least for the studies of *P. interpunctella* (Gage 1998) and *P. rapae* (Wedell and Cook 1999a), larger females were shown to have higher mating frequency and shorter periods of nonreceptivity respectively, so that the risk of sperm competition with larger females is greater. Although this is the case envisaged by the version of Galvarni and Johnstone's (1998) model in which fitness gain follows a sigmoidal function, it is difficult to separate empirically the impact of sperm competition risk per se and variance in female quality on male ejaculation characteristics. Wedell's (1998) study of the katydid *Coptaspis* sp. 2 showed that males allocate sperm in relation to female size, but only when females had not mated previously. Males increased the amount of sperm to a level equal for females of all size classes when females had mated previously. These results suggest that the engagement in sperm competition may have a greater effect on male ejaculation strategies than variation in female quality.

The contrast between Gage's (1998) results and those of Simmons and Kvarnemo (1997) shown in fig. 7.8 is interesting because it highlights the difference between predictions based on risk and on intensity of sperm competition (Parker 1998). In *P. interpunctella*, larger females are more likely to mate twice than are smaller females (females rarely mate more than twice; Gage 1998) so the risk of sperm competition increases with female size. In *K. nartee*, females can mate as often as four times, thereby increasing the intensity of sperm competition and greatly reducing female reproductive value. Thus, while ejaculate expenditure is predicted to increase with female size in *P. interpunctella* it should decrease with female size in *K. nartee*. The prospective models of Galvarni and Johnstone (1998) and the empirical studies discussed above suggest that a thorough examination of cryptic male choice should yield some interesting results.

7.4 Cryptic Female Choice

Currently theoretical models assume that females do not impinge on the ejaculation strategies of males. In a prospective discussion Parker (1998) suggested that female influences that bias paternity in favor of particular males could favor increased ejaculate expenditure by disfavored males. If favoring particular males at fertilization has some adaptive value to females, then females might be expected to retaliate by increasing the loading against the disfavored male to counter his increased expenditure. Parker (1998) suggested that such games were best modeled as evolutionary "arms races" between males and females.

Greef and Parker (2000) have recently examined how one potential mechanism of sperm selection, spermicide by females, could impact on male ejaculation strategies. The predictions generated by their models varied widely, depending on the mechanism of spermicide. The first model considered was one in which females killed a fixed number of sperm per unit time. Since two males mate at different times, different numbers of sperm from each male will be killed due to the delay between the first mating and the second. If all males are likely to suffer from reduced numbers of sperm within the female sperm stores, then an arms race can become established, where escalation in spermicide by females favors an escalation in sperm allocation by males (see also Birkhead et al. 1993). This is because all sperm produced above those killed by the female will enter the fertilization raffle and thereby contribute to male fertilization success. The same prediction was true for models in which the female selectively killed a fixed number of sperm from one particular male, thereby favoring his competitor.

The second model assumed that a fixed proportion of sperm were killed. In this case the response to spermicide depended on the interval between matings. If there was no delay between matings, and males were not affected differentially then there should be no increase in sperm allocation in response to spermicide by females. An increase in the delay between matings however, should result in a decreasing investment in sperm allocation. This is because all sperm produced by a male are subject to killing, and the delay between matings meant that one male's ejaculate was reduced in size, effectively decreasing the intensity of sperm competition. Reduced sperm allocation was also predicted when females selectively killed sperm from one male, provided that males were equally likely to be in the disfavored role (in this case having their sperm killed).

These widely differing predictions prevent any generalization concerning what males should do when females manipulate ejaculates. Biologically plausible arguments can be made for either model (Greef and Parker 2000) and the empiricist is faced with having to gain an understanding of the interactions between ejaculates and female reproductive tracts before these models can be evaluated. Currently we have only begun to examine female influences in sperm utilization (chapter 9). This area represents a challenge for future developments in sperm competition research.

7.5 Summary

It is not yet widely recognized that ejaculates are costly for males to produce and can limit the number of females a male is able to inseminate. The costs of ejaculate production should favor plasticity in the amount of ejaculate allocated to each mating, thereby maximizing male lifetime reproductive success. Recent theoretical advances in sperm competition have modeled ejaculation strategies as an evolutionary game between two players who compete for fertilizations. Fitness gain is assumed to be directly proportional to the relative numbers of sperm each male has in the female's reproductive tract. Across species, increased probability of double mating by females should favor increased expenditure on the ejaculate by males, a prediction for which there is some evidence. Within species, if males have information regarding the current risk of sperm competition they should increase their expenditure on the ejaculate when that risk is high. There is a growing body of empirical evidence for apparent adaptive plasticity in the amount of sperm transferred to females in conditions of high sperm competition risk. When a female mates with more than two males, the intensity of sperm competition is increased and so the reproductive value of that female is decreased. Under conditions of increased sperm competition intensity, theory suggests that males should reduce their expenditure on the ejaculate and conserve resources for females of greater value. There is currently little empirical data for this prediction. The general influence of female quality on ejaculation strategies has just recently received theoretical attention, and some empirical studies suggest that males may allocate their ejaculates based on female quality. Cryptic male choice may thus prove an important process in sexual selection. Ejaculate manipulation by females is also expected to influence how much a male should expend on his ejaculate. Current models predict that males could counter female manipulation by increasing ejaculate expenditure, but they could equally respond by decreasing ejaculate expenditure. Strategic ejaculation is a relatively new area in sperm competition research and many more empirical studies are required to test the assumptions and predictions of the growing number of theoretical models.

8 Sperm in Competition II: Sperm Morphology

8.1 Introduction

Insect sperm have exhibited a remarkably rapid and divergent evolution. In stark contrast to the typical image of a tadpolelike cell, insect sperm cells can be aflagellate and immotile (Dallai et al. 1975), or they can have a multitude of flagella (fig. 8.1). In species of dytiscid beetles sperm associate in pairs, while in some locusts and fishflies hundreds of sperm are bound by their heads in a single spermatodesm, disassociating only after they arrive at the site of storage (Szöllösi 1974; Hayashi 1998). In the Tettigoniidae, sperm heads are equipped with hooklike processes at their apex, the shape and elaboration of which vary markedly between species (Baccetti 1987). Sperm length can vary considerably. In some species sperm are relatively short while in others they can reach gigantic proportions; the sperm of *Scatophaga stercoraria* are just 0.2 mm in length (Ward and Hauschteck-Jungen 1993) but in *Drosophila bifurca* the 58 mm long sperm is 35 times the length of the male producing it (Pitnick et al. 1995b). In other taxa sperm are polymorphic. Thirteen species in the *obscura* group of *Drosophila* have both long and short spermatozoa whose lengths can differ by up to a factor of 6.5 in the case of *D. azteca* (Joly and Lachaise 1994). The pentatomid bug *Arvelius albopunctatus* has three sperm lengths (Schrader and Leuchtenberger 1950), while the hymenopteran wasp *D. fuscipennis* has as many as five different sperm morphs that vary in size and shape (Lee and Wilkes 1965). Sperm dimorphism is an almost universal phenomenon in the Lepidoptera. One sperm morph, the "eupyrene" sperm, is the usual nucleated sperm that fertilizes eggs. The other, the "apyrene" sperm, is smaller and lacks any genetic material. Apyrene sperm are produced in large numbers, often comprising in excess of 90% of the total sperm numbers ejaculated (Meves 1902; Silberglied et al. 1984; Gage and Cook 1994).

The rapid and divergent nature of spermatozoon evolution has been recognized by spermiocladists for many years (Afzelius 1975; Baccetti 1987; Jamieson 1987). Indeed, sperm morphology is utilized for the construction of insect phylogenies, often when whole organism morphology is unable to resolve ancestral relationships between taxa (Jamieson 1987). As with genitalia (chapter 3), the rapid and divergent nature of sperm evolution is indicative of intense selection on sperm morphology. Indeed, although he did not

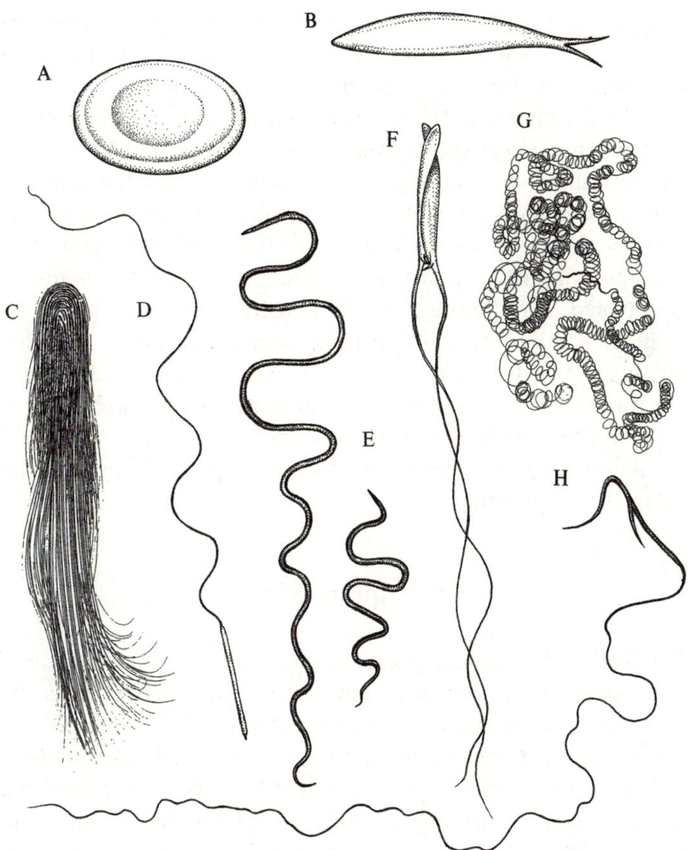

Figure 8.1 An illustration of the degree of interspecific variation in sperm morphology across insects: (A) From the order Protura, the sperm of *Eosentonon transitorium* is an immotile disk that forms a cup shape after ejaculation (Jamieson 1987); (B) the fishlike sperm of the dipteran *Telmatoscopus albipuntus* (Jamieson 1987); (C) sperm bundle of the fishfly *Parachauliodes japonicus* (Hayashi 1998); (D) 1 mm sperm of the dung beetle *Onthophagus taurus*; (E) apyrene (short) and eupyrene (long) sperm of *Plodia interpunctella*; (F) the paired sperm of the water beetle *Dytiscus marginalis* (Jamieson 1987); (G) giant 58 mm sperm of *Drosophila bifurca* (Pitnick et al. 1999); and (H) the hook-headed sperm of the tettigoniid *Tessellana tesselata* (Baccetti 1987).

discuss it in his original review of insect sperm competition, Parker (1970e) noted that there is every reason to suppose that selection acts on individual sperm. Whenever there is competition between sperm for access to ova, any trait of an individual sperm that enhances its success in fertilization over its competitors (be it swimming speed, longevity, or ability to fasten to and penetrate the egg) will be favored in the male producing it. Thus, Sivinski (1980, 1984) argued that sperm competition may be partly responsible for rapid and divergent evolution in sperm morphology. For example, Sivinski

(1984) suggested that sperm transferred in pairs or in groups (for examples see fig. 8.1) might cooperate in reaching the female's sperm stores, combining their propulsive forces to outswim sperm in rival ejaculates (Taylor 1952). Working with fishflies *Parachauliodes japonicus* Hayashi (1998) has recently shown that larger sperm bundles, presumed to contain more individually motile sperm, do have a greater swimming speed than small sperm bundles when swimming through a range of fluid viscosities. Outswimming rival sperm could benefit the male and favor the transfer of sperm contained within groups. Alternatively, at least in fishflies, increased swimming speed could be favored if it allowed males to transfer their sperm before females remove the externally attached spermatophore. Perhaps due to the logistics of examining sperm behavior, few studies are currently available that address the functional significance of variation in sperm morphology. In this chapter I examine a number of hypotheses put forward to explain variation in sperm morphology, and the extent to which sperm competition may be involved in sperm evolution.

8.2 Sperm Size

Parker's (1982) theoretical analysis concluded that sperm competition should favor the production of numerous tiny sperm. The analysis made two assumptions; that sperm competition conformed to a simple fair raffle in which males obtain fertilizations in direct proportion to the representation of their sperm in the female's sperm stores, and that there is a finite availability of reproductive reserves generating a trade-off between the number of sperm a male is capable of producing and the size of individual sperm. Sperm remain small and numerous because any increase in sperm size should come at a selective disadvantage due to the necessary reduction in sperm number and fertilization success. Nevertheless, it is abundantly clear that sperm are not always small and numerous.

The genus *Drosophila* exhibits greater variation in sperm length than all other animal taxa combined (Joly et al. 1991a), and Pitnick and his colleagues have utilized this variation in their studies of the evolutionary consequences of sperm length (Pitnick and Markow 1994a,b; Pitnick et al. 1995a; Pitnick 1996). In a comparative study across 11 species of *Drosophila*, Pitnick (1996) reported significant evolutionary associations between testis mass and sperm length and between testis mass and the total amount of sperm produced. Further analysis revealed that far more of the variation in testis mass is explained by variation in sperm length than by the amount of sperm produced. Thus, the evolution of long sperm appears to have been associated with an increase in the length and thus mass of testes (fig. 8.2). In accord with Parker's (1982) theoretical assumptions, there is an evolutionary trade-off between sperm length and number. After controlling for testis mass,

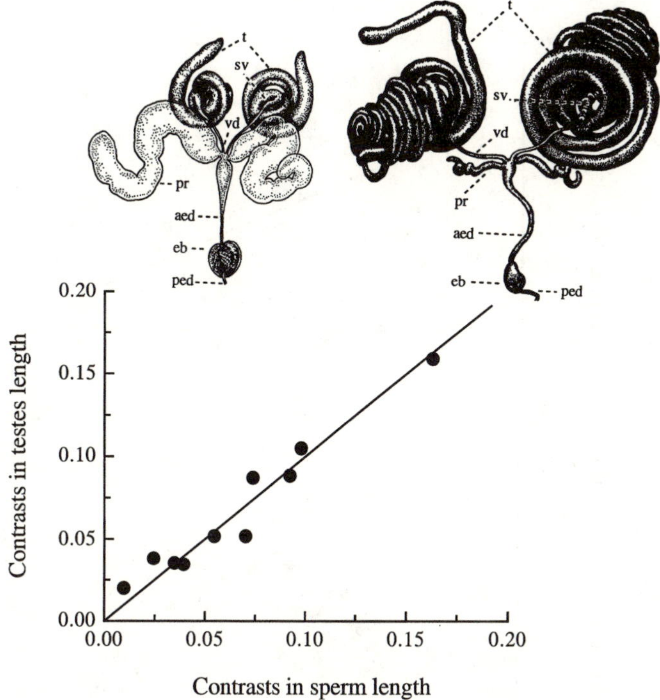

Figure 8.2 Correlated evolution of sperm length and testis length across eleven species in the genus *Drosophila*. The data represent independent contrasts in testis length and sperm length after controlling for phylogeny. An evolutionary increase in sperm length is associated with an evolutionary increase in the length of the testis, and thus in testis mass (from Pitnick 1996). The extremes of testis length variation are illustrated by the reproductive tract morphology of male *D. micromelonica* (*left*) that produce sperm 1.41 mm in length, and *D. bifurca* (*right*) that produce sperm 58.3 mm in length. t, testes; sv, seminal vesicles; vd, vas deferens; pr, paragonia; aed, anterior ejaculatory duct; eb, ejaculatory bulb; ped, posterior ejaculatory duct. (From Patterson 1943)

Pitnick (1996) found a negative evolutionary association between the number of sperm produced and the length of sperm (fig. 8.3). The inevitable consequence of this trade-off is that the males of long-sperm-producing species appear to be severely sperm limited (Pitnick 1993; Pitnick and Markow 1994b).

Pitnick et al. (1995a) also examined the life history costs of producing long sperm across 42 species of *Drosophila*, finding an evolutionary association between the age at sexual maturity and sperm length for males but not for females (fig. 8.3). Testis growth occurs predominantly between the time of adult eclosion and sexual maturity so that species producing long sperm should require longer prereproductive periods for testis growth (Pitnick 1993). Moreover, Pitnick et al. (1996) found that species producing longer

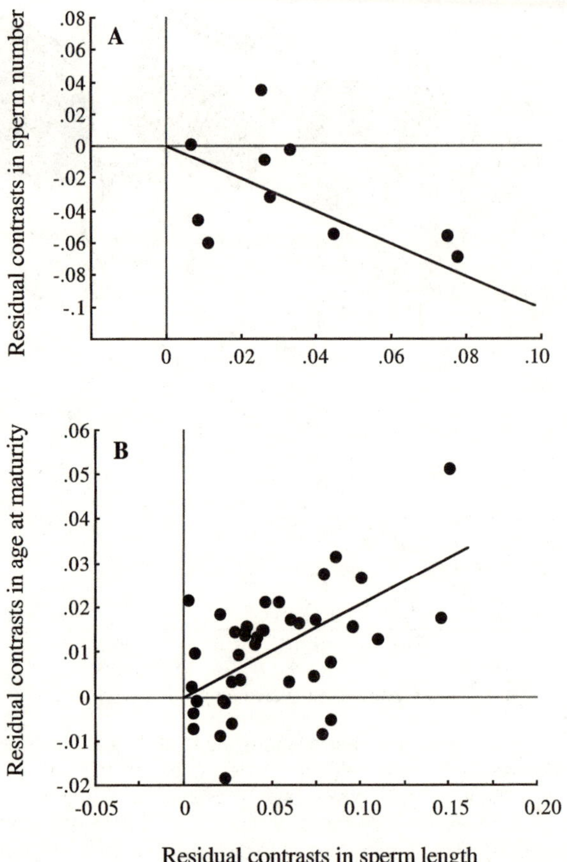

Figure 8.3 The costs of producing long sperm in the genus *Drosophila*. (A) There is an evolutionary trade-off between the number of sperm produced and sperm length. Data are independent contrasts in sperm numbers and sperm length, each controlled for their positive relationships with testis mass (from Pitnick 1996). (B) In order to produce longer sperm, males must invest more in testes production, extending the prereproductive period. Thus, there is an evolutionary association between age at maturity and sperm length. Data are independent contrasts in age at maturity and sperm length derived from a comparative analysis across forty-two species of *Drosophila*, controlling for the positive relationships of each variable with male body size (from Pitnick et al. 1995a).

sperm invested relatively more in the development of their testes than species producing short sperm; the allometric relationship between body mass and testis mass was greater than 1.0 so that large species have larger testes than might be expected for their body size. Testis mass represents 9.2% of male body mass in *D. pachea* and the resources required to sustain this investment mean that males must spend 30–50 % of their adult life in a prereproductive state (Pitnick 1993), some 10 days longer than females require to produce eggs. Furthermore, because males are limited in the number of sperm they can produce, they submaximally inseminate females so that female reproductive success is a positive function of the number of matings they can obtain (Pitnick 1993). In natural populations these two factors generate a skew in the operational sex ratio toward an excess of females (Pitnick 1993) so that the production of giant sperm should, theoretically, be associated with a reversal in the direction of sexual competition (section 1.2).

In an intraspecific study of the costs of long sperm in *D. hydei*, Pitnick and Markow (1994a) found that the costs of producing long sperm can be offset by increased male body size; larger males had longer testes and produced a greater number of sperm. In consequence, larger males were able to inseminate a greater number of females, transfer more sperm to each female, and thus produce a greater number of progeny than smaller males. Moreover, the age at sexual maturity was lower for larger males. Thus we might expect selection to favor an increase in male body size to offset the costs of increased sperm length, and indeed across species there is a positive evolutionary association between body mass and sperm length (Pitnick 1996). As pointed out by Pitnick (1996), however, such a relationship is equally expected if selection for increased sperm length was effective in large-bodied species because of their ability to withstand the costs of increased testis investment.

Giant sperm are not unique to *Drosophila*, being reported from a range of taxa including the coleopterans *Divales bipustulatus* (Mazzini 1976) and *Ptinella aptera* (Taylor 1982), the hemipteran *Notonecta glauca* (Afzelius et al. 1976), and the lepidopteran *Xenosoma geometrina* (Morrow 2000). Nevertheless, the selective advantages required to counter the obvious costs associated with sperm elongation are poorly understood. Currently there are two schools of thought. First, selection may favor increased male investment in zygote production via resources carried by the spermatozoa. Giant sperm frequently contain large mitochondrial derivatives containing protein that could be absorbed and utilized by the zygote (Perotti 1973; Afzelius et al. 1976; Sivinski 1980, 1984). The elongated tails of *Drosophila* sperm can be found throughout development, in the anterior end of the embryo and enveloped by the developing midgut. During development proteins are slowly stripped from the sperm tail, which is finally eliminated from the body soon after hatching when the larva first defecates (Pitnick and Karr 1998).

Parker's (1982) analysis concluded that in the absence of sperm competition an increase in sperm size at the expense of sperm number could be favored via the benefits of zygote nutrition. Recently Pitnick et al. 2001 examined the evolutionary responses in male *D. melanogaster* to the experimental removal of sperm competition through enforced monogamy. In accord with Parker's (1982) theoretical analysis, the males in one of two experimentally monogamous lines did have an increase in sperm length and both lines had decreased sperm numbers compared with promiscuous controls. All else being equal, competition between the sperm of different males is arguably reduced in *D. pseudoobscura* and *D. melanogaster* because of displacement and incapacitation of previously stored sperm (Price et al. 1999). The phylogeny of drosophilids places *D. pseudoobscura* and *D. melanogaster* ancestral to the giant-sperm-producing *nannoptera* and *hydei* species groups (Karr and Pitnick 1996) so that reduced sperm competition in these ancestral species could favor increased sperm provisioning. Bressac et al. (1995) favor this scenario, based on the observation that the tails of giant sperm enter the egg, and that for individual sperm paternity assurance increases with sperm length. However, Pitnick et al. (1995b) show that in many species only a fraction of the sperm tail enters the egg, for *D. bifurca* less than 5%. In a phylogenetic analysis of sperm-egg interactions, Karr and Pitnick (1996) have shown that sperm gigantism has evolved independently of sperm tail entry into the egg, evidence that paternal provisioning via the sperm is not the selective advantage behind sperm length evolution in *Drosophila*. Furthermore, as noted above, sperm limitation resulting from increased sperm length is associated with submaximal insemination and increased remating by females (Pitnick and Markow 1994b). Concomitant is an increase in multiple paternity and thus direct competition between the sperm of multiple males; thus, from the limited data available, P_2 seems to decline with increasing sperm length across *Drosophila* so that increased length is associated with more intense competition between the sperm of individual males and sperm numbers are critical in fertilization success (fig. 8.4). Parker's analysis shows that even an extremely low level of sperm competition, such as that seen in *D. melanogaster* and *D. pseudoobscura,* should counter any increase in sperm length. Moreover, a copulating male's sperm are actively involved in the sperm displacement process (Price et al. 1999; Chapman et al. 2000; Gilchrist and Partridge 2000), making sperm numbers important for male fitness even though sperm competition per se is reduced. Indeed, in the monogamous lines of Pitnick et al. 2001, males were less successful in gaining fertilizations as second males than were males from promiscuous control lines, suggesting that increased sperm length at the expense of reduced sperm numbers did reduce male fitness as predicted by Parker (1982). It therefore seems unlikely that sperm length has increased in *Drosophila* under selection for male paternal investment. Pitnick and Karr (1998) suggest that the processing of sperm tail fragments during *Drosophila* embryogenesis is more likely to represent a mechanism by which intragenomic conflict from paternal mitochondrial DNA is avoided.

The second school of thought is that sperm length may be subject to directional selection via sperm competition. There are at least four hypotheses for how sperm length may contribute to success in sperm competition: (i) swimming speed; studies from a number of taxa suggest that sperm motility may be a positive function of sperm size or length allowing larger

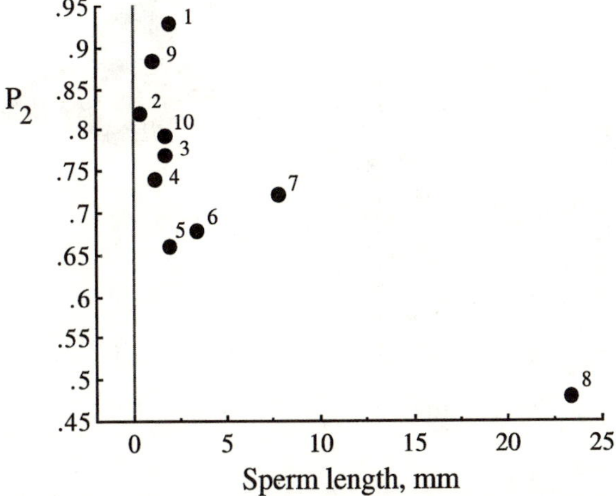

Figure 8.4 Comparison between patterns of sperm utilization and sperm length in *Drosophila*. It is now known that high values of P_2 (the proportion of offspring sired by the second male to mate) are generated via displacement of previously stored sperm from the seminal receptacle in some *Drosophila*. The limited data available suggest that the degree of sperm displacement declines with increasing sperm length, so that actual competition between sperm from different males is greater for species with long sperm. Contrary to observation, competition between sperm should maintain sperm small and numerous. More data are required, particularly from species with giant sperm, before a formal comparative analysis can be performed to rigorously test the apparent association. 1, *D. melanogaster*, 2, *D. pseudoobscura*; 3, *D. teissieri*; 4, *D. simulans*; 5, *D. mojavensis*; 6, *D. montana*; 7, *D. littoralis*; 8, D. hydei; 9, *D. mauritiana*; 10, *D. sechellia*. (Sperm lengths from Joly et al. 1991a and Pitnick et al. 1995a; P_2 values from table 2.3)

sperm to outcompete their smaller rivals in the race for fertilizations (Bressac et al. 1991b; Gomendio and Roldan 1991; Radwan 1996; LaMunyon and Ward 1998); (ii) survival; longer sperm may have more resources that enhance their longevity in storage (Parker 1993); (iii) blocking; once in storage, longer sperm may be better able to block the storage organs and thereby prevent sperm from future males competing for fertilizations (Dybas and Dybas 1981; Briskie and Montgomerie 1992; Ladle and Foster 1992; Briskie et al. 1997); and (iv) resistance; longer sperm may be better able to resist displacement once they have gained access to the sperm stores (Dybas and Dybas 1981; Sivinski 1984).

There is evidence that sperm size increases with sperm competition risk in mammals (Gomendio and Roldan 1991), birds (Briskie et al. 1997), and nematodes (LaMunyon and Ward 1999), although it was found to decrease with sperm competition intensity in fishes (Stockley et al. 1997) and to be independent of sperm competition risk in bats (Hosken 1997). There are data from the insects that suggest sperm competition risk may play an important

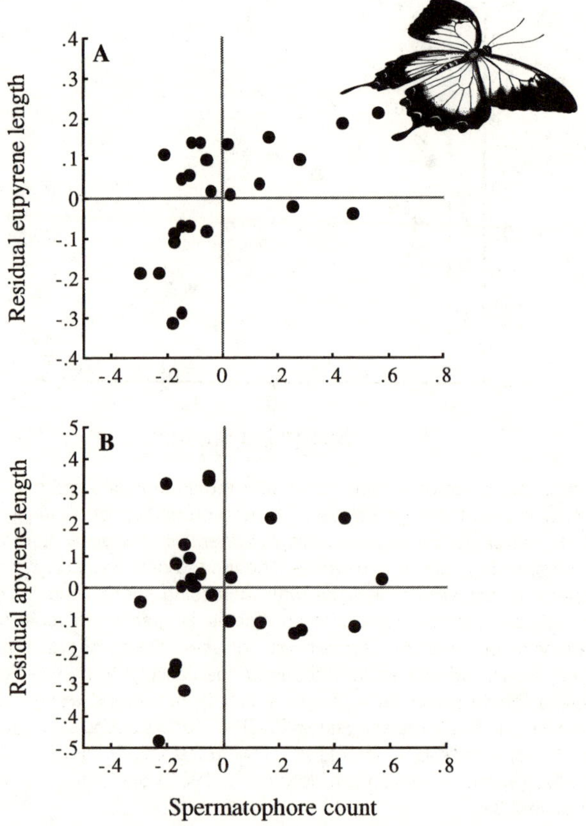

Figure 8.5 The evolutionary effect of sperm competition risk on sperm length. Associations between sperm competition risk (spermatophore counts from wild-caught females) and the lengths of eupyrene and apyrene sperm across twenty-five species of butterfly. Residual values of sperm length control for associations between length and species body size. Potentially confounding effects of common ancestry are controlled by phylogenetic subtraction using family means (see Gage 1994). Only eupyrene sperm persist in the female's sperm stores and are capable of fertilizing eggs. The data suggest that increased risk and intensity of competition among fertilizing sperm results in an increase in their length (A). Because apyrene sperm do not compete for fertilizations, they are apparently not subject to this selection pressure (B).

role in the evolution of sperm length. In Gage's (1994) comparative study of butterflies, the length of eupyrene, but not apyrene, sperm was found to increase with the degree of female multiple mating as determined from spermatophore counts of wild-caught females (fig. 8.5). Using testis size as an indirect measure of sperm competition risk, Morrow and Gage (2000) found that in moths both eupyrene and apyrene length increased with risk. Finally, within-species variation in sperm competition risk is also associated with differences in sperm length. Species of onthophagine dung beetles adopt alternative mating tactics in which horned males guard females while hornless males sneak copulations with already mated females (Emlen 1997; Moczek and Emlen 2000). The asymmetry in sperm competition inherent in alternative tactics should favor traits in sneaks that increase their success in competition (see fig. 7.4). For *Onthophagus binodis* hornless sneaks are rare and have longer sperm than horned guards. In contrast, for *O. taurus* sneaks are common so that horned guards are also subject to high risk of sperm competition. In this case there are no differences in the length of sperm between tactics (Simmons et al. 1999c). Together, these studies provide compelling evidence that selection via sperm competition can favor increased sperm length.

Although Parker's (1993) theoretical analysis concluded that in general sperm competition should not favor increased sperm length, he noted four special circumstances under which it could: (i) ejaculate mass can only increase by increasing sperm size; (ii) the competitive benefits of sperm size increase with increasing numbers of sperm in competition; (iii) sperm size affects sperm survival and sperm competition risk increases with female remating interval; and (iv) sperm size increases competitive ability at the expense of survivorship but sperm competition risk decreases with female remating interval. If it is the case that the competitive benefits of sperm size increase with increasing numbers of sperm in competition, we should only expect to see sperm gigantism in species with high female mating frequency and, most importantly, where females store and mix sperm from all of their mating partners. That is, increased sperm size would not be predicted for species with mechanisms of sperm displacement. These criteria appear to be met in the giant-sperm-producing drosophilids, since in these groups females exhibit sperm mixing (Pitnick and Markow 1994b). Species with sperm precedence could be selected to have increased sperm size under Parker's (1993) third exception. Since sperm competition increases as sperm precedence mechanisms break down, increased sperm survival, and thus sperm size, could be favored. A number of studies show that actual competition between sperm becomes increasingly important (P_2 approaches 0.5) with time since the final copulation (table 2.3). Nevertheless, there are currently no data concerning the required relationship between sperm length and longevity.

Parker's (1993) theoretical analysis assumed that sperm compete directly for access to ova under a "raffle" mechanism that was loaded by a competitive weight dependent on sperm size. But sperm size may affect fertilization success in ways other than through direct competition for ova. In their study of beetles belonging to the genus *Bambara*, Dybas and Dybas (1981) noted a strong positive correlation between the size of sperm and the dimensions of the female's reproductive tract. The diameter of the sperm in the *B. invisibilis* group approximates the diameter of the female's spermathecal duct and the length of sperm approximates both the length of the duct and the length of the spermathecal lumen. Consequently, only one spermatozoon can pass to the spermatheca at any one time. Sperm are stored side by side within the spermatheca and because of their size very few sperm can be stored. Dybas and Dybas (1981) noted that often sperm can be seen protruding from the female's genital opening and postulated that the large size of sperm could function in blocking the female's sperm storage organs to future males (see also Sivinski 1980, 1984; Ladle and Foster 1992) or to prevent sperm from being displaced from storage once they have gained entry. Both mechanisms could feasibly confer a selective advantage in sperm competition and both predict the coevolution between sperm morphology and female reproductive tract morphology seen across species of *Bambara*. The analysis presented by Dybas and Dybas (1981) did not control for phylogenetic relationships between the species examined. However, recent studies of stalk-eyed flies (Presgraves et al. 1999), *Drosophila* (Pitnick et al. 1999), and moths (Morrow and Gage 2000) have reported similar patterns of coevolution between female reproductive tract dimensions and sperm length, illustrating that female influences are an important selective pressure on the evolution of sperm morphology (fig. 8.6).

Pitnick et al. (1999) reviewed four hypotheses that might predict covariation between sperm length and female reproductive tract morphology. The first, for which there is no empirical evidence, was that sperm length may evolve under selective pressures other than sperm competition, such as natural selection for male nutrient investments, and female tract morphology tracks evolutionary change in sperm length to ensure efficient storage and utilization of sperm. This hypothesis would predict that females with short sperm storage organs, relative to the average length of sperm produced, should have a lower fertilization efficiency than females with organs close to the mean length of sperm produced by males. The remaining hypotheses all involve aspects of the sperm competition hypotheses outlined above. The first of these considered sexual conflict over sperm usage. Selection could favor the evolution of long sperm if these sperm prevented rival sperm gaining access to the sperm stores (cf. Dybas and Dybas 1981) or if long sperm prevent incoming sperm from gaining precedence by stratification effects within the sperm stores (Briskie and Montgomerie 1992). Females may respond to such adaptations in males by increasing the length of their sperm storage organs so as to allow access of further ejaculates. Counteradaptation of this nature might be expected, for example, where females gain benefits from greater control over paternity (Briskie and Montgomerie 1992) or when mixtures of ejaculates from different males facilitate increased female fitness via the avoidance of genetic incompatibility (e.g., Zeh and Zeh 1997; Tregenza and Wedell 1998; Watson 1998). The product of such conflict would be an evolutionary arms race between sperm length and sperm storage organ morphology that resulted in a correlated evolution of the two traits. Pres-

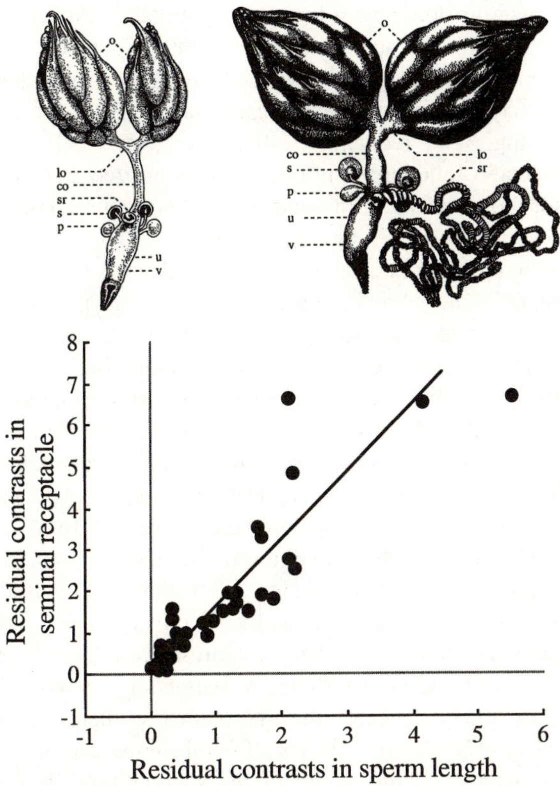

Figure 8.6 Correlated evolution of sperm length and female reproductive tract morphology across forty-five species of *Drosophila*. The data represent residual values of independent contrasts in the length of the seminal receptacle (the principal site of sperm storage and utilization) and sperm length taken from separate regressions of each variable on body size. Thus, an evolutionary increase in seminal receptacle length, beyond that expected for an evolutionary increase in female body size, is positively associated with an evolutionary increase in sperm length, beyond that expected from an increase in male body size (from Pitnick et al. 1999). The extremes of seminal receptacle length are illustrated by the diagrams of the reproductive tracts of *D. pseudoobscura* (*left*) and *D. bifurca* (*right*). *D. pseudoobscura* has a seminal recepticle 0.41 mm in length, compared with 81.67 mm in *D. bifurca*. o, ovaries; lo, lateral oviduct; co, common oviduct; u, uterus; v, vagina; s, spermatheca; sr, seminal receptacle; p, parovarium. (From Patterson 1943)

graves et al.'s (1999) study of stalk-eyed flies showed that an evolutionary shift in the site of sperm storage from the relatively small spermatheca to the relatively large ventricle receptacle was associated with an increase in sperm length. However, the notion that sperm length increases in order to block access to the sperm stores is not supported by available data from *Drosophila*; counter to the blocking hypothesis, the females of species with the longest sperm tend to store sperm from multiple males (Pitnick and Markow 1994b). The few studies of sperm utilization that are available for the genus *Drosophila* are at least consistent with the resistance hypothesis, in that longer sperm may be better able to resist displacement once in storage, thereby increasing P_1 or reducing P_2 (fig. 8.4). Nevertheless, many more studies of parentage following multiple matings are required in this genus, particularly for those species with giant sperm, before a reliable comparative analysis of this hypothesis can be performed.

The remaining hypotheses considered by Pitnick et al. (1999) incorporated aspects of female choice. Sperm storage organs might increase in length if this enabled only males with longer sperm to gain access to the sites of sperm storage and so obtain fertilizations. Coevolution between sperm length and sperm storage organ morphology could arise via a Fisherian process if, for example, assortative fertilizations resulted from a tendency for females with longer sperm storage organs to store longer sperm. Alternatively, if sperm length were correlated with some aspect of offspring performance, then a good genes process could operate (see section 1.1 for a discussion of the various models of preference evolution). Because long sperm are costly to produce (Pitnick et al. 1995a; Pitnick 1996), females could discriminate against males unable to bear these costs by selectively storing only the longest sperm. Currently there are no data to address these hypotheses. There is some suggestion from the yellow dung fly *Scatophaga stercoraria* that long sperm are more likely to enter the spermathecae than are short sperm (Otronen et al. 1997). Across species of moth, eupyrene sperm length is positively associated with the length of the spermathecal duct but not the dimensions of the spermatheca (Morrow and Gage 2000). The match between sperm length and duct length could be interpreted as a mechanism by which females could filter sperm; with female-controlled movement of sperm from storage to site of fertilization, sperm the length of the duct are likely to arrive ahead of sperm shorter than the duct, resulting in assortative fertilization. Of course such patterns could arise because of the competitive superiority of longer sperm, the tendency for females to selectively store or use longer sperm, or a combination of both selective processes. For example, Morrow and Gage (2000) argue that selection may favor sperm that increase to an equivalent length to the spermathecal duct if this enables them to migrate more rapidly from storage, gaining greater progressive forces because of the increased contact between tail and duct wall. Unfortunately, until we know more about the function of female reproductive tracts and the behavior of sperm within these tracts interpretations must remain speculative.

Within-species variation in sperm length is widespread in insects (Beatty and Sidhu 1967; Ward and Hauschteck-Jungen 1993; Pitnick and Markow 1994b; Sait et al. 1998; Simmons et al. 1999c) as well as other animal taxa (Ward 1998a) and can have a heritable basis (Joly et al. 1995; Ward 1998a).

Yet little attention has been paid to the evolutionary significance of sperm length variation within species. We need more studies on the influence of sperm length on aspects of sperm function such as motility, survival, storage, and fertilization success, before we can fully appreciate the selective pressures that influence individual sperm.

8.3 Sperm Polymorphism

Sperm polymorphisms have been reported from a diversity of insect taxa (section 8.1). Sperm dimorphism occurs throughout the order Lepidoptera (Jamieson 1987; Friedländer 1997), where the ejaculate contains two types of spermatozoa, apyrene and eupyrene. Apyrene sperm are typically one-third to one-half the length of eupyrene sperm (Gage and Cook 1994; He et al. 1995a) and because they lack a nucleus they are incapable of fertilizing eggs (Friedländer and Gitay 1972; Friedländer 1997). Sperm dimorphisms are also common in the *obscura* group of *Drosophila* (Joly and Lachaise 1994) and have recently been reported in several genera of stalk-eyed flies (Presgraves et al. 1997, 1999). Although dimorphic in the length of the head and tail, unlike the dimorphic sperm in Lepidoptera, both short and long sperm types possess a nucleus (Hauschteck-Jungen and Rutz 1983; Presgraves et al. 1997).

The functional significance of sperm polymorphism has been the subject of debate for many years and there are several competing hypotheses that have been generated. Some of these were developed specifically in relation to the apyrene/eupyrene dimorphism seen in Lepidoptera but all may be relevant to sperm polymorphisms in general: (1) Cohen (1977) suggested that "aberrant" sperm arise as a consequence of errors during meiosis due to chiasmata dysfunction and have no adaptive value; (2) Katsuno (1977b) suggested that the apyrene sperm of Lepidoptera facilitate the emigration of eupyrene sperm from the testes to the vasa deferentia; (3) it has been suggested that nonfertilizing sperm may assist fertilizing sperm in their migration to the sperm storage organ(s) of the female (Iriki 1941; Holt and North 1970; Friedländer and Gitay 1972); (4) nonfertilizing sperm might provide nourishment to the fertilizing sperm, the female, or the zygote and thereby represent male parental investment (Riemann and Gassner 1973; Sivinski 1980, 1984); (5) different morphs of sperm may be involved in fertilization at different times following insemination (Sivinski 1980, 1984); (6) nonfertilizing sperm may function in displacing or inactivating rival fertilizing sperm present in the female's sperm storage organ(s) (Silberglied et al. 1984); and (7) nonfertilizing sperm may represent "cheap filler" designed to stimulate the female reproductive tract and delay remating, thereby facilitating the avoidance of sperm competition from future rivals (Silberglied et al. 1984).

Developmental Errors

The error hypothesis is difficult to reconcile with the highly controlled and dichotomous nature of spermatogenesis (Friedländer 1997). In the Lepidoptera, cysts of spermatocytes begin to be produced midway through larval development, undergoing meiosis and elongation in the last larval stage. These spermatocytes are destined to produce eupyrene sperm. Apyrene spermatogenesis begins much later and appears to be triggered by a reduction in juvenile hormone titer in the hemolymph at pupation (Leviatan and Friedländer 1997). Spermatocysts produced after this time undergo shortened meiotic phases, with subsequent loss of nuclei. All spermatocysts produced after pupation thereby produce apyrene sperm. If random errors were responsible for sperm polymorphisms we should expect to see a mixture of sperm types arising from individual cysts, rather than specialization in the type of sperm produced at different stages of the life history. Likewise in *Drosophila* and stalk-eyed flies, sperm develop in discrete bundles containing either long or short sperm (Beatty and Sidhu 1970; Policansky 1970; Presgraves et al. 1997), suggesting that the development of alternative sperm morphs is precisely programmed rather than the product of random error. Errors can occur during spermatogenesis and are often related to environmental and/or genetic perturbations. Nevertheless, in sperm dimorphic species the evidence suggests that errors are specific to particular sperm morphs. Only eupyrene spermatogenesis is disrupted by sublethal rearing temperatures in *Plodia interpunctella* and *Ephestia cautella* (Lum 1977) or gamma irradiation in *E. cautella* (Riemann 1973). In *Drosophila* and stalk-eyed flies, meiotic drive systems result in the degeneration of *Y*-bearing long sperm but short sperm are unaffected (Bircher et al. 1995; Presgraves et al. 1997). These data, together with the apparent facultative adjustment in ratios of sperm morphs ejaculated (see below), are inconsistent with the notion that the apyrene sperm of Lepidoptera or the short sperm of *Drosophila* and stalk-eyed flies are the product of developmental errors.

Sperm Transport

The second hypothesis, that apyrene sperm somehow assist the transport of eupyrene sperm within the male, also seems unlikely, given that they are not activated until the time of ejaculation (Sheperd 1975; Herman and Peng 1976; Leopold 1976; Katsuno 1978), that is, after both apyrene and eupyrene sperm have migrated to the vasa deferentia. Moreover, apyrene sperm are transferred in large quantities to the female reproductive tract, which would seem unnecessary if they functioned only in the male reproductive tract.

The third hypothesis, that nonfertilizing sperm assist the movement of fertilizing sperm within the female's reproductive tract, derives from the observation that apyrene sperm are activated within the male reproductive tract on ejaculation and are highly motile when they are ejaculated (Sheperd 1975). In contrast, eupyrene sperm are transferred in bundles and are relatively immotile. The bundles are surrounded by the motile apyrene sperm as the entire ejaculate is transported to the spermatheca, where the two morphs separate and the eupyrene bundles finally separate into individual sperm (Holt and North 1970; Friedländer and Gitay 1972; Ferro and Akre 1975; Katsuno 1978; He et al. 1995a). Katsuno (1977a) provided compelling evidence for a role of apyrene sperm in facilitating eupyrene sperm release in an experimental manipulation of ejaculate features of the moth *Bombyx mori*. He examined the separation of eupyrene sperm bundles in three environments, in seminal fluid with no apyrene sperm, in the absence of seminal fluid and with inactive apyrene sperm, and in the presence of both seminal fluid and active apyrene sperm. Only eupyrene bundles in the last treatment separated successfully, suggesting that active apyrene sperm are essential for this process. While compelling, the sperm transport hypothesis itself does not account for the general sustained motility of apyrene sperm after their arrival in the spermatheca, or their long-term storage; sperm transport and eupyrene separation takes a matter of hours while apyrene sperm can persist in the female tract for several days to weeks after copulation (Holt and North 1970; Cook and Gage 1995; He et al. 1995a; Cook and Wedell 1999). Moreover, Proshold et al. (1975) found that the defective eupyrene sperm of *Heliothis subflexa* x *H. virescens* hybrids failed to reach the spermatheca despite the fact that these hybrids produced normal apyrene sperm. This strongly suggests that eupyrene sperm are responsible for their own transport within the female. Storage of apyrene sperm can be highly specialized. The spermatheca in Noctuidae is divided into two compartments, the ultriculus and the lagena (Drummond 1984). Epyrene sperm are stored in the ultriculus, distal to the lagena where apyrene sperm are stored (Holt and North 1970; He et al. 1995a). These facts seem difficult to reconcile with a simple eupyrene transport hypothesis for apyrene sperm.

Paternal Provisioning

The nutrient hypothesis for alternative sperm morphs suffers from the same problems as those outlined for the evolution of sperm length in section 8.2 above; specifically, any increased investment in zygotes via nutrient-carrying sperm is likely to cost males in terms of their ability to produce fertilizing sperm. With even low levels of multiple mating by females, which is all the more likely if males provide valuable resources for female reproduction, sperm competition should prevent the evolution of paternal investment via sperm (Parker 1984). There are additional factors that need to be considered. Female butterflies and moths derive nutrients from the seminal fluid of the ejaculate. Nutrients are incorporated into the somatic tissues of females and into developing eggs, thereby contributing to female survival and lifetime reproductive success (Rutowski et al. 1987; Dussourd et al. 1991; Kaitala and Wiklund 1994; Bissoondath and Wiklund 1995; Karlsson 1996; Wedell 1996; Oberhauser 1997). It therefore seems unlikely that apyrene sperm should serve a nutritive function when other aspects of the ejaculate serve this purpose. Seminal products are also translocated into developing eggs in *Drosophila* (Pitnick et al. 1991, 1997). Furthermore, in the sperm dimorphic species *D. obscura* and *D. pseudoobscura* males regurgitate crop contents to feed females during courtship (Steele 1986), making a further nutritive pathway via specialized sperm morphs seem unnecessary. Snook and Markow (1996) actually tested the nutrient hypothesis in *D. pseudoobscura*. Short sperm begin to break down and disappear from the female's reproductive tract 6 h after copulation. Using males labeled with radioactive isotopes, Snook and Markow (1996) found that the accumulation of male-derived amino acids within female tissues occurred before short sperm began to disintegrate. This observation is strong evidence against the notion that nutrients are derived from the breakdown of short sperm.

Sperm Competition

The remaining hypotheses all have their basis in sperm competition. The first of these was proposed by Sivinski (1980, 1984) and rests on the assumption that all sperm are fertilization competent. For example, different sperm morphs might represent a mixed strategy in sperm competition; long sperm could be better able to block access to sperm stores or resist displacement, while short sperm, because of their greater numbers, might be better able to outnumber rival sperm. Thus, different sperm may be adapted to compete in different regions of the reproductive tract or at different times after copulation (Sivinski 1980, 1984). Clearly this cannot explain the occurrence of apyrene sperm in Lepidoptera, which are not capable of fertilizing eggs (Friedländer and Gitay 1972). Neither can it explain the occurrence of dimorphisms in *Drosophila*. Although the two size classes of sperm are produced and transferred to the female's reproductive tract in almost equal numbers, only long sperm persist in the sperm storage organs. Moreover, only long sperm participate in fertilization (Snook et al. 1994; Bressac and Hauschteck-Jungen 1996; Snook and Karr 1998), falsifying the assumption that all sperm morphs are fertilization competent. It therefore seems highly unlikely that alternative sperm morphs represent alternative means of gaining fertilizations.

In their early discussion of the functional significance of apyrene sperm in the Lepidoptera, Silberglied et al. (1984) presented two hypotheses based on sperm competition. They argued that apyrene sperm may represent a type of worker morph that is dedicated to either an offensive or defensive role in sperm competition, allowing their eupyrene counterparts to gain fertilizations. They envisage two ways in which this could be achieved: by displacing sperm stored by the female from previous matings (their elimination hypothesis) and/or by preventing further matings (their prevention hypothesis). Silberglied et al. (1984) argued that in Lepidoptera the last male to mate with a female sires the majority of offspring and suggested that apyrene sperm may be responsible for the removal, inactivation, or destruction of sperm from previous males, generating the last-male advantage. However, it is clear from the studies in table 2.3 that last-male sperm priority is not the norm in Lepidoptera; as frequently it is the first male that fathers the majority of offspring. Across Lepidoptera, the mean value of P_2 is only 0.67 ± 0.05 and within species a bimodal distribution of P_2 values with peaks at the extremes of zero and 1 is common. These data do not provide general support for sperm displacement, let alone a role for apyrene sperm in such a process.

Etman and Hooper (1979) dissected female *Spodoptera litura* at various time intervals after copulation and examined their spermathecae for the presence of sperm. For initial matings, sperm first appeared in the spermatheca 45 min after spermatophore transfer. Following the second mating, however, sperm were lost from the spermatheca during the first 45 min after spermatophore transfer and then reappeared 60 min later. These data support the notion of sperm displacement and led Etman and Hooper (1979) to the conclusion that last-male sperm precedence occurred. However, that the spermatheca is devoid of sperm for over 30 min shows that the apyrene sperm of the second male cannot be directly responsible for this displacement. Etman and Hooper (1979) concluded that some unknown physiological mechanism of sperm expulsion was responsible.

Backcrosses between hybrids of *H. subflexa* and *H. virescens* produce sterile males due to morphogenic defects in eupyrene sperm. Eupyrene sperm from hybrid males fail to reach the spermatheca, hence male sterility (Proshold and LaChance 1974; Proshold et al. 1975). Apyrene sperm, however, appear normal (Richard et al. 1975). Pair et al. (1977) used these males in competitive situations with normal males to assess sperm utilization patterns. Normal males had an estimated P_2 of 0.76 while there was a reduction in fertility of females remated to backcross hybrids that predicted a P_2 of 0.83 had these males also transferred normal eupyrene sperm. Sperm counts from spermathecae of doubly mated females showed that after mating to backcross hybrid second males, 88% of the eupyrene sperm previously stored was lost from the female's spermatheca. Consistent with this interpretation, LaMunyon (2000) found both single- and double-mated females to have the same number of eupyrene sperm stored in their spermathecae. These data might suggest a role for apyrene sperm in sperm displacement. However, given the results of Etman and Hooper (1979), those obtained by Pair et al. (1977) and LaMunyon (2000) could also be explained by some physiological process in the female, triggered by the transfer of a complete spermatophore, regardless of the type of sperm contained within it.

The numerical fluctuations in short and long sperm within the sperm storage organs of *Drosophila pseudoobscura* are at least consistent with a role for short sperm in sperm displacement. Short and long sperm are transferred in approximately equal proportions to the female's uterus. Two hours after copulation, short sperm outnumber long sperm in the spermatheca and the ventral receptacle (fig. 8.7), suggesting that they migrate more rapidly to the sperm storage organs than do long sperm. Nevertheless, the number of long sperm increases rapidly so that by 12 h they outnumber short sperm. Twenty-four to thirty-six hours after copulation the sperm stores contain mainly long sperm. Although Snook (1998) rejected the possibility, the fluctuations in short and long sperm numbers within the female's sperm storage organs are at least consistent with a scenario in which short sperm migrate to the sperm stores early to affect displacement of rival sperm prior to the arrival of the fertilization-competent long sperm.

We have already seen how males are capable of adjusting ejaculate features in response to the perceived risks of sperm competition (chapter 7). Within-species plasticity may therefore provide insight into the functional significance of nonfertilizing sperm. If nonfertilizing sperm morphs functioned in sperm displacement, the numbers ejaculated might be expected to increase with the number of rival sperm that must be displaced. Although *Plodia interpunctella* are clearly capable of adjusting the numbers of sperm ejaculated in relation to sperm competition risk (Cook and Gage 1995), they do not increase the numbers of apyrene sperm transferred to females with larger stores of eupyrene sperm (see fig. 7.7) suggesting that apyrene sperm are not involved in displacement. Likewise, experiments with *D. pseudoobscura* failed to find an effect of female mating status on the relative numbers of short and long sperm transferred (Snook 1998). However, consistent with a sperm displacement function, males of the butterfly *Pieris rapae* do appear to transfer more apyrene sperm when competing with a large previous ejaculate than when competing with a small previous ejaculate (Wedell and Cook 1999a).

It has also been suggested that nonfertilizing sperm may interact directly with rival sperm, incapacitating them so that their fertilizing siblings will have a greater probability of being used for fertilization (Silberglied et al. 1984; Baker and Bellis 1988). Although incapacitation is a potential alternative route to sperm displacement for gaining last-male priority at fertilization, there is currently no evidence to suggest that sperm themselves are involved in such a process. Recently Kura and Nakashima (2000) examined the theoretical conditions necessary for the evolution of "soldier sperm" that specialize in killing rival sperm. In their models they assumed that sperm compete numerically according to Parker's (1982) fair raffle and contrasted the fitness of males producing soldiers versus those not producing soldiers. Kura and Nakashima (2000) conclude that soldier sperm can be favored even when the number of rival sperm killed per soldier is less than 1, and

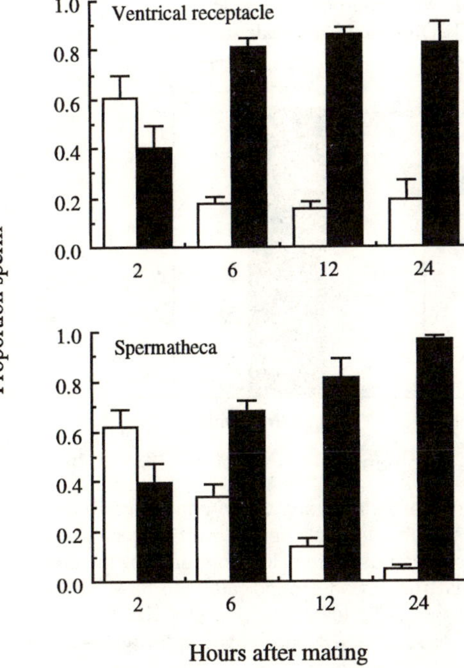

Figure 8.7 Numerical fluctuations of short (open bars) and long (solid bars) sperm within the sperm storage organs of *Drosophila pseudoobscura*. Initially, short sperm predominate in the sperm storage organs but rapidly become outnumbered by long sperm. Short sperm are lost from the sperm storage organs and play no role in fertilization. The patterns are consistent with a function for small sperm in initial displacement of rival long sperm from storage, prior to the arrival of their long-morph "siblings." (Data from Snook et al. 1994)

that the evolution of soldier sperm becomes more likely as this "killing ratio" increases. The increase in numbers of soldier sperm should be checked when the numbers of fertilizing sperm become so scarce as to compromise fertilization efficiency. Kura and Nakashima's (2000) analysis also found that soldier sperm should be smaller than fertilizing sperm, as observed in the apyrene sperm of Lepidoptera and the short sperm of drosophilids and stalk-eyed flies.

There is just one study that has attempted to examine the fertilization consequences of sperm dimorphism. Unlike the discrete sperm morphs seen in members of the subgroup *obscura*, dimorphism in *D. teissieri* is characterized by a continuous distribution of sperm lengths with two major peaks. Moreover, there is remarkable geographic variation, with some populations showing monomorphic sperm and others dimorphic sperm. Joly et al. (1991b) used this variation in experiments to investigate the influence of sperm dimorphism on patterns of sperm utilization. They crossed females from sperm monomorphic populations with both a sperm monomorphic male from their own population and a sperm dimorphic male from a different population. The experiment was repeated using females from dimorphic populations. They found no significant female effect on P_2 but a significant effect due to males and a significant male by female interaction (fig. 8.8). When females from sperm dimorphic populations were tested there was no

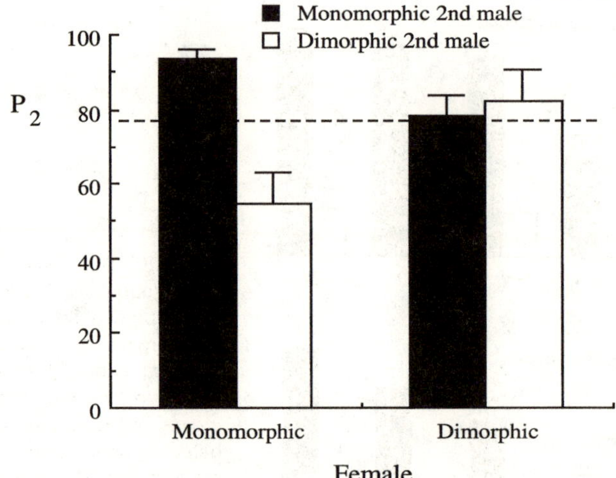

Figure 8.8 Outcome of sperm competition experiments in which both a sperm monomorphic and a sperm dimorphic male *Drosophila teissieri* were allowed to copulate with a single female. The experiment was repeated using females from each of the sperm monomorphic and dimorphic populations. The dotted line indicates the mean value of P_2 across all treatments; error bars represent ± SE (data from Joly et al. 1991b). Dimorphic males appear to be at a selective disadvantage when mating with females from the monomorphic population. Unfortunately, however, these experiments are confounded by homogamy and by the possibility of evolutionary coadaptations between males and females within populations.

difference in the proportion of offspring sired by the second male when he was either sperm monomorphic or sperm dimorphic. However, when females from sperm monomorphic populations were used, sperm dimorphic males had a significant disadvantage. We know that in other sperm dimorphic *Drosophila* short sperm are not fertilization competent (Snook et al. 1994; Snook and Karr 1998) so that the reduced paternity of sperm dimorphic males in competition with monomorphic males may arise because sperm dimorphic males have relatively fewer fertilizing sperm. This raises the questions as to how sperm dimorphism could arise in a sperm monomorphic population, and how it could be maintained? However, the sperm utilization patterns seen in *D. teissieri* could equally be due to homogamy, that is, the tendency for sperm from the female's own population to be more successful in fertilization. Thus, across all possible combinations, there is a tendency for the second male to father the majority of offspring (fig. 8.8). Homogamy results in second males from the female's own population gaining more fertilizations than would be expected from sperm competition while second males from different populations obtain fewer fertilizations than expected. Moreover, the type of interaction observed in Joly et al.'s (1991b) study of *D. teissieri* has recently been reported from crosses between isogenic lines of

D. melanogaster, and is thought to arise from an evolutionary conflict of interest between males and females over seminal-product-mediated sperm displacement (Clark et al. 1999; see section 11.3). It is not possible to disentangle these effects from those due to possible sperm dimorphism effects in the experiments of Joly et al. (1991b).

The storage of sperm in a spermatheca is clearly an important feature in the development of sperm dimorphisms in stalk-eyed flies. The ancestral state in this group is one in which there are three functional spermathecae and dimorphism in sperm length (Presgraves et al. 1999). An evolutionary loss of sperm storage function by the spermathecae is associated with a loss of sperm dimorphism. Moreover, there is an evolutionary association between short sperm length and spermathecal volume, but not long sperm length. These two lines of evidence strongly suggest that short sperm function is localized to the spermathecae (Presgraves et al. 1999). Although there is some suggestion from *Drosophila* that this function may involve sperm displacement, the evidence is limited and equivocal.

Silberglied et al.'s (1984) second hypothesis proposed that apyrene sperm served the role of avoiding sperm competition by preventing the female from remating. Female Lepidoptera become unreceptive to further males after mating, and the duration of the refractory period has been linked to the presence of a spermatophore in the corpus bursae (Sugawara 1979) and the presence of motile sperm in the bursae and/or spermatheca (Taylor 1967; Riddiford and Ashenhurst 1973; Thibout 1975; LaChance et al. 1978). Thus, Silberglied et al. (1984) suggested that apyrene sperm may represent "cheap filler" that delays remating by the female and thereby reduces the risk of future sperm competition. Apyrene sperm are transferred in large numbers, representing over 90% of the total number of sperm transferred (e.g., Cook and Gage 1995; He et al. 1995a), and are highly motile in the reproductive tract (Sheperd 1975; Katsuno 1978). Males could benefit by producing apyrene rather than eupyrene sperm because the costs of cell growth and synapsis necessary for the production of effective nucleated cells are bypassed. Certainly nonfertilizing sperm can be produced more rapidly than fertilizing sperm in both Lepidoptera (Friedländer 1997) and *Drosophila* (Snook 1998).

Variation in ejaculate features within species of Lepidoptera provide some evidence in favor of the prevention hypothesis. In the moths *P. interpunctella* and *Pseudaletia separata*, increased rearing density had the effect of increasing the numbers of apyrene sperm produced and ejaculated by adults (Gage 1995; He and Miyata 1997). Rearing density was related to the risk of sperm competition because females emerging from high-density populations mated more often than those emerging from low-density populations. This could be taken as evidence in favor of the prevention hypothesis because, with increased female polyandry, there would be a selective advantage to males able to delay remating via the transfer of apyrene sperm. It could also be taken as evidence for the displacement hypothesis, however, because increased risk of sperm competition would mean that males are more likely to be called upon to displace eupyrene sperm from rival males. Counter to this latter hypothesis, Cook and Gage (1995) have shown that, unlike eupyrene numbers, apyrene sperm numbers do not vary with sperm competition risk (see fig 7.7). Rather, males transfer more apyrene sperm to young females than they do to old females. Cook and Gage (1995) interpret this result as evidence for a role of apyrene sperm in delaying the onset of sexual receptivity; old females are unlikely to remate so that the transfer of apyrene sperm to old females would be unnecessary.

Similar evidence in favor of the prevention hypothesis comes from Wedell and Cook's work on *P. rapae* (Cook and Wedell 1996; Wedell and Cook 1999a,b). The interval between copulation and remating is influenced by the size of the spermatophore placed in the corpus bursae (Sugawara 1979). Males are able to transfer large spermatophores on their first mating but, because of the depletion of seminal fluids following mating, they can only produce small subsequent spermatophores. Cook and Wedell (1996) found that, although the spermatophores of recently mated males were smaller than those of their unmated counterparts, the numbers of apyrene and eupyrene sperm contained within them were considerably greater. When males are unable to avoid sperm competition via the production of a large spermatophore they may adopt an alternative tactic of increasing receptivity-inhibiting apyrene sperm. Direct evidence for this claim comes from work on the congener *P. napi* (Cook and Wedell 1999). As in *P. rapae*, female remating interval is influenced by the size of the spermatophore received (Kaitala and Wiklund 1994, 1995; Wiklund and Kaitala 1995) and recently mated males produce smaller spermatophores (Bissoondath and Wiklund 1996a). Cook and Wedell (1999) allowed virgin females to mate with either a virgin male or a recently mated male and then gave females the opportunity to remate daily for a period of 10 days. Females that failed to remate were found to have more apyrene sperm in their sperm stores than those that remated, while the numbers of eupyrene sperm present in each group did not vary (fig. 8.9). Moreover, for those females that did remate, the time to remating was positively correlated with the number of apyrene sperm in storage but not with the number of eupyrene sperm. This experiment provides good evidence that apyrene sperm function in delaying female remating and thereby avoiding sperm competition from future rivals.

Bressac et al. (1991a) compared the mating patterns of three species of *Drosophila*, two with monomorphic sperm and one with dimorphic sperm. As predicted from the prevention hypothesis, females of the dimorphic species *D. affinis* mated less frequently than the sperm monomorphic species *D. latifasciaeformis* and *D. littoralis*. However, there are many other possible variables that could explain the observed differences in female remating tendency, as well as phylogenetic effects that are not considered in this compar-

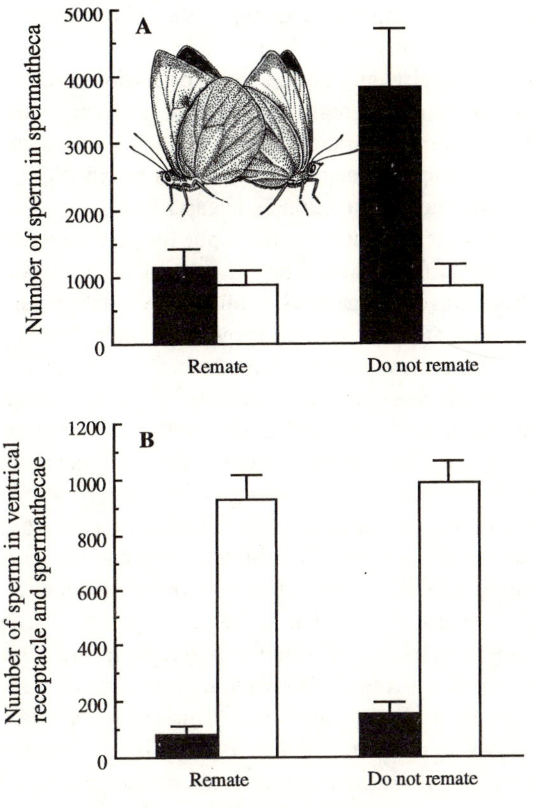

Figure 8.9 Testing the function of nonfertilizing sperm in Lepidoptera and *Drosophila*. (A) Apyrene sperm stored in the spermatheca of the butterfly *Pieris napi* are associated with a loss of sexual receptivity. The data are the mean (± SE) number of apyrene (solid bars) and eupyrene (open bars) sperm recovered from the spermatheca 10 days after the initial mating, or for those females that remated immediately on remating, before rival sperm had had the opportunity to arrive in the spermatheca. Within the remating group, the time to remating was also correlated with the numbers of apyrene but not eupyrene sperm (from Cook and Wedell 1999). (B) Neither short sperm (solid bars) or long sperm (open bars) appear to be related to sexual receptivity in *Drosophila pseudoobscura* (from Snook 1998).

ison. Snook (1998) performed an experiment similar to that of Cook and Wedell (1999) using the sperm dimorphic *D. pseudoobscura*. Although the numbers of short sperm in the ventricle receptacle and spermathecae were higher in nonreceptive females, the difference was not significant (fig. 8.9). One striking difference between the lepidopteran study and the *Drosophila* study that can be drawn from fig. 8.9 is the obvious difference in the relative proportions of nonfertilizing sperm in storage between the two taxa. Nonfertilizing sperm are not a major component of the sperm stores in *Drosophila*. Furthermore, in *Drosophila* short sperm have a lower beat frequency and wave propagation velocity, and unlike long sperm do not increase their motility on entering the female's reproductive tract (Bressac et al. 1991b). Both of these factors contrast markedly with what we know of lepidopteran apyrene sperm and support the conclusion drawn by Snook (1998) that short sperm in *Drosophila* do not function as cheap filler.

The various hypotheses that have been put forward to explain the evolution of sperm polymorphisms need not be mutually exclusive. For example, sperm competition favors correlated evolutionary changes in both apyrene and eupyrene sperm (Gage 1994; Morrow and Gage 2000), and we have seen how male butterflies increase the numbers of apyrene and eupyrene sperm when the immediate risk and intensity of sperm competition is high (Cook and Wedell 1996; Wedell and Cook 1999a,b). If apyrene sperm function in separating eupyrene sperm bundles, increased eupyrene transfer for sperm competition purposes could require greater numbers of apyrene sperm simply to separate the greater number of bundles or assist in their transport to the sperm stores. It is entirely possible that apyrene sperm could function both in separating eupyrene bundles and in reducing female receptivity to future matings. Nevertheless, what is clear from the evidence discussed above is that apyrene sperm are subject to selection under sperm competition. The evidence for a role of *Drosophila* short sperm in sperm competition is less convincing. Sperm dimorphism in *Drosophila* is limited to the *obscura* group but it is clear that this is not simply due to phylogenetic inertia. Snook (1997) used an autoregressive comparative method to analyze sperm length variation across 10 species from the *obscura* group, finding that while 22% of the variation in long sperm length was due to phylogeny variation in short sperm length was not. This analysis shows that short sperm are highly labile evolutionarily, and must have evolved, and be currently maintained, by some adaptive advantage. Clearly short sperm do not serve a nutritive or receptivity inhibition function, and more studies are required on their potential impact on sperm displacement.

8.4 Intraejaculate Sperm Competition

Within an ejaculate individual sperm can be seen as siblings who share some proportion of their genome. Because they are not genetically identical we might expect competition among them to fertilize available ova (Sivinski 1980, 1984). For example, all else being equal, a random mutation in the germ line that resulted in some sperm being better able to fertilize eggs should rapidly go to fixation at the expense of sperm not carrying the mutation. However, sperm in competition with their siblings could result in a reduction in the effectiveness of the ejaculate as a whole, if for example rogue sperm somehow disabled their weaker siblings. Thus, intraejaculate sperm competition could prove maladaptive from the perspective of the male progenitor. Sivinski (1980, 1984) suggested that the absence of phenotypic expression (haploid control) may be an adaptation to suppress the potential deleterious effects of intraejaculate competition, a suggestion supported by Haig and Bergstrom's (1995) theoretical analysis. Nevertheless, studies of vertebrates show that haploid expression can be an important component of spermatogenesis and sperm function (Erickson 1990), so that it should not be dismissed entirely.

Parker and Begon's (1993) models predict a conflict between parental and gametic interests over sperm size and number that is dependent on the degree of interejaculate sperm competition. If sperm size contributed to fertilization success under intraejaculate sperm competition, selection could favor a general increase in sperm length at the expense of sperm number. However, with interejaculate sperm competition, males with longer sperm would be disfavored because of their lower sperm counts, thereby preventing elongation of sperm. Thus Parker and Begon (1993) concluded that sperm morphology is likely to be a reflection of the resolution of conflict between haploid and diploid interests.

Because the haploid genomes of sperm can vary considerably in their genetic relatedness, Haig and Bergstrom (1995) suggest that intraejaculate sperm competition may be a significant obstacle to the evolution of sperm polymorphisms where different morphs have different functions. If, for example, helper sperm sacrificed their own prospects of fertilization for the sake of their siblings, those helpers that cheated would be over-represented in the successful sperm of a male's ejaculate so that the helping trait would soon be lost. Kura and Nakashima's (2000) theoretical analysis also showed that soldier sperm are less likely to evolve when there is an influence of the sperm genome on specialization. One solution to this problem, analogous to parent-offspring conflict, could be the disabling of helper sperm by the diploid parent, as occurs by the deletion of nuclear material during the development of apyrene sperm (Friedländer 1997). Haig and Bergstrom (1995) suggest that because the sperm produced by haplodiploid species are genetically identical, haplodiploid species might be more likely to exhibit sperm polymorphisms because of the greatly reduced intraejaculate sperm competition (see also Kura and Nakashima 2000). Interestingly, the haplodiploid hymenopteran *D. fuscipennis* has a total of five sperm morphs (Lee and Wilkes 1965). A detailed comparative study focusing on the degree of sperm polymorphism in diploid versus haplodiploid species is required to test Haig and Bergstrom's (1995) hypothesis.

8.5 Summary

Insect sperm exhibit rapid and divergent evolution, a pattern consistent with intense selection on sperm morphology. Competition amongst sperm is likely to favor adaptations that enhance fertilization potential. Despite the theoretical prediction that sperm should be small and numerous, a number of studies suggest that sperm competition is associated with an increase in sperm length. In fact a number of taxa exhibit sperm gigantism. Studies of giant-sperm-producing drosophilids suggest that increased sperm length is costly although the selective advantage for the maintenance of this trait remains elusive. Nevertheless it seems clear that giant sperm do not contribute to zygote nutrition. Sperm length and female reproductive tract morphology often show strongly correlated evolutionary change. Such patterns are consistent with a number of hypotheses, including an advantage in sperm competition via the blocking of sperm storage organs to sperm of future rivals or the resistance to sperm displacement. Correlated evolution is also consistent with a number of models of female choice. Unfortunately, we have little knowledge of the role sperm length plays in functional aspects of sperm biology such as motility and survival, or of the behavior of sperm within female reproductive tracts.

Some taxa produce both fertilizing and nonfertilizing sperm. In Lepidoptera, males adjust the numbers of fertilizing (eupyrene) and nonfertilizing (apyrene) sperm in relation to the risk of sperm competition. There is good evidence that apyrene sperm play a role in reducing female receptivity to further matings, thereby avoiding sperm competition from rival males. However, evidence for the functional significance of sperm dimorphisms in other taxa is less convincing. Despite their obvious importance, few studies in sperm competition have focused on observations of sperm dynamics or on selection acting on sperm morphology. An increased effort in this area should provide us with a greater understanding of the selective pressures responsible for the rapid and divergent nature of spermatozoon evolution.

9 Ejaculate Manipulation: Mechanisms of Female Choice

9.1 Introduction

Sperm competition has traditionally been viewed as an extension of male-male competition, occurring during and after copulation and favoring adaptations in males for achieving greater numbers of fertilizations within the competitive arena of the female's reproductive tract (Parker 1970e). Nevertheless, Parker (1970e) was explicit in his statement that "the female cannot be regarded as an inert environment in and around which this form of adaptation evolves." The storage, maintenance, and utilization of sperm is ultimately under the control of females and, as Lloyd (1979) first pointed out, female choice might also be expected to continue during and after copulation, especially in species where males monopolize females and prevent them from choosing before copulation. Lloyd (1979) thus suggested that females may manipulate ejaculates, selectively storing, using, or digesting them, depending on the characteristics of the copulating male. Multiple mating, coupled with an ability of females to store sperm in complex sperm storage organs, has thus been interpreted as an adaptation in females for mate choice (Lloyd 1979; Walker 1980; Thornhill and Alcock 1983; Sivinski 1984; Simmons 1987b; Eberhard 1996). Nevertheless, the influence of females over sperm utilization patterns, and the resulting selection on males through this avenue of female choice, has received relatively little attention in comparison with aspects of male-male competition examined in previous chapters of this volume. The initial bias in empirical investigation toward male influences reflects two issues. First, the adaptive significance of female choice has been the center of controversy since Darwin (1871) originally raised the idea. Initially biologists could not understand how preferences for often bizarre male traits could bestow a selective advantage on the choosing female (reviewed in Otte 1979; and in Andersson 1994). Once plausible models for preference evolution were derived, resistance persisted because biologists could not and perhaps still do not fully appreciate how genetic variance underlying preferred traits could be maintained in the face of directional selection. Second, male-male competition is often conspicuous and intense and can readily generate nonrandom mating without the active participation of females. In contrast, mechanisms of female choice are often subtle. When based on the same traits that function to increase a male's success in competition, as female preferences so often are (Berglund et al. 1996), ascribing variance in nonrandom mating to either process becomes difficult. Thus, behavioral ecologists were rightly cautious of ascribing nonrandom mating patterns to female choice (Partridge and Halliday 1984; Partridge 1994). Only very recently in the history of sexual selection have workers been able to establish, using rigorous experimental methods, that female choice is a significant selection pressure (Andersson 1994). Imagine then the problems in attributing nonrandom paternity to male and/or female influences, when we are unable to observe processes occurring within the female's reproductive tract. For example, a given male may achieve a disproportionate number of fertilizations simply because some characteristic of his ejaculate, such as the number of sperm or the swimming speed of sperm, allows them to reach the site of fertilization first, so that females play no role in generating nonrandom paternity (Partridge 1994). On the other hand, ejaculate performance may be environment dependent so that female influences over ejaculate performance of different males contribute to their fertilization success. Identifying, let alone disentangling, such processes is daunting. Nevertheless, it is essential if we are to assess the significance of female choice during and after copulation. Sperm competition as a discipline has had a far briefer history than premating sexual selection and, given the above, it is not surprising that attention was initially focused predominantly on aspects of male-male competition.

Knowlton and Greenwell (1984) provided the first theoretical analysis of the influence of female interests in their analysis of the evolution of sperm competition avoidance mechanisms in males. They assumed that mechanisms of sperm competition avoidance were costly to females, an assumption that now has empirical support (e.g., Chapman et al. 1995). Knowlton and Greenwell's (1984) models recognized that selection on females could prevent the evolution of sperm competition avoidance mechanisms in males, but the evolutionary outcome of sexual conflict over sperm competition avoidance was likely to reflect the relative costs for females and the benefits for males. In many cases the costs for females will be typically less than the benefits for males (Knowlton and Greenwell 1984). Further, the costs of sperm competition avoidance for females may even be outweighed by female benefits. Such benefits could include the potential for females to determine paternity of their offspring by a variety of means outlined below. Alternatively, if sperm aging resulted in a depression in fertility and/or zygote fitness, sperm displacement would constitute a significant direct advantage for females so that their interests need not conflict with those of males (Stockley and Simmons 1998; Reinhardt et al. 1999). Proximate mechanisms for the avoidance of sperm competition may also provide significant benefits to females (Waage 1984; Wilcox 1984; Fincke 1986; Tsubaki et al. 1994).

Thus Knowlton and Greenwell (1984) concluded that the resolution of sexual conflict would probably favor the evolution of sperm competition avoidance mechanisms in males.

Thornhill (1983, 1984a) defined female-influenced processes occurring during and/or after copulation that bias offspring production toward one male rather than another as cryptic female choice. The processes are cryptic in the sense that they occur within the body of the female and cannot be observed directly. Eberhard (1996) looked at current literature on reproductive biology from the female's perspective to illustrate more than 20 different potential mechanisms by which female behavior, morphology, and physiology could directly or indirectly contribute to variation in nonrandom paternity. Fifteen of these mechanisms were seen as applicable to insects (table 9.1) and can be classified into five main categories; female influence over remating, the transfer of sperm during mating, transport and storage of sperm, sperm utilization, and investment in offspring. Differential investment in offspring is not relevant to the current discussion of insect sperm competition and will not be considered further. For a general discussion of the phenomena the reader is directed to Burley (1988) and Eberhard (1996), and for specific examples in insects to Thornhill (1983, 1984a), Hughes (1985), Simmons (1987a), and Wedell (1996). Many arguments that were put forward by Eberhard (1996) for female "control" over paternity can be countered by alternative arguments for male "control" (see, for example, arguments for the evolution of male genitalia in chapter 3). Interpretations of male versus female control can rarely be more than a point of view, neither of which can be said to be right or true. Moreover, use of the term "control" implies that one or other sex has absolute authority over processes that by their nature require two participating individuals. Thus, I agree with Pitnick and Brown (2000) that use of the subjective term "control" should be avoided. Nevertheless, looking at existing literature from a different perspective is of considerable value because it generates new and alternative hypotheses to those generally advocated. Female influences have thus been discussed in previous chapters in this volume. In this chapter I explicitly examine the available evidence for female influence over sperm transfer, storage, and utilization as potential mechanisms of female choice via ejaculate manipulation.

9.2 Influence over Remating

Female influence over remating is not strictly a mechanism of cryptic female choice since mating is directly observable from behavior. Nevertheless, it is a means by which females can exploit the mechanism of sperm competition in mate selection. Thornhill and Alcock (1983) suggested that multiple mating by females may represent what they termed "genetic-benefit polyandry."

Table 9.1
Proposed mechanisms for cryptic female choice in insects.*

Influence remating
1. Remate with additional male
2. Impede plugging of reproductive tract
3. Impede plug removal by new males
Influence sperm transfer
4. Prevent intromission
5. Terminate copulation
6. Impede sperm transfer through changes in reproductive tract morphology
7. Remove spermatophore of current male
8. Resist manipulations necessary for spermatophore function
Influence sperm storage
9. Discard sperm of current male after copulation
10. Discard sperm from previous male
11. Fail to transport sperm to storage organ(s)
12. Move stored sperm to site where it can be removed by copulating male
Sperm selection
13. Bias use of stored sperm
Differential reproductive investment
14. Reduce rate of oviposition
15. Invest less in offspring

*For a more complete discussion of these potential mechanisms, see Eberhard 1996, 1997.

Females may remate to replace the sperm of a previous genetically inferior mate, thereby continually improving on their reproductive success as they encounter new males in the population (see also Walker 1980). Such a mechanism relies on the ability of females to utilize predominantly the sperm of the superior male when she finds one, or to continue to use sperm from previous mates if she does not. In many odonates females have two distinct oviposition strategies; they either copulate with the male resident on the territory before oviposition or they oviposit without copulation (Waage 1979a; Koenig 1991). In *Calopteryx splendens xanthostoma* females that oviposit without copulation do so by actively rejecting the copulation attempts of the territory owner (Siva-Jothy and Hooper 1995). DNA analysis of the sperm within the sperm storage organs of the female reproductive tract revealed that there was a greater genetic diversity of sperm in both the spermathecae and bursae of females that oviposit without copulating (Siva-Jothy and Hooper 1995). The lower genetic diversity of sperm in copulating females arises because territory owners displace sperm from the bursa copulatrix prior to delivering their own ejaculate (Siva-Jothy and Hooper 1995) thereby gaining 98% of fertilizations (Hooper and Siva-Jothy 1996). Siva-Jothy and Hooper (1996) were able to attribute offspring produced by females to the sperm store from which sperm were drawn to fertilize eggs. In accord with their previous study (Siva-Jothy and Hooper 1995), they found that when females copulated prior to oviposition, they used sperm predominantly from the bursa copulatrix so that the copulating male sired the majority of offspring. When they oviposited without copulation, females drew sperm predominantly from the spermathecae, using sperm from previous mates. These data show how females can control paternity via selective remating. Whether they gain fitness benefits from doing so remains to be seen.

With mechanisms of sperm displacement or sperm precedence such as those found in odonates, it is a relatively simple step to regulate paternity via remating; females simply allow a male to displace sperm from their stores or to place his sperm in a position where they will be utilized. Nevertheless, even with sperm mixing females can influence paternity through prolonged mating associations (e.g., Simmons 1986, 1987b; Müller and Eggert 1989; Lissemore 1997). Prolonged mating associations occur in the field cricket *Gryllus bimaculatus* (Simmons 1986). Females are attracted to the acoustic signals produced by males. On her arrival in a male's burrow he will begin to court the female. Females will mount the courting male who attaches an externally positioned spermatophore. The amount of time females spend in a male's burrow depends on a number of factors, including male body size (Simmons 1986), degree of relatedness to the female (Simmons 1989, 1990b, 1991a), and the levels of infection by a protozoan gut parasite (Simmons 1990c). Females who remain with males will copulate repeatedly, accepting multiple ejaculates (Simmons 1986, 1991a). Because sperm are utilized in direct proportion to their numerical representation within the spermatheca, multiple mating with the same male has the effect of biasing paternity in favor of preferred males (Simmons 1987b). Female choice in this species appears to confer indirect benefits on the choosing female's offspring; they have an increased developmental speed and survival (Simmons 1987a) and males preferred as mates sire sons who are themselves preferred as mates (Wedell and Tregenza 1999).

Eberhard (1996) suggested that females could control remating via cryptic means, for example, by accepting or rejecting mating plugs from their last partner (for a discussion of the evolution of mating plugs see section 4.2). He illustrated this possibility with Lorch et al.'s (1993) work on stalk-eyed flies *Cyrtodiopsis whitei*. In this species sperm are transferred packaged in a spermatophore which appears to serve the dual function of plugging the female's reproductive tract. Copulations with recently mated females are terminated after a matter of seconds, without the successful transfer of an ejaculate. Only after the spermatophore has been ejected are subsequent copulations successful. There is considerable variation in the timing of spermatophore ejection and, although there is no evidence of such, females could act in a selective manner by retaining the spermatophores of desirable males for longer than those of undesirable males.

9.3 Influence over Sperm Transfer

A more cryptic means of influencing a male's fertilization success is via control over the number of sperm he is able to transfer to the reproductive tract. Eberhard (1996) noted five means by which females could influence sperm transfer (table 9.1), and there is good evidence in the literature that females do indeed have considerable control over this aspect of mating.

Returning to the reproductive behaviour of *G. bimaculatus* (section 9.2), in addition to mating repeatedly with preferred males, females have been shown to have considerable control over the number of sperm they receive from each copulation. After males have attached the spermatophore they enter a period during which they "guard" the female while sperm are transferred to her reproductive tract, a process that takes up to 60 min (see also section 5.4). During this period the female may repeatedly attempt to remove the spermatophore and if she is successful will thereby terminate sperm transfer. Thornhill and Alcock (1983) suggested that, by persistently attempting to remove spermatophores, females could ensure insemination only by high-quality males able to prevent them from removing spermatophores. The amount of sperm transferred by a male would represent the outcome of sexual conflict over spermatophore attachment, with females able to exert their influence over low-quality but not high-quality males, resulting in a biased paternity toward high-quality males (Simmons 1991b; Zuk and Simmons 1997). The duration of the guarding period in *G. bimaculatus* increases with male body size so that females leave the spermatophores of large males attached for longer (Simmons 1986). Females will also leave the spermatophores of large males attached for longer in the absence of a guarding male, indicating that female assessment of male quality is not dependent on their ability to subvert the male's attempts to prevent spermatophore removal (Simmons 1986; fig. 9.1). Moreover, Simmons (1987b) showed that interference with sperm transfer had a direct impact on parentage; premature removal of spermatophores resulted in a reduced proportion of offspring sired at the time of fertilization (fig 9.1). Thus in this species of cricket, females exert their influence over paternity by remating and by controlling the amount of sperm transferred.

Control over sperm transfer is not limited to species with externally attached spermatophores. Sperm transfer in the midge *Culicoides mellitus* is via an internally positioned spermatophore (Linley 1981). Females become increasingly unreceptive to mating as they age and will kick out with their hind legs at males attempting to gain genital contact (Linley and Mook 1975). Unmated females quickly cease resistance on intromission. However, previously mated females will continue their defensive kicking during copulation. Linley and Hinds (1975b) found that the number of sperm transferred to females decreased by up to 21% depending on the female's previous mat-

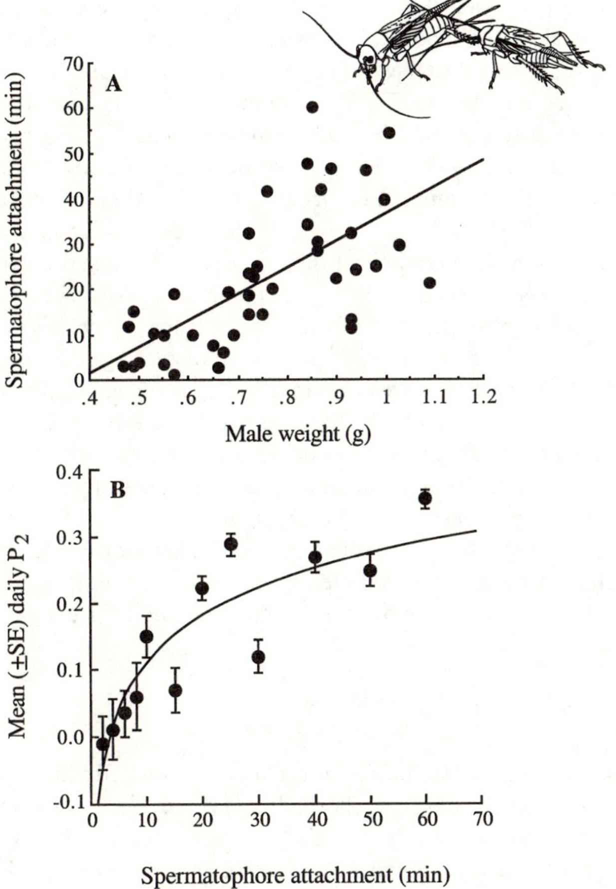

Figure 9.1 Control over insemination and paternity in the field cricket *Gryllus bimaculatus*. Sperm take around 60 min to be transferred from the externally attached spermatophore to the female's spermatheca. Following copulation females reach back and remove the spermatophore with their mandibles before consuming it. (A) In the absence of males, females will remove the spermatophore of small males earlier than those of large males ($F_{(1,\ 41)} = 32.49$, $P < 0.001$; Simmons, previously unpublished data). When males are present they will attempt to prevent females from removing spermatophores, but the same relationship persists (Simmons 1986). (B) The duration of spermatophore attachment is related to the number of sperm transferred and thus to the outcome of sperm competition. Males whose spermatophores are attached for longer obtain a greater proportion of fertilizations than males whose spermatophores are removed. Coupled with multiple mating with large males, female control over insemination biases paternity in favor of large males (see Simmons 1986; 1987). (Vignette from Loher and Rence 1978, with permission of Blackwell Wissenschafts-Verlag)

ing history and the total time she spent kicking during copulation. Experimental removal of the hind legs of twice-mated females resulted in males transferring more sperm to these manipulated females than to unmanipulated controls (Linley and Mook 1975). A corollary of female resistance and reduced sperm transfer is that the second male to mate obtains just 30% of fertilizations (Linley 1975). These experiments provide evidence that females can influence the amount of ejaculate transferred during copulation. It has not been established, however, whether females resist some males more than others, as would be required if this behavior represented cryptic female choice. The observed changes in sperm numbers are also consistent with strategic ejaculation by the male. We saw in chapter 7 how males might be expected to decrease the number of sperm with increasing intensity of sperm competition (Parker et al. 1996) or with decreased female quality (Galvarni and Johnstone 1998). Females do store sperm from more than one mating (Linley and Hind 1975a) and studies of parentage suggest that sperm mix in storage (Linley 1975). Thus, an alternative explanation to female influence is that males use female resistance to copulation as a cue to the current intensity of sperm competition and reduce the size of their ejaculates in relation to the decline in fitness gain expected from sperm competition. Removal of female hind limbs would remove the stimulus required for male assessment of females.

Nuptial Feeding

The evolution of nuptial feeding in insects has been directly linked to female manipulation of sperm transfer. Thornhill's (1976b, 1983) work with hanging flies *Hylobittacus apicalis* and *Harpobittacus apicalis* was the first to demonstrate the influence females can have over insemination. Males offer their mates a prey item on which to feed during copulation. Females prefer males with larger prey items, rejecting males offering small prey items prior to copulation. However, females can also determine the duration of copulation, terminating copulations with males offering small prey items before insemination is completed (see fig. 6.8). Female control over copulation duration appears to be a general phenomenon in insects where males provide prey items at copulation (see also Thornhill 1979; Svensson et al. 1990). In at least one of these species, *Panorpa vulgaris*, the outcome of sperm competition conforms to a mechanism of random sperm mixing so that copulation duration, and thus the number of sperm transferred, is directly related to a male's paternity expectation (Thornhill and Sauer 1991).

Nuptial feeding is particularly prevalent in the tettigoniids where, like the gryllids discussed earlier, sperm are transferred via an externally attached spermatophore. Male tettigoniids also provide a gelatinous mass, known as the spermatophylax, that is attached to the sperm-containing ampulla of the spermatophore and transferred to the female at copulation. As in crickets, female tettigoniids reach back after copulation to remove the spermatophore. However, they must first remove and consume the spermatophylax before they can gain access to the sperm-containing ampulla. It has been suggested that sexual conflict over sperm transfer, such as that seen in *G. bimaculatus*, may have been the selective pressure responsible for the evolutionary origin of these ejaculate protection devices (Boldyrev 1915; Thornhill 1976c). Given female manipulation of sperm transfer, there would be an immediate selective advantage associated with any trait in males that served to prevent females from manipulating sperm transfer. The ejaculate protection hypothesis thus predicts that males should provide enough material to distract the female until sperm have been successfully transferred. This prediction has been borne out for a number of species of tettigoniid (Wedell and Arak 1989; Simmons and Gwynne 1991; Reinhold and Heller 1993; Simmons 1995a; see fig. 9.2), as well as for the gryllid *Gryllodes sigillatus*, in which there has been convergent evolution for a similar spermatophore protection device (Sakaluk 1984). Moreover, the size of the spermatophylax, and thus time available for sperm transfer, is positively associated with fertilization success (Sakaluk 1986; Wedell 1991; Sakaluk and Eggert 1996; Sakaluk 1997). The ejaculate protection hypothesis is also supported by two independent comparative analyses across 16 species (Wedell 1993b) and 46 species (Vahed and Gilbert 1996) of tettigoniid. Vahed and Gilbert's (1996) analysis shows that evolutionary increases in the number of sperm transferred at copulation, and the mass of the sperm-containing ampulla of the spermatophore, are associated with evolutionary increases in the amount of spermatophylax material provided to the female (fig. 9.2). Moreover, the chemical composition of the spermatophylax suggests that it has evolved to exploit a preexisting sensory system in the female. Heller et al. (1998) working with five species of phaneropterid tettigoniid, and Warwick (1999) working with *G. sigillatus*, found that the spermatophylax was composed almost exclusively of free amino acids, in concentrations above those found in normal food. Chemical composition varied little across species of phaneropterid and was independent of dietary composition in *G. sigillatus*, suggesting that the amino acid content is important in function. Importantly, the amino acid content was highly imbalanced, with very low levels of essential amino acids but high levels of glycine and proline in *G. sigillatus* and glycine and glutamine in phaneropterids. Free amino acids are powerful phagostimulants (Cook 1977; Reinecke 1985) and Warwick (1999) found that females initiate feeding on spermatophylaxes and spend longer feeding on them than they do normal food. Sakaluk (2000) has recently shown that spermatophore attachment can be extended in three species of gryllid that do not provide a spermatophylax meal by providing them with a spermatophylax derived from *G. sigillatus*. The spermatophylax thus invokes the same general gustatory response in females of species in which males do not provide a spermatophylax. These data strongly imply that spermatophylax composition has

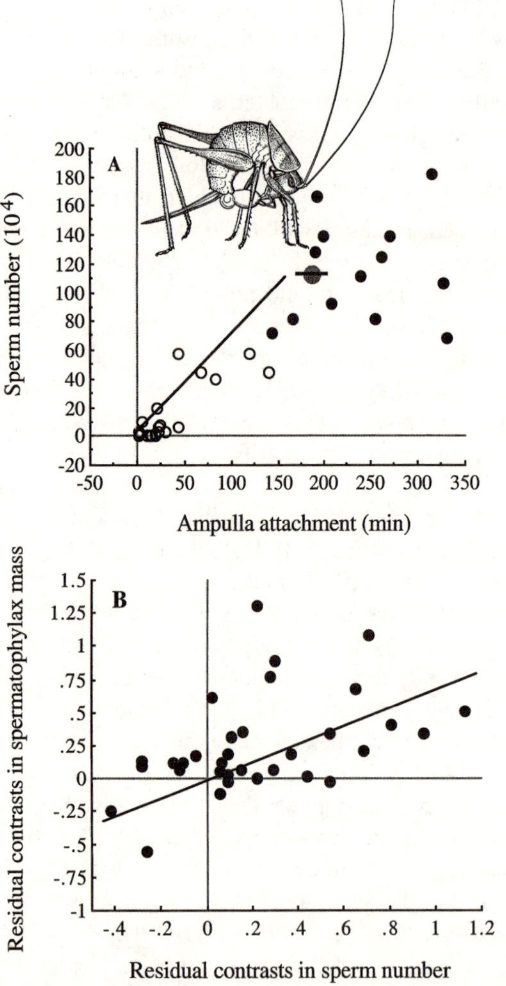

Figure 9.2 Male tettigoniids have regained control over sperm transfer via the evolution of an accessory gland product, the spermatophylax, that is attached to the ampulla of the spermatophore and removed by the female when she attempts to remove the spermatophore. (A) the amount of spermatophylax material is just sufficient to allow the transfer of the ejaculate from the ampulla. The data show the amount of sperm transferred to the spermatheca of female *Requena verticalis* in relation to the duration of ampulla attachment. Open circles are experimental females whose ampulla were removed with forceps. Closed circles are females who removed the ampulla themselves after a normal mating. The large shaded circle is the mean (± SE) time for females to consume the spermatophylax. The spermatophylax is fully consumed at a time when no further sperm are transferred from the ampulla (data from Simmons 1995a). (B) Across species of tettigoniid, an evolutionary increase in the number of sperm contained within the ampulla of the spermatophore is associated with an evolutionary increase in the mass of spermatophylax material provided to the female. The data are independent contrasts from a comparative analysis across thirty-five species of tettigoniid (redrawn from Vahed and Gilbert 1996).

evolved to exploit the female's feeding response. Thus, cryptic female choice, in this case through interference with sperm transfer, appears to have favored the evolution of a physiological response in males, the synthesis of a spermatophylax by the accessory glands, that increases the probability of successful sperm transfer and fertilization success. The original context of spermatophore removal in the tettigoniids may have been a mechanism to enhance offspring fitness through female choice, as in *G. bimaculatus* (Simmons 1986, 1987a). The habit of removing and eating spermatophores is ancestral to the evolution of the spermatophylax (Gwynne 1997). The sexual conflict that ensued has probably limited the extent to which females can now determine paternity via this route.

As with lepidopteran seminal fluids (section 4.3), this adaptation for sperm transfer in males can in some cases provide incidental benefits to females. In some cases the nutrients and/or moisture obtained through the consumption of the spermatophylax can increase the number of offspring produced or their viability and growth (Gwynne 1984a, 1988a; Simmons 1990a; Ivy et al. 1999; Reinhold 1999). In at least one species, *Requena verticalis*, the spermatophylax appears to have been subject to further elaboration via natural selection for male parental investment; males appear capable of adjusting the nutrient composition of the spermatophylax as well as increasing its size beyond that necessary for ejaculate transfer when the benefits of investing resources in females are high (Gwynne 1986; Simmons 1995a,b; Simmons et al. 1999a). When they do invest parentally, they increase the numbers of sperm in the ejaculate, which assures them of paternity of the offspring in which they invest (Simmons and Achmann 2000). Wedell's (1994) comparative analysis suggests that subsequent natural selection for increased nutritive function of the spermatophylax may have occurred several times in tettigoniid evolution.

The evolutionary conflict of interest over sperm transfer that seems to drive the evolution of the spermatophylax in tettigoniids and some gryllids, and the donation of prey items in hanging flies and scorpion flies, may be generally responsible for the evolution of adaptations that distract females during copulation. In a number of spermatophore-donating orthopterans, males feed females with glandular secretions (Brown 1997) and even parts of their own body (Johnson et al. 1999) during and after copulation. Even sexual cannibalism may have its origin in facilitating sperm transfer (Andrade 1996). But in some taxa males can be severely limited in their ability to provide nuptial gifts and under these circumstances selection appears to have favored adaptations in males for sexual coercion (Thornhill 1980a, 1984b; Thornhill and Sauer 1991; Sakaluk et al. 1995). In the scorpion flies, males possess a dorsal clamp with which they grip the female's wings during copulation. Normally, males will offer a prey item to females in order to achieve a copulation that is optimal for them. Under these conditions the dorsal clamp seems to play no role in extending copula duration. However, if males are unable to obtain prey items, they will produce salivary masses that they offer as alternative gifts to females. Moreover, males who are unable to manufacture salivary masses will attempt to copulate using the clamp alone (Thornhill 1980a). Females prefer males providing prey items over those providing salivary masses and will attempt to terminate copulations when there is no prey item provided. Under these circumstances males with a functional dorsal clamp are able to extend the duration of copula beyond that of males with an experimentally manipulated, nonfunctional clamp (Thornhill 1984b; Thornhill and Sauer 1991). This example clearly illustrates the intensity of sexual conflict over reproduction and the importance of female control over reproductive events in generating selection on males. It also raises an interesting alternative explanation for the rapid and divergent evolution of insect genitalia. Richards (1927) and Thornhill and Alcock (1983) suggested that the often complex male genitalia may act as holdfast devices to avoid takeover from rival males during copulation. Like the dorsal clamp of scorpion flies, complex genitalia may equally represent adaptations to circumvent termination of copulation by the female before an amount of sperm has been transferred that is optimal from the male's perspective.

9.4 Influence over Sperm Storage

In some species males ejaculate directly into the female's sperm storage organ(s) (Gregory 1965; Ono et al. 1989; Gack and Peschke 1994; Tadler 1999). More typically, however, males deliver their ejaculate into the vagina or bursa copulatrix (an area of the reproductive tract internal to the genital opening of the female) and sperm are subsequently transported to the sperm storage organ(s) by muscular contractions of the female's reproductive tract (Linley 1981; Heming-Van Battum and Heming 1986; Bloch Qazi et al. 1996; Kaufmann 1996; Lachmann 1997; Bloch Qazi et al. 1998; Simmons and Achmann 2000; Hosken and Ward 2000). Variation in the numbers of sperm transferred to the sperm stores by females could have major implications for male fertilization success, and represent a mechanism of cryptic female choice.

Female influence over sperm entry and exit from the spermatheca has been examined in two species of beetle in which the spermatheca has associated musculature, implying female involvement in sperm storage and release (Villavaso 1975a; Rodriguez 1994b). By cutting the spermathecal muscle of female boll weevils *Anthonomus grandis*, Villavaso (1975a) showed that a functional spermathecal muscle was not necessary for sperm storage; sperm either enter the spermatheca due to the fluid pressures involved in ejaculation and/or they swim there under their own control. The spermathecal muscle in chrysomelid beetles *Chelymorpha alternans* does influence the uptake of sperm (Rodriguez 1994b), and in both species it controls the exit of sperm from the spermatheca. Females with cut spermathecal muscles were unable to utilize sperm stored in their spermathecae and thus laid infertile eggs. In *A. grandis* paternity by second males was reduced from 66% to 22% by cutting the spermathecal muscle (Villavaso 1975a). These data suggest that, as with female dung flies (see section 6.2), female *A. grandis* may be complicit in sperm displacement by second males. However, whether females show selective cooperation with males dependent on mate phenotype has not been established. Moreover, subsequent experiments using females that had had their spermathecae surgically removed failed to support the original findings; the last male to copulate fertilized 66% of eggs with normal females and 62% of eggs with females that had no spermatheca (Nilakhe and Villavaso 1979) suggesting that the spermatheca plays no role in sperm displacement by males.

The overt expulsion of sperm from the female's reproductive tract following copulation is suggestive of cryptic female choice (Eberhard 1996). For example, in the chrysomelid *C. alternans*, Rodriguez (1995) found that females would sometimes, but not always, expel a droplet of sperm from their reproductive tract after copulation. The male has an elongated flagellum, averaging 24 mm in length, that is threaded through the spermathecal duct of the female and often into the spermatheca (Eberhard 1996). Sperm flow from the spermatophore in the bursa copulatrix along the length of the flagellum and into the spermatheca. Thus, the flagellum in this species plays an important role in the transportation of sperm to the sperm storage organ. The length of the male flagellum was shorter in copulations with sperm ejection, than in copulations without sperm ejection and experimental shortening of the flagellum resulted in an increased probability of sperm ejection (Rodriguez 1995). The ejection of sperm was associated with fewer sperm being stored in the spermatheca. Moreover, among sperm ejecting females, those mated with intact males had more sperm in their spermatheca after copulation than those mating with manipulated males. It therefore seems that males may have some influence over sperm storage via the entry of the flagellum into the spermatheca. Nevertheless, experimentally cutting the spermathecal muscle of females also affected the probability of sperm ejection, suggesting that females also influence sperm storage (Rodriguez 1995). Sperm utilization experiments showed that sperm mix in storage and that a male's fertilization success is a positive function of his flagellum length, relative to the lengths of his competitors flagella (Rodriguez 1994a). Thus, selection through increased success in sperm competition should favor increased flagellum length in males. The source of that selection is the morphology of the female's tract (the length of the spermathecal duct) and the tendency for females to eject sperm that are not placed into the spermatheca. But the selection occurs in the context of sperm competition, so male and female influences seem equally important. It is not known if there is covariation between flagellum length and spermathecal duct length across species in this genus. However, it might be expected if males and females are in a coevolutionary arms race to gain control over transport of sperm into the sperm stores.

Copulatory Courtship

Eberhard (1991, 1994, 1996) suggested that females may assess males during copulation and recognized the widespread occurrence of "copulatory courtship" as a possible avenue by which males might attempt to persuade females to transport their sperm to the sperm stores and thereby increase their individual reproductive success. In his review, Eberhard (1994) found that 81% of 131 insect species exhibited behavior during copulation that could be interpreted as copulatory courtship. He utilized strict conservative criteria in the interpretation of behaviors. For example, he did not include behaviors directed at other individuals attempting takeover (section 5.5). Rather, he included repeated behaviors by the male directed toward females, such as tapping, stroking, or rubbing of the body with the antennae, legs, or genitalia. In support of a courtship function for behavior delivered during copulation, the behavior is often decoupled from ejaculation, occurring either before or after transfer of the ejaculate (Otronen 1990; Eberhard 1996; Simmons et al. 2000). Moreover, Eberhard (1996) noted that copulatory courtship in beetles of the genus *Macrohaltica* is species specific, as would be expected if it were under sexual selection (fig. 9.3). Importantly, Eberhard (1994) noted that definitive proof that such behavior functions as copulatory courtship requires evidence that the behavior is related to male fertilization success.

Copulatory courtship has been reported in solitary bees and appears to be involved in the loss of sexual receptivity in female *Centris pallida* (Alcock et al. 1976; Alcock 1979a; Alcock and Buchmann 1985). Males search for females at emergence sites and dig into the ground to reach emerging females. Intromission and insemination are brief, following which the male performs a series of stroking movements with its legs and antenna and produces an acoustic signal or "buzzing." Copulatory buzzing appears widespread in solitary bees and wasps (see Larsen 1986; Simmons et al. 2000). Alcock and Buchmann (1985) found that females who had been experimentally separated from their mate before the onset of copulatory courtship were more likely to mate a second time than were females who received copulatory courtship. Simmons et al.'s (2000) study of *Amegilla dawsoni* suggests that copulatory courtship may stimulate females to transport sperm from the bursa copulatrix to the spermatheca. However, unlike for *C. pallida*, experimental removal of copulatory courtship had no influence on a female's immediate probability of remating.

Our most detailed understanding of the adaptive significance of copulatory courtship comes from Otronen's work with the carrion fly *Dryomyza anilis*. Females arrive at fresh carcasses or droppings to feed and lay their eggs.

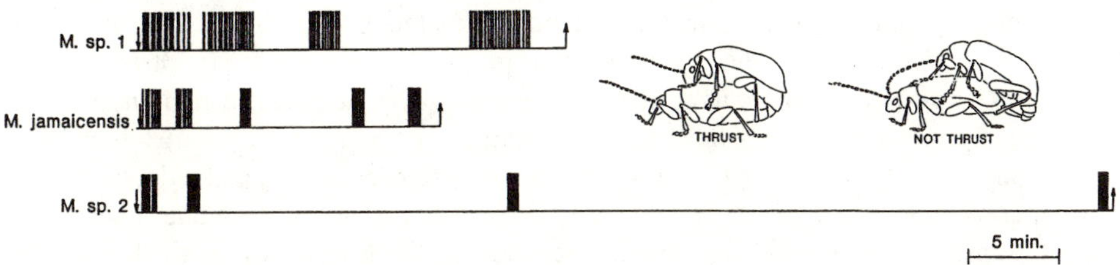

Figure 9.3 Species specificity in the patterns of genitalic thrusting in three species of *Macrohaltica* chrysomelid beetles. Each vertical bar represents a thrust, in which the male pushes his genitalia deep into the female; the width of the bar is the duration of the thrust. Intromission begins with the downward-pointing arrows, and ends with the upward arrows. (From Eberhard 1996)

Males establish territories at these resources and soon capture incoming females (Otronen 1984). The pair leave the resource to copulate in the surrounding vegetation. While mounted the male holds the female around the wings and abdomen with his hind legs. Mating follows a complex series of events. Intromission and sperm transfer occur during the first minute, after which the male withdraws his aedeagus. Copulatory courtship then ensues as the male performs rhythmic tapping of the female's external genitalia with his genital claspers and simultaneously squeezes the female's abdomen with his hind limbs. Tapping movements last around 1.5 s and are repeated on average 22 times. After the male has performed a number of tapping sequences the pair return to the resource where the female expels a droplet of sperm from her reproductive tract, before beginning to lay a batch of eggs. The pair then leave the resource and engage in a second copulation bout (intromission and tapping sequence), before returning for a second bout of oviposition. This continues for up to six copulation and oviposition bouts before the female finally leaves the resource. The majority of eggs are laid in oviposition bouts occurring late in the copulation bout sequence. The first copulation bout is characterized by an average of ten tapping sequences, while subsequent copulation bouts have about five tapping sequences (Otronen 1990).

Otronen and Siva-Jothy (1991) used radioactively labeled amino acid mixtures to label ejaculates and observe their distribution within the female's reproductive tract and within the droplet of sperm discarded prior to oviposition. Females ejected about the same volume of sperm prior to oviposition as they received from a single copulation. In matings where one male was labeled they found that males delivered their ejaculate into the bursa copulatrix. However, most of those sperm were ejected in the droplet; 10–26 % stayed in the bursa but only about 0.7–6 % of labeled ejaculate was retained within the spermathecae. In double matings, where each male was labeled with a different isotope, they found that an increase in the number of tapping sequences performed by the second male resulted in a decrease in the representation of his ejaculate in the discarded droplet, and an increase in his representation in the ejaculate retained in the bursa copulatrix. Otronen (1997a) subsequently counted the numbers of sperm in different regions of the reproductive tract. The female has three spermathecae, a singlet that connects to the bursa copulatrix via its own spermathecal duct, and a doublet that consists of two separate spermathecae that share a common duct. Otronen (1997a) found that tapping resulted in the transfer of sperm from the bursa to the singlet spermatheca, and a decrease in the number of sperm ejected in the droplet. Moreover, sperm were drawn from the singlet spermatheca for fertilization. Together, these studies suggest that sperm are ejaculated into the bursa copulatrix and are exchanged between the bursa and the singlet spermatheca, a pattern that is remarkably similar to that found in the yellow dung fly (see section 6.2). The data show that copulatory courtship, in this case the male's tapping sequence, is critical for the transport of sperm within the female's reproductive tract. Not surprisingly, Otronen (1990) found that a male's fertilization success increased with the number of tapping sequences in a copulation bout (fig. 9.4), the number of tapping movements within a tapping sequence (Otronen 1997b), and the number of copulation bouts (Otronen 1994b).

Although there is considerable within-male variation in copulatory courtship performance, there is significantly greater variation between males in key aspects such as the number of tapping movements per sequence, the number of tapping sequences, and realized fertilization success (Otronen 1997b). Thus, there is the potential for female-imposed selection, based on differences in copulatory courtship and fertilization success, to drive evolutionary change in males. Some of this variation is related to male size, with larger males having a greater number of tapping movements per tapping sequence (Otronen 1997b) and a higher fertilization success (Otronen 1994a). Moreover, selection appears to be acting on male genitalic morphology. The male has two genital claspers on each side of the body. One pair is significantly larger than the other and they differ in their patterns of asymmetry. The large clasper is bilaterally symmetrical while the small clasper exhibits directional asymmetry, with the right hand small clasper being larger than the left hand small clasper. Together, the two pairs of claspers form the functional unit used to tap the female's external genitalia. Otronen (1998) found that males with greater directional asymmetry in small clasper length had a higher fertilization success than males with symmetrical small claspers. Interestingly, the female's sperm storage organs are positioned asymmetrically within the abdominal cavity. Females utilize sperm from the singlet spermatheca, tapping increases the number of sperm transferred to the singlet spermatheca, and the singlet spermatheca lies on the female's right hand side (Otronen and Siva-Jothy 1991). It is the right hand small clasper that is enlarged.

It might be argued that female *D. anilis* have control over male reproduc-

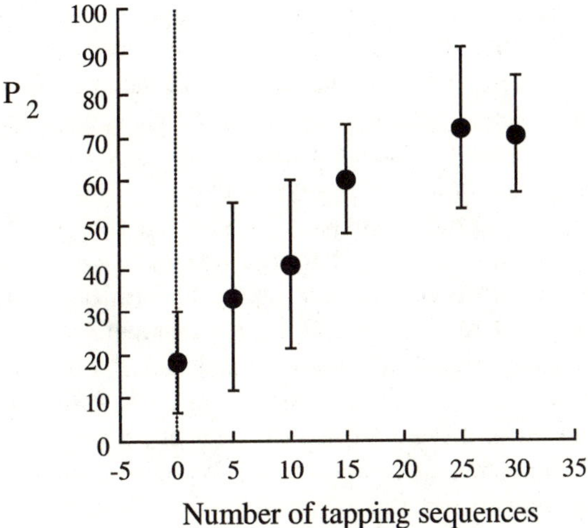

Figure 9.4 Copulatory courtship facilitates cryptic female choice in the fly *Dryomyza anilis.* After transferring sperm to the female's bursa copulatrix, the male taps the female's external genitalia with his genital claspers and squeezes her abdomen with his hind limbs. Tapping mobilizes sperm within the female's reproductive tract, increasing the proportion of the male's sperm retained in the sperm storage organs and decreasing the proportion in a sperm droplet that is ejected prior to oviposition. The fertilization success of males thereby increases with the number of tapping sequences performed in a copulation bout. Moreover, males with more asymmetrical small claspers and with shorter small claspers have higher fertilization success. (See Otronen 1990; Otronen 1998)

tive success (Eberhard 1996). However, males also adjust their behavior in a manner that is adaptive from their own perspective. Otronen (1990) found that males would terminate copulation sequences early with immature females or with females with few eggs to be laid. She also found that in natural matings the most important variables affecting male copulatory courtship behavior were the number of eggs a female carried and her body size. The time males spent tapping and the number of copulation bouts both increased with the number of eggs a female contained. These data show that males invest more time and effort in females who offer a greater fitness gain, a situation analogous to that found in yellow dung flies (section 6.2). They also stress the importance of considering both male and female perspectives. Otronen's work provides strong support for cryptic female choice in its broadest sense; female processes generate selection on males that favor the evolution of behavior and morphology that increases their fertilization success. The alternative view, that selection due to sperm competition favors these behaviors in males, is equally true although clearly female influences play an important role in the selection process.

Very recently Edvardsson and Arnqvist (2000) have provided evidence that copulatory courtship is also an important aspect of fertilization success in red flour beetles *Tribolium castaneum*. The outcome of sperm competition in this species is highly variable and 58% of the variation in P_2 values can be attributed to differences between females (Lewis and Austad 1990), suggesting that females may influence paternity. One of the problems with correlational studies such as those described above for *D. anilis* is that variation in copulatory courtship may be associated with some third variable, such as sperm numbers, sperm viability, or fertilization capacity, that could equally account for the observed correlated variation in fertilization success. However, Edvardsson and Arnqvist's (2000) study manipulated the female's perception of copulatory courtship but not the male's performance of the behavior. During copulation the male rubs the lateral edges of the female's elytra with the tarsi of his mid or hind legs. The behavior is performed in bouts of one to a few strokes per bout and can involve the use of one or two legs simultaneously. Edvardsson and Arnqvist (2000) shortened the tarsi of experimental males so that they could not reach the lateral edges of the female's elytra. Nevertheless, they continued to perform the stroking movements. Edvardsson and Arnqvist (2000) recorded the rate of stroking movements during copulations by both manipulated and control males. The manipulation did not affect the numbers of sperm transferred to females. However, it had a significant effect on realized fertilization success (fig. 9.5). Fertilization success increased with the rate of leg rubbing performed during copulation by control males. Importantly, this relationship was absent from manipulated males, whose mates were unable to perceive the copulatory courtship. Taken together with the known female control over transportation of sperm from the bursa to the spermatheca during copulation (Bloch Qazi et al. 1998), these data provide unequivocal evidence that copulatory courtship can play a critical role in male fertilization success (Eberhard 1996). But whether variation in male paternity is adaptive from the female's perspective remains to be seen. It is possible that assortative fertilization between females that respond to male copulatory courtship and males able to provide high levels of copulatory courtship results from a Fisherian process. Alternatively, males with more vigorous copulatory courtship could sire offspring of superior viability so that females obtain indirect fitness benefits for their offspring. Lewis and Austad (1994) have found a positive association between male olfactory attractiveness and P_2, and Yan and Stevens (1995) found that parasitic infection reduces P_2. These two male traits could be relevant to indirect mechanisms of preference evolution. It may be that males able to invest more resources into pheromone production are also able to invest more in copulatory courtship, which has been shown to entail nontrivial energetic costs at least for spiders (Watson and Lighton 1994). Similarly, males with parasitic infections are likely to be in poor condition and unable to engage in energetic activities such as copulatory courtship. Reduced levels of copulatory courtship would result in reduced sperm storage and paternity (fig. 9.5). With heritable resistance to parasite infection (Ham-

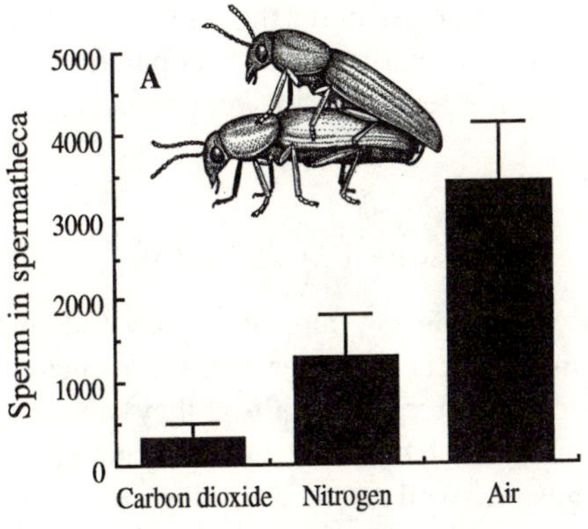

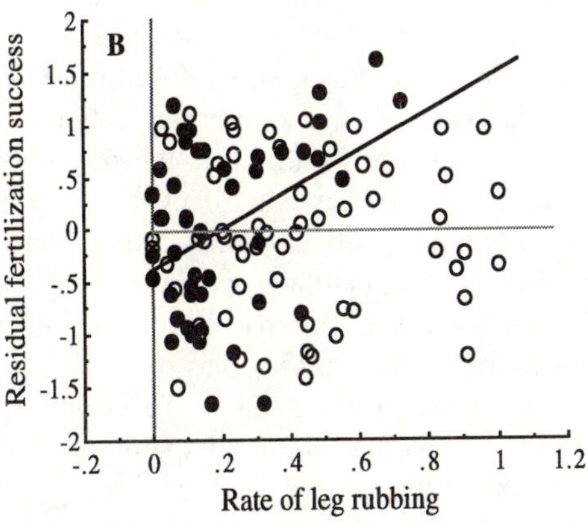

Figure 9.5 The influence of copulatory courtship on male fertilization success in *Tribolium castaneum.* (A) Females control the transport of sperm from the bursa copulatrix to the spermatheca. When kept under CO_2 or nitrogen anesthesia following intromission, sperm are not transferred to the spermatheca. CO_2 has no influence on sperm motility, but does inhibit muscular contraction of the female's reproductive tract (from Bloch Qazi et al. 1998). (B) Males rub their mid and hind tarsi against the lateral edges of the female's elytra during copulation. The rate of leg rubbing is positively related to a male's fertilization success in unmanipulated matings (solid circles). However, when male tarsi are shortened so that females are no longer able to perceive the male rubbing action, fertilization success is no longer related to rubbing rate (open circles). It seems that copulatory courtship influences the transfer of sperm from the bursa copulatrix to the spermatheca from which they are drawn during fertilization (see Edvardsson and Arnqvist 2000).

ilton and Zuk 1982) or genetic variance in male attractiveness, cryptic female choice could contribute indirectly to offspring fitness. What is clear from the example of *T. casteneum* is that selection on males to avoid sperm competition and selection on males through cryptic female choice can operate hand in hand. Haubruge et al.'s (1999) study showed how selection has favored adaptations in male genitalia for the removal of rival sperm from the bursa copulatrix so that only sperm from the copulating male are in a position to be transferred by the female to the spermatheca.

Sensory Exploitation

Many movements occurring during copulation may function in the removal of rival sperm. In some respects, however, these movements may function in a manner analogous to copulatory courtship, if they influence a male's ability to remove sperm. In most insects the female reproductive tract is involved in the processing of eggs, as well as in receiving, transporting, and storing sperm; the egg must pass from the ovaries, be provisioned with accessory secretions, held in the correct position for fertilization, and then passed out of the female onto a suitable substrate. It is not surprising to find that the female's tract has an often complex proprioreceptive system (Okelo 1979; Siva-Jothy 1987a; Miller 1990; Sugagawa 1993). Miller (1987b, 1990, 1991) suggested that copulating male damselflies and libellulid dragonflies might stimulate the campaniform sensilla that line the female's genital tract, and which control the fertilization reflex that affects the ejection of sperm stored in the sperm storage organs. Artificial stimulation of the sensillae results in the reflex contraction of the spermathecal muscle (Miller 1990) and it is entirely plausible that male genitalia may have evolved to exploit this sensory system. Thus a preexisting sensory bias in females, favored by natural selection for efficient fertilization, may favor the evolution of aedeagus morphology that induces the ejection of stored sperm to a site where it can be removed by the copulating male. Córdoba-Aguilar (1999) found evidence for just such an effect in the damselfly *Calopteryx haemorrhoides asturica*. As in most species of *Calopteryx* studied to date (section 3.3), sperm are removed from the bursa copulatrix and the spermatheca during stage 1 of copulation. Sperm are removed from the bursa by the distal projections of the aedeagus and their associated spines. Nevertheless, as in other species, no structure on the aedeagus appears to be able to enter the spermathecae. Females exhibit natural asymmetry in the numbers of sensilla present on the left and right lateral vaginal plates and in natural copulations Córdoba-Aguilar (1999) found that more sperm were lost from the spermatheca on the side with most sensilla, suggesting that stimulation of these sensilla may indeed facilitate ejection of sperm from the spermathecae. Thus, Córdoba-Aguilar (1999) experimentally stimulated females using dissected aedeagi. He removed the aedeagus heads to remove any possible variation in sperm removal function. The experimental aedeagi differed only in their width. The rationale was that wider aedeagi should create greater distension of the lateral plates during mechanical stimulation. As predicted, Córdoba-Aguilar (1999) found that after artificial stimulation of the female's tract fewer sperm remained in the spermathecae of females stimulated with the wider aedeagi.

A similar phenomenon has been suggested to occur in the tettigoniid *Metaplastes ornatus* (von Helversen and von Helversen 1991). Here male and female are both cryptically and overtly involved in the sperm removal process. The male does not possess an aedeagus. Rather, he has a modified subgenital plate that is inserted into the female's genital chamber. The male then pulls backward and forward, withdrawing the subgenital plate about halfway on each movement. The flanks of the subgenital plate have sharp spurs that leave scars on the internal walls of the female's genital chamber. The movements result in the ejection of sperm from the spermatheca into the genital chamber (fig. 9.6). After approximately 30 min the male everts the genital chamber upon which the female bends back and consumes the sperm mass. Only then will the male transfer the typical spermatophore, with its associated spermatophylax. von Helversen and von Helverson (1991) suggest that the mechanical stimulation mimics the passage of eggs through the genital chamber, which stimulates ejection of sperm for fertilization. The damselfly and tettigoniid examples again show how female processes can be involved in the evolution of traits in males that enhance their success in sperm competition. In both cases, the female response is probably naturally selected in the context of fertilization efficiency so that the female "preferences" are of no indirect benefit (Kirkpatrick and Ryan 1991). Females may nevertheless gain direct benefits from cooperating with sperm removal; access to oviposition sites in the case of *Calopteryx* and potential nutrient benefits from the spermatophylax in *Metaplastes*.

Female involvement in sperm displacement has also been implicated in species without mechanisms of mechanical stimulation. Female yellow dung flies must transport sperm from the bursa copulatrix to the spermatheca in order for displacement to take place (Simmons et al. 1999b; section 6.2). Likewise, Wilson et al. (1997) uncovered a female influence over sperm displacement in the bean weevil *Callosobruchus maculatus*. There is good evidence in *C. maculatus* that the last male to mate obtains the majority of fertilizations by a mechanism of sperm displacement (Eady 1994b, 1995). Although the precise mechanism by which displacement is achieved is not yet known, ejaculate size does seem to be important (Eady 1995). Wilson et al. (1997) reasoned that, if variation in sperm displacement were largely female determined, patterns of sperm utilization should be more similar among closely related females than among unrelated females. They examined the repeatability of P_2 values for male pairs mated to a group of females that were either full sisters or unrelated (both groups of females were unrelated to the males used to assess P_2). They found that P_2 values were highly repeatable across full sisters but not across unrelated females, implying that female genotype may be important in determining a male's success in sperm displacement. However, there are a number of conclusions that

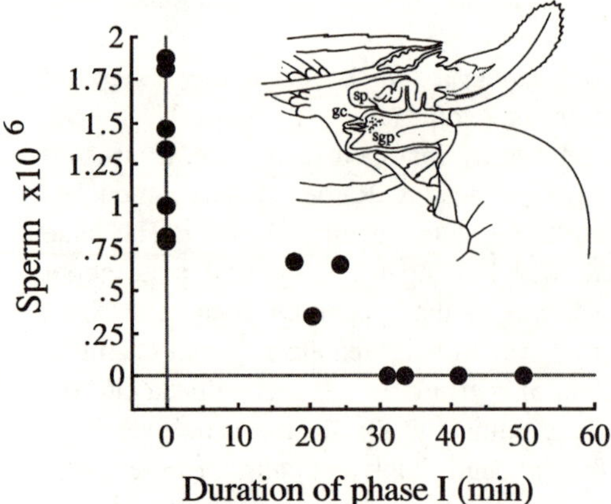

Figure 9.6 Mechanical stimulation of the female's reproductive tract during the first phase of copulation induces the release of sperm from the spermatheca of female *Metaplastes ornatus*. The male inserts his subgenital plate (sgp) into the female's genital chamber (gc) and rhythmically withdraws and reinserts it for a period averaging 31 min. During this mechanical stimulation, sperm are ejected from the spermatheca (sp) and are consumed by the female when the male everts the female's genital chamber on withdrawal. The male transfers his spermatophore in phase II of copulation. (After von Helversen and von Helversen 1991)

could be drawn, depending on an understanding of the mechanisms behind sperm storage and utilization. High repeatability across full sisters could arise if sisters were similar with respect to the size of their sperm storage organs (see also Kempenaers et al. 2000; Pitnick and Brown 2000). Spermathecal size shows positive variation with female body size in a variety of insect species (e.g., Gage 1998; Parker et al. 1999) and not surprisingly Parker et al. (1999) showed that the displacement of rival sperm was negatively dependent on female size in yellow dung flies because larger females with their larger spermathecae store more sperm (Ward 1998b). Female size is heritable in *C. maculatus* (Fox and Savalli 1998) so that full sisters would be phenotypically more similar than unrelated females. Thus the repeatability of P_2 values across full sisters may simply reflect consistency in the dimensions of their reproductive tracts rather than any differential acceptance of sperm from particular males. Wilson et al. (1997) found that repeatability in P_2 was greater in full sisters when the copulating males were related than when they were unrelated. Based on this apparent interaction they argued that a male's ability to achieve sperm displacement depends on his ability to stimulate sperm uptake into the spermatheca, which in turn depends on the interacting genotypes of both male and female. However, Savalli and Fox (1998a) have recently shown that ejaculate size is heritable and we know that it contributes to variation in P_2 (Eady 1995). Values of P_2 are therefore expected to be more similar among males who are related than among those who are not, irrespective of any female influence. Unfortunately, until more is known of the mechanisms underlying sperm displacement in this species Wilson et al.'s (1997) results will remain equivocal.

9.5 Sperm Selection

Sperm selection occurs when a female has a random mixture of sperm from a number of males stored in her sperm storage organ(s) and she selectively utilizes those from a particular individual at the time of fertilization (Simmons and Siva-Jothy 1998). Sperm selection is distinct from other mechanisms of cryptic female choice described above in that females must have sperm from a number of males in storage in order for them to be able to choose among them at the time of fertilization. It is thus a process that occurs after receipt and transfer of sperm to the sperm storage organ(s). This is an important distinction because it is confusion over exactly what constitutes sperm selection that lies at the heart of recent debate over how best to examine the phenomenon (regarding Birkhead's 1998 criteria for demonstrating what he termed female sperm choice, and Eberhard 2000, see recent comments by Pitnick and Brown 2000). Some define any female-mediated process that results in nonrandom paternity as female sperm choice (Eberhard 2000; Pitnick and Brown 2000). I have discussed above how females can bias paternity toward a given male by influencing sperm transfer and/or storage. But biased use of stored sperm is the ultimate in cryptic mechanisms for female choice, was seen as a distinct mechanism in Eberhard's (1996) original review (p. 167), and is the mechanism over which there is most controversy, hence Birkhead's (1998) attempt to clarify the means by which it could be demonstrated. Accuracy in terminology is important if a unified approach to examining phenomena is to be achieved and is why Simmons and Siva-Jothy (1998) offered a classification and definition of previously undefined terms used in the sperm competition literature (section 2.2). The term coined by Birkhead (1998), "female sperm choice," has been confused with choice by the female at a number of levels. Hence the necessity to be clear regarding exactly what is at issue. Here I specifically examine the final avenue by which females can influence a male's fertilization success, by selecting sperm of a particular genotype from those available in their sperm stores.

Potential Mechanisms

Let us first consider the potential avenues by which females could select amongst the sperm available in their sperm stores. Females of a number of insect taxa have multiple sperm storage organs; odonates use the bursa copulatrix and two spermathecae (e.g., Waage 1984; Siva-Jothy 1987a), in *Drosophila* there is the seminal vesicle and two spermathecae (Pitnick et al. 1999), and in dung flies (Ward 1993) and carrion flies (Otronen and Siva-Jothy 1991) there are the singlet and doublet spermathecae. Hellriegel and

Ward (1998) developed a theoretical model of sperm storage and utilization for systems with single versus multiple sites of sperm storage, and showed how, theoretically, females could have greater scope for selecting among potential sires if they stored sperm from different males in different sperm storage organ(s). The findings of Siva-Jothy and Hooper (1995, 1996) provide empirical support for Hellriegel and Ward's (1998) suggestion; when females did not mate prior to oviposition, they were found to fertilize 37% of their eggs using bursal sperm (from their last copulation) and 54% of their eggs using spermathecal sperm (stored from previous copulations). Thus, females could potentially exercise sperm selection if they were able to partition the ejaculates from different males within their reproductive tract.

During fertilization the insect egg is passed through the reproductive tract and precisely aligned with the ducts that lead to the sperm storage organs. Sperm gain entrance to the egg through one or more narrow channels in the chorion, called micropyles. They are released from the sperm storage organs at the time when the micropyles are aligned with the opening of their ducts (Davey 1985a; Sander 1985), a process that appears to be facilitated by muscular contractions of the sperm stores and/or their ducts induced by mechanical stimulation from the egg (Miller 1990; Sugagawa 1993). The transportation of sperm from the sperm stores to the micropyle is thus determined by the female; denervation of the spermatheca of the grasshopper *Shistocerca vaga* reduces fertilization success to less than 50% (Okelo 1979) and cutting of the spermathecal muscle in *A. grandis* and *C. alternans* results in females being unable to fertilize their eggs (Villavaso 1975a; Rodriguez 1994b). Local attraction of sperm to eggs may be facilitated by species-specific chemoattractants and receptor molecules identified on the surface of egg and sperm, respectively (Vacquier 1998). The specificity of egg-sperm signaling systems could potentially facilitate selection of compatible sperm within species. Sperm enter the egg via the micropyles where they encounter the vitelline membrane. The acrosome reaction releases lytic material that presumably aids in the penetration of the vitelline membrane. In the case of houseflies, secretions from the female accessory gland are required to trigger the acrosome reaction (Degrugillier and Leopold 1976). The effectiveness of female accessory secretions in initiating the acrosome reaction may vary with sperm genotype, resulting in sperm selection. Complex intracellular interactions between sperm and maternal cytoplasm are required to bring the maternal and paternal pronuclei together for syngamy (Karr 1991). Selective transport of paternal pronuclei within the egg could be a means by which females, or rather their eggs, bias paternity.

Conspecific Sperm Precedence

There are a number of sperm competition studies of insects where variation in P_2 is indicative of homogamy (see table 2.3; Hewitt et al. 1989; Bella et al. 1992; Gregory and Howard 1994; Robinson et al. 1994; Price 1997; for a recent review see Howard 1999). When females are mated to a conspecific and a heterospecific male, a significant tendency for eggs to be fertilized by the conspecific male overlies the basic pattern of sperm utilization. In many of these studies P_2 is intermediate, suggesting the extensive sperm mixing required for sperm selection. The bias toward conspecific sperm is akin to the loaded raffle mechanism of sperm competition proposed by Parker et al. (1990) in that the conspecific male's sperm have an advantage over and above their numerical representation. Some potential mechanisms by which loading could occur are discussed above; it may be that eggs resist penetration by heterospecific sperm or conspecific sperm may be competitively superior to heterospecific sperm. An interaction between these mechanisms is also possible in that a competitive disadvantage for heterospecific sperm may arise because of a lack of adaptation to the environment in the female's reproductive tract. In the ladybirds *Epilachna vigintioctomaculata* and *E. pustulosa,* heterospecific sperm appear to be incapacitated by the female's reproductive tract (Katakura 1986). In evolutionary terms females can ensure fertilization by genetically compatible mates by keeping the environment in the reproductive tract within narrow physiological limits.

In the ground crickets *Allenomobius fasciatus* and *A. socius*, females mate multiply. They store the sperm from multiple males who generally share in paternity (Gregory and Howard 1994, 1996). However, in their hybridization studies Gregory and Howard (1994) found that there was almost total superiority of conspecific sperm when in competition with heterospecific sperm, irrespective of mating order. Heterospecific sperm are capable of fertilizing eggs; females do produce hybrid offspring when multiply mated to heterospecific males. Thus, postinsemination barriers to fertilization exist in these species (Howard et al. 1998). Moreover, the barriers appear to depend on the numbers of sperm present in the sperm storage organ (Howard and Gregory 1993). The sperm number dependency might imply that the barrier is determined by the competitive abilities of sperm since increasing sperm numbers of heterospecific males can overcome the homospecific advantage. This example illustrates that sperm selection can have significant evolutionary implications, at least at the level of species isolation.

A similar phenomenon was reported for the flour beetles *Tribolium castaneum* and *T. freemani* (Robinson et al. 1994; Wade et al. 1994). In *T. castaneum* there is a temporary second-male advantage at fertilization due to sperm precedence effects (Schlager 1960; Lewis and Jutkiewicz 1998). Robinson et al. (1994) found that female *T. castaneum* mated first to conspecific males followed by *T. freemani* males failed to exhibit the usual second-male sperm precedence. Conversely, female *T. castaneum* mated first to *T. freemani* males followed by conspecific males showed a much more rapid shift to second-male sperm utilization than expected from conspecific males. Wade et al. (1994) and Robinson et al. (1994) suggest that, as with the ground crickets discussed above, a postinsemination barrier to fertilization was important in reproductive isolation. However, given the recent work of Edvardsson and Arnqvist (2000), this is unlikely to involve sperm selection. Rather, copulatory courtship in *Tribolium* appears to be required for the transportation of sperm to the sperm stores. Given that patterns of copulatory courtship are species specific (Eberhard 1993, 1996), it is more likely that *T. freemani* males are unable to stimulate *T. castaneum* females to transport their sperm to the spermatheca.

Homogamy can potentially have its effects at a lower taxonomic level, isolating strains or geographically separated races of the same species. This is illustrated by recent work with *Drosophila. Drosophila* seminal products play a critical role in the storage of sperm by females, and in the displacement and incapacitation of rival sperm (section 4.3). Moreover, there is remarkably strong evolutionary coadaptation between male seminal products and female responses to them (Rice 1996, 1998a). It is not surprising to find, then, that there is rapid divergence in the molecular composition of *Drosophila* seminal products, and that they exhibit species-specific activity (Fuyama 1983; Stumm-Zollinger and Chen 1988; see review in Eberhard 1996 and in section 4.3). Thus, Price (1997) found that variation in seminal fluid activity across conspecific and heterospecific matings of *Drosophila* was responsible for high conspecific sperm displacement and low heterospecific sperm displacement; only the seminal products of conspecific males are able to displace previously stored sperm from the female's sperm storage organs (see also Price et al. 2000).

Homogamy can also be an important aspect in determining paternity in crosses between reproductively isolated populations (Joly et al. 1991b; see fig. 8.8). Within-species variation in paternity has been studied in detail by Clark and his colleagues, who examined the influence of male and female genotypes on sperm displacement in *D. melanogaster*. They found significant variation between laboratory strains in the sperm displacement abilities of males that was associated with variation at accessory gland protein loci (Clark et al. 1995). They also found significant genetic variation across female strains that influenced sperm displacement ability of males (Clark and Begun 1998). Importantly, using pairwise tests with isogenic female lines, they have shown a significant male by female interaction; the degree of sperm displacement achieved by a particular male genotype is dependent on the genotype of the female with which he is mating (Clark et al. 1999). This work supports the early findings of DeVries (1964), who found significant variation in the numbers of sperm in storage after crosses between different strains. These data provide good evidence that male and female influences are both critical in determining the degree of sperm displacement in *Drosophila*. Pitnick and Brown (2000) suggest that demonstration of a male by female interaction provides good evidence for female sperm choice. In its broadest sense this is true. Nevertheless, one must consider the mechanisms of sperm transfer, storage, and utilization before conclusions regarding mechanisms behind nonrandom paternity can be made (Simmons and Siva-Jothy 1998). As discussed above, sperm selection occurs when females choose the sperm of one male from an available set of sperm stored in the sperm storage organs. In the case of *Drosophila*, accessory gland products displace and incapacitate sperm stored from previous males, so that nonrandom paternity reflects random usage of sperm by females following displacement (Price et al. 1999; Gilchrist and Partridge 2000). The result of Clark et al. (1999) provides good evidence for female involvement in sperm displacement as discussed in section 9.4 above. It is not evidence for a mechanism of sperm selection. Likewise, Lewis and Austad (1990) found a significant male by female interaction in their study of paternity that utilized crosses between two laboratory strains of *T. castaneum*. Divergence in copulatory courtship between strains would account for this interaction. Like *Drosophila*, female *Tribolium* are clearly involved in determining the extent of sperm storage, but there is no evidence for sperm selection after storage. It may be that sperm selection can bias paternity toward conspecifics over and above that expected from sperm displacement. Such effects can only be examined after controlling for the effects of conspecific sperm transfer and/or displacement.

Intraejaculate Sperm Selection

There are a number of situations in which female fitness might best be served by producing offspring of one particular sex. For example, a local scarcity of one sex or sex-biased effects of resource competition can favor the facultative production of the opposite sex (e.g., Hamilton 1967; Werren and Charnov 1978; Charnov 1982; King 1996). This has proved particularly relevant to haplodiploid species such as parasitoid wasps (Godfray 1994), in which females can adjust sex ratio by laying fertilized or unfertilized eggs. Nevertheless, there is an increasing body of evidence to suggest that females in heterogametic taxa can also facultatively adjust the sex ratio of their offspring at the time of fertilization when the fitness returns for producing sons and daughters vary (Gunnarsson and Andersson 1996; Svensson and Nilsson 1996; Komdeur et al. 1997). In heterogametic insects, sex ratio biases have been associated with a delay between copulation and fertilization (James 1937; Werren and Charnov 1978; Stockley and Simmons 1998). Delays between copulation and fertilization may represent a proximate signal to females of a local paucity of males and thus the potential benefits of biasing their brood sex ratio toward sons. It is not known how females are able to distinguish between male-producing and female-producing sperm. Sivinski (1984) suggests that a mechanism could be available if sex chromosomes differ in size, resulting in slight phenotypic differences in sperm size, but also notes that attempts to separate male- and female-producing sperm have yielded mixed results. Mange (1970) found that female *D. melanogaster* mated with young males produced predominantly male offspring in their initial broods, while females mated to old males produced predominantly female offspring. Females left to exhaust their sperm supplies produced overall an equal number of sons and daughters, so that males must have transferred equal numbers of male-producing and female-producing sperm. Mange (1970) therefore concluded that the most likely mechanism for the observed changes in sex ratio production was sperm selection by the female. He suggested that storage of male sperm in the ventricle receptacle and female sperm in the spermathecae could facilitate this selection (sensu Hellriegel and Ward 1998), since sperm are used from the spermathecae only once ventricle sperm are exhausted (Gromko et al. 1984a). A similar phenomenon has been reported recently in the tettigoniid *Poecilimon veluchianus* (Reinhold 1996a). In the case of *P. veluchianus*, females produce predominantly female offspring when mated with young males and switch to producing male offspring when mated with old males. Both Mange (1970) and Reinhold (1996a) were able to show that the sex ratio was independent of female age. However, the adaptive significance of variation in sex ratio is unclear. Reinhold (1996b) suggests that it may be adaptive to produce members of one sex when there is sex differential migration that generates a bias in the local sex ratio. Female *P. veluchianus*, but not males, do become increasingly likely to emigrate as they age (Reinhold 1996a).

Perhaps the best example of intraejaculate sperm selection comes from the early studies of Childress and Hartl (1972). When females of the *D. melanogaster* strain carrying the chromosomal translocation $T(1;4)B^s$ were mated once, they produced a decreasing proportion of offspring that were the products of the $B^s + 4$ bearing sperm with time since mating. Childress and Hartl (1972) interpreted this result as selective use of sperm that is dependent on exposure of the reproductive tract to $B^s + 4$ bearing sperm; females essentially learn to discriminate against $B^s + 4$ sperm. They consider various alternative arguments for the observed "brood" effects, including sperm competition. Different sperm genotypes may have different viabilities and/or motilities (Gromko et al. 1984a). Childress and Hartl (1972) argued that if $B^s + 4$ sperm were less competitive than other sperm types, their representation in offspring should increase with time as the more competitive genotypes are used up. However, the reverse result was obtained. The result could also be explained if there were nonrandom loss of sperm from the female's reproductive tract, for example, if $B^s + 4$ sperm had shorter longevity than other sperm genotypes, or their competitive ability declined with age. If this were the case, identical patterns should be obtained when females are remated after exhausting their current sperm supply. However, when Childress and Hartl (1972) remated females the discrimination against $B^s + 4$ sperm was even greater than that seen in once-mated females, suggesting that females "remembered" the disfavored genotype, supporting their conclusion that females were capable of sperm selection.

While intraejaculate sperm selection can contribute to female fitness, it will have little impact on male fitness via sexual selection because the sperm used are always segregants of the male's genotype. Nevertheless, these examples show that females have the potential to bias sperm use toward sperm carrying particular genotypes and thus the potential for biased sperm use to influence sexual selection.

Sperm Selection and Male Phenotype

Two studies have claimed to have found evidence for sperm selection based on male phenotype. In the first, LaMunyon and Eisner (1993) observed that in double matings of the moth *Utetheisa ornatrix* larger males obtained the majority of fertilizations, irrespective of their position in the mating sequence, and concluded that females were responsible for generating variation in nonrandom paternity. Sperm utilization follows the typical lepidopteran pattern in which P_2 is distributed bimodally with peaks at zero and 1; when the first male is principal sire P_2 is zero and when the second male is principal sire P_2 is 1. LaMunyon and Eisner (1993) argue that the bimodal distribution of P_2 values is a manifestation of cryptic female choice. However, when interpreted within a framework of mechanism of sperm transfer and storage, these data provide little evidence for sperm selection. Simmons and Siva-Jothy (1998) raised a number of alternative hypotheses that could generate a bimodal distribution in P_2 values. The pattern could arise due to a combination of plugging of the female tract by the first spermatophore and sperm displacement and/or precedence by second males who were successful in circumventing the plug, or when either first or second male failed to successfully position its spermatophore within the reproductive tract so that sperm were not transferred. Neither of these explanations seems likely in the case of *U. ornatrix*, however, because LaMunyon and Eisner (1993) rejected any females from their analyses in which either first- or second-male spermatophores appeared not to have successfully drained. Nevertheless, LaMunyon and Eisner (1994) later discovered that it was spermatophore size that determined principal sire in this system. Larger spermatophores contain more sperm (He and Miyata 1997), so that paternity advantages might be expected for males with larger spermatophores without sperm selection by females. Indeed, a number of studies of butterflies and moths have revealed that ejaculate size is critical in determining a male's success in sperm competition (Bissoondath and Wiklund 1997; Cook et al. 1997; Wedell and Cook 1998). In Wedell and Cook's (1998) study of *Pieris rapae* paternity was found to be related to the weight of ejaculate remaining in the spermatophore; the spermatophores of principal sires were drained to a greater extent than spermatophores of nonsires. Why the spermatophores of some males are not fully drained is unknown. Given that females are responsible for rupturing the spermatophore in the corpus bursae (Sugawara 1979), it could represent female discrimination, but this would be via an influence over sperm storage rather than sperm selection. On the other hand, poor drainage could result from relatively poor spermatophore positioning or sperm function. A principal sire effect can also arise when sperm do not mix randomly within the sperm stores (see, for example, section 2.5 and Harvey and Parker 2000).

Working with yellow dung flies, Ward (1993) examined the distribution of sperm in the three spermathecae after single matings. Not surprisingly, he found the number of spermathecae containing sperm increased with copula duration and the size of the copulating male; both of these variables are correlated with the amount of sperm transferred (Parker et al. 1990; Simmons and Parker 1992; Parker and Simmons 1994; Ward 1998b; Parker and Simmons 2000). Ward (1993) also found that sperm became more dispersed among the three spermathecae as male size increased. Following double matings, and thus sperm displacement, Ward (1993) found that when large males copulated second, fewer spermathecae contained sperm and a lower proportion of sperm were contained in the singlet spermatheca than when small males copulated second. Although these data illustrate extreme variability in the distribution of sperm among the three sperm storage organs, patterns appear to be inconsistent. For example, Ward (1998b) subsequently found that the proportion of sperm stored in the singlet spermatheca increased with the size of the second male, and Hellriegel and Bernasconi (2000) found that a greater proportion of second-male sperm were stored in the singlet spermatheca, irrespective of the size of the second male. Hellriegel and Bernasconi's (2000) estimate of the proportion of second-male sperm stored was inconsistent with the proportion estimated from radioisotope-labeled ejaculates (Simmons et al. 1999b), and with the proportion of eggs fertilized by the second male (Parker 1970f; Simmons and Parker 1992; Simmons et al. 1996). Otronen et al. (1997) have found significant female by spermatheca (singlet or doublet) interactions in the numbers of sperm stored, indicating that there is significant between-female variation in which store is used for the majority of sperm. Ward (2000) was able to select for increased numbers of spermathecae (a low frequency of females from natural populations have four, rather than the typical three spermathecae). Females from his selected lines had a greater total sperm storage volume and Ward (2000) found that males had a lower P_2 when mating with large, four-spermathecae females than with small four-spermathecae females or with three-spermathecae females. He concluded that the greater number of spermathecae gave females enhanced ability to select sperm, hence reducing the second-male effect. However, we have already seen in section 6.2 how larger volumes of sperm are more difficult for males to displace, so that P_2 declines with sperm storage volume (Parker et al. 1999), irrespective of any active mechanism for sperm selection among females.

Ward (1993) performed an experiment in which he allowed two males to copulate with a single female for a fixed time period and assessed their paternity using allozyme markers. He found that large males attained a higher P_2 than did small males and concluded that females were exercising sperm selection, favoring large males as mates by drawing sperm from spermathecae that stored predominantly large male sperm. However, Simmons and Parker (1992), Ward (1993), and Parker and Simmons (1994, 2000) have shown that large males have a higher constant rate of sperm transfer and displacement than small males. Simmons et al. (1999b) have shown that the mechanism of sperm transfer and displacement conforms to one in which there is volumetric displacement of previously stored sperm coupled with instantaneous random sperm mixing (section 6.2). Following sperm displacement, sperm are assumed to be used at random from the female's sperm storage organs so that there is no sperm selection. Because large males have a higher rate of sperm displacement they are expected to attain a higher P_2, even without sperm selection on the basis of male size. Simmons et al. (1996) assessed a given male's success in fertilization as a second male with replicate females that had experienced first males of different sizes. They found no significant variance in P_2 among females but significant variance among males that was related to male size; as in Ward's (1993) study, large males gained a higher P_2 than small males (fig. 9.7). After calculating the P_2 expected on the basis of size-dependent rates of sperm transfer and displacement and subtracting this from the observed value of P_2, variance in residual P_2 was unrelated to male size and showed no significant variation among males or females (fig. 9.7). Thus, sperm utilization from the female's sperm storage organs was random with respect to male size once the numbers of sperm transferred and displaced were controlled. In general, these types of experimental approaches, coupled with an understanding of sperm competition mechanisms, are essential for interpreting nonrandom patterns of sperm utilization (Simmons et al. 1996). Ward (1993) did find brood effects in females mated with small second males; P_2 increased across three successive clutches when females were not allowed to remate between clutches. Such brood effects are at least consistent with the notion that increasing numbers of sperm from small males were utilized, perhaps as sperm from large first males became exhausted. Sperm selection could thereby refine fertilization biases already imposed via male effects.

During his sperm utilization studies Ward (1998b) discovered that allele frequencies in offspring were often nonrandom. Females taken from a natural population and allowed to oviposit in the laboratory without mating produced offspring of different genotypes at the phosphoglucomutase (PGM) locus depending on ambient conditions; in "sunlight" (illumination from a strong light bulb) females produced predominantly offspring of the PGM_3

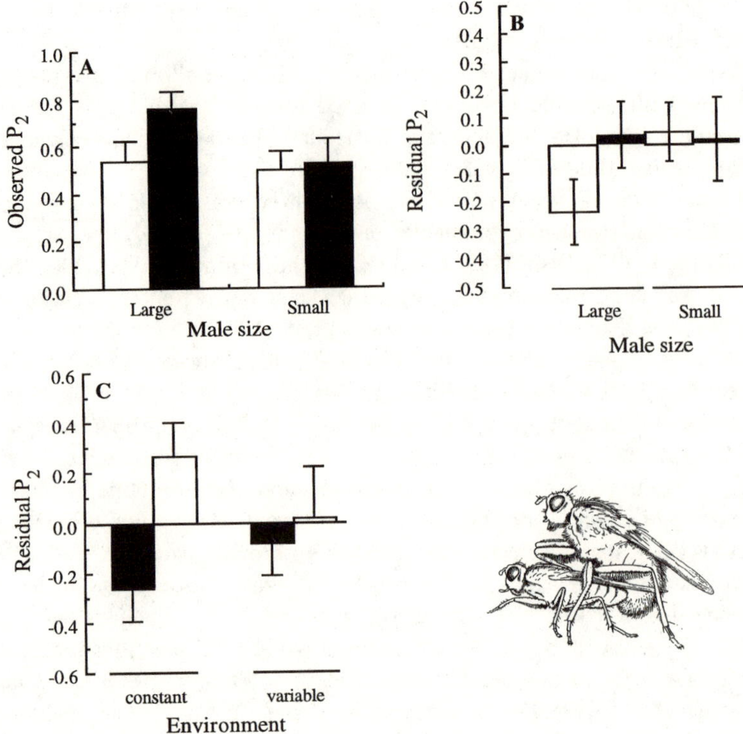

Figure 9.7 Distinguishing between sperm displacement and sperm selection in the yellow dung fly *Scatophaga stercoraria*. (A) The outcome of double matings for large and small males assessed with replicate females that differed in their previous mating experience; females mated first with a large male (open bars) or first with a small male (solid bars). Large males achieve a higher proportion of fertilizations than small males, irrespective of the size of the female's first mate. This has been taken as evidence for sperm selection (Ward 1993). However, large males have a higher constant rate of sperm displacement. (B) After controlling for the proportion of offspring fertilized by the second male expected from size-dependent sperm displacement, there is no significant influence of male size on paternity (from Simmons et al. 1996). Nevertheless, underlying the strong male influence on paternity in yellow dung flies, is evidence of sperm selection. (C) Females can increase the fitness of their offspring by producing homozygotes when environmental conditions are constant and heterozygotes when conditions are variable. After controlling for sperm displacement effects, the paternity of the last male to copulate is dependent on the environment in which females deposit their eggs and on the genotype of the last male to copulate, relative to the female's own genotype (last male genotype: different, solid bars; the same, open bars) (from Ward 2000).

genotype (the number refers to the relative electophoretic mobility of the allele) while in "shade" (no bulb) they produced offspring of predominantly PGM_2 genotype. These data imply that females are able to select sperm on the basis of their PGM genotype. Ward (1998b) reasoned that sunlight would represent a more variable environment, since pats in the direct sun would increase in temperature greatly during the day while those in the shade would not. He examined the development of larvae under variable and constant temperature regimes and found that PGM genotype had a significant influence on the performance of female larvae; heterozygotes achieved a higher pupal weight in the variable environments while homozygotes achieved a higher pupal weight in a constant environment. There was a similar trend in male larvae in the constant environment but not the variable environment. These data suggest that females could potentially increase the fitness of their offspring if they were able to adjust their PGM genotype dependent on the location of dung pats in which they were laying. To test for this, Ward (2000) allowed homozygous females reared in either constant or variable environments to mate with two homozygous males that were either the same genotype or a different genotype from themselves. Ward (2000) predicted that females reared in a constant environment should fertilize their eggs with the sperm of males with the same genotype as themselves in order to produce homozygous offspring who fare better in constant conditions. Conversely, females reared in variable conditions should fertilize their eggs with sperm from males of a different genotype from themselves in order to produce heterozygous offspring that perform better in variable conditions. In this study, Ward (2000) controlled for variation in P_2 expected due to variation in copula duration and controlled the size of males across females from different environments. As predicted, copula duration had a large influence on sperm utilization, males that copulated for longer displacing more sperm and obtaining more fertilizations. However, after removing this effect, Ward (2000) found subtle differences in fertilization success that were due to male genotypes, for females from a constant environment P_2 was higher when the last male was of the same genotype than when he was of a different genotype, as predicted. The converse prediction, that P_2 with females from a variable environment should be greater when the second male was of a different genotype was not supported, although P_2 for different genotype males was greater with variable than with constant environment females. These data provide compelling support for the notion that, underlying a predominantly male-driven pattern of sperm utilization, females can have some influence over paternity even though, because of sperm displacement, their contribution to nonrandom paternity is relatively weak.

Ransford (1997) has examined the relationship between premating female choice and cryptic female choice via sperm selection in the two-spot ladybird *Adalia bipunctata*. In some populations of two-spot ladybird there are two morphs, the typical red morph as well as a melanic morph. Females have been shown to have a genetically based mating preference that favors melanic males during precopulatory sexual selection (Majerus et al. 1982, 1986; O'Donald and Majerus 1992). Ransford (1997) established independent breeding lines of ladybirds taken from natural populations in which both phenotypes were present. He examined variation between lines in the strength of female preference for melanic males. He also examined variation between lines in paternity when females were mated first to typical males and then to a melanic male. He found no significant covariation between premating preferences for melanic males and cryptic female choice as revealed by biases in paternity toward melanic males. Importantly, Ransford (1997) also established the mechanism of sperm transfer, storage, and utilization in this species; P_2 increases with copula duration and conforms with a mechanism of sperm competition in which the last male to copulate displaces sperm at a constant rate from the female's sperm storage organs during copulation (see table 2.5). After controlling for variation in sperm displacement due to copula duration, Ransford (1997) similarly found no covariation between premating preferences for melanic males and the paternity of melanic males (fig. 9.8). Thus, female choice via sperm selection was not predicted on the basis of known female preferences for melanic males.

Sperm Selection and Multiple Mating

Maternal filtering of sires is well known in plants, and fertilization by multiple pollen donors has been associated with increased maternal fitness; mixed pollen loads have been shown to increase the total weight of mature seeds produced (for a review see Delph and Havens 1998). Sivinski (1980, 1984) suggested that female incitement of sperm competition through multiple mating could evolve under similar selection pressures. Thus, multiple mating by females could be the visible manifestation of cryptic female choice via sperm selection. By acting as an arena for male sperm competition, females could ensure that only the most competitive sperm fertilize their eggs; sperm selection is achieved by sperm competition. An important feature of this argument is that sperm must compete for access to fertilizations, so that females must store the sperm from multiple males within their sperm storage organs in order to ensure they compete. The potential for incitement of sperm competition via multiple mating to facilitate sperm selection will be reduced or absent when adaptations arise in males to avoid sperm competition, for example, through the removal and/or incapacitation of rival sperm.

The theoretical considerations underlying the evolution of cryptic female preferences differ little from those of classical precopulatory female preferences. Theoretical models fall into two broad categories: (i) direct models, whereby females choose males that provide superior resources or parenting ability; and (ii) indirect models, whereby the benefits of choice are bestowed on the female's offspring (section 1.1). Cohen (1967, 1973) has argued that errors during crossover leave few sperm with intact chromosomes so that many sperm within an ejaculate will be infertile. Multiple mating by females may thus protect them against infertile sperm. However, because infertile sperm are unlikely to be competitive, such direct benefits are not relevant to

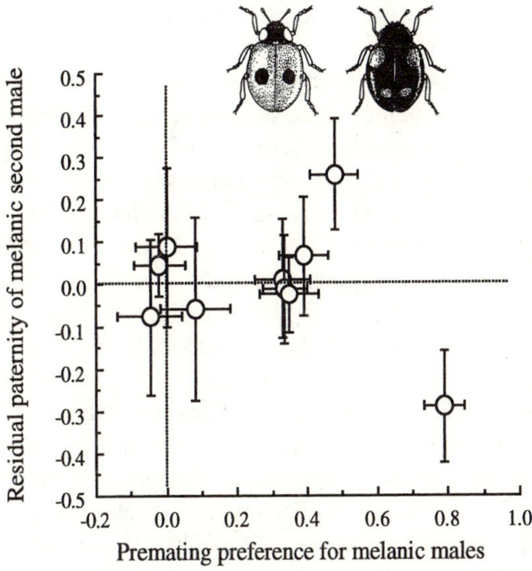

Figure 9.8 Examination of mate choice and sperm selection in two-spot ladybirds *Adalia bipunctata.* Two morphs occur, the typical red morph and the melanic morph. Females show a genetically based mating preference for melanic males. Across ten independent lines there is significant variation in the strength of preference for melanic males. There was no significant variation across these lines in the proportion of offspring sired by melanic males, either before or after controlling for sperm displacement. Thus, there was no covariation across lines between premating preferences for melanic males and cryptic female choice based on male morph. The data show the mean (± SE) residual proportion of offspring sired by melanic second males in competition with typical first males, after controlling for variation in sperm displacement arising due to variation in copula duration, and the mean (± SE) strength of preference for melanic males. They show that females who prefer melanic males as mates do not preferentially select melanic male sperm to fertilize their eggs. There was no covariation even before controlling for sperm displacement, showing melanic males are not favored by other potential mechanisms of cryptic female choice such as influence over the displacement process. (Data from Ransford 1997)

arguments of sperm selection. There are two principal indirect models of preference evolution that could be relevant to sperm selection; Fisher's runaway and viability models (section 1.1; Andersson 1994). Under the Fisherian model females might benefit from incitement of sperm competition because, by ensuring that their eggs are fertilized by competitively superior sperm, they will produce sons who will themselves have competitively superior sperm (Sivinski 1984; Harvey and May 1989; Keller and Reeve 1995). The process could favor adaptations in sperm morphology, such as sperm length, if some aspect of the female, such as the length of her spermathecal ducts, resulted in longer or shorter sperm being more successful in achieving fertilizations (Pitnick et al. 1999).

Curtsinger (1991) formally modeled what he termed the sexy-sperm hypothesis with a two-locus genetic model; one locus determined sperm competitiveness while the other locus determined female mating behavior (in this case multiple mating, which facilitates sperm selection). Positive linkage disequilibrium (a genetic correlation between multiple mating in females and sperm competitiveness in males) was required for the sexy-sperm process to work. This condition is of course typical for all indirect models of premating preference evolution. Curtsinger (1991) found that a slight cost associated with multiple mating was sufficient for the mechanism to decay. However, he assumed that sperm competitive ability rapidly attained a selective equilibrium whereby no further increase in competitiveness could accrue. As Keller and Reeve (1995) point out, it is becoming increasingly clear that fitness variation can be maintained in natural populations (Rice 1988; Pomiankowski et al. 1991; Pomiankowski and Møller 1995; Rowe and Houle 1996). For example, cycles of antagonistic coadaptation between males and females are equally intense to the host-parasite cycles predicted to maintain genetic variance in male secondary sexual traits, and have been demonstrated to maintain variation in traits important in sperm competition (Rice 1996, 1998a). Genetic variation in traits such as testis size, ejaculate size, sperm length, and success in sperm competition appears widespread (Beatty and Sidhu 1970; Service and Fales 1993; Reinhold 1994; Clark et al. 1995; Service and Vossbrink 1996; Hughes 1997; Radwan 1998; Savalli and Fox 1998a; Pitnick and Miller 2000). Thus, Keller and Reeve (1995) argued that a Fisherian process could indeed favor the evolution of multiple mating in females through a number of mechanisms of cryptic female choice, including sperm selection.

Sivinski (1980, 1984) suggested that viability selection could favor the evolution of multiple mating in females, if sperm competitive ability were correlated with offspring competitive ability. Females who mated multiply would produce offspring of higher viability than those who mated singly, because sperm competition would ensure that only the most competitive sperm fertilized their eggs. They could even heighten the competition by providing chemically challenging environments within which sperm are forced to compete (Sivinski 1984; Birkhead et al. 1993). Yasui (1997) presented such a "good sperm" model for the evolution of multiple mating by females. He concluded that female multiple mating could be maintained if the sperm competitive ability of a male is influenced by his general viability and when mutations affecting viability are deleterious, rather than advantageous. Madsen et al. (1992) favored a viability model of sperm selection in interpreting their finding that female adders that copulate with more males produce fewer stillborn young. However, as pointed out by Parker (1992b), such an interpretation requires a correlation between sperm success and offspring viability. That is, we need proof that stillborn young are fathered by males whose sperm fare poorly under sperm competition. Studies of insects and arachnids have recently revealed that multiple mating by females can increase the viability of their young (Zeh 1997; Tregenza and Wedell 1998; Watson 1998). Accordingly, females appear to favor multiple matings with different males to multiple matings with the same male (Bateman 1998; Zeh et al. 1998; Archer and Elgar 1999), encouraging interejaculate rather than intraejaculate sperm competition. The only empirical test of the viability model for sperm selection is that of Simmons (in press), who examined the correlation between sperm competitive success and offspring performance in the Australia field cricket *Teleogryllus oceanicus*. Like Tragenza and Wedell's (1998) study of *G. bimaculatus*, Simmons found that females mated to more than one male had a higher hatching success than did females mating the same number of times with a single male. However, Simmons (in press) found no evidence to suggest that the increased hatching success was due to greater fertility of one particular male. More importantly, males who were successful in sperm competition did not produce offspring who were of superior viability (fig. 9.9). Neither was there a correlation between a male's success in sperm competition and the performance of his surviving offspring measured by development rate or final adult size. Thus, the viability model for multiple mating and sperm selection does not gain empirical support from this study.

There are a number of concerns regarding the general feasibility of a viability model for sperm selection via multiple mating. First, fitness genetically mediated via the haploid genotype of sperm may not translate into fitness in the diploid organism where the phenotype is the product of a combination of genes from sperm and egg and their interacting effects (Sivinski 1984). Second, the fitness of a diploid organism could be inversely related to sperm competitive ability where parasitic organisms enhance their own reproduction at the expense of their host. In *T. confusum*, for example, males infected with the intracellular parasite *Wolbachia pipientis* have a competitive advantage over uninfected males during sperm competition; when females mate with both an infected and a pharmacologically cured male, the infected male's sperm invariably gain the majority of fertilizations in the lottery that follows (Wade and Chang 1995). *Walbachia* species have been termed cytoplasmic incompatibility microorganisms because females mated with infected males produce fewer offspring. Wade and Chang (1995) concluded that, despite the deleterious effects on female fitness, competitive superiority of infected sperm accelerates the spread of *Walbachia* through the host population.

Genetic incompatibility has been argued to be an important selective agent behind the evolution of female multiple mating (Zeh and Zeh 1996, 1997). As well as microorganisms like *Walbachia*, meiotic drive systems, such as sex ratio distorters, can reduce female fitness by rendering certain combinations of maternal and paternal haplotypes incompatible within a developing embryo (Zeh and Zeh 1996). A common source of incompatibility may arise

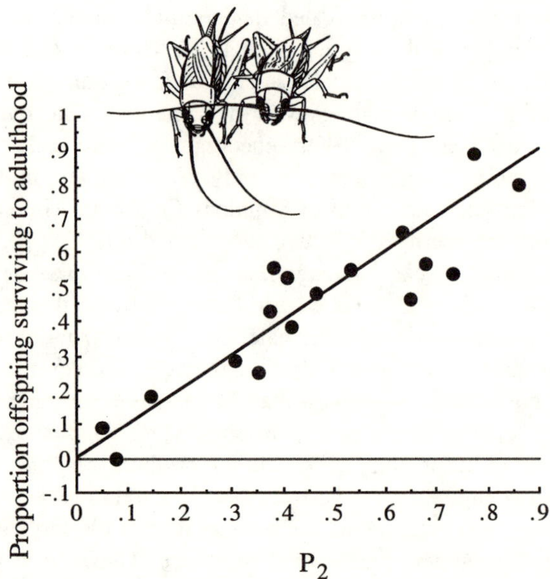

Figure 9.9 A test of the "good-sperm" model of sperm selection in *Teleogryllus oceanicus*. Given a correlation between sperm competitiveness and offspring viability, females could gain indirect genetic benefits for their offspring by mating with multiple males and inciting sperm competition. Only competitive sperm would fertilize their eggs. The good-sperm model of sperm selection therefore predicts that a male's success in sperm competition should be correlated with the viability of the offspring he sires. The data show the proportion of offspring hatching from eggs that are fathered by the second male to mate with a doubly mated female, plotted against the proportion of that male's offspring surviving to adulthood. If males successful in sperm competition sire more viable offspring, we would expect the representation of successful male's offspring to be relatively greater in the adult population than those of unsuccessful males. The line of expected contribution to the adult population given a male's fertilization success is shown. The expected line shown does not differ from the observed line in intercept or slope. Contrary to the good-sperm hypothesis, males successful in sperm competition did not have a greater representation of offspring in the adult population than expected from the proportion of offspring they sired. (Simmons, in press; vignette from Loher and Rence 1978, with permission of Blackwell Wissenschafts-Verlag).

when there is a high risk of matings between close relatives (Tregenza and Wedell 2000). In their extreme, incompatible combinations can result in early embryonic failure and reduced offspring production for females. Sublethal effects could also arise where incompatibility results in offspring sterility, or in offspring with inferior developmental and/or adult performance. By mating with multiple males, females could ensure that they had a mixture of sperm genotypes available for fertilization, and thereby select compatible sperm for fertilization (Zeh and Zeh 1997). This hypothesis seems at odds with Wade and Chang's (1995) finding that infected male *T. confusum* have superior fertilization success. However, as Zeh and Zeh (1997) point out, uninfected male *T. confusum* could have had inferior fertilization success because of unknown effects on sperm production and/or function of the antibiotics used to cure them. Evidence in favor of the genetic incompatibility hypothesis comes from studies of the sex ratio distorter driving elements in *Drosophila*. Because the meiotic drive system functions by destroying developing spermatids carrying a particular sex chromosome, males with the drive have reduced sperm numbers and reduced success in sperm competition (Wu 1983). In the wasp *Nasonia vitripennis*, the selfish *B* chromosome destroys other paternal chromosomes in fertilized eggs, rendering the zygote haploid and so biasing the sex ratio toward an excess of males. Females mated singly to carriers of selfish *B* chromosomes or noncarriers will utilize all of the sperm stored in their sperm storage organs. However, females mated with both *B* chromosome carriers and normal males will exhibit a biased use of normal male sperm. This could arise through sperm selection by females, or if *B*-chromosome-carrying sperm were of inferior competitive ability, or if males carrying *B* chromosomes produce fewer sperm (*N. vitripennis* exhibit high levels of sperm mixing, required for sperm to compete and for sperm selection; Holmes 1974; Beukeboom 1994). In both cases described above, females that mated with just one male would risk producing biased sex ratios when their mates were affected by the driving system. With multiple matings, males affected by a drive system would gain few fertilizations, thereby increasing female fitness.

An important distinction between Zeh and Zeh's (1996, 1997) genetic incompatibility hypothesis and the Fisherian and viability models for the evolution of sperm selection via multiple mating is that the effects of interacting paternal and maternal genotypes are nonadditive. Thus, sperm selection based on compatibility should not result in directional sexual selection on males because individual males who are reproductively successful with females of one genotype will be unsuccessful with another. For example, in the case of yellow dung flies (fig. 9.7) sperm selection by females appears to be driven by the production of offspring with genotypes best suited to development under current environmental conditions. Thus, sperm selection based on the PGM genotype depends critically on the female's own genotype, so that a male who is discriminated against by one female will be selected by another. The net action of selection acting on males or rather their PGM locus will be zero. A second example of apparent sperm selection based on genetic compatibility comes from Stockley's (1999) work with *G. sigillatus*. Stockley (1999) mated females with both a sibling and an unrelated male to see if females showed preferential usage of sperm from nonrelatives to avoid inbreeding. It has been established that female crickets utilize other mechanisms of cryptic female choice to avoid sibling matings (Simmons 1989, 1991a). When the first male to mate was the female's sibling, his paternity tended to be less when the second male was unrelated to the female than when the second male was also a sibling. Unfortunately, Stockley's (1999) test suffered from a rather small sample size and was therefore weak; there was no significant difference between sibling and unrelated second males in their fertilization success. Nevertheless, the results contribute to an emerging pattern that, in general, cases that provide evidence for sperm selection all involve aspects of genetic compatibility (Bishop 1996; Bishop et al. 1996; Olsson 1996; Olsson et al. 1997; Palumbi 1999).

Criteria for Demonstrating Sperm Selection

Birkhead (1998) offered a number of criteria required to demonstrate sperm selection. He argued that it was necessary to demonstrate that there was variation in fertilization success and that some of this variation was due to differences between males and some was due to differences between females. Birkhead (1998) outlined two experimental designs that could be used to examine these criteria. Several authors have criticized Birkhead's protocol (Eberhard 2000; Kempenaers et al. 2000; Pitnick and Brown 2000). Eberhard (2000) was primarily concerned that Birkhead's discussion might be viewed as relevant to all mechanisms of cryptic female choice described in table 9.1. However, Birkhead was clear that his discussion was focused only on one mechanism of cryptic female choice, what I refer to here as sperm selection. All contributors to the debate noted that a lack of variation due to females is not evidence against sperm selection, because all females may be congruent on what constitutes a good male. This will be particularly true when sperm selection is based on some aspect of sperm competitive ability, such as sperm length or fertilization capacity, or on a correlate of sperm fitness such as offspring viability. Neither can variation due to females be taken as evidence in favor of sperm selection, since between-female variation can arise, as we have seen above, due to morphological differences or similarities between females. Moreover, variation in paternity among females may reflect cryptic male choice, where males invest more heavily in gaining fertilizations with females of high quality (Galvarni and Johnstone 1998; Pitnick and Brown 2000; section 7.3). Pitnick and Brown (2000) assert that interactions between male and female in determining fertilization success can be indicative of sperm selection. However, interaction effects are only likely to be important in cases of genetic compatibility, which themselves are unlikely to fuel sexual selection acting on males.

I consider that there are two critical areas which must be addressed if we are to provide unequivocal evidence for sperm selection. First we must demonstrate the mechanisms involved in sperm transfer, storage, and utilization (see also Simmons and Siva-Jothy 1998). In general, species with little or no sequence effects and high variance in paternity are those most likely to have the potential for sperm selection (Sivinski 1984; Simmons and Siva-Jothy 1998). Quite simply, if females do not have a random mix of sperm in their sperm stores from a number of males, they will be unable to choose among males via sperm selection. Only once we have a clear understanding of mechanisms can we begin to ascribe variance in paternity due to males and females to a process of sperm selection, as opposed to differential sperm transfer or transportation and storage. The second avenue is to test models of preference evolution from a quantitative genetic perspective. All models of preference evolution predict a genetic correlation between the preference in females and the trait in males. Thus, if a species truly has a mechanism of sperm selection, we should expect to see genetic covariance between males and females in the trait that is subject to selection and female preference for that trait. If the trait is some aspect of sperm morphology such as sperm length, and long spermathecal ducts allow females to utilize longer sperm for fertilization, then we should expect a genetic correlation between sperm length in males and spermathecal duct length in females. If females select competitive sperm by multiple mating, we should expect to see a genetic correlation between female multiple mating behavior and male sperm competitiveness. If, on the other hand, sperm length or competitiveness were favored purely through sperm competition, we should not expect genetic correlations between these traits in males and aspects of female behavior or morphology. Quantitative genetic approaches such as these offer a powerful way with which to unravel even the most cryptic mechanism of sexual selection (Partridge 1994).

9.6 Summary

Females cannot be regarded as passive vehicles within which males compete for fertilizations. Ultimately it is females that control the storage and utilization of sperm so that they have the potential to exercise choice over which males fertilize their eggs. Nevertheless, because of the difficulties involved in examining processes occurring within the female's reproductive tract, female influences are poorly documented. Females may exercise what has become known as cryptic female choice at a number of key stages of reproduction. They can determine whether they remate, how much sperm they receive from a copulating male, whether they transport that sperm to their sperm storage organs, whether they allow males to displace sperm, and, finally, which of the stored sperm they actually utilize at fertilization. The order of these stages represents an increasing level of crypsis and thus difficulty in unequivocally demonstrating male and female influences.

Female influence over remating is not truly cryptic as mating is readily observable from behavior. Numerous studies have shown how females can influence the amount of sperm received during copulation in a manner that favors particular individuals as sires. Adaptation in males to circumvent female control of sperm transfer is seen in some taxa where males provide nuptial food gifts that dissuade females from removing spermatophores or terminating copulation. The storage of sperm in the sperm storage organ(s) is ultimately under female control, and sperm that are not transferred to the sperm stores can be ejected after copulation. The widespread occurrence of copulatory courtship in insects has been proposed to function in persuading females to store sperm after ejaculation. Recent studies of changes in the distribution of ejaculates within the female's reproductive tract and the effects of copulatory courtship on paternity provide compelling evidence in favor of this avenue of cryptic female choice. However, the behavior of males during copulation need not be adaptive from the female's perspective, and there is some evidence to suggest that males have evolved to exploit existing sensory biases in females that facilitate the ejection of sperm from the sperm storage organ(s).

Perhaps the most controversial area of cryptic female choice, and indeed the most cryptic of processes, is the notion that females can select among multiple genotypes of sperm within their sperm stores and bias paternity in favor of certain males. Potential mechanisms could include the partitioning of sperm among different sperm storage sites, the selective transport of sperm from the stores to the site of fertilization, or biased fertilization due to egg-sperm interactions. Evidence that sperm selection can be a potential avenue of female choice comes from studies of homogamy in which conspecific sperm are used in preference to heterospecific sperm, and from studies of facultative adjustment of offspring sex ratios. Nevertheless, partitioning variance in fertilization success between males and females is proving extremely difficult. The most promising approach is to determine the mechanisms behind sperm transfer and storage and to partition variation in paternity expected on the basis of these mechanisms. Quantitative genetic approaches to test predictions underlying preference evolution will represent a powerful test of cryptic mechanisms of female choice. Currently very few studies provide unequivocal support for the notion that females can bias paternity by sperm selection. Those studies that do support the hypothesis all involve aspects of genetic compatibility. This may have important implications for female fitness, but is unlikely to impose sexual selection on males, because males who are successful in gaining fertilizations with one female will be unsuccessful with another.

Female influences over sperm competition have only recently received the attention they deserve. Currently only a handful of studies provide evidence that female influences are able to bias paternity in favor of particular males. As with any newly advanced paradigm, there is a tendency for researchers to interpret data in line with new and popular ideas. Currently we know too little about the physiology and behavior of sperm, the female reproductive tract, and the interactions between them. Until we are better equipped to address these issues, caution should be exercised over claims arising from observed patterns of nonrandom fertilization and the roles of males and females in generating these patterns.

10 Social Insects

10.1 Introduction

Multiple mating and sperm competition has unusual evolutionary implications in social insects because of their mechanism of sex determination. All Hymenoptera exhibit haplodiploidy in which males are the product of unfertilized eggs (haploid) while females are the product of fertilized eggs (diploid). Haplodiploidy has profound effects on the degree of relatedness between individuals. Each sperm contains all of the male's genes while eggs have on average a 50% chance of containing a given gene from the female. Thus, female offspring are more closely related to their sisters than would be the case for a diploid species because full sisters will have a probability of 0.75 of sharing genes via common ancestry due to the complete contribution from their father. Conversely, females are less related to their brothers than would be the case for diploid species because they only have a probability of 0.25 of sharing genes in common derived from their mother. Moreover, females are more closely related to their sisters than they would be to their own offspring (0.75 versus 0.50) so that their evolutionary fitness should best be served by producing sisters rather than offspring. Thus Hamilton (1964a, 1964b) proposed that haplodiploidy facilitated the evolution of reproductive altruism among hymenopteran females, explaining both the prevalence of eusociality in the Hymenoptera and the restriction of worker behavior to females (see also Trivers and Hare 1976). Eusociality has multiple evolutionary origins within the haplodiploid Hymenoptera, the bees, ants, and wasps (Wilson 1971; Crozier and Pamilo 1996), but is rare in diploid taxa (Choe and Crespi 1997b), being reported in termites (Shellman-Reeve 1997), one genus of beetle (*Austroplatypus incompertus*, Kent and Simpson 1992), one genus of thrips (*Oncothrips*, Crespi 1992), and a few genera of aphid (Aoki 1982). Unlike Hymenoptera, diploid eusocial species have not been subject to intense theoretical and empirical research and they will not be dealt with further in this chapter. When I refer to social insects I refer strictly to the social Hymenoptera, the bees, ants, and wasps.

Multiple mating, or more specifically multiple paternity, is expected to reduce the level of relatedness between females within social insect colonies. For example, if a female mates with two males, each receiving equal paternity, the average relatedness between her female offspring will be 0.50, rather than 0.75 with single paternity. The greater the number of males sharing paternity the lower the average relatedness between sisters. Of course, the relatedness between mother and offspring remains unchanged. Multiple paternity should therefore prevent the evolution of altruistic behavior between sisters, and thus eusociality (Hamilton 1964b,a; Wilson 1971; Hamilton 1972; Trivers and Hare 1976; Pamilo 1991c). Nevertheless, reports of multiple mating are widespread in social insects (Page and Metcalf 1982; Cole 1983; Starr 1984; Crozier and Page 1985; Page 1986; Keller and Reeve 1994; Boomsma and Ratnieks 1996) so that sperm utilization following multiple mating is likely to be of fundamental importance to the social structure of insect colonies, imposing selection on reproductive males and females, and on workers. In this chapter I discuss aspects of sperm competition that are likely to influence the evolution of social insect behavior. Colony relatedness can be influenced by factors other than multiple mating, such as multiple maternity within colonies (so-called polygyny). I restrict my discussion to the effects of multiple mating and paternity. For a complete analysis of social evolution in insects the reader is directed to the excellent works of Wilson (1971), Bourke and Franks (1995), and Crozier and Pamilo (1996).

10.2 Multiple Mating and Multiple Paternity

It has long been known that honey bees mate with a number of different males during a mating flight (Taber 1955; Taber and Wendel 1958) so that they represented a paradox in discussions of social evolution (Hamilton 1964b; Wilson 1971; Trivers and Hare 1976). An early solution to this paradox was the claim that sperm from different males clumped in storage and only a single male fertilized offspring until his sperm were exhausted (Trivers and Hare 1976; Alexander and Sherman 1977; Charnov 1978). Sperm precedence resulting from clumping could, theoretically, maintain high genetic relatedness amongst sisters (Boomsma and Ratnieks 1996). However, Crozier and Brüchner (1981) quite rightly challenged the validity of the sperm clumping claim. From examination of the nonsocial insect literature it is clear that where sperm precedence does occur it is very short lived due to subsequent mixing of sperm within the female sperm storage organ(s) (see table 2.3). Studies of naturally mated honey bees have since shown that the sperm of different males are utilized at random from the female sperm stores, with little or no evidence for the extreme clumping required to maintain sister relatedness at 0.75 (Page and Metcalf 1982; Laidlaw and Page 1984; Moritz 1986; Estoup et al. 1994; Haberl and Tautz 1998). Thus, early reviews of sperm utilization in social insects gathered evidence against the paradigm that social insects were essentially monandrous and attempted to develop adaptive scenarios for the evolution of multiple mating (Page and Metcalf 1982; Cole 1983; Starr 1984; Crozier and Page 1985; Page 1986).

The particular focus of these reviews was not the incidence of multiple mating per se, but the extent to which multiple mating resulted in multiple paternity, since it is only the effectiveness of mating that should impact on social evolution. Thus previous reviews categorized data on queen mating frequency as observational, dissection and counts of sperm in storage, and visual or allozyme genetic markers to assign paternity. Page (1986) suggested that observational data are likely to give a biased estimate of the incidence of genetic polyandry because, although males may attempt to mate with any female they encounter, females may be successful in resisting insemination. In the most recent review, Boomsma and Ratnieks (1996) show that within the ants (for which most data are available) the documentation of polyandry is significantly associated with the type of data collected; multiple mating is more likely to be reported from observational studies than from dissections of sperm stores or use of genetic markers. Dissection studies can confirm multiple insemination only when females are able to store more sperm in their sperm stores than a single male can transfer (e.g., Reichardt and Wheeler 1996). However, sperm counts are not informative if a single male can transfer as much or more sperm than an average female can store. This is because females may mate with several males but only store a part of the sperm received from each (e.g., Page 1986). Furthermore, studies of nonsocial insects show quite clearly that males can use large ejaculates to flush out sperm stored from previous males (Otronen and Siva-Jothy 1991; Eady 1995; section 2.5) and thereby gain the majority of fertilizations. Similarly, studies using genetic markers to assign paternity show the end product of interactions between males in sperm competition. Thus, if mixed paternity is revealed it can be inferred that females are multiply mated. However, single paternity can result from single mating by females or from multiple mating and the operation of mechanisms in males for the removal or incapacitation of rival sperm. Female odonates, for example, exercise multiple mating but have almost single paternity within a clutch due to the mechanical removal of sperm (table. 2.3). Thus, the fact that observational studies point to a higher degree of multiple mating does not necessarily advance a false impression of this behavior in social insects. Social insect biologists have been concerned principally with examining the influence of mixed paternity on colony relatedness so that it is the outcome of processes of sperm transfer, storage, and utilization that is important from their perspective (Boomsma and Ratnieks 1996). However, as with nonsocial insects, a complete understanding of these underlying processes will be necessary if we are to fully appreciate the evolutionary consequences of multiple mating and sperm competition within social insects. Selection via sperm competition in nonsocial insects favors adaptations in males to ensure complete fertilization success (Parker 1970e). In social insects there may also be selection for complete fertilization success for a given male, but in this instance selection may act via the female to ensure colony relatedness and queen reproductive success, as well as via the male to ensure paternity. Thus, queen and male interests may coincide. Alternatively, if multiple paternity has advantages for social life, queen and male interests may conflict. The social insects may therefore represent one of the most promising groups within which to examine female influences over sperm transfer, storage, and utilization (cf. chapter 9).

While studies of sperm utilization in social insects have not examined processes occurring between copulation and offspring production, they represent our most extensive knowledge of multiple paternity within natural populations. Studies are now available from 74 species of eusocial Hymenoptera, predominantly from ants, on the incidence of multiple paternity. The data are provided in table 10.1 at the end of this chapter, which represents an extension of the review provided by Boomsma and Ratnieks (1996). Since Boomsma and Ratnieks (1996) review there has been an explosion of studies of multiple paternity facilitated by the advent of molecular DNA technology. Additional information on the incidence of multiple mating obtained from observational studies of social insects can be obtained from earlier reviews (Page and Metcalf 1982; Cole 1983; Starr 1984; Crozier and Page 1985; Page 1986; Keller and Reeve 1994). The data show considerable variation in the proportion of females in natural populations that produce progeny of mixed paternity. Nevertheless, the general pattern appears to be consistent with kin selection theory, within species eusocial Hymenoptera females show a low probability of producing offspring of multiple paternity (table 10.1). Considerably more data are available for the ants and the general trend in the data for this taxon is illustrated graphically in figure 10.1A. It is also possible to examine patterns of sperm utilization in females that have mated more than once. Unlike in nonsocial insects, contributions of males following double matings cannot be assigned to either the first or second male because experimental matings have never been performed. Rather, paternity is inferred from genetic data. Thus values of P_2 or P_1 cannot be calculated. In this area of evolutionary biology the variable that examines relative paternity of different males is known as paternity skew, calculated as the sum of the squared proportions of offspring sired by each male. Paternity skews are presented for doubly mated females in table 10.1 and the possible causes discussed in section 10.3. Among females that do produce offspring sired by two different males, mixed paternity is often the result, although equally as often one particular male is over-represented (variation in paternity skew is illustrated for ants in fig. 10.1B). Published values of paternity skew are undoubtedly underestimates because of problems associated with limited numbers of offspring genotyped (nonsampling error) and because of limited variation in the genetic markers used (nondetection error) (see Pedersen and Boomsma 1999). Thus the true paternity skew is likely to be greater than that shown in table 10.1 and fig. 10.1B. The consequence of low proportions of multiple paternity broods and high paternity skew in mixed broods is to keep the effective paternity frequency close to 1. Boomsma and Ratnieks (1996) calculated an average effective paternity frequency for ants

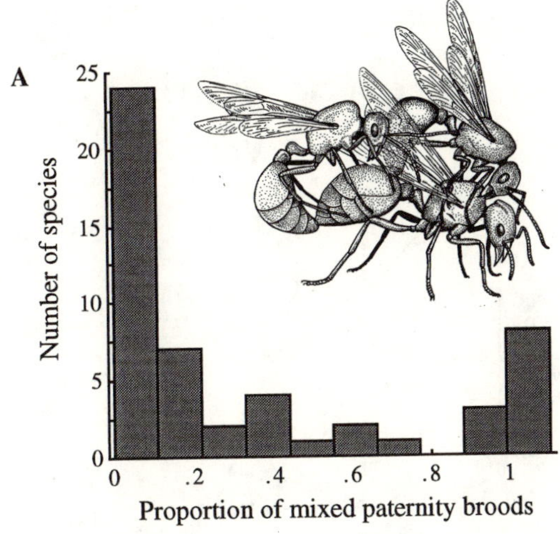

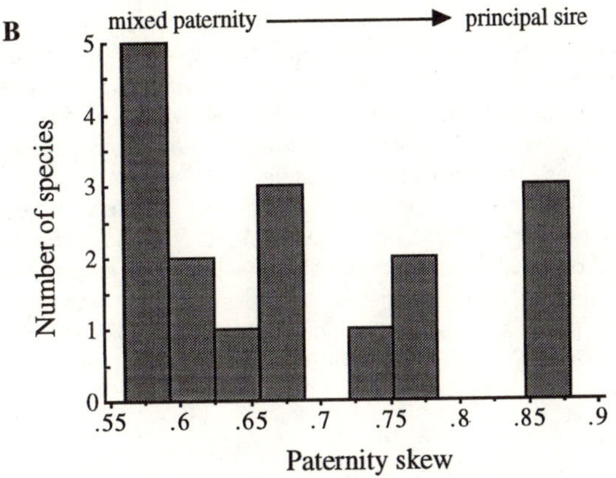

Figure 10.1 Incidence of multiple paternity in ants. (A) Frequency distribution of the proportion of females found to produce broods of mixed paternity across thirty-two species of ants. In general, single paternity is the most common finding. (B) Frequency distribution of paternity skew for doubly mated female ants. Paternity skew is calculated as the ΣP_i^2, where P is the proportion of offspring sired by the ith male (Starr 1984). With double mating, a skew of 0.50 indicates complete mixed paternity, while skews tending toward 1.0 indicate that one male gains the majority or all fertilizations (data from table 10.1). All species studied show paternities skewed toward one male, although to varying degrees.

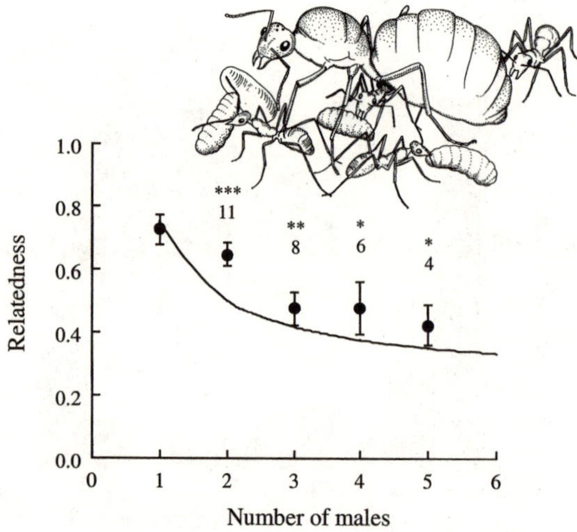

Figure 10.2 The effect of multiple paternity on average relatedness between siblings within colonies of *Formica truncorum*. The solid line illustrates the decline in relatedness when all copulating males obtain an equal proportion of fertilizations. The data are pedigree relatedness estimates with 95% confidence intervals for progenies from laboratory colonies founded by queens mated with 1–5 males. Numbers of colonies are given above each mean value. If all males contributed equally, multiple paternity would reduce average colony relatedness to a level below 0.5 so that workers would be more closely related to their own offspring and would have increased fitness by ceasing to rear sisters. Paternity skew in doubly mated females is 0.56, so that although there is predominance of one male, the second male obtains a large proportion of fertilizations. Nevertheless, paternity skew is sufficient to prevent colony relatedness falling to the level expected with equal paternity, as indicated by the significant deviations from expected relatedness. (* $P < 0.05$, ** $P < 0.01$, *** $P < 0.001$; redrawn from Sundström 1993)

of ≤1.16, which reduced the relatedness among colony members by just 8%. Although the increased data set for ants now yields a somewhat higher arithmetic mean effective paternity frequency of 1.37 ± 0.11, the distribution remains highly skewed with over 70% of species falling between 1.0 and 1.3 (table 10.1). Nevertheless, it is clear that multiple paternity occurs to some degree in 61% of species studied. Multiple paternity does reduce colony relatedness across species (see Boomsma and Ratnieks 1996), although paternity skew can moderate this effect (fig. 10.2). It is also clear that some species of social insect have both a high probability of multiple mating, and a high representation of different fathers in mixed broods (fig. 10.1). Specifically, high genetic polyandry and low intracolony relatedness appear common in three taxonomic groups, the vespine wasps (*Vespula* and some *Dolichovespula*), the honey bees (*Apis*), and the leafcutter ants (*Acromyrmex* and *Atta*) (table 10.1 and Boomsma and Ratnieks 1996).

10.3 Social Consequences of Sperm Competition

Trivers and Hare (1976) used Fisher's (1930) sex ratio theory to show how, within social insect colonies where female workers help in reproduction, asymmetries in relatedness between sisters and their brothers can lead to conflict with queens over sex allocation. Fisher's (1930) theory states that the population sex ratio investment is at an evolutionary equilibrium when the fitness payoffs from producing sons and daughters are equal. Thus, in the simplest case when sons and daughters are equally costly to produce and yield the same fitness returns in terms of their reproductive success, a female should invest equally in sons and daughters so that the population sex ratio is 1:1. The sex ratio is evolutionarily stable because the fitness payoffs associated with sons and daughters are frequency dependent and determined by the population sex ratio. Thus, in a female-biased population the fitness payoffs for producing sons will exceed that for daughters so that selection will favor an increased investment in sons. As the number of males in the population increases, so the payoff for producing sons decreases as they begin to compete for the relatively decreasing numbers of females. Once the 1:1 sex ratio is restored, selection will no longer favor increased investment in sons because the fitness payoff for sons no longer exceeds that for daughters and will become lower than that for daughters if the sex ratio should become male biased. Deviations from a 1:1 sex ratio can arise for various reasons. For example, if daughters cost twice as much to produce as sons and yield the same fitness return, a reproducing female will be selected to invest twice as much in sons so that her fitness return per unit investment remains the same for either sex.

In diploid and haplodiploid taxa, reproducing females are equally related to their sons and daughters (each share, on average, 0.50 of their genes) and the population sex ratio should equilibrate at 1:1. However, when female workers control reproduction, Trivers and Hare (1976) showed that there will be conflict over sex allocation. In a colony headed by a singly mated queen, workers are related to their sisters by 0.75 and to their brothers by 0.25. There is therefore a relatedness asymmetry of 3:1. From a worker's perspective, the fitness payoff of producing a female is three times that of producing a male so that workers should invest more in sisters than in brothers. If workers control reproduction, the population sex ratio is expected to become biased toward an excess of females, equilibrating at a ratio of 3:1 females to males. Although at this ratio females will have just one-third the mating success of males, this is compensated by the threefold relatedness-based fitness return for every female offspring that actually mates. Thus, because of relatedness asymmetry, queens prefer a 1:1 sex ratio while workers prefer a 3:1 sex ratio, resulting in queen-worker conflict over sex allocation (for more detailed and/or mathematical treatments of sex ratio theory, see Trivers and Hare 1976; Benford 1978; Pamilo 1982a, 1991a; Bourke and Franks 1995, and references therein).

Multiple mating by queens and multiple paternity within colonies reduces the relatedness asymmetry between workers and female offspring (i.e., among sisters) so that workers should favor a less female-biased sex ratio, thereby reducing queen-worker conflict. For example, if all females mate twice and there is equal paternity between sires, relatedness between workers falls to 0.5 (the average of 0.75 for full sisters and 0.25 for half sisters). The value of females for workers is reduced to being only twice that of males so that the stable sex ratio for workers will be 2:1 females to males. The higher the queen mating frequency the closer the workers' stable sex ratio, and the population sex ratio, approaches the queen's optimum of 1:1. Reduced queen-worker conflict has been suggested as one selective advantage for multiple mating in eusocial Hymenoptera (Starr 1984; Moritz 1985; Pamilo 1991b; Ratnieks and Reeve 1992; Queller 1993; Ratnieks and Boomsma 1995). However, multiple mating can only be at a selective advantage where females gain an immediate fitness benefit associated with the behavior (Ratnieks and Boomsma 1995).

One method by which an immediate fitness benefit to multiple mating can accrue is via facultative worker policing (Starr 1984; Woyciechowski and Lomnicki 1987; Ratnieks 1988). Although workers in most species of eusocial Hymenoptera do not mate, some can produce unfertilized eggs that produce males. With single queen mating, workers are more closely related to males produced by their full sisters (0.375) than they are to those produced by their mother (0.25) so that they should be in conflict with the queen over whose sons to rear. With multiple mating this conflict is removed because workers will be on average more closely related to their brothers than to their sisters' sons, leading to worker policing in which workers kill the eggs laid by their sisters (Ratnieks and Visscher 1989). Foster and Ratnieks (2000) found a positive correlation between worker relatedness and the proportion of males produced by workers in the wasp *Dolichovespula saxonica*; worker policing in colonies headed by multiply mated queens reduced male production by workers. Thus, multiple mating can provide immediate fitness benefits to females and, in contrast to the early concerns of Trivers and Hare (1976), can actually promote cooperation and social evolution (Foster and Ratnieks 2000).

An alternative mechanism by which females can gain immediate fitness returns through multiple mating is where mechanisms exist for facultative sex allocation by workers. If multiple mating varies between individuals within eusocial species, as appears to be the case (table 10.1), and workers within colonies are able to assess their own queen's mating frequency and control sex allocation, it has been shown that sex ratios between colonies should vary although the population sex ratio remains constant (Boomsma and Grafen 1990, 1991). Imagine a species in which a small proportion p of queens has mated twice while $1-p$ have mated just once. The p colonies will have a relatedness asymmetry 2:1 (workers related by 0.5 to female offspring and 0.25 to male offspring) and $1-p$ colonies will have the usual relatedness asymmetry of 3:1. When p is small the population sex ratio should reflect the preferred sex ratio of the more numerous $1-p$ colonies, which will produce an excess of females. The relatedness asymmetry of the p colonies will therefore be below the average relatedness asymmetry reflected in the population-wide sex ratio, so that for them sons will have a greater fitness return than daughters; despite their lower relatedness to brothers, workers can expect a greater fitness return from brothers because of their relatively greater mating success in the female-biased population. Thus, selection should favor the production of only male offspring by workers in the p colonies. Boomsma and Grafen (1990, 1991) refer to this as split sex ratios (fig. 10.3). As p increases in evolutionary time, the pure male production of p colonies will affect the population-wide sex ratio so that the $1-p$ colonies will adjust their sex ratio to maintain the population sex ratio bias that is stable for them, until a point is reached where they produce only females (fig. 10.3). When p exceeds a critical frequency (33% of the population's colonies are headed by doubly mated queens) the population-wide sex ratio falls below the relatedness asymmetry for the p colonies ($1-p$ colonies now produce all females and can no longer compensate for the male production of p colonies). Females begin to increase in value so that the p colonies begin to produce mixed broods. Eventually, when p approaches 1.0, all colonies produce a sex ratio of 2:1 dictated by their relatedness asymmetry (fig. 10.3). If workers cannot assess the relatedness asymmetry within their colonies, then selection should favor all colonies responding to the population average relatedness asymmetry and produce a sex ratio that equals this population average; sex ratios should not be split (Boomsma and Grafen 1990; Ratnieks 1990a; Boomsma and Grafen 1991; Ratnieks 1991).

Van der Have et al. (1988) found that worker investment in males and females in the ant *Lasius niger* was associated with differences in the population sex ratio and multiple mating, and Murakami et al. (2000) found a negative association between queen mating frequency and the proportion of males produced across seven species of attine ant. Split sex ratio theory (Boomsma and Grafen 1990, 1991) specifically predicts that colonies headed by multiply mated queens should invest predominantly in male production while colonies headed by singly mated queens should invest mostly in female production. Sundström (1994) examined the sex ratio of reproductives produced by colonies of the ant *Formica truncorum* over a four-year period. She found that 43% of queens were multiply mated in this species and that multiple mating reduces within-colony relatedness asymmetry (fig. 10.2). In accord with split sex ratio theory, colonies headed by a multiply mated queen invested in male production while colonies headed by singly mated queens invested in female production (fig. 10.4). This pattern has been main-

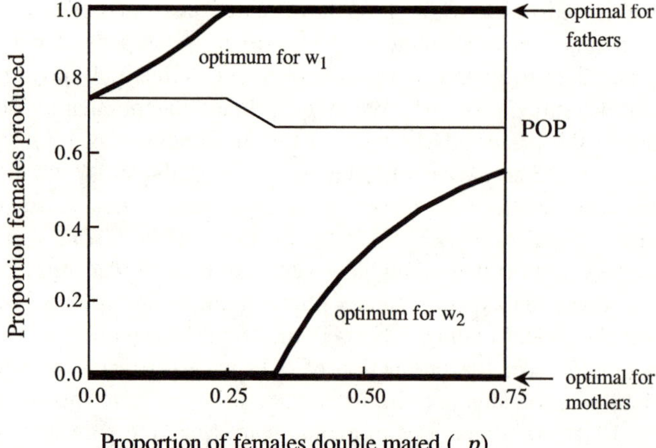

Figure 10.3 Split sex ratio theory in eusocial insects (Boomsma and Grafen 1991). The stable sex ratio in terms of the proportion of females produced is plotted against the proportion of queens in the population that have mated with two males, *p*. The optimal sex ratio for workers in a colony with a single-mated queen is given by w_1 and the optimal sex ratio for workers in a colony with a double-mated queen is given by w_2. The population sex ratio is given by POP. When $p < 0.25$, colonies with single-mated queens will balance the all-male production of colonies with double-mated queens by increasing female production, and the populationwide sex ratio would remain 3:1. When $0.25 \leq p \leq 0.33$, colonies headed by single-mated queens will produce all females and those headed by double-mated queens will produce all males. The population sex ratio will represent a weighted average between the relatedness asymmetries of double- and single-mated colonies. When $p > 0.33$, colonies with double-mated queens balance the all-female production of colonies with single-mated queens, and the population sex ratio will be 2:1. (Modified from Boomsma 1996)

tained by these colonies for over six years (Sundström and Ratnieks 1998). Sundström et al. (1996) later found split sex ratios in a second species, *F. exsecta*. More importantly, there were no differences in colony sex ratios of singly and multiply mated queens when sex ratio was assessed at the egg stage, as expected from the queen's preferred 1:1 sex ratio. Changes toward female-biased sex ratios in singly mated colonies occurred between the egg stage and adult emergence. The data strongly support the conclusion of worker control, suggesting that workers eliminate males when the female is singly mated and their preferred sex ratio is 3:1 or even more female biased. Sundström and Ratnieks (1998) showed how doubly mated queen *F. truncorum* have a 37% fitness advantage over singly mated queens, providing empirical support for theoretical models that predict selection for multiple mating under queen-worker conflict over sex allocation (Boomsma and Grafen 1991; Ratnieks and Boomsma 1995).

Relatedness-induced worker-controlled split sex ratios have been reported from 17 species of eusocial insect (e.g., Mueller 1991; Queller et al. 1993b;

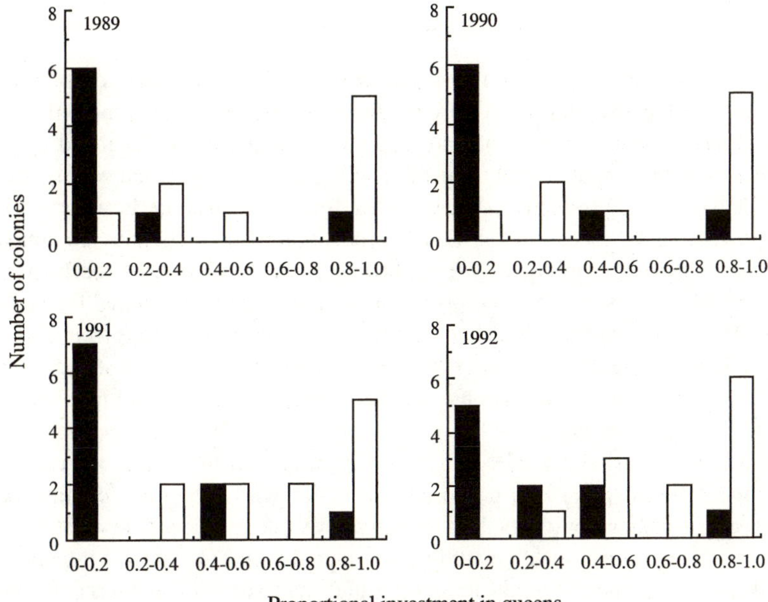

Figure 10.4 Split sex ratios in the ant *Formica truncorum*. The frequency distributions show the colony sex ratio investment in females, calculated as the proportion dry weight of female pupae to male pupae, for 17–22 colonies sampled in each of 4 years. The distributions are bimodal, as predicted by Boomsma and Grafen's (1991) split sex ratio theory. Colonies headed by a single-mated queen (open bars) specialized on female production, while colonies headed by double-mated queens (solid bars) specialized on male production. Queen mating frequency was estimated using allozyme variation at 3–4 polymorphic loci. (Redrawn from Sundström 1994)

Chan and Bourke 1994; Packer and Owen 1994; Evans 1998; see reviews in Queller and Strassmann 1998; Chapuisat and Keller 1999). Nevertheless, there are anomalies. Brown and Keller (2000) failed to find differences in relatedness asymmetry between male- and female-producing polygynous colonies of *F. exsecta*, and Helms (1999) found split sex ratios in populations of the ant *Pheidole desertorum*, which appears to be monandrous (see also Pamilo and Seppä 1994; Aron et al. 1999). In *P. desertorum* it was queen control rather than worker control that resulted in biased sex allocation in male specialist colonies (Helms 2000), which might explain why multiple mating has not arisen in this species. In conclusion, worker-controlled split sex ratios can provide the requisite fitness gain necessary for the evolution of multiple mating by queens (Ratnieks and Boomsma 1995), and further study is required to assess the general applicability of split sex ratio theory to the evolution of multiple mating in social insects.

Male-Queen Conflict

The occurrence of split sex ratios presents an immediate selection pressure on males. Because males can only realize reproductive success through their daughters (remember that sons are haploid, arising from unfertilized eggs), males that have mated with queens heading colonies producing only sons will have zero reproductive success. Thus, male interests are best served by single mating by queens and all-female production by their colonies (Boomsma 1996). All else being equal, we might expect the evolution of mechanisms for the avoidance of sperm competition such as male mate choice for virgin females (section 5.6), or mechanisms for preventing the female from remating (chapters 4 and 5). However, where female interests are best served by reducing relatedness asymmetry within colonies through multiple mating (Ratnieks and Boomsma 1995), or via other fitness benefits (section 10.4), there is likely to be sexual conflict over remating so that male interests are not met. Boomsma (1996) proposed that males who exercised "continence" in sperm transfer could obtain a selective advantage over males that either rejected nonvirgin females or fully inseminated them. Consider a population with low frequency of multiple mating and split sex ratios. A second male who fully inseminates a queen will have zero reproductive success due to colony specialization on males. However, a male who reduced his ejaculate size so that there is a high paternity skew in favor of the first male will reduce the relatedness asymmetry in resulting colonies only marginally so that the colonies are less likely to specialize on male production (Boomsma 1996). Of course the male will gain lower fertilization success than his sperm competition rival, but his fitness return will be greater than if he fully inseminated the queen so that her workers produced only males (fig. 10.5). Boomsma's (1996) model showed that the evolution of strategic ejaculation by males is dependent on the frequency of double-mating females in the population; for the specific conditions analyzed by Boomsma (1996), strategic ejaculation should be favored between a probability of queen double-mating of 0.29 and 0.48, a range encompassing observed double mating frequencies in natural populations (table 10.1). The model assumes that males are able to assess female mating status, an assumption that has good empirical support from nonsocial insects (section 5.6). It also assumes that workers are able to recognize variation in relatedness among their nest mates, an assumption for which there is also some support (Breed and Bennett 1987; Waldman et al. 1988; Crozier and Pamilo 1996).

Boomsma's (1996) model of strategic ejaculation has interesting parallels with those developed for nonsocial insects (chapter 7). Galvarni and Johnstone's (1998) analysis predicted that males should vary ejaculate size in relation to female quality. One aspect of female quality is of course the fitness returns expected for male investment in terms of fertilization gain. With split sex ratios, mated females are of reduced value because of worker

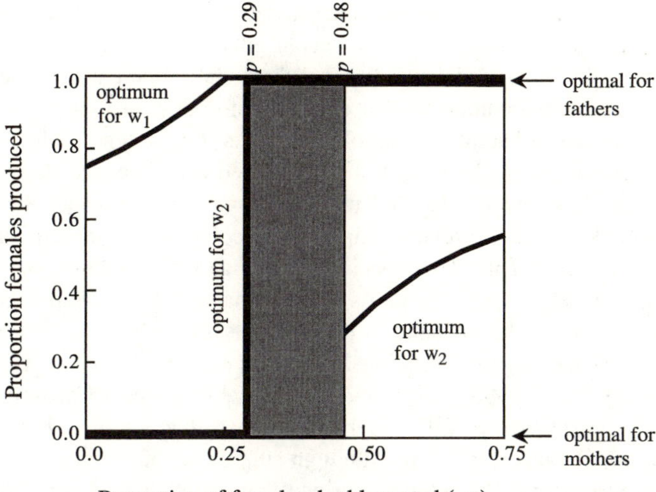

Figure 10.5 Conditions under which reduced insemination by males as an adaptation to multiple mating by queens can spread. The figure shows the split sex ratio optima for workers in relation to the proportion of double-mated queens in the population, p (see also fig. 10.3); w_1 is the optima for workers of a single-mated queen, w_2 for workers of a double-mated queen with equal insemination by two males, and w_2' for workers of a double-mated queen where the second male reduces ejaculate size. The sex ratio optima are equal to paternal fitness. The example assumes that the second male reduces sperm numbers to just 18% of the first male's contribution, so that the effective paternity frequency is 1.33, the nest relatedness is reduced to 0.625, and relatedness asymmetry is 2.5:1 instead of 2:1 with full insemination. When $p < 0.29$, the optimal sex ratio of both w_2 and w_2' workers is all male so that selection would not favor males who reduce ejaculate size. Rather, males should reject mated females. When $0.29 \leq p \leq 0.49$, w_2' workers should produce all females favoring males who transfer an ejaculate 18% the size of the first male's ejaculate. The proportion of females produced by w_2 workers rises with increasing p. A threshold is reached, at 0.48 for the parameters given above, where the reduced fertilization success of ejaculate-reducing males results in a fitness return equal to that expected for fully inseminating males. Beyond this point, males reducing ejaculate size will be disfavored because their fertilization gain will be less than that expected from full insemination and worker adjustment of sex ratio. (Modified from Boomsma 1996)

manipulation of sex ratios. Thus, both Boomsma (1996) and Galvarni and Johnstone's (1998) models converge on the same solution, reduction of ejaculate expenditure with low-quality females. In Boomsma's (1996) models, the proportion of females in a population that are doubly mated is equivalent to Parker et al.'s (1997) sperm competition risk. In Parker et al.'s (1997) analysis increased risk favored increased expenditure on the ejaculate as males competed in a fertilization raffle (section 7.2). There is also a positive association between ejaculate size and risk of double mating predicted from Boomsma's (1996) models but for different reasons; ejaculate size should decrease with decreasing sperm competition risk because split sex ratios result in zero reproductive success for second males with large ejaculates.

Boomsma's (1996) models predict that there should be a negative association between the population-wide probability of double mating and paternity skew; when the probability of double mating is low, second males with reduced ejaculate size will be favored and reduced ejaculate size should result in paternity skew toward the first male to mate. Boomsma and Sundström (1998) found just such a relationship in their comparative analysis of seven species of *Formica* ants. Two species, *F. rufa* and *F. sanguinea*, have been shown not to have split sex ratios (Boomsma and Sundström 1998) and arguably should not be included in this analysis. Nevertheless, even without these species the negative relationship seems strong (fig. 10.6). Boomsma and Van der Have (1998) also found a negative relationship between paternity skew and frequency of double mating across 36 colonies of *Lasius niger* from six populations (fig. 10.6). Although supportive of the male continence hypothesis, data from 11 colonies derived from another population that was characterized by a considerably higher frequency of multiple mating did not conform with the expected pattern. Moreover, *L. niger* from two populations (one Swiss and one Swedish) did not appear to exhibit split sex ratios and there was no evidence to suggest that workers regulated sex allocation in response to queen mating frequency (Fjerdingstad et al. MS).

An alternative mechanism by which paternity skew can arise is via sperm clumping (see section 10.2 above and Boomsma and Sundström 1998). Under male-queen conflict, the reproductive interests of all males that mate with a single female are best served by patterns of sperm utilization in which there is a principal sire, since this will maximize relatedness asymmetry that favors female production by colonies (Boomsma 1996). Sperm clumping could effectively allow males to "take turns" in siring reproductive daughters produced by the colony. Harvey and Parker's (2000) analysis shows how principal sires can occur when two males' sperm remain in large packets within the sperm stores of the female. Sundström and Boomsma (2000) examined the long-term patterns of sperm utilization by doubly mated queens in *F. truncorum*. Averaged across nine years, they found that two males received the same proportional representation in offspring that was close to 0.5 (i.e., two males had equal long-term fertilization success). This is strong evidence against the notion of male continence. However, Sundström and Boomsma (2000) found that there was significant heterogeneity in paternity skew between different cohorts of offspring, with one male gaining predominantly more offspring than the other in any given cohort. This pattern is consistent with moderate sperm clumping (Harvey and Parker 2000). Importantly, they found that colonies with higher average cohort-specific relatedness asymmetry produced more female-biased sex ratios. Thus, selection can favor males whose sperm remain clumped within the female's sperm store, rather than mixing with rival sperm and competing for

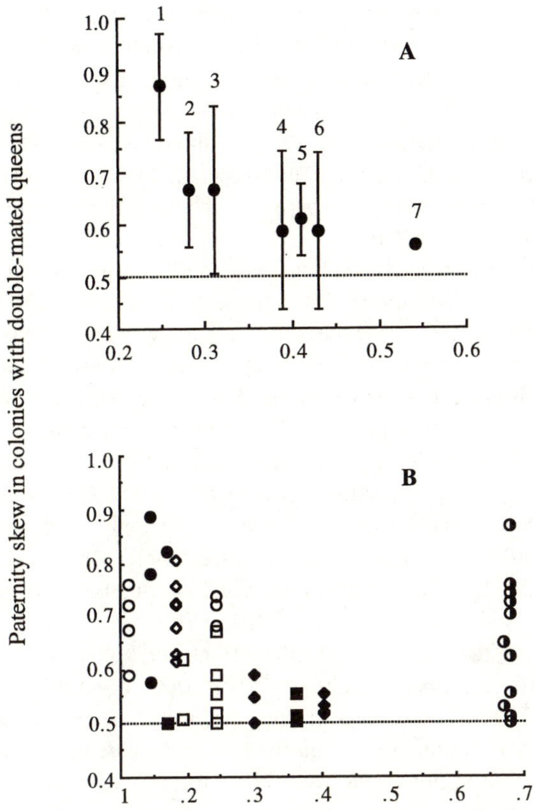

Figure 10.6 Empirical support for queen-male conflict over sperm allocation in ants. Strategic ejaculation under worker-controlled split sex ratios predict a strong paternity skew against ejaculate-reducing males. Moreover, paternity skew should decline with increased frequency of queen double mating because male continence will provide lower fitness returns as workers in colonies headed by double-mated queens begin to produce females. (A) Across seven species of *Formica*, paternity skew (sum of the squared proportions of offspring sired by each of two males ± SD) declines with the populationwide estimate of the proportion of queens double mated. The relationship is supported by phylogenetic comparisons based on ten possible phylogenies of the species used. 1, *F. transkaucasica*; 2, *F. fusca*; 3, *F. pressilabris*; 4, *F. exsecta*; 5, *F. truncorum*; 6, *F. rufa*; 7, *F. sanguinea* (after Boomsma and Sundström 1998). (B) Observed values of paternity skew decline with the proportion of double-mated queens across thirty-six colonies of *Lasius niger* from six populations that exhibit double mating. A value of 0.5 represents equal paternity for both sires, or no skew in paternity. However, paternity skew in a further population with mating frequencies greater than 2 do not conform with the observed pattern. Colonies from each population are depicted with different symbols. (Modified from Boomsma and Van der Have 1998)

fertilizations. Sperm mixing appears to be more complete in *F. exsecta* where there are no cohort differences in paternity contribution (Keller et al. 1997). In *F. exsecta* one male consistently fathered more offspring than the other, and Keller et al. (1997) suggest that this was most likely due to one male having more sperm in the sperm stores than the other. More studies such as these will be required to examine the general significance of patterns of sperm mixing versus variation in sperm numbers for the evolution of multiple mating under split sex ratios.

Unfortunately, experimental matings have not been performed to examine the mechanisms behind observed paternity skew in ants. It could arise due to strategic ejaculation and subsequent mixing of sperm predicted by male-queen conflict over sperm allocation (Boomsma 1996). Alternatively, it could arise because of first- or second-male sperm precedence or because of sperm displacement. Boomsma and Sundström (1998) raise several objections to a second-male sperm precedence/displacement scenario. First, there are no obvious genitalic structures in ants that could facilitate sperm displacement. Second, males emerge with a fixed supply of sperm and their testes subsequently degenerate (Heinze and Hölldobler 1993) so that they are sperm limited. Such a physiology might preclude the use of large quantities of ejaculate to flush rival sperm from the sperm stores. Third, selection is expected to favor the evolution of adaptations for second-male sperm displacement and/or precedence only when the risks of multiple mating are high (Parker 1970e). In the case of ants, paternity skew is associated with low risk of sperm competition. Fourth, first-male paternity is a rarely reported phenomenon in nonsocial insects, but it has been repeatedly found in nonsocial Hymenoptera where female multiple mating is rare (e.g., Wilkes 1966; see table 2.3; Allen et al. 1994; El Agoze et al. 1995). Ultimately studies of ejaculate sizes and sperm utilization for first and second males in social insects are necessary tests of Boomsma's (1996) theory of strategic ejaculation under split sex ratios. The long-term study of Sundström and Boomsma (2000) represents an excellent approach to this problem.

10.4 Alternative Hypotheses for Multiple Mating

We have seen above how multiple mating by queens can be favored by the reduction in queen-worker conflict over sex allocation when there are split sex ratios. The selective advantage associated with reduced relatedness asymmetry is unlikely to favor extreme multiple mating (Queller 1993), however, since as mating frequency increases the workers' stable population sex ratio will result in a decreasing female bias, which would decrease the payoffs for queens from multiple mating (Ratnieks and Boomsma 1995). The available data suggest that queen mating frequencies are indeed low. However, as noted above, there are taxa in which mating frequencies are extremely high. A number of potential benefits to multiple mating have been proposed for nonsocial insects (table 1.1). Some of these benefits cannot apply to eusocial insects while those outlined below may be more relevant to social insects than to nonsocial insects (Crozier and Page 1985; see table 2 in Boomsma and Ratnieks 1996).

Sperm Limitation

Unlike nonsocial insects, social insects mate during a brief prereproductive period prior to the establishment of colonies, usually during a single nuptial flight. Sperm acquired during the nuptial flight must be sufficient to last the queen her entire reproductive lifespan, which in some ants can be as long as 30 years (Pamilo 1991d). Moreover, social insect colonies can be extremely large. Thus Cole (1983) suggested that increased queen fecundity may favor multiple mating among social insects, noting a positive association between multiple mating and colony size across 14 species of ant. Starr (1984) and Crozier and Page (1985) among others have criticized Cole's (1983) hypothesis on the grounds that males should evolve sperm loads adequate to meet their conspecific females' needs. Implicit in this criticism is the established dogma that sperm are cheap to produce, which later research has shown not to be the case (section 7.1). Males may often be limited in their sperm supplies (Heinze and Hölldobler 1993). Indeed, the work of Heinze et al. (1998) suggests a trade-off between dispersal and sperm production in *Cardiocondyla* ants. Typically in ants the testes degenerate prior to emergence and males invest in large flight muscle and sensory structures that allow them to disperse and seek out females on nuptial flights. In species of *Cardiocondyla*, however, there exists a class of wingless, workerlike "ergatoid" males who do not disperse but seek out copulations with young queens in the maternal nest. The testes of ergatoid males do not degenerate but continue to produce sperm throughout their lives. Moreover, they have larger ejaculates than winged males. Similar trade-offs between dispersal and reproduction have been revealed in both male and female nonsocial insects (Roff 1984; Roff and Fairbairn 1993; Sakaluk 1997). Thus, energetic constraints may prevent dispersing males from providing females with a sufficient supply of sperm. Fortelius et al. (1987) have found dimorphisms in dispersal activity and allocation to flight in *F. exsecta* and *F. sanguinea*. The same phenomenon appears to occur in *F. paralugubris*; some individuals disperse while others remain close to the natal nest (Cherix et al. 1991), and Chapuisat (1998) examined the mating frequency of dispersing and nondispersing queens. If energetic investment in dispersal resulted in reduced sperm transfer by males and queens remated to ensure an adequate sperm supply, we might expect dispersing queens to have a higher effective mating frequency than nondispersing queens. Contrary to the prediction, Chapuisat (1998) found an equal effective mating frequency for dispersing and nondispersing queens, suggesting that trade-offs between dispersal and sperm production are unlikely to drive female remating at least in the species of *Formica* that have been studied thus far.

Tschinkel (1987) found a positive relationship between the number of ovarioles (indicative of the lifetime number of eggs a female can potentially produce) and the number of sperm stored in the sperm storage organs of 25 ant species. The relationship was positively allometric, indicating that females of more fecund species stored relatively more sperm than those of less fecund species. It should be noted that the relationship observed by Tschinkel (1987) does not control for phylogeny so that the two variables may be unrelated in evolutionary terms. Nevertheless, if female ants mated only once, this relationship would suggest that male sperm loads have evolved to match their females sperm demands (cf. the criticism of Starr 1984; Crozier and Page 1985). However, female ants can mate more than once (table 10.1) so that the relationship could also be interpreted as females themselves collecting and storing as much sperm as they are likely to require during their lifetimes.

It is interesting to note that Tschinkel (1987) excluded from his analysis two apparent outliers, *Atta texana* and *Acromyrmex versicolor*, that had higher sperm counts than would be expected for the number of ovarioles. These leafcutter ants are now known to exhibit high effective mating frequencies (table 10.1). In their study of *A. colombica*, Fjerdingstad and Boomsma (1998) found a positive relationship between queen mating frequency and the numbers of sperm stored. A conservative estimate of the number of sperm an average queen would require over the minimum observed lifespan of ten years was calculated as 80–270 million. To obtain this many sperm a female would need to mate at least twice, indicating that multiple mating to avoid sperm limitation may well be an adaptive strategy in *A. colombica*. Indeed, Murakami et al. (2000) found positive associations between the number of oocytes per queen, colony size, and mating frequency across eight species of attine ants.

Finally, Boomsma and Ratnieks (1996) tested the sperm limitation hypothesis using available data on ants. In support of Cole (1983) they found a significant positive relationship between insemination frequency and colony size across 11 species of monogynous ant. Although this analysis also lacks phylogenetic control, the relationship held when analysis was performed at the genus level. However, it was not present in polygynous ants. The latter result is perhaps not surprising given that colonies with multiple queens can grow to large size because of the combined efforts of multiple queens and are less likely to be constrained by the sperm supplies of individual queens. Schmid-Hempel (1998) has since confirmed the relationship between colony size and multiple mating across 34 species of ants using appropriate comparative analysis by independent contrasts. Thus there is support for the role of sperm limitation in favoring the evolution of multiple mating in ants.

Sperm limitation cannot explain multiple mating in honey bees. During mating flights female *Apis* can mate with up to 53 different males (table 10.1). Page (1986) provides a comprehensive review of the mating biology of *Apis*. Sperm are delivered to the bursa copulatrix and the oviducts where they are mixed, by both the explosive nature of male ejaculation and the muscular contractions of the female's reproductive tract. Sperm are gradually transferred to the spermatheca over a period of 24 h. Although each male delivers about 6 million sperm, the female stores only 5.3–5.7 million, the remainder being ejected from the female's reproductive tract. Although patterns of sperm utilization show that females store a mixture of sperm from multiple males (table 10.1), the ejection of excess sperm clearly shows that multiple mating does not function in increasing the numbers of sperm stored (Page 1986).

Genetic Variance

The remaining hypotheses for the evolution of multiple mating relate to the potential fitness benefits associated with increased genetic diversity within social insect colonies. Cole and Wiernasz (1999) found that the growth rates of colonies of *Pogonomyrmex occidentalis* were negatively correlated with colony relatedness. Increased growth rate influenced colony survival and the onset of reproduction, leading to a 35-fold increase in fitness of fast-growing colonies. Cole and Wiernasz (2000) concluded that this fitness advantage favored multiple mating in this species. Crozier and Page (1985) suggested that genetic variance hypotheses represent the most likely general explanation for multiple mating in social insects and much work has focused on empirical and theoretical evaluation of these hypotheses.

Haplodiploid organisms require a different mechanism of sex determination than diploid ones (Cook 1993; Cook and Crozier 1995; Crozier and Pamilo 1996). Sex determination seems to be controlled by one or more loci with multiple alleles. Individuals that are heterozygous at a sex determining locus are female while those who are homozygous or in the case of haploid individuals hemizygous are males. If by chance a female should mate with a male who carried a sex determining allele that was identical to one of her own, half of her eggs fertilized by that male would develop into diploid males because they would be homozygous at the sex determining locus. Diploid males have been reported from a number of social insects (Crozier and Pamilo 1996) and since they are sterile they seriously reduce colony fitness (Page and Metcalf 1982; Ross and Fletcher 1986). With outbreeding and random mating, the probability of mating with a male carrying the same allele will be $2/n$ where n is the number of alleles at the sex determining locus. If a female mated twice, utilized both males' sperm, and one of the males shared an allele in common with her, only 25% of her diploid brood would be sterile males. The probability of mating with a single male having a common allele is higher than the probability of mating with two males both with alleles in common, such that multiple mating can reduce the risk of diploid male production for individual queens (Page and Metcalf 1982; Crozier and Page 1985; Pamilo et al. 1994). Avoidance of diploid male production is one way in which multiple mating can reduce the risks of genetic incompatibility (Zeh and Zeh 1996, 1997). However, the benefits to be gained from multiple mating depend on the stage at which colony reproduction occurs; singly mating queens are predicted to have greater fitness returns when reproduction occurs early in colony growth while doubly mating queens have greater fitness returns when reproduction occurs near the peak of colony growth (Crozier and Page 1985; Pamilo et al. 1994). Moreover, Ratnieks (1990b) showed how the detection and early removal of diploid males could favor multiple mating in queens while nonremoval could favor single mating. Thus, although diploid males represent a significant colony mortality factor in fire ants, *Solenopsis invicta*, effective queen mating frequency is close to 1 (table 10.1). Reproduction occurs during the early phase of colony growth and diploid males are not removed (Ross and Fletcher 1986). In contrast honey bees, *Apis*, remove diploid males as young larvae, reproduce late in colony growth, and have the highest known effective mating frequency of all social insects (Ratnieks 1990b) (table 10.1). The fact that diploid male production can favor single as well as multiple mating might suggest that it is unlikely to be a generally applicable model for the evolution of multiple mating in social insects (see Sherman et al. 1988). However, data on the prevalence of diploid male production and colony growth curves for many more species will be required before this can be confirmed.

It has been suggested that increased genetic variance can potentially protect colonies from invasion by parasites and pathogens that would otherwise reduce their fitness (Hamilton 1987; Sherman et al. 1988; Keller and Reeve 1994; Schmid-Hempel 1998). Multiple mating is one means by which genetic variation within colonies can be increased. Schmid-Hempel and his colleagues have provided good evidence for such effects in their studies of bumble bees *Bombus terrestris*. Shykoff and Schmid-Hempel (1991b,a) found that the prevalence of infection of the trypanosome *Crithidia bombi* was greater within experimental groups of closely related workers than in groups of unrelated workers (but see Schmid-Hempel and Schmid-Hempel 1993). Liersch and Schmid-Hempel (1997) manipulated within-colony genetic variance by cross fostering broods between colonies, finding that experimentally induced genetic variance reduced parasite loads in free-living colonies. Finally, Baer and Schmid-Hempel (1999) experimentally manipulated the levels of polyandry by means of artificial insemination. They found the colonies headed by queens inseminated by a mixture of sperm from four unrelated males had lower parasite loads and fewer species of parasite infecting their colonies than did queens inseminated by a mixture of sperm from four sibling males. Nevertheless, despite the potential fitness benefits associated with multiple mating, *B. terrestris* queens mate only once (Estoup et al. 1995). Single mating by queen bumble bees may be the consequence of male adaptation to sperm competition, since males transfer a mating plug which, at least following artificial insemination, prevents sperm from entering the spermatheca (Duvoisin et al. 1999). In contrast, Page et al. (1995) found no effect of genetic variation on parasite loads among worker honey bees *A. mellifera*, despite the fact that honey bees have extreme levels of multiple mating (see Kraus and Page 1998; and the response of Sherman et al. 1998).

In their comparative study of 24 species of ant, Schmid-Hempel and Crozier (1999) found a positive relationship between parasite richness (the number of parasite species infecting a given host species) and within-colony relatedness. The relationship was robust to control for phylogenetic relationships between the species involved. However, there was no difference in parasite richness between single-mating and multiple-mating species. Another route to genetic variance is via polygyny. However, there was no difference in parasite richness between monogynous and polygynous species. Fjerdingstad et al. (1998) found no evidence for survival benefits associated with high genetic diversity in colonies of *A. colombica*. The data currently available for ants would therefore suggest that, although the combined effects of polyandry and polygyny on colony genetic variability can be beneficial in terms of decreased susceptibility to parasites, there is little to suggest that multiple mating has arisen under selection for parasite resistance. In accord with this conclusion, with the exception of the leafcutters, monogynous ants have very low effective paternity frequencies and high degrees of intracolonial relatedness (table 10.1).

As with diploid male production, Schmid-Hempel (1998) notes that the benefits of multiple mating derived from parasite resistance depend strongly on the relationship between fitness and the size of the colony at the time of reproduction; when fitness exhibits a convex function with colony size, selection through parasites or pathogens should favor single mating rather than multiple mating. The fitness function for *B. terrestris* is indeed a convex function, perhaps explaining why bumble bees mate only once despite the potential benefits of polyandry. As with arguments based on diploid male production, the fact that parasites and pathogens can favor both single and multiple mating means that data on colony growth curves from many more species will be required to assess the general applicability of the hypothesis for the evolution of multiple mating in social insects.

The final reason why high genetic variance within colonies may be adaptive is because it can potentially result in a more efficient division of labor and thus colony productivity (Starr 1984; Crozier and Page 1985; Fuchs and Moritz 1999). There is now good evidence that, as well as strong environmental effects (Wheeler 1986), behavioral specialization within insect colonies can also have a genetic basis. In *A. mellifera*, specializations on guarding of the nest entrance, removal of corpses, grooming of nest mates, and feeding of nest mates are all genetically determined (Frumhoff and Baker

1988; Robinson and Page 1988). There is also genetic variance for preferred foraging distance, for resource type (pollen or nectar), and for host plant (Robinson and Page 1989; see review of Page and Robinson 1991; Oldroyd et al. 1992, 1993). In ants, genetic variation is responsible for behavioral specialization on guarding, patrolling, and foraging (Stuart and Page 1991; Snyder 1992; Carlin et al. 1993; Snyder 1993) and the ratio of large to small castes within colonies has a genetic component (Fraser et al. 2000). Multiple mating could thereby provide queens with a diversity of worker genotypes allowing each patriline (offspring of one male) within a colony to specialize on a given task (Oldroyd et al. 1994b). Studies of honey bees (Oldroyd et al. 1992; Fuchs and Schade 1994; Page et al. 1995; Giray et al. 2000) provide evidence that increased genetic variance can increase colony performance. It is interesting to note that extreme levels of multiple mating are generally confined to the honey bees, the vespine wasps, and the leafcutter ants, all of which have behaviorally and morphologically distinct queen and worker castes (Wilson 1971; Hölldobler and Wilson 1990). Hamilton (1964b) noted that if multiple mating arose after the evolution of a worker caste, multiple paternity and reduced colony relatedness would no longer pose a threat to colony cohesiveness because sterile workers would be unable to desert the colony to reproduce on their own (although workers can sometimes produce unfertilized eggs, and thus males, in bees and wasps worker policing generally prevents this behavior; Ratnieks 1988; Oldroyd et al. 1994a; Foster and Ratnieks 2000). Thus, we might generally expect to see a correlated evolution of caste specialization and multiple mating; as caste specialization increases, the social costs of multiple paternity and reduced colony relatedness are reduced, allowing multiple mating to evolve, reinforced by the benefits of improved specialization and any other adaptive benefits that multiple mating may yield. In support of the hypothesis, Murakami et al. (2000) have found that body-size-related caste specialization is associated with higher mating frequencies across eight species of attines, but age-related specialization is not. The hypothesis requires that single mating is the ancestral state, for which there is evidence in both vespine wasps (Foster et al. 1999) and attine ants (Villesen et al. 1999).

The various hypotheses based on benefits of genetic variation would all be weakened if there was evidence that sperm from multiple males were not mixed within the female's spermatheca so that only one male achieved paternity within a cohort (Sherman et al. 1988). With the exception of the leafcutters, the high paternity skews in table 10.1 suggest that genetic variance arguments are less relevant to multiple mating in the majority of ants. Genetic variance models are supported by the patterns of mating, sperm storage, and utilization in honey bees. So-called drone congregation areas contain males derived in equal proportions from surrounding colonies (Baudry et al. 1998). Thus, the males available for mating represent the total amount of genetic variation available to a female at the time of her nuptial flight. If genetic variability in sperm is beneficial for females they should not only mate multiply but also mate with as many different males as possible. Indeed, field studies have found that the diversity among patrilines is as great as that within the drone congregations (Baudry et al. 1998). Following mating, the male's genitalia break off and remain within the female, the so-called mating sign. This has previously been interpreted as a mating plug designed to prevent additional males from inseminating the female (Boorman and Parker 1976; Thornhill and Alcock 1983). However, recent work has shown that, rather than preventing remating by the female, the mating sign actually makes the queen more attractive to males within the congregation (Koeniger 1990). Additional males readily remove the mating sign and copulate with the queen. Koeniger (1990) suggested that the mating sign could make it easier for future males to copulate with the female. He interpreted his results from a male perspective, seeking an advantage for males that make their mates attractive to others. However, such a scenario seems unlikely given that multiple mating and increased sperm competition intensity greatly reduces the fitness returns for individual males. Viewed from the female perspective, however, the mating sign makes more sense. Under genetic variance benefit models, females that are able to facilitate remating by different males will be favored. Thus, the retention of a current male's genitalia would serve two beneficial purposes; it would avoid rematings with the same male and attract different males for mating. The valvefold of the female's internal genitalia (Page 1986) could allow her to hold the male's genitalia in position after intromission so that they are pulled from his abdomen after copulation. Once females have collected sperm from multiple males, they actively mix them within the oviducts before transferring a sample to the spermatheca for storage so that the effective paternity frequency is almost equal to the mating frequency (Page 1986; table 10.1). Thus, the mating behavior of honey bees strongly suggests that mixed paternity is beneficial for females and that selection has shaped female behavior to maximize the genetic diversity of her work force.

10.5 Summary

The evolution of social behavior in insects is thought to have been facilitated by haplodiploidy, which makes females more closely related to their sisters than to their own offspring. However, multiple mating and mixed paternity represent a challenge to kin selection theory because they will reduce the relatedness among the female work force. Multiple paternity is generally low due to low frequencies of queen remating and high paternity skew favors only one of the queen's mates. Nevertheless, extremely high effective paternities are found in three taxa, the honey bees, vespine wasps, and leafcutter ants.

Multiple mating can have a significant influence on social evolution. With single mating, there is a queen-worker conflict over sex allocation; queens prefer a 1:1 sex ratio while workers favor the production of sisters over brothers in the ratio of 3:1. Reduced relatedness asymmetry between sisters and brothers achieved through multiple mating can alleviate queen-worker conflict and result in split sex ratios; with low probability of double mating, colonies headed by doubly mated queens should be selected to specialize on the production of males to exploit the female-biased sex ratio produced by the work force of colonies headed by single-mated queens. Male production is against paternal interests because haplodiploidy results in male fitness being served only through the production of daughters. The resulting queen-male conflict over sperm allocation may favor the strategic allocation of sperm depending on female mating history; males who submaximally inseminate already mated queens may be favored since the resulting paternity skew may not reduce within-colony relatedness to the extent that workers specialize on producing males. An alternative route for the avoidance of reduced relatedness asymmetry arising with multiple mating is via sperm clumping and high cohort-specific relatedness asymmetry. There is some evidence that paternity skew is related to the probability of queen double mating. There is also some evidence for moderate sperm clumping. However, studies of sperm allocation and first- versus second-male paternity are generally lacking in social insects.

A number of alternative hypotheses for the evolution of multiple mating have been proposed. Because queens can only mate during the early part of their reproductive lives, sperm supply may be a constraint driving the evolution of multiple mating. There is some evidence that ants with large colonies mate more than once to attain a sufficient supply of sperm for colony growth. Alternatively, increased genetic variance within colonies may be favored via the avoidance of genetic incompatibility (diploid male production) and/or increased resistance to parasites and pathogens. However, both of these pressures can favor multiple or single mating, depending on how fitness scales with colony growth and the timing of reproduction. Where task specialization has a genetic basis, multiple mating can potentially improve colony performance because of a more efficient division of labor. Multiple mating is unlikely to represent an evolutionary cost in highly derived taxa where there is a high level of caste differentiation, because sterile workers cannot abandon the colony to reproduce independently. The three taxa that exhibit extreme levels of multiple mating are highly advanced in terms of caste differentiation.

The studies of multiple mating and sperm utilization in nonsocial and social insects have advanced in parallel with little or no cross fertilization. The work of social insect biologists provides our best knowledge of the extent of multiple paternity in natural populations while the work of nonsocial insect biologists offers our best knowledge of the mechanisms behind sperm utilization and sperm competition. Integrating these areas of research should be fruitful.

Table 10.1

Patterns of multiple insemination and sperm utilization in social insects.

Species	Study Method	Paternity or Insemination Frequency[a] [Range]	Proportion Multiple Paternity	Paternity Skew[b] [Range]	Effective Paternity Frequency[c] [Range]	Source
VESPIDAE						
Dolichovespula maculata media	M	1	0.00		1	Thorén 1998 in Foster et al. 1999
Dolichovespula norwegica saxonica	M	[4–11]	1.00		4.81	Thorén 1998 in Foster et al. 1999
Dolichovespula norwegica arenaria	M	[1–3]			1.09	Thorén 1998 in Foster et al. 1999
Parachartergus colobopterus	M	1	0.00			Goodnight et al. 1996
Polistes annularis	M	1.05 [1–2]	0.05			Peters et al. 1995
Polistes bellicosus	M	1	0.00		1	Field et al. 1998
Polistes chinensis antennalis	A	[1–2]	≤0.20	≥0.63		Miyano and Hasegawa 1998
Polistes jadwigae	A	1				Tsuchida 1994
Polistes metricus	A	≥2		≥0.82		Metcalf and Whitt 1977
Polistes variatus	A	≥2		0.82 [0.68–0.98]		Metcalf 1980
Ropalidia marginata	A	2.0 [1–3]				Muralidharan et al. 1986
Vespa crabro	M	1.4 [1–3]	0.36	≥0.73 [0.51–0.91]	1.11	Foster et al. 1999
Vespula rufa rufa	M	≥2	1.00			Thorén et al. 1995
Vespula squamosa[d]	A	>4.08 [2–6]	1.00		5.5	Ross and Carpenter 1991

Table 10.1 (*cont.*)

Species	Study Method	Paternity or Insemination Frequency[a] [Range]	Proportion Multiple Paternity	Paternity Skew[b] [Range]	Effective Paternity Frequency[c] [Range]	Source
Vespula vulgaris maculifrons[d]	A	>2.20 [1–4]	0.90		9.5	Ross and Carpenter 1991
APIDAE						
Apis andreniformis	M	13.5 [10–20]	1.00		9.1 [8.4–10.8]	Oldroyd et al. 1997
Apis dorsata	M	30.2 [19–53]	1.00		25.6 [14.1–44.1]	Moritz et al. 1995
Apis florea	M	8.0 [5–14]	1.00		5.65 [3.3–9.4]	Oldroyd et al. 1995
Apis mellifera	M	13.8 [7–20]	1.00		12.4 [6.6–17.9]	Estoup et al. 1994
Bombus hortorum	M	1	0.00		1	Schmid-Hempel and Schmid-Hempel 2000
Bombus hypnorum	DM	[2–4]	0.66		1.14 [1–1.46]	Estoup et al. 1995; Schmid-Hempel and Schmid-Hempel 2000
Bombus lapidarius	DM	1	0.00		1	Estoup et al. 1995; Schmid-Hempel and Schmid-Hempel 2000
Bombus lucorum	DM	1	0.00		1	Estoup et al. 1995; Schmid-Hempel and Schmid-Hempel 2000

Bombus pascuorum	M	1	0.00		1	Schmid-Hempel and Schmid-Hempel 2000
Bombus pratorum	DM	1	0.00		1	Estoup et al. 1995; Schmid-Hempel and Schmid-Hempel 2000
Bombus sicheli	M	1	0.00		1	Schmid-Hempel and Schmid-Hempel 2000
Bombus terrestris	DM	1	0.00		1	Estoup et al. 1995; Schmid-Hempel and Schmid-Hempel 2000
Melipona subnitida	A	1	0.00		1	Contel and Mestiner 1974; Contel and Kerr 1976
Plebia droryana	A	1	0.00		1	Machado et al. 1984
HALICTIDAE						
Augochlorella striata	M	1	0.00		1	Mueller et al. 1994
Halictus ligatus	A	≥1				Richards et al. 1995
Lasioglossum zephyrum	A	[1–2]	≤0.23			Crozier et al. 1987; Kukuk et al. 1987
FORMICIDAE[e]						
Myrmicinae						
Acromyrmex echinatior	M	2.53 [1–4]	0.92	≥0.62	2.23 [1.0–3.1]	Bekkevold et al. 1999
Acromyrmex octospinosus	M	6.1 [4–10]	1.00		3.93 [2–6]	Boomsma et al. 1999; Murakami et al. 2000
Acromyrmex versicolor	DM	c.3			3.26	Reichardt and Wheeler 1996
Aphaenogaster rudis[f]	A	1	≤0.05		≤1.02	Crozier 1973; Crozier 1974
Apterostigma collare	M	1	0.00		[1.03–1.06]	Villesen et al. 1999
Apterostigma mayri	M				≈1[h]	Murakami et al. 2000
Atta colombica	M	2.6 [1–5]	0.92	≥0.63 [0.51–0.87]	2.31 [1.0–5.7]	Fjerdingstad et al. 1998; Murakami et al. 2000

Table 10.1 (*cont.*)

Species	*Study Method*	*Paternity or Insemination Frequency*[a] *[Range]*	*Proportion Multiple Paternity*	*Paternity Skew*[b] *[Range]*	*Effective Paternity Frequency*[c] *[Range]*	*Source*
Atta sexdens	M	2.7 [2–3]	1.00	0.52	2.6 [1.91–2.72]	Fjerdingstad and Boomsma 2000
Cyphomyrmex costatus	M				≈1[h]	Murakami et al. 2000
Cyphomyrmex longiscapus	M	1	0.00		[1.03–1.04]	Villesen et al. 1999
Cyphomyrmex rimosus	M				≈1[h]	Murakami et al. 2000
Harpagoxenus sublaevis[f]	A	1	≤0.05		≤1.02	Bourke et al. 1988
Leptothorax pergandei[f]	A	[1–2]	0.08	≥0.85	1.04	Heinze et al. 1995
Myrmica lobicornis	M	1			1	Seppä 1994
Myrmica punctiventris	AM	1				Snyder and Herbers 1991; Herbers and Mouser 1998
Myrmica rubra[f]	A	[1–2]	0.18	≥0.73	1.21	Seppä 1994b in Boomsma and Ratnieks 1996
Myrmica ruginodis[f]	A	1.2 [1–3]	0.15	≥0.68	1.07	Seppä 1994
Myrmica sulcinodis	A	1.38 [1–2]	0.35	0.61 [0.50–0.87]	1.19[g]	Pedersen and Boomsma 1999
Myrmica tahoensis	M	1	0.00			Evans 1993; Evans 1995
Myrmicocrypta ednaella	M	1	0.00		[1.05–1.07]	Villesen et al. 1999; Murakami et al. 2000
Sericomyrmex amabalis	M				≈2[h]	Murakami et al. 2000

Solenopsis geminata[f]	A	1	≤0.04		<1.02	Ross et al. 1987; Ross et al. 1988
Solenopsis invicta[f]	A	1	≤0.02		≤1.02	Ross and Fletcher 1985
Solenopsis richteri[f]	A	1	≤0.03		≤1.02	Ross et al. 1987; Ross et al. 1988
Trachymyrmex isthmicus	M				≈1[h]	Murakami et al. 2000
Dolichoderinae						
Conomyrma insana[f]	A	1	≤0.19		≤1.09	Berkelhamer 1984
Linepithema humile (*Iridomyrmex humilis*)	M	1	0.00		1	Keller and Passera 1992; Krieger and Keller 2000
Formicinae						
Camponotus ligniperdus	M	1.02 [1–2]	0.02	≥0.56	1.02	Gadau et al. 1998
Camponotus consobrinus	M	1	0.00			Crozier et al. 1999
Colobopsis nipponicus[f]	A	1	≤0.08		1.03	Hasegawa 1994
Formica aquilonia[f]	A	[1–3+]	0.57	≥0.64	1.48	Pamilo 1993
Formica exsecta[f]	A	[1–2]	0.11	≥0.78	1.16	Pamilo and Rosengren 1984
Formica fusca	A	1.24 [1–2]	0.29	≥0.67 [0.50–0.80]		Boomsma and Sundström 1998
Formica paralugubris	M	1.25 [1–3]	0.22		1.13	Chapuisat 1998
Formica pressilabris[f]	M	[1–2]	0.15	≥0.67	1.12	Pamilo 1982b
Formica rufa	A	1.48 [1–3]	0.44	≥0.59 [0.40–0.90]		Boomsma and Sundström 1998
Formica sanguinea[f]	A	[1–2]	0.08	≥0.56	1.31	Pamilo and Varvio-Aho 1979; Pamilo 1982b
Formica transkaucasica[f]	A	[1–2]	0.03	>0.87	1.03	Pamilo 1982b
Formica truncorum	AM	1.97 [1–3]	0.31	≥0.62	1.43	Sundström 1993; Sundström and Ratnieks 1998
Lasius flavus[f]	A	[1–2]	≤0.39		≤1.20	Boomsma et al. 1993

Table 10.1 (*cont.*)

Species	*Study Method*	*Paternity or Insemination Frequency*[a] *[Range]*	*Proportion Multiple Paternity*	*Paternity Skew*[b] *[Range]*	*Effective Paternity Frequency*[c] *[Range]*	*Source*
Lasius neglectus[f]	A	[1–2]	≤0.63		≤1.36	Boomsma et al. 1990
Lasius niger	A	[1–3]	0.12	>0.58 [0.40–0.88]	1.16 [1.0–1.4]	Boomsma and van der Have 1998

Notes: A, allozyme variation; M, molecular techniques including mini and microsatellites, and randomly amplified polymorphic DNA; D, dissection and sperm counts.

[a]The observed number of male genotypes represented in the offspring produced or in the females sperm stores, irrespective of their relative contributions.

[b]Paternity skew is calculated as ΣP_i^2 where P_i is the proportion of offspring sired by the *i*th male (Starr 1984). Values are given for cases in which females have mated with two males. Thus, a paternity skew of 0.5 would reflect equal paternity for both males. A value of 0.6 reflects a moderate bias (c. 75%) toward one male, and a value of 0.9 almost complete paternity for one of the males. The values are calculated either from data provided in the source papers without correction for nonsampling detection error or taken directly from source papers. Many of the values are undoubtedly underestimates of the true paternity skew because Pamilo's (1993) correction, which has been routinely applied in published estimates, has been found not to apply to isolated paternity classes (see Pedersen and Boomsma 1999). Hence, true values are ≥ to those quoted.

[c]The populationwide effective paternity frequency is the reciprocal of the paternity skew summed across all colonies (including those with single paternity). It is the mean number of males that actually contribute to offspring production and is the value that will directly influence genetic relatedness among workers. The values presented are taken directly from source papers so that the methods of calculation may differ. In some cases the ranges presented will be overestimates because of the problems inherent in applying the non-detection error of Pamilo (1993) to individual colonies (Pedersen and Boomsma 1999).

[d]Lower values were incorrectly reported by Ross (1986): 7.14 for *V. maculifrons* and 3.33 for *V. squamosa*.

[e]Boomsma and Ratnieks (1996) provide data on fifteen additional species of ant for which data are available, which indicate mating frequency but for which estimates of actual proportion of females multiply mating or the effective paternity frequency are not possible. Nine of these are single mating species and six are double or multiple mating species.

[f]Values in body of table from Boomsma and Ratnieks (1996) calculations. To provide conservative estimates of the proportion of females multiply mated and effective paternity frequency, they assumed that when no colony was found to have a multiply mated queen, the next colony sampled would have, thereby attempting to correct for nondetection errors in small samples.

[g]Because females were mated with two related males, the genetic effect of multiple mating yielded an effective paternity frequency of just 1.06.

[h]Implied from colony level relatedness.

11 Broader Significance

11.1 Sperm Competition and Sexual Selection

The aim of this volume has been to show that sperm competition is a powerful force in sexual selection, shaping aspects of an organism's reproductive behavior, morphology, and physiology. Students of sexual selection typically focus on secondary sexual traits, such as the peacock's tail, and how these traits contribute to variance in male reproductive success. It is clear from preceding chapters in this volume that the rapid and divergent evolution of traits such as seminal fluid, genitalia, or sperm more than rival the variation seen in secondary sexual traits. Moreover, variation in these traits clearly has considerable influence on variation in male reproductive success and is the focus of intense sexual selection. Nevertheless, sperm competition was omitted from a recent review of the discipline (Andersson 1994). Variation in reproductive success can arise at a number of levels, including differential mating success, postmating events associated with sperm competition and/or sperm selection, and postfertilization events such as differential parental investment and/or abortion (Møller 1998). Yet differential reproductive success is often accredited to variation in male mating success and the traits that contribute to variation in mating success, without due consideration of processes occurring after mating. Such a focus might be acceptable where females are genetically monogamous; however, the advent of molecular techniques for assigning parentage has taught us that monogamy is the exception rather than the rule. In fact, accrediting variation in reproductive success to traits that enhance male mating success can be misguided and result in errors in identifying processes of sexual selection (see also Birkhead and Møller 1998b). For example, male body size is widely recognized as a trait important in generating variation in male reproductive success (Andersson 1994) particularly in insects (Thornhill and Alcock 1983; Choe and Crespi 1997a). McLain (1991) partitioned size-dependent variation in male reproductive success between different components of fitness in stink bugs, *Nezara viridula*. Standardized selection differentials revealed that the intensity of sexual selection acting on male body size was an order of magnitude greater via sperm competition than via male mating success. Without an insight into sperm competition in this species, an examination of male reproductive success would have concluded erroneously that selection on male body size results from its association with competitive mating success. Differential fertilization success may enhance selection acting via differential mating success where the secondary sexual traits responsible for variation in mating success are also associated with enhanced success in sperm competition. Male flour beetles *Tribolium castaneum* and arctuid moths *Utetheisa ornatrix* with more attractive pheromones achieve a higher paternity than do males with less attractive pheromones (Dussourd et al. 1991; Lewis and Austad 1994), while male stalk-eyed flies *Cyrtodiopsis whitei* with wide eyespans achieve a higher paternity than do males with narrow eyespans under conditions where they have an equal mating frequency (Burkhardt et al. 1994). On the other hand, sperm competition could ameliorate the effects of premating competition among males. Where females benefit from multiple mating and the mixing of sperm within their spermathecae, they may choose to produce offspring with a variety of different males, not just those individuals who are successful in competition or most attractive (Simmons 1991a). Moreover, Danielsson (2000) has recently shown that, despite the apparent selective advantage of large body size in male contest competition among water striders *Gerris lacustris*, there is no net selection for increased male size because selection under sperm competition acts antagonistically; large males mate more often but as a consequence they suffer sperm depletion and reduced fertilization success when in competition with small males. Clearly a greater understanding and integration of the processes occurring between mating and offspring production will provide us with a broader understanding of sexual selection as an evolutionary process.

11.2 Life History Evolution

Studies of life history evolution suggest that selection under sperm competition can result in significantly correlated responses in life history traits. Populations of *Drosophila melanogaster* subjected to long-term selection for delayed or accelerated senescence show correlated responses in performance in sperm competition (Service and Fales 1993). Specifically, females mated to males from lines selected for late-life fitness were less likely to remate and when they did so were less likely to fertilize eggs using sperm of the second male than females mated to males from populations selected for early-life fitness. Experimental removal of selection arising from sperm competition by enforcing monogamous mating in *D. melanogaster* had the effect of increasing larval development rate and decreasing a male's ability to gain fertilizations when in the position of second male to mate (Holland and Rice 1999; Pitnick et al. 2001). We have already seen how increased sperm length across species of *Drosophila* is positively associated with increases in the period of prereproductive adult development (Pitnick et al. 1995a; section 8.2). Pitnick and Miller (2000) have recently shown how selection on increased testis length in *Drosophila hydei* resulted in positive correlated responses in egg-to-adult development time, adult thorax length, and posteclosion maturation time after just 11–12 generations of selection. These studies of *Drosophila* suggest that, more generally, trade-offs between investment in traits important in sperm competition and life history traits such as the age or size at reproduction may be important in shaping a species' life history.

The phenomenon of protandry, where the peak emergence of males occurs before that of females, is particularly common in insects (Wiklund and Fagerstrom 1977; Thornhill and Alcock 1983) and has received considerable theoretical attention (Fagerström and Wiklund 1982; Bulmer 1983; Iwasa et al. 1983; Parker and Courtney 1983; Zonneveld 1992, 1996). Protandry seems to be prevalent in species were females rarely mate more than once on emergence, so that males who emerge before the period of female emergence have a selective advantage over later emerging males because of an increased probability of encountering unmated females. Protandry seems to be the rule in solitary Hymenoptera, which, like many social Hymenoptera (chapter 10), have a low incidence of remating (Thornhill and Alcock 1983; Alcock and Buchmann 1985; Simmons et al. 2000). In at least one species of tettigoniid, *Requena verticalis*, there is a strong first-male advantage in sperm competition so that despite polyandry females are genetically monandrous (Simmons and Achmann 2000), and males emerge some five weeks earlier than females to increase their chances of encountering unmated females (Simmons et al. 1994). In monandrous butterflies of the genus *Heliconius* protandry is extreme in that males emerge and copulate with females while they are still in the pupal stage (Deinert et al. 1994). Protandry is widespread in butterflies and, although many species are monandrous, protandry is also a characteristic of polyandrous species. Wiklund and Forsberg (1991) argued that selection via sperm competition in polyandrous species should act as a selective pressure to reduce the degree of protandry. Their argument was based on the assumption that large male body size represents a selective advantage in sperm competition, and that there is a trade-off between adult body size and development time. There is support for the first of these assumptions. Larger male butterflies are capable of producing larger spermatophores, which induce longer periods of unreceptivity in females, and higher fertilization success when sperm are in competition with those of rival males (Oberhauser 1989; Wiklund and Kaitala 1995; Bissoondath and Wiklund 1996a, 1997; Wedell and Cook 1998). Thus selection via sperm competition should favor increased male body size. Indeed, in his theoretical analysis Parker (1992c) showed how in the absence of sperm competition male dwarfism might be favored, but with increasing sperm competition male size should increase to approach that of females. Parker (1992c) analyzed patterns of sexual size dimorphism in fish and found that, across species, sperm competition was associated with increases in male body size relative to females, as predicted. Under the assumption of a trade-off between body size and development time, a correlated response to selection for increased male body size would be an extension of male development time and thus a reduction in the degree of protandry (Wiklund and Forsberg 1991). In support of their argument, Wiklund and Forsberg (1991) presented comparative evidence for an increase in sexual size dimorphism and the degree of polyandry across 11 species of pierid and 12 species of satyrid butterflies; males are typically smaller than females in monandrous species but equal in body size in polyandrous species. Although this analysis did not control for phylogenetic effects, an appraisal of the data suggests that it may be robust (Nylin and Wedell 1994). Thus sperm competition may indeed favor increased male body size. Wiklund and Forsberg's (1991) sperm competition hypothesis predicted, then, that polyandrous species should have a lower degree of protandry than monandrous species. There is weak support for such a relationship in species with diapause development but not for species with direct development. It seems that the assumption of a trade-off between size and development may not hold, with males achieving greater size through increased rates of development. Subsequent theoretical treatments suggest that, while polyandry might be expected to reduce the degree of protandry, the levels of polyandry seen across butterflies are too low to relax selection for protandry (Zonneveld 1992, 1996). Moreover, most theoretical treatments of protandry assume that all matings are equally valuable to males because of last male sperm precedence. As we have seen, for butterflies last-male sperm precedence is not an acceptable assumption (table 2.3). With polyandry, not all matings will be equally valuable (see Zonneveld 1992 for an analysis of protandry in which matings with already mated females were considered to decline in value). In the tettigoniid *Decticus verrucivorus*, females lay approximately 30% of their total lifetime egg production during the refractory period following their first mating (Wedell 1992). On remating, females lay fewer eggs and males obtain paternity of a fraction of these, depending on the number of competing ejaculates. Thus, virgin females are of premium reproductive value and *D. verrucivorus* is strongly protandrous (Wedell 1992). The same pattern is true for polyandrous butterflies; egg production declines and the intensity of sperm competition increases with female age (e.g., Drummond 1984; Rutowski et al. 1987; Svärd and Wiklund 1988a; Oberhauser 1997). The same is true for fishflies (Hayashi 1999a). Thus, selection via sperm competition should act to maintain protandry in polyandrous species. Only in highly polyandrous species, where females do not refrain from mating for significant periods of reproduction following mating, should the reproductive value of virgin females decline to a level where selection on protandry will be relaxed.

Sexual dimorphism in body size is ubiquitous in animal taxa, and has been the focus of considerable attention (reviewed in Andersson 1994). A number of hypotheses have been proposed to explain sexual size dimorphism (Fairbairn 1990). Perhaps the earliest was that propounded by Darwin (1871), who interpreted larger female size as a response to fecundity selection; where the relationship between size and fecundity is greater in females than males, selection should be greater in females leading to a female-bias sexual size dimorphism (see Wiklund and Karlsson 1988; Sivinski and Dodson 1992). Fecundity selection in females may be balanced by sexual selection in males if male size contributes to fitness via competitive mating success (e.g., Ward 1988). However, the association of decreasing sexual size dimorphism with polyandry noted in butterflies (Wiklund and Forsberg 1991) may also result from sexual selection on males due to sperm competition. Evidence for such an effect also comes from Wedell's (1997) comparative study of 20 species of tettigoniid. Wedell (1997) found a positive association between the degree of sexual size dimorphism and the amount of sperm transferred by males at mating. Although this analysis was not performed within a phylogenetic context, there was no significant variation in either variable attributable to subfamily grouping and the relationship persisted when analysis was conducted at the genus level. Ejaculate size is know to be subject to directional selection under sperm competition for at least two species of tettigoniid (Wedell 1991; Gwynne and Snedden 1995; Simmons and Achmann 2000) and is positively correlated with male size (Wedell 1992, 1993a; Simmons 1995b; Simmons and Kvarnemo 1997). The data on butterflies (Wiklund and Forsberg 1991) and tettigoniids (Wedell 1997) thus strongly suggest that sexual selection via sperm competition can contribute to observed variation in sexual size dimorphism in insects.

Finally, adaptations to sperm competition in males have been proposed to contribute to the evolution of phenotypic plasticity. We have seen how selection for the avoidance of sperm competition can result in adaptations in male seminal fluids that influence female remating (see section 4.3). Thus in butterflies selection has favored increased ejaculate mass because of its effects on delayed remating by females (table 4.1). A consequence of male ejaculate investment is the ability of females to utilize nutrients from the seminal fluid in reproduction (Boggs and Watt 1981; Rutowski et al. 1987; Wiklund et al. 1993; Wedell 1996; Oberhauser 1997). Leimar et al. (1994) proposed that in species, such as many butterflies, where all nutrients for reproduction must be gathered in the larval stage, the additional avenue for resource acquisition for females via male ejaculates should influence patterns of life history. Specifically, as resources become limited, females of species with no access to resources as adults should extend their larval development to attain increased size, and thus fecundity, at reproduction. Species in which females can acquire resources from males at mating, however, can afford to maintain their larval development time and thus emerge earlier, compensating for poor larval resources through increased acquisition of ejaculates. Thus, Leimar et al. (1994) predicted that variation in female body size should increase with the amount of ejaculate transferred by conspecific males, a relationship that was present across 16 species of butterflies and 12 phylogenetically independent contrasts derived from these species. The hypothesis also predicts that the females of species in which males provide large ejaculates (polyandrous species) should exhibit greater phenotypic plasticity in response to larval food stress than the females of species in which males provide small ejaculates (monandrous species). In contrast, the reverse should be true for males; males of polyandrous species should exhibit less phenotypic plasticity than females, increasing development time to maintain body size because size-dependent ejaculate size contributes to reproductive success. Males of monandrous species should show greater phenotypic plasticity than females because ejaculate size does not contribute to their reproductive success. As predicted, Leimar et al. (1994) showed that female size decreased to a greater extent than male size in the polyandrous *Pieris napi*, while Karlsson et al. (1997) showed that male size decreased to a greater extent than female size in the monandrous *Pararge aegeria.*

The studies described above provide compelling evidence that sexual selection via sperm competition can have evolutionary consequences far beyond the context of sperm transfer and utilization.

11.3 Speciation

A recurring theme in this volume has been that sexual conflict can frequently arise over sperm transfer and utilization. Sexual conflict will arise whenever the reproductive interests of males and females differ (Parker 1979). In the context of sperm competition, males will be under strong selection to achieve fertilization and, as we have seen, adaptations to achieve fertilization may not be in the females' best interests; in *Drosophila*, accessory gland proteins that influence female remating and the displacement and incapacitation of rival sperm impose increased mortality on females (section 4.3). Thus selection for increased sperm competition ability in males via the manipulation of female physiology is likely to generate counterselection on females to overcome the deleterious effects of male accessory gland products, resulting in antagonistic coevolution between genes associated with male and female fitness (Holland and Rice 1998). There is good empirical evidence for such antagonist coevolution in *Drosophila.* Experimental arrest of female evolution resulted in an increase in male sperm competition ability which was associated with an elevated mortality for female lines prevented from coevolving with males (Rice 1996, 1998a). Experimental removal of sperm competition by enforced monogamy resulted in both an increased mortality for monogamous females when mated to control polygamous males (Holland and Rice 1999), and a decreased fertilization success for monogamous males competing with polygamous control males for fertilizations (Pitnick et al. 2001). Antagonistic coevolution should generate extremely rapid and divergent evolution in proteins involved in fertilization, a pattern supported by recent molecular studies from a range of animal taxa including insects (Palumbi and Metz 1991; Aguadé et al. 1992; Vacquier and Lee 1993; Metz and Palumbi 1996; Tsaur et al. 1998; Wyckoff et al. 2000).

It has been recognized for some time that coevolution between male secondary sexual traits and female mating preferences can generate premating reproductive isolation and fuel speciation (reviewed in Andersson 1994; Sætre et al. 1997; Price 1998). More recently, the findings from *Drosophila* have prompted theoretical analyses of the potential for antagonistic coevolution under sperm competition to generate postmating reproductive isolation that leads to speciation (Alexander et al. 1997; Holland and Rice 1998; Parker and Partridge 1998; Rice 1998b; Gavrilets 2000). Antagonistic coevolution between males and females will continue when populations of species become separated, either due to changing geography or because of dispersal to new regions, but they are likely to diverge phenotypically and genetically in the absence of gene flow, making them genetically incompatible. Thus, coevolution between male genitalia and female reproductive tracts (Eberhard 1985; Arnqvist 1998), male sperm morphology and female sperm storage organs (Pitnick et al. 1999), or male seminal products and female physiology (Rice 1996) could result in barriers to fertilization when populations are brought back together; so called allopatric speciation. The occurrence of conspecific sperm precedence or homogamy lends support to such postmating reproductive barriers (see examples in table 2.3; reviewed by Howard 1999) and evidence for its involvement in speciation comes from studies of sperm competition between races of two species of grasshopper, *Chorthippus parallelus* and *Podisma pedestris*. Both species have two races that meet at hybrid zones. In both species when females mate with a male from their own-race and one from the other race, own race males have a higher fertilization success than would be expected from the mechanism of sperm competition involved (Hewitt et al. 1989; Bella et al. 1992). Interactions between male and female genotypes are critical in determining fertilization success amongst experimentally isolated populations of *Drosophila melanogaster* (Clark et al. 1999).

If antagonistic coevolution due to sperm competition is a general driving force in insect speciation, we would predict that evolutionary lineages in which there is sperm competition should have a greater species diversity than evolutionary lineages in which there is no sperm competition. Arnqvist et al. (2000) have recently tested this prediction in a comparative analysis of insects. They performed a series of phylogenetic contrasts between the numbers of extant species in insect clades were females typically mate with many different males so that there is the potential for antagonistic coevolution in the context of fertilization, and clades where females typically mate only once. They found that clades where females mate with many males had four times as many species as those where females mate only once. They also examined two potentially confounding factors that could explain species richness, trophic ecology and geographic range. The conclusion that species richness was greater in clades with multiple mating was strengthened by controlling for these effects. Thus, in general, the data for insects do appear to support the notion that sperm competition can result in antagonistic co-evolution between males and females that ultimately leads to speciation.

Parker's (1979) original analysis and that of Parker and Partridge (1998) argued that sexual conflict should often result in females acting as a selective force for speciation, but males acting as a force against it. For example, in meetings between sibling species, it will pay males to mate but females not to mate, depending on the relative difference in parental investments between the sexes (Parker 1979). Thus, whether sexual conflict fuels speciation or not depends on who wins the conflict. The patterns of speciation seen in insects might suggest that it is females who are ahead in this particular evolutionary chase.

11.4 Concluding Remarks

Thirty years have passed since Parker (1970e) first proposed that sperm competition should be an important agent in sexual selection. Many of the early years in this discipline's history were devoted to understanding the evolutionary consequences of sperm competition for male reproductive biology. We are now beginning to come to grips with the techniques necessary to examine the evolutionary consequences for females and the complex interactions between males and females at the gametic level. More than in any other discipline within behavioral ecology, students of sperm competition have grasped the importance of a multidisciplinary approach to research and the discipline has made huge strides forward via the integration of physiology, quantitative and molecular genetics, evolutionary and behavioral ecology, and comparative morphology. This correspondence between disciplines is recent, perhaps accounting for the almost explosive number of discoveries over the last few years. We are now realizing that selection due to sperm competition can have far-reaching consequences for an organism's evolution. Nevertheless, some areas are in need of greater effort. In particular, behavioral and physiological aspects of sperm form and function are poorly understood, as are the interactions between sperm, the female reproductive tract, and ultimately the egg. Without doubt the study of sperm competition has come of age, and the rapid progress currently being made promises to answer many of the questions left unanswered by this volume.

References

Achmann, R., K.-G. Heller, and J. T. Epplen. 1992. Last-male sperm precedence in the bushcricket *Poecilimon veluchianus* (Orthoptera, Tettigoniidae) demonstrated by DNA fingerprinting. *Mol. Ecol.* 1: 47–54.

Adler, P. H., and C. R. L. Adler. 1991. Mating behavior and the evolutionary significance of mate guarding in three species of crane flies (Diptera: Tipulidae). *J. Insect Behav.* 4: 619–632.

Ae, S. A. 1979. The phylogeny of some *Papilio* species based on interspecific hybridization data. *Syst. Entomol.* 4: 1–16.

Afzelius, B. A. 1975. *The Functional Anatomy of the Spermatozoon.* Pergamon Press, Oxford.

Afzelius, B. A., B. Bacetti, and R. Dallai. 1976. The giant spermatozoa of *Notonecta. J. Submicr. Cytol.* 8: 149–161.

Aguadé, M., N. Miyashita, and C. H. Langley. 1992. Polymorphism and divergence in the *Mst26A* male accessory gland region in *Drosophila. Genetics* 132: 755–770.

Aigaki, T., I. Fleischmann, P. S. Chen, and E. Kubli. 1991. Ectopic expression of sex peptide alters reproductive behavior of female *Drosophila melanogaster. Neuron* 7: 557–563.

Aiken, R. B. 1992. The mating behaviour of a boreal water beetle *Dytiscus alaskanus* (Coleoptera: Dytiscidae). *Ethol. Ecol. Evol.* 4: 245–254.

Alcock, J. 1979a. The evolution of intraspecific diversity in male reproductive strategies in some bees and wasps. In M. S. Blum and N. A. Blum, eds., *Sexual Selection and Reproductive Competition in the Insects*, pp. 381–402. Academic Press, London.

Alcock, J. 1979b. Multiple mating in *Calopteryx maculata* (Odonata: Calopterygidae) and the advantage of non-contact guarding by males. *J. Nat. Hist.* 13: 439–446.

Alcock, J. 1982. Post-copulatory mate guarding by males of the damselfly *Hetaerina vulnerata* Selys (Odonata: Calopterygidae). *Anim. Behav.* 30: 99–107.

Alcock, J. 1983. Mate guarding and the acquisition of new mates in *Calopteryx maculata* (P. De Beauvois) (Zygoptera: Calopterygidae). *Odonatologica* 12: 153–159.

Alcock, J. 1987. Male reproductive tactics in the libellulid dragonfly *Paltothemis lineatipes*: Temporal partitioning of territories. *Behavior* 103: 157–173.

Alcock, J. 1991. Adaptive mate-guarding by males of *Ontholestes cingulatus* (Coleoptera: Staphylinidae). *J. Insect Behav.* 4: 763–772.

Alcock, J. 1992. The duration of strong mate guarding by males of the libellulid dragonfly *Paltothemis lineatipes*: proximate causation. *J. Insect Behav.* 5: 507–515.

Alcock, J. 1994. Postinsemination associations between males and females in insects: The mate-guarding hypothesis. *Ann. Rev. Entomol.* 39: 1–21.

Alcock, J., and S. L. Buchmann. 1985. The significance of post insemination display by male *Centris pallida* (Hymenoptera: Anthophoridae). *Z. Tierpsychol.* 68: 231–243.

Alcock, J., and A. Forsyth 1988. Post-copulatory aggression toward their mates by males of the rove beetle *Leistotrophus versicolor* Coleoptera staphylinidae. *Behav. Ecol. Sociobiol.* 22: 303–308.

Alcock, J., C. E. Jones, and S. L. Buchmann. 1976. Location before emergence of the female bee *Centris pallida* by its male (Hymenoptera: Anthophoridae). *J. Zool.* 179: 189–199.

Alexander, R. D. 1964. The evolution of mating behaviour in arthropods. In K. C. Highnam, ed., *Insect Reproduction*, pp. 78–94. Symp. Roy. Entomol. Soc. Lond., no. 2.

Alexander, R. D. 1974. The evolution of social behavior. *Ann. Rev. Ecol. Syst.* 5: 325–383.

Alexander, R. D., and D. Otte. 1967. The evolution of genitalia and mating behavior in crickets (Gryllidae) and other Orthoptera. *Misc. Pub. Mus. Zool. Univ. Mich.* 133: 1–59.

Alexander, R. D., and P. W. Sherman. 1977. Local mate competition and parental investment in social insects. *Science* 196: 494–500.

Alexander, R. D., D. C. Marshall, and J. R. Cooley. 1997. Evolutionary perspectives on insect mating. In J. C. Choe and B. J. Crespi, eds., *The Evolution of Mating Systems in Insects and Arachnids*, pp. 4–31. Cambridge University Press, Cambridge, UK.

Allen, G. R., D. J. Kazmer, and R. F. Luck. 1994. Post-copulatory male behaviour, sperm precedence and multiple mating in a solitary parasitoid wasp. *Anim. Behav.* 48: 635–644.

Alonso-Pimentel, H., and D. R. Papaj. 1996. Operational sex ratio versus gender density as determinants of copulation duration in the walnut fly, *Rhagoletis juglandis* (Diptera: Tephritidae). *Behav. Ecol. Sociobiol.* 39: 171–180.

Alonso-Pimentel, H., and D. R. Papaj. 1999. Resource presence and operational sex ratio as determinants of copulation duration in the fly *Rhagoletis juglandis*. *Anim. Behav.* 57: 1063–1069.

Alonso-Pimentel, H., L. P. Tolbert, and W. B. Heed. 1994. Ultrastructural examination of the insemination reaction in *Drosophila*. *Cell Tissue Res.* 275: 467–479.

Amano, H., and K. Hayashi. 1998. Costs and benefits for water strider (*Aquarius paludum*) females of carrying guarding, reproductive males. *Ecol. Res.* 13: 263–272.

Andersen, N. M. 1994. The evolution of sexual size dimorphism and mating systems in water striders (Hemiptera, Gerridae): A phylogenetic approach. *Écoscience* 1: 208–214.

Andersen, N. M. 1997. A phylogenetic analysis of the evolution of sexual dimorphism and mating systems in water striders (Hemiptera: Gerridae). *Biol. J. Linn. Soc.* 61: 345–368.

Andersson, J., A. -K. Borg-Karlson, and C. Wiklund. 2000. Sexual cooperation and conflict in butterflies: A male-transferred anti-aphrodisiac reduces harassment of recently mated females. *Proc. Roy. Soc. Lond. B* 267: 1271–1275.

Andersson, M. 1982. Female choice selects for extreme tail length in a widowbird. *Nature* 299: 818–830.

Andersson, M. 1986. Evolution of condition-dependent sex ornaments and mating preferences: Sexual selection based on viability differences. *Evolution* 40: 804–816.

Andersson, M. 1994. *Sexual Selection*. Princeton University Press, Princeton, NJ.

Andrade, M. C. B. 1996. Sexual selection for male sacrifice in the Australian redback spider. *Science* 271: 70–72.

Andrés, J. A., and A. Cordero Rivera. 2000. Copulation duration and fertilization success in a damselfly: An example of cryptic female choice? *Anim. Behav.* 59: 695–703.

Aoki, S. 1982. Soldiers and altruistic dispersal in aphids. In M. D. Breed, C. D. Michener, and H. E. Evans, eds., *The Biology of Social Insects*, pp. 154–158. Westview Press, Boulder, CO.

Archer, M. S., and M. A. Elgar. 1999. Female preference for multiple partners: Sperm competition in the hide beetle, *Dermestes maculatus* (De Geer). *Anim. Behav.* 58: 669–675.

Arnaud, L., and E. Haubruge. 1999. Mating behaviour and male mate choice in *Tribolium castaneum* (Coleoptera, Tenebrionidae). *Behaviour* 136: 67–77.

Arnold, S. J., and D. Duvall. 1994. Animal mating systems: A synthesis based on selection theory. *Am. Nat.* 143: 317–348.

Arnqvist, G. 1988. Mate guarding and sperm displacement in the water strider *Gerris lateralis* Schumm. (Heteroptera: Gerridae). *Freshw. Biol.* 19: 269–274.

Arnqvist, G. 1989a. Multiple mating in a water strider: Mutual benefits or intersexual selection. *Anim. Behav.* 38: 749–756.

Arnqvist, G. 1989b. Sexual selection in a water strider: The function, mechanism of selection and heritability of a male grasping apparatus. *Oikos* 56: 344–350.

Arnqvist, G. 1992a. Pre-copulatory fighting in a water strider: Inter-sexual conflict or mate assessment? *Anim. Behav.* 43: 559–568.

Arnqvist, G. 1992b. Spatial variation in selective regimes: Sexual selection in the water strider, *Gerris odontogaster*. *Evolution* 46: 914–929.

Arnqvist, G. 1997a. The evolution of animal genitalia: Distinguishing between hypotheses by single species studies. *Biol. J. Linn. Soc.* 60: 365–379.

Arnqvist, G. 1997b. The evolution of water strider mating systems: Causes and consequences of sexual conflicts. In J. C. Choe and B. J. Crespi, eds., *The Evolution of Mating Systems in Insects and Arachnids*, pp. 146–163. Cambridge University Press, Cambridge, UK.

Arnqvist, G. 1998. Comparative evidence for the evolution of genitalia by sexual selection. *Nature* 393: 784–786.

Arnqvist, G., and I. Danielsson. 1999a. Copulatory behavior, genital morphology, and male fertilization success in water striders. *Evolution* 53: 147–156.

Arnqvist, G., and I. Danielsson. 1999b. Postmating sexual selection: The effects of male body size and recovery period and egg production rate in a water strider. *Behav. Ecol.* 10: 358–365.

Arnqvist, G., and R. Thornhill. 1998. Evolution of animal genitalia: Patterns of phenotypic and genotypic variation and condition dependence of genital and non-genital morphology in water striders (Heteroptera: Gerridae: Insecta). *Genet. Res., Camb.* 71: 193–212.

Arnqvist, G., M. Edvardsson, U. Friberg, and T. Nilsson. 2000. Sexual conflict promotes speciation in insects. *Proc. Natl. Acad. Sci. USA* 97: 10460–10464.

Arnqvist, G., R. Thornhill, and L. Rowe. 1997. Evolution of animal genitalia: Morphological correlates of fitness components in a water strider. *J. Evol. Biol.* 10: 613–640.

Aron, S., E. Campan, J. J. Boomsma, and L. Passera. 1999. Relatedness and split sex-ratios in the ant *Pheidole pallidula*. *Ethol. Ecol. Evol.* 11: 209–227.

Arthur, B. I., E. Hauschteck-Jungen, R. Nöthiger, and P. I. Ward. 1998. A female nervous system is necessary for normal sperm storage in *Drosophila melanogaster*: A masculinized nervous system is as good as none. *Proc. Roy. Soc. Lond. B* 265: 1749–1753.

Asada, N., and O. Kitagawa. 1988. Formation and the inhibition of reaction plug in mated *Drosophila*: Study of a primitive defense reaction. *Dev. Comp. Immunol.* 12: 521–530.

Aspi, J. 1992. Incidence and adaptive significance of multiple mating in females of two boreal *Drosophila virilis* group species. *Ann. Zool. Fenn.* 29: 147–159.

Avise, J. C. 1993. *Molecular Markers, Natural History and Evolution.* Chapman and Hall, London.

Baccetti, B. 1987. Spermatozoa and phylogeny in orthopteroid insects. In B. Baccetti, ed., *Evolutionary Biology of Orthopteroid Insects*, pp. 12–112. Ellis Horwood, Chichester, UK.

Backus, V. L., and W. H. Cade. 1986. Sperm competition in the field cricket *Gryllus integer* (Orthoptera: Gryllidae). *Fla. Entomol.* 69: 722–728.

Baer, B., and P. Schmid-Hempel. 1999. Experimental variation in polyandry affects parasite loads and fitness in a bumble-bee. *Nature* 397: 151–154.

Bairati, A. 1968. Structure and ultrastructure of the male reproductive system of *Drosophila melanogaster* Meig. 2. The genital duct and accessory glands. *Monit. Zool. Ital.* 2: 105–182.

Baker, R. R., and M. A. Bellis. 1988. "Kamikaze" sperm in mammals? *Anim. Behav.* 36: 937–938.

Baker, R. R., and M. A. Bellis. 1995. *Human Sperm Competition.* Chapman and Hall, London.

Bakker, T. C. M. 1993. Positive genetic correlation between female preference and preferred male ornament in sticklebacks. *Nature* 363: 255–257.

Ball, M. A., and G. A. Parker. 1998. Sperm competition games: A general approach to risk assessment. *J. Theor. Biol.* 194: 251–262.

Ball, M. A., and G. A. Parker. 2000. Sperm competition games: A comparison of loaded raffle models and their biological implications. *J. Theor. Biol.* 206: 487–506.

Banks, M. J., and D. J. Thompson. 1985. Lifetime mating success in the damselfly *Coenagrion puella. Anim. Behav.* 33: 1175–1183.

Barrows, E. M., and G. Gordh. 1978. Sexual behavior in the Japanese beetle, *Popillia japonica* and comparative notes on sexual behavior of other scarabs (Coleoptera: Scarabaeidae). *Behav. Biol.* 23: 341–354.

Bartlett, A. C., E. B. Mattix, and N. M. Wilson. 1968. Multiple matings and use of sperm in the boll weevil, *Anthonomus grandis. Ann. Entomol. Soc. Am.* 61: 1148–1155.

Barton Browne, L., P. H. Smith, A. C. M. Van Gerwen, and C. Gillott. 1990. Quantitative aspects of the effect of mating on readiness to lay in the Australian sheep blowfly *Lucilia cuprina. J. Insect Behav.* 3: 637–646.

Basolo, A. L. 1990. Female preference predates the evolution of the sword in swordtail fish. *Science* 250: 808–810.

Basolo, A. L. 1995a. A further examination of a pre-existing bias favouring a sword in the genus *Xiphophorus. Anim. Behav.* 50: 365–375.

Basolo, A. L. 1995b. Phylogenetic evidence for the role of a pre-existing bias in sexual selection. *Proc. Roy. Soc. Lond. B* 259: 307–311.

Bateman, A. J. 1948. Intrasexual selection in *Drosophila*. *Heredity* 2: 349–368.

Bateman, P. W. 1998. Mate preference for novel partners in the cricket *Gryllus bimaculatus*. *Ecol. Entomol.* 23: 473–475.

Baudry, E., M. Solignac, L. Garnery, M. Gries, J.-M. Cornuet, and N. Koeniger. 1998. Relatedness among honeybee (*Apis mellifera*) of a drone congregation. *Proc. Roy. Soc. Lond. B* 265: 2009–2014.

Baumann, H. 1974. Biological effects of paragonial substances PS1 and PS2 in female *Drosophila funebris*. *J. Insect Physiol.* 20: 2347–2363.

Beatty, R. A., and N. S. Sidhu. 1967. Spermatozoan nucleus length in three strains of *Drosophila melanogaster*. *Heredity* 22: 65–82.

Beatty, R. A., and N. S. Sidhu. 1970. Polymegaly of spermatozoan length and its genetic control in *Drosophila* species. *Proc. Roy. Soc. Edinb. Sect. B* 71: 14–28.

Bekkevold, D., J. Frydenberg, and J. J. Boomsma. 1999. Multiple mating and facultative polygyny in the Panamanian leafcutter ant *Acromyrmex echinatior*. *Behav. Ecol. Sociobiol.* 46: 103–109.

Bella, J. L., R. K. Butlin, C. Ferris, and G. M. Hewitt. 1992. Asymmetrical homogamy and unequal sex ratio from reciprocal mating-order crosses between *Chorthippus parallelus* subspecies. *Heredity* 68: 345–352.

Benford, F. A. 1978. Fisher's theory of the sex ratio applied to the social Hymenoptera. *J. Theor. Biol.* 72: 701–727.

Benken, T., A. Knaak, C. Gack, M. Eberle, and K. Peschke. 1999. Variation of sperm precedence in the rove beetle *Aleochara curtula* (Coleoptera: Staphylinidae). *Behaviour* 136: 1065–1077.

Bentur, J. S., K. Dakshayani, and S. B. Mathad. 1977. Mating induced oviposition and egg production in the crickets, *Gryllus bimaculatus* De Geer and *Plebeiogryllus guttiventris* Walker. *Z. Ang. Entomol.* 84: 129–135.

Berglund, A., A. Bisazza, and A. Pilastro. 1996. Armaments and ornaments: An evolutionary explanation of traits of dual utility. *Biol. J. Linn. Soc.* 58: 385–399.

Berkelhamer, R. C. 1984. An electrophoretic analysis of queen number in three species of dolichoderine ants. *Insect Soc.* 31: 132–141.

Bertram, M. J., G. A. Akerkar, R. L. Ard, C. Gonzalez, and M. Wolfner. 1992. Cell type specific gene expression in the *Drosophila melanogaster* male accessory gland. *Mechs. Dev.* 38: 33–40.

Bertram, M. J., D. M. Neubaum, and M. F. Wolfner. 1996. Localization of the *Drosophila* male accessory gland protein Acp36De in the mated female suggests a role in sperm storage. *Insect Biochem. Mol. Biol.* 26: 971–980.

Beukeboom, L. W. 1994. Phenotypic fitness effects of the selfish B chromosome, paternal sex ratio (PSR) in the parasitic wasp *Nasonia vitripennis*. *Evol. Ecol.* 8: 1–24.

Bhaskaran, G., S. P. Sparagana, K. H. Dahm, P. Barrera, and K. Peck. 1988. Sexual dimorphism in juvenile hormone synthesis by corpora allata and in juvenile hormone acid methyltransferase activity in corpora allata and accessory sex glands of some Lepidoptera. *Int. J. Invert. Reprod. Dev.* 13: 87–100.

Bircher, U., H. Jungen, R. Burch, and E. Hauschteck-Jungen. 1995. Multiple morphs of sperm were required for the evolution of the Sex Ratio trait in *Drosophila*. *J. Evol. Biol.* 8: 575–588.

Birkhead, T. R. 1995. Sperm competition: Evolutionary causes and consequences. *Reprod. Fert. Dev.* 7: 755–775.

Birkhead, T. 1998. Cryptic female choice: Criteria for establishing female sperm choice. *Evolution* 52: 1212–1218.

Birkhead, T. R., and J. D. Biggins. 1998. Sperm competition mechanisms in birds: Models and data. *Behav. Ecol.* 9: 253–260.

Birkhead, T. R., and A. P. Møller. 1992. *Sperm Competition in Birds: Evolutionary Causes and Consequences*. Academic Press, London.

Birkhead, T. R., and A. P. Møller. 1998a. *Sperm Competition and Sexual Selection*. Academic Press, London.

Birkhead, T. R., and A. P. Møller. 1998b. Sperm competition, sexual selection and different routes to fitness. In T. R. Birkhead and A. P. Møller, eds., *Sperm Competition and Sexual Selection*, pp. 757–781. Academic Press, London.

Birkhead, T. R., A. P. Møller, and W. J. Sutherland. 1993. Why do females make it so difficult for males to fertilize their eggs? *J. Theor. Biol.* 161: 51–60.

Birkhead, T. R., G. J. Wishart, and J. D. Biggins. 1995. Sperm precedence in the domestic fowl. *Proc. Roy. Soc. Lond. B* 261: 285–292.

Bishop, J. D. D. 1996. Female control of paternity in the internally fertilizing compound ascidian *Diplosoma listerianum*. I. Autoradiographic investigation of sperm movements in the female reproductive tract. *Proc. Roy. Soc. Lond. B* 263: 369–376.

Bishop, J. D. D., C. S. Jones, and L. R. Noble. 1996. Female control of paternity in the internally fertilizing compound ascidian *Diplosoma listerianum*. II. Investigation of male mating success using RAPD markers. *Proc. Roy. Soc. Lond. B* 263: 401–407.

Bissoondath, C. J., and C. Wiklund. 1995. Protein content of spermatophores in relation to monandry/polyandry in butterflies. *Behav. Ecol. Sociobiol.* 37: 365–371.

Bissoondath, C. J., and C. Wiklund. 1996a. Effect of male mating history and body size on ejaculate size and quality in two polyandrous butterflies, *Pieris napi* and *Pieris rapae* (Lepidoptera: Pieridae). *Funct. Ecol.* 10: 457–464.

Bissoondath, C. J., and C. Wiklund. 1996b. Male butterfly investment in successive ejaculates in relation to mating system. *Behav. Ecol. Sociobiol.* 39: 285–292.

Bissoondath, C. J., and C. Wiklund. 1997. Effect of male body size on sperm precedence in the polyandrous butterfly *Pieris napi* L. (Lepidoptera: Pieridae). *Behav. Ecol.* 8: 518–523.

Blay, S., and B. Yuval. 1997. Nutritional correlates of reproductive success of male Mediterranean fruit flies (Diptera: Tephritidae). *Anim. Behav.* 54: 59–66.

Bloch Qazi, M. C., J. R. Aprille, and S. M. Lewis. 1998. Female role in sperm storage in the red flour beetle, *Tribolium castaneum*. *Comp. Biochem. Physiol. A* 120: 641–647.

Bloch Qazi, M. C., J. T. Herbeck, and S. M. Lewis. 1996. Mechanisms of sperm transfer and storage in the red flour beetle (Coleoptera: Tenebrionidae). *Ann. Entomol. Soc. Am.* 89: 892–897.

Boggs, C. L. 1990. A general model of the role of male-donated nutrients in female insects' reproduction. *Am. Nat.* 136: 598–617.

Boggs, C. L., and W. B. Watt. 1981. Population structure of pierid butterflies. IV. Genetic and physiological investment in offspring by male *Colias*. *Oecologia* 50: 320–324.

Boiteau, G. 1988. Sperm utilization and post-copulatory female-guarding in the Colorado potato beetle, *Leptinotarsa decemlineata*. *Entomol. Exp. Appl.* 47: 183–188.

Boldyrev, B. T. 1915. Contributions à l'étude de la structure des spermatophores et des particularites de la copulation chez Locustodea et Gryllidea. *Horae. Soc. Entomol. Ross.* 41: 1–245.

Bookstein, F. L., B. Chernoff, R. Elder, J. Humphries, G. Smith, and R. Strauss. 1985. *Morphometrics in Evolutionary Biology.* Special Publication 15, Academy of Natural Sciences of Philadelphia.

Boomsma, J. J. 1996. Split sex ratios and queen-male conflict over sperm allocation. *Proc. Roy. Soc. Lond. B* 263: 697–704.

Boomsma, J. J., and A. Grafen. 1990. Intraspecific variation in ant sex ratios and the Trivers-Hare hypothesis. *Evolution* 44: 1026–1034.

Boomsma, J. J., and A. Grafen. 1991. Colony-level sex ratio selection in the eusocial hymenoptera. *J. Evol. Biol.* 4: 383–407.

Boomsma, J. J., and F. L. W. Ratnieks. 1996. Paternity in eusocial Hymenoptera. *Phil. Trans. Roy. Soc. Lond. B* 351: 947–975.

Boomsma, J. J., and L. Sundström. 1998. Patterns of paternity skew in *Formica* ants. *Behav. Ecol. Sociobiol.* 42: 85–92.

Boomsma, J. J., and T. M. Van der Have. 1998. Queen mating and paternity variation in the ant *Lasius niger*. *Mol. Ecol.* 7: 1709–1718.

Boomsma, J. J., A. H. Brouwer, and A. J. Van Loon. 1990. A new polygynous *Lasius* species (Hymenoptera: Formicidae) from central Europe. II. Allozymatic confirmation of species status and social structure. *Insect. Soc.* 37: 363–375.

Boomsma, J. J., E. J. Fjerdingstad, and J. Frydenberg. 1999. Multiple paternity, relatedness and genetic diversity in *Acromyrmex* leaf-cutter ants. *Proc. Roy. Soc. Lond. B* 266: 249–254.

Boomsma, J. J., P. J. Wright, and A. H. Brouwer. 1993. Social structure in the ant *Lasius flavus*: Multi-queen nests or multi-nest mounds? *Ecol. Entomol.* 18: 47–53.

Boorman, E., and G. A. Parker. 1976. Sperm (ejaculate) competition in *Drosophila melanogaster*, and the reproductive value of females to males in relation to female age and mating status. *Ecol. Entomol.* 1: 145–155.

Borgia, G. 1981. Mate selection in the fly *Scatophaga stercoraria*: Female choice in a male-controlled system. *Anim. Behav.* 29: 71–80.

Boucher, L., and J. Huignard. 1987. Transfer of male secretions from the spermatophore to the female insect in *Caryedon serratus* (Ol.): Analysis of the possible trophic role of these secretions. *J. Insect Physiol.* 33: 949–957.

Bourke, A. F. G., and N. R. Franks. 1995. *Social Evolution in Ants*. Princeton University Press, Princeton, NJ.

Bourke, A. F. G., T. M. van der Have, and N. R. Franks. 1988. Sex ratio determination and worker reproduction in the slave-making ant *Harpagoxenus sublaevis*. *Behav. Ecol. Sociobiol.* 23: 233–245.

Bownes, M., and L. Partridge. 1987. Transfer of molecules from ejaculate to females in *Drosophila melanogaster* and *Drosophila pseudoobscura*. *J. Insect Physiol.* 33: 941–947.

Bradbury, J. W., and M. B. Andersson. 1987. *Sexual Selection: Testing the Alternatives*. Wiley, Chichester, UK.

Breed, M. D., and B. Bennett. 1987. Kin recognition in highly eusocial insects. In D. J. C. Fletcher and C. D. Michener, eds., *Kin Recognition in Animals*, pp. 243–285. Wiley, Chichester, UK.

Brenner, R. R., and A. Bernasconi. 1989. Prostaglandin biosynthesis in the gonads of the hematophagous insect *Triatoma infestans*. *Comp. Biochem. Physiol. B* 93: 1–4.

Bressac, C., and E. Hauschteck-Jungen. 1996. *Drosophila subobscura* females preferentially select long sperm for storage and use. *J. Insect Physiol.* 42: 323–328.

Bressac, C., A. Fleury, and D. Lachaise. 1995. Another way of being anisogamous in *Drosophila* subgenus species: Giant sperm, one-to-one gamete ratio, and high zygote provisioning. *Proc. Natl. Acad. Sci. USA* 91: 10399–10402.

Bressac, C., D. Joly, J. Devaux, and D. Lachaise. 1991a. Can we predict the mating pattern of *Drosophila* females from the sperm length distribution in males? *Experientia* 47: 111–114.

Bressac, C., D. Joly, J. Devaux, C. Serres, D. Feneux, and D. Lachaise. 1991b. Comparative kinetics of short and long sperm in sperm dimorphic *Drosophila* spp. *Cell Motil. Cytoskel.* 19: 269–274.

Briskie, J. V., and R. Montgomerie. 1992. Sperm size and sperm competition in birds. *Proc. Roy. Soc. Lond. B* 247: 89–95.

Briskie, J. V., R. Montgomerie, and T. R. Birkhead. 1997. The evolution of sperm size in birds. *Evolution* 51: 937–945.

Brooks, R., and V. Couldridge. 1999. Multiple sexual ornaments coevolve with multiple mating preferences. *Am. Nat.* 154: 37–45.

Brower, J. H. 1975. Sperm precedence in the Indian meal moth, *Plodia interpunctella*. *Ann. Entomol. Soc. Am.* 68: 78–80.

Brown, W. D. 1997. Courtship feeding in tree crickets increases insemination and female reproductive life span. *Anim. Behav.* 54: 1369–1382.

Brown, W. D., and L. Keller. 2000. Colony sex ratios vary with queen number but not relatedness asymmetry in the ant *Formica exsecta*. *Proc. Roy. Soc. Lond. B* 267: 1751–1757.

Brown, W. D., and R. Stanford. 1992. Male mating tactics in a blister beetle (Coleoptera, Meloidae) vary with female quality. *Can. J. Zool.* 70: 1652–1655.

Bryan, J. H. 1968. Results of consecutive matings of female *Anopheles gambiae* species *B* with fertile and sterile males. *Nature* 218: 489.

Bullini, L., M. Coluzzi, and A. P. Bianchi Bullini. 1976. Biochemical variants in the study of multiple insemination in *Culex pipiens* L. (Diptera, Culicidae). *Bull. Entomol. Res.* 65: 683–685.

Bulmer, M. G. 1983. Models for the evolution of protandry in insects. *Theor. Pop. Biol.* 23: 314–322.

Burke, T. 1989. DNA fingerprinting and other methods for the study of mating success. *TREE* 4: 139–144.

Burkhardt, D., I. de la Motte, and K. Lunau. 1994. Signalling fitness: Larger males sire more offspring. Studies of the stalk-eyed fly *Cyrtodiopsis whitei* (Diopsidae, Diptera). *J. Comp. Physiol. A* 174: 61–64.

Burley, N. 1988. The differential allocation hypothesis: An experimental test. *Am. Nat.* 132: 611–628.

Burns, J. M. 1968. Mating frequency in natural populations of skippers and butterflies as determined by spermatophore counts. *Proc. Natl. Acad. Sci. USA* 61: 852–859.

Burpee, D. M., and S. K. Sakaluk. 1993. Repeated matings offset costs of reproduction in female crickets. *Evol. Ecol.* 7: 240–250.

Burt, A. 1989. Comparative methods using phylogenetically independent contrasts. *Oxf. Surv. Evol. Biol.* 6: 33–53.

Butlin, R. K., C. W. Woodhatch, and G. M. Hewitt. 1987. Male spermatophore investment increases female fecundity in a grasshopper. *Evolution* 41: 221–225.

Calos, J. B., and S. K. Sakaluk. 1998. Paternity of offspring in multiply-mated female crickets: The effect of nuptial food gifts and the advantage of mating first. *Proc. Roy. Soc. Lond. B* 265: 2191–2195.

Campanella, P. J., and L. Wolf. 1974. Temporal lek as a mating system in a temperate zone dragonfly (Odonata: Anisoptera): I. *Plathemis lydia*. *Behaviour* 51: 49–87.

Carayon, J. 1966. Traumatic insemination and the paragenital system. In R. L. Usinger, ed., *Monograph of Cimicidae*, pp. 81–166. Entomological Society of America, Baltimore.

Carayon, M. J. 1974. Insemination traumatique héterosexuelle et homosexuelle chez *Xylocoris maculipennis* (Hem. Anthocoridae). *C. R. Acad. Sci. Paris Série D* 278: 2803–2806.

Carlberg, U. 1983. Copulation in *Extatosoma tiaratum* (MacLeay) (Insecta: Phasmida). *Zool. Anz.* 210: 340–356.

Carlberg, U. 1987a. Mate choice, sperm competition and storage of sperm in *Baculum* sp. 1 (Insects: Phasmida). *Zool. Anz.* 219: 182–196.

Carlberg, U. 1987b. Reproduction behavior of *Extatosoma tiaratum* (MacLeay) (Insecta: Phasmida). *Zool. Anz.* 219: 331–336.

Carlin, N. F., H. K. Reeve, and S. P. Cover. 1993. Kin discrimination and division of labour among matrilines in the polygynous ant, *Camponotus planatus*. In L. Keller, ed., *Queen Number and Sociality in Insects*, pp. 362–401. Oxford University Press, Oxford.

Carpenter, J. E. 1992. Sperm precedence in *Helicoverpa zea* (Lepidoptera: Noctuidae): Response to a substerilizing dose of radiation. *J. Econ. Entomol.* 85: 779–782.

Carroll, S. P. 1988. Contrasts in reproductive ecology between temperate and tropical populations of *Jadera haematoloma*, a mate-guarding Hemipteran (Rhopalidae). *Ann. Entomol. Soc. Am.* 81: 54–63.

Carroll, S. P. 1991. The adaptive significance of mate guarding in the soapberry bug, *Jadera haematoloma* (Hemiptera: Rhopalidae). *J. Insect Behav.* 4: 509–530.

Carroll, S. P. 1993. Divergence in male mating tactics between two populations of the soapberry bug. I. Guarding versus nonguarding. *Behav. Ecol.* 4: 156–164.

Carroll, S. P., and C. Boyd. 1992. Host race radiation in the soapberry bug: Natural history, with the history. *Evolution* 46: 1052–1069.

Carroll, S. P., and P. S. Corneli. 1995. Divergence in male mating tactics between two populations of the soapberry bug. II. Genetic change and the evolution of a plastic reaction norm in a variable social environment. *Behav. Ecol.* 6: 46–56.

Carroll, S. P., and J. E. Loye. 1990. Male-biased sex ratios, female promiscuity, and copulatory mate guarding in an aggregating tropical bug, *Dysdercus bimaculatus*. *J. Insect Behav.* 3: 33–48.

Cavalloro, R., and G. Delrio. 1974. Mating behavior and competitiveness of gamma-irradiated olive fruit flies. *J. Econ. Entomol.* 67: 253–255.

Chan, G. L., and A. F. G. Bourke. 1994. Split sex ratios in a multiple-queen ant population. *Proc. Roy. Soc. Lond. B* 258: 261–266.

Chapman, T., Y. Choffat, W. E. Lucas, E. Kubli, and L. Partridge. 1996. Lack of response to sex peptide results in increased cost of mating in dunce *Drosophila melanogaster* females. *J. Insect Physiol.* 42: 1007–1015.

Chapman, T., J. Hutchings, and L. Partridge. 1993. No reduction in the cost of mating for *Drosophila melanogaster* females mating with spermless males. *Proc. Roy. Soc. Lond. B* 253: 211–217.

Chapman, T., L. F. Liddle, J. M. Kalb, M. F. Wolfner, and L. Partridge. 1995. Cost of mating in *Drosophila melanogaster* females is mediated by male accessory gland products. *Nature* 373: 241–244.

Chapman, T., T. Miyatake, H. K. Smith, and L. Partridge. 1998. Interactions of mating, egg production and death rates in females of the Mediterranean fruit fly, *Ceratitis capitata. Proc. Roy. Soc. Lond. B* 265: 1879–1894.

Chapman, T., D. M. Neubaum, M. F. Wolfner, and L. Partridge. 2000. The role of male accessory gland protein Acp36DE in sperm competition in *Drosophila melanogaster. Proc. Roy. Soc. Lond. B* 267: 1097–1105.

Chapman, T., S. Trevitt, and L. Partridge. 1994. Remating and male-derived nutrients in *Drosophila melanogaster. J. Evol. Biol.* 7: 51–69.

Chapuisat, M. 1998. Mating frequency of ant queens with alternative dispersal strategies, as revealed by microsatellite analysis of sperm. *Mol. Ecol.* 7: 1097–1105.

Chapuisat, M., and L. Keller. 1999. Testing kin selection with sex allocation data in eusocial Hymenoptera. *Heredity* 82: 473–478.

Charlesworth, B. 1987. The heritability of fitness. In J. W. Bradbury and M. B. Andersson, eds., *Sexual Selection: Testing the Alternatives*, pp. 21–40. Wiley, Chichester, UK.

Charnov, E. L. 1976. Optimal foraging: The marginal value theorem. *Theor. Pop. Biol.* 9: 129–136.

Charnov, E. L. 1978. Sex ratio selection in eusocial Hymenoptera. *Am. Nat.* 112: 317–326.

Charnov, E. L. 1982. *The Theory of Sex Allocation.* Princeton University Press, Princeton, NJ.

Charnov, E. L., and G. A. Parker. 1995. Dimensionless invariants from foraging theory's marginal value theorem. *Proc. Natl. Acad. Sci. USA* 92: 1446–1450.

Chen, P. S. 1984. The functional morphology and biochemistry of insect male accessory glands and their secretions. *Ann. Rev. Entomol.* 29: 233–255.

Chen, P. S. 1996. The accessory gland proteins in male *Drosophila*: Structural, reproductive, and evolutionary aspects. *Experientia* 52: 503–510.

Chen, P. S., and J. Balmer. 1989. Secretory proteins and sex peptides of the male accessory gland in *Drosophila sechellia. J. Insect Physiol.* 35: 759–764.

Chen, P. S., E. Stumm-Zollinger, T. Aigaki, J. Balmer, M. Bienz, and P. Bohlen. 1988. A male accessory gland peptide that regulates reproductive behaviour of female *D. melanogaster. Cell* 54: 291–298.

Cheng, K. M., and P. B. Siegel. 1990. Quantitative genetics of multiple mating. *Anim. Behav.* 40: 406–407.

Cherix, D., D. Chautems, D. J. C. Fletcher, W. Fortelius, G. Gris, L. Keller, L. Passera, R. Rosengren, E. L. Vargo, and F. Walter. 1991. Alternative reproductive strategies in *Formica lugubris* Zett. (Hymenoptera Formicidae). *Ethol. Ecol. Evol.* 1: 61–66.

Childress, D., and D. L. Hartl. 1972. Sperm preference in *Drosophila melanogaster. Genetics* 71: 417–427.

Choe, J. C., and B. J. Crespi. 1997a. *The Evolution of Mating Systems in Insects and Arachnids.* Cambridge University Press, Cambridge, UK.

Choe, J. C., and B. J. Crespi. 1997b. *The Evolution of Social Behavior in Insects and Arachnids*. Cambridge University Press, Cambridge, UK.

Civetta, A. 1999. Direct visualization of sperm competition and sperm storage in *Drosophila. Current Biol.* 9: 841–844.

Clark, A. G., and D. J. Begun. 1998. Female genotypes affect sperm displacement in *Drosophila. Genetics* 149: 1487–1493.

Clark, A. G., M. Aguade, T. Prout, L. G. Harshman, and C. H. Langley. 1995. Variation in sperm displacement and its association with accessory gland protein loci in *Drosophila melanogaster. Genetics* 139: 189–201.

Clark, A. G., D. J. Begun, and T. Prout. 1999. Female × Male interactions in *Drosophila* sperm competition. *Science* 283: 217–220.

Clark, S. J. 1988. The effects of operational sex ratio and food deprivation on copulation duration in the water strider *Gerris remigis* Say. *Behav. Ecol. Sociobiol.* 23: 317–322.

Clarke, C. A., and P. M. Shepard. 1962. Offspring from double matings in swallow butterflies. *Entomologist* 95: 199–203.

Clutton-Brock, T. H., and G. A. Parker. 1992. Potential reproductive rates and the operation of sexual selection. *Q. Rev. Biol.* 67: 437–456.

Cobbs, G. 1977. Multiple insemination and male sexual selection in natural populations of *Drosophila pseudoobscura. Am. Nat.* 111: 641–656.

Cochran, D. G. 1979. A genetic determination of insemination frequency and sperm precedence in the German cockroach. *Entomol. Exp. Appl.* 26: 259–266.

Coffelt, J. A. 1975. Multiple mating by *Lasioderma serricorne* (F.)—Effects on fertility and fecundity. In *Proceedings of the First International Working Conference on Stored Product Entomology (1974)*, pp. 549–553. Savannah, Georgia.

Cohan, J. 1969. Why so many sperms? An essay on the arithmetic of reproduction. *Sci. Prog., Oxf.* 57: 23–41.

Cohen, J. 1967. Correlation between sperm "redundancy" and chiasmata frequency. *Nature* 215: 862–863.

Cohen, J. 1973. Cross-overs, sperm redundancy and their close association. *Heredity* 31: 408–413.

Cohen, J. 1977. *Reproduction*. Butterworths, London.

Cole, B. J. 1983. Multiple mating and the evolution of social behaviour in the Hymenoptera. *Behav. Ecol. Sociobiol.* 12: 191–201.

Cole, B. J., and D. C. Wiernasz. 1999. The selective advantage of low relatedness. *Science* 285: 891–893.

Colegrave, N., T. R. Birkhead, and C. M. Lessells. 1995. Sperm precedence in zebra finches does not require special mechanisms of sperm competition. *Proc. Roy. Soc. Lond. B* 259: 223–228.

Coleman, S., B. Drahn, G. Petersen, J. Stolorov, and K. Kraus. 1995. A *Drosophila* male accessory gland protein that is a member of the serpin superfamily of proteinase inhibitors is transferred to females during mating. *Insect Biochem. Mol. Biol.* 25: 203–207.

Conner, J. K. 1995. Extreme variability in sperm precedence in the fungus beetle, *Bolitotherus cornutus* (Coleoptera Tenebrionidae). *Ethol. Ecol. Evol.* 7: 277–280.

Conrad, K. F., and G. Pritchard. 1990. Pre oviposition mate guarding and mating behavior of *Argia vivida* (Odonata: Coenagrionidae). *Ecol. Entomol.* 15: 363–370.

Contel, E. P. B., and W. E. Kerr. 1976. Origin of males in *Melipona subnitida* estimated from data of an isozymic polymorphic system. *Genetica* 46: 271–279.

Contel, E. P. B., and M. A. Mestiner. 1974. Esterase polymorphisms at two loci in the social bee. *J. Heredity* 65: 349–352.

Convey, P. 1989. Post-copulatory guarding strategies in the non territorial dragonfly *Sympetrum sanguineum* Müller (Odonata: Libellulidae). *Anim. Behav*. 37: 56–63.

Cook, A. G. 1977. Nutrient chemicals as phagostimulants for *Locusta migratoria*. *Ecol. Entomol.* 2: 121–131.

Cook, D. F. 1990. Differences in courtship, mating and postcopulatory behavior between male morphs of the dung beetle *Onthophagus binodis* Thunberg (Coleoptera: Scarabaeidae). *Anim. Behav*. 40: 428–436.

Cook, D. F. 1992. The effect of male size on receptivity in female *Lucilia cuprina* (Diptera: Calliphoridae). *J. Insect Behav.* 5: 365–374.

Cook, J. M. 1993. Sex determination in the Hymenoptera: A review of models and evidence. *Heredity* 71: 421–435.

Cook, J. M., and R. H. Crozier. 1995. Sex determination and population biology in the Hymenoptera. *TREE* 10: 281–286.

Cook, P. A., and M. J. G. Gage. 1995. Effects of risks of sperm competition on the numbers of eupyrene and apyrene sperm ejaculated by the male moth *Plodia interpunctella* (Lepidoptera: Pyralidae). *Behav. Ecol. Sociobiol.* 36: 261–268.

Cook, P. A., and N. Wedell. 1996. Ejaculate dynamics in butterflies: A strategy for maximizing fertilization success? *Proc. Roy. Soc. Lond. B* 263: 1047–1051.

Cook, P. A., and N. Wedell. 1999. Non-fertile sperm delay female remating. *Nature* 397: 486.

Cook, P. A., I. F. Harvey, and G. A. Parker. 1997. Predicting variation in sperm precedence. *Phil. Trans. Roy. Soc. Lond. B* 352: 771–780.

Cooper, G., P. W. H. Holland, and P. L. Miller. 1996a. Captive breeding of *Ischnura elegans* (Vander Linden): Observations on longevity, copulation and oviposition (Zygoptera: Coenagrionidae). *Odonatologica* 25: 261–273.

Cooper, G., P. L. Miller, and P. W. H. Holland. 1996b. Molecular genetic analysis of sperm competition in the damselfly *Ischnura elegans* (Vander Linden). *Proc. Roy. Soc. Lond. B* 263: 1343–1349.

Cordero, A. 1990. The adaptive significance of the prolonged copulations of the damselfly, *Ischnura graellsii* (Odonata: Coenagrionidae). *Anim. Behav*. 40: 43–48.

Cordero, A., and P. L. Miller. 1992. Sperm transfer, displacement and precedence in *Ischnura graellsii* (Odonata: Coenagrionidae). *Behav. Ecol. Sociobiol.* 30: 261–267.

Cordero, A., S. Santolamazza Carbone, and C. Utzeri. 1995. Male disturbance, repeated insemination and sperm competition in the damselfly *Coenagrion scitulum* (Zygoptera: Coenagrionidae). *Anim. Behav*. 49: 437–449.

Cordero, A., S. Santolamazza Carbone, and C. Utzeri. 1998. Mating opportunities and mating costs are reduced in androchrome female damselflies, *Ischnura elegans* (Odonata). *Anim. Behav*. 55: 185–197.

Cordero, C. 1995. Ejaculate substances that affect female insect reproductive physiology and behavior—honest or arbitrary traits. *J. Theor. Biol.* 174: 453–461.

Cordero, C. 1996. On the evolutionary origin of nuptial seminal gifts in insects. *J. Insect Behav*. 9: 969–974.

Cordero, C. 1998. Chemical ornaments of semen. *J. Theor. Biol.* 192: 581–584.

Córdoba-Aguilar, A. 1999. Male copulatory sensory stimulation induces female ejection of rival sperm in a damselfly. *Proc. Roy. Soc. Lond. B* 266: 779–784.

Coulhart, M. B., and R. S. Singh. 1988. Differing amounts of genetic polymorphism in testes and male accessory glands of *Drosophila melanogaster* and *D. simulans*. *Biochem. Genet.* 26: 153–164.

Craig, G. B. 1967. Mosquitoes: Female monogamy induced by male accessory gland substance. *Science* 156: 1499–1501.

Crespi, B. J. 1992. Eusociality in Australian gall thrips. *Nature* 359: 724–726.

Crocker, G., and T. Day. 1987. An advantage to mate choice in the seaweed fly, *Coelopa frigida*. *Behav. Ecol. Sociobiol.* 20: 295–301.

Crozier, R. H. 1973. Apparent differential selection at an isozyme locus between queens and workers of the ant *Aphaenogaster rudis*. *Genetics* 73: 313–318.

Crozier, R. H. 1974. Allozyme analysis of reproductive strategy in the ant *Aphaenogaster rudis*. *Isozyme Bull.* 7: 18.

Crozier, R. H., and D. Brückner. 1981. Sperm clumping and the population genetics of hymenoptera. *Am. Nat.* 117: 561–563.

Crozier, R. H., and R. E. Page. 1985. On being the right size: Male contributions and multiple mating in social hymenoptera. *Behav. Ecol. Sociobiol.* 18: 105–115.

Crozier, R. H., and P. Pamilo. 1996. *Evolution of Social Insect Colonies: Sex Allocation and Kin Selection*. Oxford University Press, Oxford.

Crozier, R. H., B. E. Kaufmann, M. E. Carew, and Y. C. Crozier. 1999. Mutability of microsatellites developed for the ant *Camponotus consobrinus*. *Mol. Ecol.* 8: 271–276.

Crozier, R. H., B. H. Smith, and Y. C. Crozier. 1987. Relatedness and population structure of the primitively eusocial bee *Lasioglossum zephyrum* (Hymenoptera: Halictidae) in Kansas. *Evolution* 41: 902–910.

Crudgington, H. S., and M. T. Siva-Jothy. 2000. Genital damage, kicking and early death. *Nature* 407: 855–856.

Curtis, C. F. 1968. Radiation sterilization and the effect of multiple mating of females in *Glossina austeni*. *J. Insect Physiol.* 14: 1365–1380.

Curtsinger, J. W. 1991. Sperm competition and the evolution of multiple mating. *Am. Nat.* 138: 93–102.

Dadd, R. H. 1981. Essential fatty acids for mosquitoes, and other insects, and vertebrates. In G. Bhaskaran, S. F. Friedman, and J. G. Rodriguez, eds., *Current Topics in Insect Endocrinology and Nutrition*, pp. 189–214. Plenum, New York.

Dallai, R., B. Baccetti, F. Bernini, E. Bigliardi, A. G. Burrini, F. Giusti, M. Mazzini, V. Pallini, T. Renieri, F. Rosati, G. Selmi, and M. Vegni. 1975. New models of aflagellate arthropod spermatozoa. In B. A. Afzelius, ed., *The Functional Anatomy of the Spermatozoon*, pp. 279–287. Pergamon Press, Oxford.

Dambach, M., and U. Beck. 1990. Mating in the scaly cricket *Cycloptiloides canariensis* (Orthoptera: Gryllidae: Mogoplistinae). *Ethology* 85: 289–301.

Dame, D. A., and H. R. Ford. 1968. Multiple mating of *Glossina morsitans* Westw. and its potential effect on the sterile male technique. *Bull. Entomol. Res.* 58: 213–219.

Danielsson, I. 2000. Antagonistic pre- and post-copulatory sexual selection on male body size in a water strider (*Gerris lacustris*). *Proc. Roy Soc. Lond. B* 268: 77–81.

Danielsson, I., and C. Askenmo. 1999. Male genital traits and mating interval affect male fertilization success in the water strider *Gerris lacustris*. *Behav. Ecol. Sociobiol.* 46: 149–156.

Darwin, C. 1859. *On the Origin of Species by Means of Natural Selection.* John Murray, London.

Darwin, C. 1871. *The Descent of Man and Selection in Relation to Sex.* John Murray, London.

Davey, K. G. 1985a. The female reproductive tract. In G. A. Kerkut and L. I. Gilbert, eds., *Comprehensive Insect Physiology, Biochemistry and Pharmacology*, pp. 15–36. Pergamon Press, Oxford.

Davey, K. G. 1985b. The male reproductive tract. In G. A. Kerkut and L. I. Gilbert, eds., *Comprehensive Insect Physiology, Biochemistry and Pharmacology*, pp. 1–14. Pergamon Press, Oxford.

Davey, K. G., and G. F. Webster. 1967. The structure and secretion of the spermatheca of *Rhodnius prolixus* Stal.: A histochemical study. *Can. J. Zool.* 45: 653–657.

Dean, J. M. 1981. The relationship between lifespan and reproduction in the grasshopper *Melanoplus*. *Oecologia* 49: 385–388.

Degrugillier, M. E., and R. A. Leopold. 1976. Ultrastructure of sperm penetration of housefly eggs. *J. Ultrastruct. Res.* 56: 312–325.

Deinert, E. I., J. T. Longino, and L. E. Gilbert. 1994. Mate competition in butterflies. *Nature* 370: 23–24.

de Jong, P. W., P. M. Brakefield, and B. P. Geerinck. 1998. The effect of female mating history on sperm precedence in the two-spot ladybird, *Adelia bipunctata* (Coleoptera, Coccinellidae). *Behav. Ecol.* 9: 559–565.

de Jong, P. W., M. D. Verhoog, and P. M. Brakefield. 1993. Sperm competition and melanic polymorphism in the 2-spot ladybird, *Adalia bipunctata* (Coleoptera, Coccinellidae). *Heredity* 70: 172–178.

Delisle, C. L., and M. Hardy. 1997. Male larval nutrition influences the reproductive success of both sexes of the Spruce Budworm, *Choristoneura fumiferana* (Lepidoptera: Tortricidae). *Funct. Ecol.* 11: 451–463.

Delph, L., and K. Havens. 1998. Pollen competition in flowering plants. In T. R. Birkhead and A. P. Møller, eds., *Sperm Competition and Sexual Selection*, pp. 147–174. Academic Press, London.

Destephano, D. B., and U. E. Brady. 1977. Prostaglandin and prostaglandin synthetase in the cricket, *Acheta domesticus*. *J. Insect Physiol.* 23: 905–911.

De Villiers, P. S., and S. A. Hanrahan. 1991. Sperm competition in the Namib desert beetle, *Onymacris unguicularis*. *J. Insect Physiol.* 37: 1–8.

DeVries, J. K. 1964. Insemination and sperm storage in *Drosophila melanogaster*. *Evolution* 18: 271–282.

Dewsbury, D. A. 1982. Ejaculate cost and male choice. *Am. Nat.* 119: 601–610.

Dewsbury, D. A. 1984. Sperm competition in muroid rodents. In R. L. Smith, ed., *Sperm Competition and the Evolution of Animal Mating Systems*, pp. 547–571. Academic Press, London.

Dickinson, J. L. 1986. Prolonged mating in the milkweed leaf beetle *Labidomera clivicollis clivicollis* (Coleoptera: Chrysomelidae): A test of the "sperm-loading" hypothesis. *Behav. Ecol. Sociobiol.* 18: 331–338.

Dickinson, J. L. 1988. Determinants of paternity in the milkweed leaf beetle. *Behav. Ecol. Sociobiol.* 23: 9–19.

Dickinson, J. L. 1995. Trade-offs between post-copulatory riding and mate location in the blue milkweed beetle. *Behav. Ecol.* 6: 280–286.

Dickinson, J. L., and R. L. Rutowski. 1989. The function of the mating plug in the chalcedon checkerspot butterfly. *Anim. Behav.* 38: 154–162.

Downes, J. A. 1970. The feeding and mating behaviour of the specialised Empidinae (Diptera); observations on four species of *Rhamphomyia* in the high arctic and general discussion. *Can. Entomol.* 102: 769–791.

Droney, D. C. 1998. The influence of the nutritional content of the adult male diet on testis mass, body condition and courtship vigour in a Hawaiian *Drosophila*. *Funct. Ecol.* 12: 920–928.

Drummond, B. A. 1984. Multiple mating and sperm competition in the Lepidoptera. In R. L. Smith, ed., *Sperm Competition and the Evolution of Animal Mating Systems*, pp. 547–572. Academic Press, London.

Dufour, L. 1848. Anatomie générale des Dipteres. *Ann. de Sci. Nat.* 1: 244–264.

Dumont, C. 1920. Note biologique sur *Rhytirrhinus surcoufi* Peyerh. *Bull. Soc. Entomol. Fr.* 1920: 119–120.

Dussourd, D. E., C. A. Harvis, J. Meinwald, and T. Eisner. 1991. Pheromonal advertisement of a nuptial gift by a male moth (*Utetheisa ornatrix*). *Proc. Natl. Acad. Sci. USA* 88: 9224–9227.

Duvoisin, N., B. Baer, and P. Schmid-Hempel. 1999. Sperm transfer and male competition in a bumblebee. *Anim. Behav.* 58: 743–749.

Dybas, L. K., and H. S. Dybas. 1981. Coadaptation and taxonomic differentiation of sperm and spermathecae in featherwing beetles. *Evolution* 35: 168–174.

Eady, P. E. 1991. Sperm competition in *Callosobruchus maculatus* (Coleoptera: Bruchidae): A comparison of two methods used to estimate paternity. *Ecol. Entomol.* 16: 45–53.

Eady, P. 1994a. Intraspecific variation in sperm precedence in the bruchid beetle *Callosobruchus maculatus*. *Ecol. Entomol.* 19: 11–16.

Eady, P. 1994b. Sperm transfer and storage in relation to sperm competition in *Callosobruchus maculatus*. *Behav. Ecol. Sociobiol.* 35: 123–129.

Eady, P. E. 1995. Why do male *Callosobruchus maculatus* beetles inseminate so many sperm? *Behav. Ecol. Sociobiol.* 36: 25–32.

Eady, P., and S. Tubman. 1996. Last-male sperm precedence does not break down when females mate with three males. *Ecol. Entomol.* 21: 303–304.

Eberhard, W. G. 1982. Beetle horn dimorphism: Making the best of a bad lot. *Am. Nat.* 119: 420–426.

Eberhard, W. G. 1985. *Sexual Selection and Animal Genitalia*. Harvard University Press, Cambridge, MA.

Eberhard, W. G. 1991. Copulatory courtship and cryptic female choice in insects. *Biol. Rev.* 66: 1–31.

Eberhard, W. G. 1992. Species isolation, genital mechanics, and the evolution of species-specific genitalia in three species of *Macrodactylus* beetles (Coleoptera, Scarabeidae, Melolothinae). *Evolution* 46: 1774–1783.

Eberhard, W. G. 1993. Copulatory courtship and genital mechanisms of three species of *Macrodactylus* (Coleoptera Scarabaeidae Melolonthinae). *Ethol. Ecol. Evol.* 5: 19–63.

Eberhard, W. G. 1994. Evidence for widespread courtship during copulation in 131 species of insects and spiders, and implications for cryptic female choice. *Evolution* 48: 711–733.

Eberhard, W. G. 1996. *Female Control: Sexual Selection by Cryptic Female Choice*. Princeton University Press, Princeton, NJ.

Eberhard, W. G. 1997. Sexual selection by cryptic female choice in insects and arachnids. In J. C. Choe and B. J. Crespi, eds., *The Evolution of Mating Systems in Insects and Arachnids*, pp. 32–57. Cambridge University Press, Cambridge, UK.

Eberhard, W. G. 2000. Criteria for demonstrating postcopulatory female choice. *Evolution* 54: 1047–1050.

Eberhard, W. G., and C. Cordero. 1995. Sexual selection by cryptic female choice on male seminal products—a bridge between sexual selection and reproductive physiology. *TREE* 10: 493–496.

Eberhard, W. G., and E. E. Gutierrez. 1991. Male dimorphisms in beetles and earwigs and the question of developmental constraints. *Evolution* 45: 18–28.

Eberhard, W. G., B. A. Huber, R. L. S. Rodriguez, R. D. Briceño, I. Salas, and V. Rodriguez. 1998. One size fits all? Relationships between the size and degree of variation in genitalia and other body parts in twenty species of insects and spiders. *Evolution* 52: 415–431.

Economorpoulos, A. P., and H. T. Gordon. 1972. Sperm replacement and depletion in the spermatheca of the s and cs strains of *Oncopeltus fasciatus*. *Entomol. Exp. Appl.* 15: 1–12.

Edvardsson, M., and G. Arnqvist. 2000. Copulatory courtship and cryptic female choice in red flour beetles *Tribolium castaneum*. *Proc. Roy. Soc. Lond. B* 267: 559–563.

Ehrlich, A. H., and P. R. Ehrlich. 1978. Reproductive strategies in the butterflies: I. Mating frequency, plugging, and egg number. *J. Kan. Entomol. Soc.* 51: 666–697.

Eisner, T., and J. Meinwald. 1995. The chemistry of sexual selection. *Proc. Natl. Acad. Sci. USA* 92: 50–55.

El Agoze, M., M. Poirié, and G. Périquet. 1995. Precedence of the first male sperm in successive matings in the hymenoptera *Diadromus pulchellus*. *Entomol. Exp. Appl.* 75: 251–255.

Emlen, D. J. 1994. Environmental control of horn length dimorphism in the beetle *Onthophagus acuminatus* (Coleoptera: Scarabaeidae). *Proc. Roy. Soc. Lond. B* 256: 131–136.

Emlen, D. J. 1997. Alternative reproductive tactics and male-dimorphism in the horned beetle *Onthophagus acuminatus* (Coleoptera: Scarabaeidae). *Behav. Ecol. Sociobiol.* 41: 335–342.

Emlen, S. T., and L. W. Oring. 1977. Ecology, sexual selection, and the evolution of mating systems. *Science* 197: 215–223.

Endler, J. A., and T. McLellan. 1988. The process of evolution: Towards a newer synthesis. *Ann. Rev. Ecol. Syst.* 19: 395–421.

Engelmann, F. 1970. *The Physiology of Insect Reproduction*. Pergamon Press, New York.

Epstein, M. E. 1987. Mating behavior of *Acraea andromacha andromacha* (Fabricus) (Nymphalidae) in New Caledonia. *J. Lepid. Soc.* 41: 119–121.

Erickson, R. P. 1990. Post-meiotic gene expression. *Trends Genet.* 6: 264–269.

Estoup, A., A. Scholl, A. Pouvreau, and M. Solignac. 1995. Monandry and polyandry in bumble bees (Hymenoptera; Bombinae) as evidenced by highly variable microsatellites. *Mol. Ecol.* 4: 89–93.

Estoup, A., M. Solignac, and J.-M. Cornuet. 1994. Precise assessment of the number of patrilines and of genetic relatedness in honeybee colonies. *Proc. Roy. Soc. Lond. B* 258: 1–7.

Etman, A. A. M., and G. H. S. Hooper. 1979. Sperm precedence of the last mating in *Spodoptera litura*. *Ann. Entomol. Soc. Am.* 72: 119–120.

Evans, A. R. 1983. A study of the behaviour of the Australian field cricket *Tele-*

ogryllus commodus (Walker) (Orthoptera: Gryllidae) in field and habitat simulations. *Z. Tierpsychol.* 62: 269–290.

Evans, A. R. 1988. Mating systems and reproductive strategies in three Australian gryllid crickets: *Bobilla victoriae* Otte, *Balamara gidya* Otte and *Teleogryllus commodus* (Walker) (Orthoptera: Gryllidae: Nemobiinae; Gryllinae). *Ethology* 78: 21–52.

Evans, J. D. 1993. Parentage analysis in ant colonies using simple sequence repeat loci. *Mol. Ecol.* 2: 393–397.

Evans, J. D. 1995. Relatedness threshold for the production of female sexuals in colonies of a polygynous ant, *Myrmica tahoensis*, as revealed by microsatellite DNA analysis. *Proc. Natl. Acad. Sci. USA* 92: 6514–6517.

Evans, J. D. 1998. Parentage and sex allocation in the facultatively polygynous ant *Myrmica tahoensis*. *Behav. Ecol. Sociobiol.* 44: 23–34.

Evans, M. R., and B. J. Hatchwell. 1992a. An experimental study of male adornment in the scarlet-tufted malachite sunbird: I. The role of pectoral tufts in territorial defense. *Behav. Ecol. Sociobiol.* 29: 413–419.

Evans, M. R., and B. J. Hatchwell. 1992b. An experimental study of male adornment in the scarlet-tufted malachite sunbird: II. The role of the elongated tail in mate choice and experimental evidence for a handicap. *Behav. Ecol. Sociobiol.* 29: 421–427.

Fagerström, T., and C. Wiklund. 1982. Why do males emerge before females? Protandry as a mating strategy in male and female butterflies. *Oecologia* 52: 164–166.

Fairbairn, D. J. 1990. Factors influencing sexual size dimorphism in temperate waterstriders. *Am. Nat.* 136: 61–86.

Fairbairn, D. J. 1993. Costs of loading associated with mate-carrying in the waterstrider, *Aquarius remigis*. *Behav. Ecol.* 4: 224–231.

Falconer, D. S., and T. F. C. Mackay. 1996. *Introduction to Quantitative Genetics*. Longman, Harlow, UK.

Fan, Y., A. Rafaeli, C. Gileadi, E. Kubli, and S. W. Applebaum. 1999. *Drosophila melanogaster* sex peptide stimulates juvenile hormone synthesis and depresses sex pheromone production in *Helicoverpa armigera*. *J. Insect Physiol.* 45: 127–133.

Farias, G. J., R. T. Cunningham, and S. Nakagawa. 1972. Reproduction in the Mediterranean fruit fly: Abundance of stored sperm affected by duration of copulation, and affecting egg hatch. *J. Econ. Entomol.* 65: 914–915.

Farmer, D. C., and C. J. Barnard. 2000. Fluctuating asymmetry and sperm transfer in male decorated field crickets (*Gryllodes sigillatus*). *Behav. Ecol. Sociobiol.* 47: 287–292.

Fellowes, M. D. E., A. R. Kraaijeveld, and H. C. J. Godfray. 1999. The relative fitness of *Drosophila melanogaster* (Diptera, Drosophilidae) that have successfully defended themselves against the parasitoid *Asobata tabida* (Hymenoptera, Braconidae). *J. Evol. Biol.* 12: 123–128.

Felsenstein, J. 1985. Phylogenies and the comparative method. *Am. Nat.* 125: 1–15.

Ferguson, I. M., and D. J. Fairbairn. 2000. Sex-specific selection and sexual size dimorphism in the waterstrider *Aquarius remigis*. *J. Evol. Biol.* 13: 160–170.

Ferro, D. N., and R. D. Akre. 1975. Reproductive morphology and mechanics of mating of the codling moth, *Laspeyresia pomonella*. *Ann. Entomol. Soc. Am.* 68: 417–424.

Field, J., C. R. Solís, D. C. Queller, and J. E. Strassmann. 1998. Social and genetic structure of paper wasp cofoundress associations: Tests of reproductive skew models. *Am. Nat.* 151: 545–563.

Field, S. A., and M. A. Keller. 1993. Alternative mating tactics and female mimicry as post copulatory mate guarding behavior in the parasitic wasp *Cotesia rubecula*. *Anim. Behav.* 46: 1183–1189.

Fincke, O. M. 1984. Sperm competition in the damselfly *Enallagma hageni* Walsh (Odonata:Coenagrionidae): Benefits of multiple mating to males and females. *Behav. Ecol. Sociobiol.* 14: 235–240.

Fincke, O. M. 1986. Underwater oviposition in a damselfly (Odonata: Coenagrionidae) favors male vigilance, and multiple mating by females. *Behav. Ecol. Sociobiol.* 18: 405–412.

Fisher, R. A. 1915. The evolution of sexual preference. *Eugenics Review* 7: 184–192.

Fisher, R. A. 1930. *The Genetical Theory of Natural Selection.* Clarendon Press, Oxford.

Fjerdingstad, E. J., and J. J. Boomsma. 1998. Multiple mating increases the sperm stores of *Atta colombica* leafcutter ant queens. *Behav. Ecol. Sociobiol.* 42: 257–262.

Fjerdingstad, E. J., and J. J. Boomsma. 2000. Queen mating frequency and relatedness in young *Atta sexdens* colonies. *Insect. Soc.* 47: 1–3.

Fjerdingstad, E. J., J. J. Boomsma, and P. Thorén. 1998. Multiple paternity in the leafcutter ant *Atta colombica*—a microsatellite DNA study. *Heredity* 80: 118–126.

Fjerdingstad, E. J., P. J. Gertsch, and L. Keller. Unpublished ms. Why do some social insect queens mate with several males? Testing the sex ratio manipulation hypothesis in *Lasius niger*.

Flint, H. M., and E. L. Kressin. 1968. Gamma irradiation of the tobacco budworm: Sterilization, competitiveness, and observations on reproductive biology. *J. Econ. Entomol.* 61: 477–483.

Forbes, M. R. L., J. M. L. Richardson, and R. L. Baker. 1995. Frequency of female morphs is related to an index of male density in the damselfly, *Nehalennia irene* (Hagen). *Ecoscience* 2: 28–33.

Fortelius, W., P. Pamilo, R. Rosengren, and L. Sundström. 1987. Male size dimorphism and alternative reproductive tactics in *Formica exsecta* ants (Hymenoptera, Formicidae). *Ann. Zool. Fenn.* 24: 45–54.

Foster, K. R., and F. L. W. Ratnieks. 2000. Facultative worker policing in a wasp. *Nature* 407: 692–693.

Foster, K. R., P. Seppä, F. L. W. Ratniek, and P. A. Thorén. 1999. Low paternity in the hornet *Vespa crabro* indicates that multiple mating by queens is derived in vespine wasps. *Behav. Ecol. Sociobiol.* 46: 252–257.

Foster, S. P. 1993. Neural inactivation of sex pheromone production in mated lightbrown apple moths, *Epiphyas postvittana* (Walker). *J. Insect Physiol.* 39: 267–273.

Foster, W. 1967a. Hormonal-mediated nutritional control of sexual behavior in male dung flies. *Science* 158: 1596–1597.

Foster, W. A. 1967b. Co-operation by male protection of ovipositing female in the Diptera. *Nature* 214: 1035–1036.

Fowler, G. L. 1973. Some aspects of the reproductive biology of *Drosophila melanogaster*: Sperm transfer, sperm storage, and sperm utilisation. *Adv. Genet.* 17: 293–360.

Fowler, K., and L. Partridge. 1989. A cost of mating in female fruitflies. *Nature* 338: 760–761.

Fox, C. W. 1993a. The influence of maternal age and mating frequency on egg size and offspring performance in *Collosobruchus maculatus* (Coleoptera: Bruchidae). *Oecologia* 96: 139–146.

Fox, C. W. 1993b. Multiple mating, lifetime fecundity and female mortality of the bruchid beetle, *Callosobruchus maculatus* (Coleoptera: Bruchidae). *Funct. Ecol.* 7: 203–208.

Fox, C. W., and U. M. Savalli. 1998. Inheritance of environmental variation in body size: Superparasitism of seeds affects progeny and grandprogeny body size via a nongenetic maternal effect. *Evolution* 52: 172–182.

Fox, C. W., D. L. Hickman, E. L. Raleigh, and T. A. Mousseau. 1995. Paternal investment in a seed beetle (Coleoptera: Bruchidae): Influence of male size, age, and mating history. *Ann. Entomol. Soc. Am.* 88: 100–103.

Frankie, G. W., S. B. Vinson, and R. E. Colville. 1980. Territorial behavior of *Centris adani* and its reproductive function in the Costa Rican dry forest (Hymenoptera: Anthophoridae). *J. Kans. Entomol. Soc.* 53: 837–857.

Frankino, W. A., and S. K. Sakaluk. 1994. Post-copulatory mate guarding delays promiscuous mating by female decorated crickets. *Anim. Behav.* 48: 1479–1481.

Fraser, V. S., B. Kaufmann, B. P. Oldroyd, and R. H. Crozier. 2000. Genetic influence on caste in the ant *Camponotus consobrinus. Behav. Ecol. Sociobiol.* 47: 188–194.

Friedel, T., and C. Gillot. 1976. Male accessory gland substance of *Melanoplus sanguinipes*: Oviposition stimulant under the control of corpus allatum. *J. Insect Physiol.* 22: 489–495.

Friedel, T., and C. Gillot. 1977. Contribution of male-produced proteins to vitellogenesis in *Melanoplus sanguinipes. J. Insect Physiol.* 23: 145–151.

Friedländer, M. 1997. Control of the Eupyrene-apyrene sperm dimorphism in Lepidoptera. *J. Insect Physiol.* 43: 1085–1092.

Friedländer, M., and H. Gitay. 1972. The fate of the normal-anucleated spermatozoa in inseminated females of the silkworm *Bombyx mori. J. Morph.* 138: 121–130.

Frumhoff, P. C., and J. Baker. 1988. A genetic component to division of labour within honey bee colonies. *Nature* 333: 358–361.

Fryer, T., C. Cannings, and G. T. Vickers. 1999a. Sperm competition. I: Basic model, ESS and dynamics. *J. Theor. Biol.* 196: 81–100.

Fryer, T., C. Cannings, and G. T. Vickers. 1999b. Sperm competition. II: Post-copulatory guarding. *J. Theor. Biol.* 197: 343–360.

Fuchs, M. S., and E. A. Hiss. 1970. The partial purification and separation of the protein components of matrone from *Aedes aegypti. J. Insect Physiol.* 16: 931–939.

Fuchs, M. S., G. B. J. Craig, and D. D. Despommier. 1969. The protein nature of the substance inducing female monogamy in *Aedes aegypti. J. Insect Physiol.* 15: 701–709.

Fuchs, S., and R. F. A. Moritz. 1999. Evolution of extreme polyandry in the honeybee *Apis mellifera. Behav. Ecol. Sociobiol.* 45: 269–276.

Fuchs, S., and V. Schade. 1994. Lower performance in honeybee colonies of uniform paternity. *Apidologie* 25: 155–168.

Fuyama, Y. 1983. Species-specificity of paragonial substances as an isolating mechanism in *Drosophila. Experientia* 39: 190–192.

Gack, C., and K. Peschke. 1994. Spermathecal morphology, sperm transfer and a novel mechanism of sperm displacement in the rove beetle, *Aleochara curtula* (Coleoptera, Staphylinidae). *Zoomorphology* 114: 227–237.

Gadau, J., P. J. Gertsch, J. Heinze, P. Pamilo, and B. Hölldobler. 1998. Oligogyny by unrelated queens in the carpenter ant, *Camponotus ligniperdus*. *Behav. Ecol. Sociobiol.* 44: 23–34.

Gage, A. R., and C. J. Barnard. 1996. Male crickets increase sperm number in relation to competition and female size. *Behav. Ecol. Sociobiol.* 38: 349–353.

Gage, M. J. G. 1991. Risk of sperm competition directly affects ejaculate size in the Mediterranean fruit fly. *Anim. Behav.* 42: 1036–1037.

Gage, M. J. G. 1992. Removal of rival sperm during copulation in a beetle, *Tenebrio molitor*. *Anim. Behav.* 44: 587–589.

Gage, M. J. G. 1994. Associations between body size, mating pattern, testis size and sperm lengths across butterflies. *Proc. Roy. Soc. Lond. B* 258: 247–254.

Gage, M. J. G. 1995. Continuous variation in reproductive strategy as an adaptive response to population density in the moth *Plodia interpunctella*. *Proc. Roy. Soc. Lond. B* 261: 25–30.

Gage, M. J. G. 1998. Influences of sex, size, and symmetry on ejaculate expenditure in a moth. *Behav. Ecol.* 9: 592–597.

Gage, M. J. G., and R. R. Baker. 1991. Ejaculate size varies with socio-sexual situation in an insect. *Ecol. Entomol.* 16: 331–337.

Gage, M. J. G., and P. A. Cook. 1994. Sperm size or numbers? Effects of nutritional stress upon eupyrene and apyrene sperm production strategies in the moth *Plodia interpunctella* (Lepidoptera: Pyralidae). *Funct. Ecol.* 8: 594–599.

Galvarni, A., and R. Johnstone. 1998. Sperm allocation in an uncertain world. *Behav. Ecol. Sociobiol.* 44: 161–168.

Gangrade, G. A. 1963. A contribution to the biology of *Necroscia sparaxes* Westwood (Pasmidae: Phasmida). *Entomologist* 96: 83–93.

Garcia-Bellido, A. 1964. Das Sekret der Paragonien als Stimulus der Fekundität Bei weibchen von *Drosophila melanogaster*. *Z. Naturforsch.* 19b: 491–495.

Gavrilets, S. 2000. Rapid evolution of reproductive barriers driven by sexual conflict. *Nature* 403: 886–889.

George, J. A. 1967. Effect of mating sequence on egg-hatch from female *Aedes aegypti* (L.) mated with irradiated and normal males. *Mosquito News* 27: 82–86.

Gerhardt, U. 1913. Copulation and Spermatophoren von Grylliden und Locustiden. *Zool. Jahrb. Abt. Syst. Okol. Geogr. Tiere* 35: 461–531.

Getty, T. 1999. Chase-away sexual selection as noisy reliable signaling. *Evolution* 53: 299–302.

Giglioli, M. E. C. 1963. The female reproductive system of *Anopheles gambiae melas*. Part I. *Riv. Malar.* 42: 149–176.

Giglioli, M. E. C., and G. F. Mason. 1966. The mating plug in anopheline mosquitoes. *Proc. Roy. Entomol. Soc. Lond. A* 41: 123–129.

Gilbert, D. G. 1981. Ejaculate esterase 6 and initial sperm use by female *Drosophila melanogaster*. *J. Insect Physiol.* 27: 641–650.

Gilbert, D. G., R. C. Richmond, and K. B. Sheehan. 1981. Studies of esterase 6 in *Drosophila melanogaster*. V. Progeny production and sperm use in females inseminated by males having active or null alleles. *Evolution* 35: 21–37.

Gilbert, L. E. 1976. Postmating female odour in *Heliconius* butterflies: A male-contributed antiaphrodisiac? *Science* 193: 419–420.

Gilbert, L. I., and H. A. Schneiderman. 1961. The content of juvenile hormone and lipid in Lepidoptera: Sexual differences and developmental changes. *Gen. Comp. Endocr.* 1: 453–472.

Gilburn, A. S., S. P. Foster, and T. H. Day. 1993. Genetic correlation between a female mating preference and the preferred male character in seaweed flies (*Coelopa frigida*). *Evolution* 47: 1788–1795.

Gilchrist, A. S., and L. Partridge. 1997. Heritability of pre-adult viability differences can explain apparent heritability of sperm displacement ability in *Drosophila melanogaster*. *Proc. Roy. Soc. Lond. B* 264: 1271–1275.

Gilchrist, A. S., and L. Partridge. 1995. Male identity and sperm displacement in *Drosophila melanogaster*. *J. Insect Physiol.* 41: 1087–1092.

Gilchrist, A. S., and L. Partridge. 2000. Why it is difficult to model sperm displacement in *Drosophila melanogaster*: The relation between sperm transfer and copula duration. *Evolution* 54: 534–542.

Gillies, M. T. 1956. A new character for the recognition of nulliparous females of *Anopheles gambiae*. *Bull. Wld. Hlth. Org*. 15: 451–459.

Gillott, C. 1996. Male insect accessory glands: Functions and control of secretory activity. *Invert. Reprod. Dev.* 30: 199–205.

Ginsberg, J. R., and D. J. Rubenstein. 1990. Sperm competition and variation in zebra mating behavior. *Behav. Ecol. Sociobiol.* 26: 427–434.

Giray, T., E. Guzmán-Novoa, C. W. Aron, B. Zelinski, S. E. Fahrbach, and G. E. Robinson. 2000. Genetic variation in worker temporal polyethism and colony defensiveness in the honey bee, *Apis mellifera*. *Behav. Ecol.* 11: 44–55.

Godfray, H. C. J. 1994. *Parasitoids: Behavioral and Evolutionary Ecology*. Princeton University Press, Princeton, NJ.

Gomendio, M., and E. R. S. Roldan. 1991. Sperm competition influences sperm size in mammals. *Proc. Roy. Soc. Lond. B* 243: 181–185.

Gomendio, M., and E. R. S. Roldan. 1993. Mechanisms of sperm competition: Linking physiology and behavioural ecology. *TREE* 8: 95–100.

Gomendio, M., A. H. Harcourt, and E. R. S. Roldán. 1998. Sperm competition in mammals. In T. R. Birkhead and A. P. Møller, eds., *Sperm Competition and Sexual Selection*, pp. 667–756. Academic Press, London.

Goodnight, K. F., J. E. Strassmann, C. J. Klingler, and D. C. Queller. 1996. Single mating and its implications for kinship structure in a multiple-queen wasp, *Parachartegus colobopterus*. *Ethol. Ecol. Evol.* 8: 191–198.

Goulson, D. 1993. Variation in the genitalia of the butterfly *Maniola jurtina* (Lepidoptera: Satyrinae). *Zool. J. Linn. Soc.* 107: 65–71.

Grafen, A. 1989. The phylogenetic regression. *Phil. Trans. Roy. Soc. B* 326: 119–156.

Grafen, A. 1990a. Biological signals as handicaps. *J. Theor. Biol.* 144: 517–546.

Grafen, A. 1990b. Sexual selection unhandicapped by the Fisher process. *J. Theor. Biol.* 144: 473–516.

Grafen, A., and M. Ridley. 1983. A model of mate guarding. *J. Theor. Biol.* 102: 549–567.

Graves, J., J. Ortega-Ruano, and P. J. B. Slater. 1993. Extra-pair copulations and paternity in shags: Do females choose better males? *Proc. Roy. Soc. Lond. B* 253: 3–7.

Greef, J. M., and G. A. Parker. 2000. Spermicide by females: What should males do? *Proc. Roy. Soc. Lond. B* 267: 1759–1763.

Green, A. J. 1999. Allometry of genitalia in insects and spiders: One size does not fit all. *Evolution* 53: 1621–1624.

Gregory, G. E. 1965. The formation and fate of the spermatophore in the African migratory locust, *Locusta migratoria migratorioides* Reiche and Fairmaire. *Trans. Roy. Entomol. Soc. Lond.* 117: 33–66.

Gregory, P. G., and D. J. Howard. 1994. A postinsemination barrier to fertilization isolates two closely related ground crickets. *Evolution* 48: 705–710.

Gregory, P. G., and D. J. Howard. 1996. Multiple mating in natural populations of ground crickets. *Entomol. Exp. et Appl.* 78: 353–356.

Gromko, M. H. 1992. Genetic correlation of male and female mating frequency: Evidence from *Drosophila melanogaster*. *Anim. Behav.* 43: 176–177.

Gromko, M. H., A. Briot, S. C. Jensen, and H. H. Fukui. 1991. Selection on copulation duration in *Drosophila melanogaster*: Predictability of direct response versus unpredictability of correlated response. *Evolution* 45: 69–81.

Gromko, M. H., D. G. Gilbert, and R. C. Richmond. 1984a. Sperm transfer and use in the multiple mating system of *Drosophila*. In R. L. Smith, ed., *Sperm Competition and the Evolution of Animal Mating Syustems*, pp. 371–426. Academic Press, London.

Gromko, M. H., M. A. E. Newport, and M. G. Kortier. 1984b. Sperm dependence of female receptivity to remating in *Drosophila melanogaster*. *Evolution* 38: 1273–1282.

Gross, M. R. 1996. Alternative reproductive strategies and tactics: Diversity within sexes. *TREE* 11: 92–98.

Gunnarsson, B., and A. Andersson. 1996. Sex ratio variation in sheet-web spiders: Options for female control. *Proc. Roy. Soc. Lond. B* 263: 1177–1182.

Gwynne, D. T. 1984a. Courtship feeding increases female reproductive success in bushcrickets. *Nature* 307: 361–363.

Gwynne, D. T. 1984b. Male mating effort, confidence of paternity, and insect sperm competition. In R. L. Smith, ed., *Sperm Competition and the Evolution of Animal Mating Systems*, pp. 117–149. Academic Press, London.

Gwynne, D. T. 1986. Courtship feeding in katydids (Orthoptera: Tettigoniidae): Investment in offspring or in obtaining fertilizations? *Am. Nat.* 128: 342–352.

Gwynne, D. T. 1988a. Courtship feeding and the fitness of female katydids (Orthoptera: Tettigoniidae). *Evolution* 42: 545–555.

Gwynne, D. T. 1988b. Courtship feeding in katydids benefits the mating male's offspring. *Behav. Ecol. Sociobiol.* 23: 373–377.

Gwynne, D. T. 1990. Testing parental investment and the control of sexual selection in katydids: The operational sex ratio. *Am. Nat.* 136: 474–484.

Gwynne, D. T. 1997. The evolution of edible sperm sacs and other forms of courtship feeding in crickets, katydids and their kin (Orthoptera: Ensifera). In J. C. Choe and B. J. Crespi, eds., *The Evolution of Mating Systems in Insects and Arachnids*, pp. 110–129. Cambridge University Press, Cambridge, UK.

Gwynne, D. T., and W. J. Bailey. 1999. Female-female competition in katydids: Sexual selection for increased sensitivity to a male signal? *Evolution* 53: 546–551.

Gwynne, D. T., and L. W. Simmons. 1990. Experimental reversal of courtship roles in an insect. *Nature* 346: 172–174.

Gwynne, D. T., and A. W. Snedden. 1995. Paternity and female remating in *Requena verticalis* (Orthoptera: Tettigoniidae). *Ecol. Entomol.* 20: 191–194.

Haberl, M., and D. Tautz. 1998. Sperm usage in honey bees. *Behav. Ecol. Sociobiol.* 42: 247–256.

Hadrys, H., B. Schierwater, S. L. Dellaporta, R. Desalle, and L. W. Buss. 1993. Determination of paternity in dragonflies by random amplified polymorphic DNA fingerprints. *Mol. Ecol.* 2: 79–87.

Hafernik, J. E. J., and R. W. Garrison. 1986. Mating success and survival rate in a population of damselflies: Results at variance with theory? *Am. Nat.* 128: 353–365.

Hagan, D. V., and U. E. Brady. 1981. Effects of male photoperiod on calling, pheromone levels, and oviposition of mated female *Trichoplusia ni*. *Ann. Entomol. Soc. Am.* 74: 286–288.

Haig, D., and C. T. Bergstrom. 1995. Multiple mating, sperm competition and meiotic drive. *J. Evol. Biol.* 8: 265–282.

Halliday, T., and S. J. Arnold. 1987. Multiple mating by females: A perspective from quantitative genetics. *Anim. Behav.* 35: 939–941.

Hamilton, W. D. 1964a. The genetical evolution of social behaviour. I. *J. Theor. Biol.* 7: 1–16.

Hamilton, W. D. 1964b. The genetical evolution of social behaviour. II. *J. Theor. Biol.* 7: 17–52.

Hamilton, W. D. 1967. Extraordinary sex ratios. *Science* 156: 477–488.

Hamilton, W. D. 1972. Altruism and related phenomena, mainly in social insects. *Ann. Rev. Ecol. Syst.* 3: 193–232.

Hamilton, W. D. 1987. Kinship, recognition, disease, and intelligence: Constraints on social evolution. In Y. Itô, J. L. Brown, and J. Kikkawa, eds., *Animal Societies*, pp. 81–102. Japan Scientific Societies Press, Tokyo.

Hamilton, W. D., and M. Zuk. 1982. Heritable true fitness and bright birds: A role for parasites? *Science* 218: 384–387.

Happ, G. M. 1969. Multiple sex pheromones of the mealworm beetle *Tenebrio molitor*. *Nature* 222: 180–181.

Harcourt, A. H., P. H. Harvey, S. G. Larson, and R. V. Short. 1981. Testis weight, body weight and breeding system in primates. *Nature* 293: 55–57.

Harshman, L. G., and A. G. Clark. 1998. Inference of sperm competition from broods of field-caught *Drosophila*. *Evolution* 52: 1334–1341.

Harshman, L. G., and T. Prout. 1994. Sperm displacement without sperm transfer in *Drosophila melanogaster*. *Evolution* 48: 758–766.

Hartmann, R., and W. Loher. 1996. Control mechanisms of the behavior 'secondary defense' in the grasshopper *Gomphocerus rufus* L. (Gomphocerinae: Orthoptera). *J. Comp. Physiol. A* 178: 329–336.

Harvey, I. F., and S. F. Hubbard. 1987. Observations on the reproductive behaviour of *Orthemis ferruginea* (Fabricius) (Anisoptera: Libellulidae). *Odonatologica* 16: 1–8.

Harvey, I. F., and G. A. Parker. 2000. "Sloppy" sperm mixing and intraspecific variation in sperm precedence (P_2) patterns. *Proc. Roy. Soc. Lond. B* 267: 2537–2542.

Harvey, P. H., and R. M. May. 1989. Out for the sperm count. *Nature* 337: 508–509.

Harwalker, M. R., and G. W. Rahalkar. 1973. Sperm utilization in the female red cotton bug. *J. Econ. Entomol.* 66: 805–806.

Hasegawa, E. 1994. Sex allocation in the ant *Colobopsis nipponicus* (Wheeler). I. Population sex ratio. *Evolution* 48: 1121–1129.

Hassan, A. T. 1978. Reproductive behaviour of *Acisoma panorpoides inflatum* Selys (Anisoptera: Libellulidae). *Odonatologica* 7: 237–245.

Hassan, A. T. 1981. Coupling and oviposition behaviour in two macrodiplacinid libellulids—*Aethriamantha rezia* (Kirby) and *Urothemis assignata* Selys (Libellulidae: Odonata). *Zool. J. Linn. Soc.* 1981: 289–296.

Haubruge, E., L. Arnaud, J. Mignon, and M. J. G. Gage. 1999. Fertilization by proxy: Rival sperm removal and translocation in a beetle. *Proc. Roy. Soc. Lond. B* 266: 1183–1187.

Hauschteck-Jungen, E., and G. Rutz. 1983. Arginine-ich nucleoprotein transition occurs in the two size classes of spermatozoa of *Drosophila subobscura* males. *Genetica* 62: 25–32.

Hayashi, F. 1992. Large spermatophore production and consumption in dobsonflies *Protohermes* (Megaloptera, Corydalidae). *Jap. J. Entomol.* 60: 59–66.

Hayashi, F. 1993. Male mating costs in two insect species (*Protohermes*, Megaloptera) that produce large spermatophores. *Anim. Behav.* 45: 343–349.

Hayashi, F. 1996. Insemination through an externally attached spermatophore: Bundled sperm and post-copulatory mate guarding by male fishflies (Megaloptera: Corydalidae). *J. Insect Physiol.* 42: 859–866.

Hayashi, F. 1998. Sperm co-operation in the fishfly, *Parachauliodes japonicus*. *Funct. Ecol.* 12: 347–350.

Hayashi, F. 1999a. Ejaculate production schedule and the degree of protandry in fishflies (Megaloptera: Corydalidae). *Funct. Ecol.* 13: 178–189.

Hayashi, F. 1999b. Rapid evacuation of spermatophore contents and male post-mating behaviour in Alderflies (Megaloptera: Sialidae). *Entomol. Sci.* 2: 49–56.

Hayashi, K. 1985. Alternative mating strategies in the water strider *Gerris elongatus* (Heteroptera, Gerridae). *Behav. Ecol. Sociobiol.* 16: 301–306.

He, Y., and T. Miyata. 1997. Variations in sperm number in relation to larval crowding and spermatophore size in the armyworm, *Pseudaletia separata*. *Ecol. Entomol.* 22: 41–46.

He, Y., and Y. Tsubaki. 1991. Effects of spermatophore size on female remating in the armyworm, *Pseudaletia separata*, with special reference to larval crowding. *J. Ethol.* 9: 47–50.

He, Y., and Y. Tsubaki. 1992. Variation in spermatophore size in the armyworm, *Pseudaletia separata*, (Lepidoptera: Noctuidae) in relation to rearing density. *Appl. Entomol. Zool.* 27: 39–45.

He, Y., T. Tanaka, and T. Miyata. 1995a. Eupyrene and apyrene sperm and their numerical fluctuations inside the female reproductive tract of the armyworm, *Pseudaletia separata*. *J. Insect Physiol.* 41: 689–694.

He, Y. B., Y. Tsubaki, K. Itou, and T. Miyata. 1995b. Gamma radiation effects on reproductive potential and sperm use patterns in *Pseudaletia separata* (Lepidoptera: Noctuidae). *J. Econ. Entomol.* 88: 1626–1630.

Heady, S. E. 1993. Factors affecting female sexual receptivity in the planthopper, *Prokelisia dolus*. *Physiol. Entomol.* 18: 263–270.

Heinze, J., and B. Hölldobler. 1993. Fighting for a harem of queens: Physiology of reproduction in *Cardiocondyla* male ants. *Proc. Natl. Acad. Sci. USA* 90: 8412–8414.

Heinze, J., B. Hölldobler, and S. Trenkle. 1995. Reproductive behavior in the ant *Leptothorax (Dichothorax) pergandei*. *Insect. Soc.* 42: 309–315.

Heinze, J., B. Hölldobler, and K. Yamauchi. 1998. Male competition in *Cardiocondyla* ants. *Behav. Ecol. Sociobiol.* 42: 239–246.

Heller, K.-G., and D. v. Helverson. 1991. Operational sex ratio and individual mating frequencies in two bushcricket species (Orthoptera, Tettigoniidae, *Poecilimon*). *Ethology* 89: 211–228.

Heller, K.-G., S. Faltin, P. Fleischmann, and O. v. Helversen. 1998. The chemical composition of the spermatophore in some species of phaneropterid bushcrickets (Orthoptera: Tettigonioidea). *J. Insect Physiol.* 44: 1001–1008.

Hellriegel, B., and G. Bernasconi. 2000. Female-mediated differential sperm storage in a fly with complex spermathecae, *Scathophaga stercoraria*. *Anim. Behav.* 59: 311–317.

Hellriegel, B., and P. I. Ward. 1998. Complex female reproductive tract morphology: Its possible use in postcopulatory female choice. *J. Theor. Biol.* 190: 179–186.

Helms, K. R. 1999. Colony sex ratios, conflict between queens and workers, and apparent queen control in the ant *Pheidole desertorum*. *Evolution* 53: 1470–1478.

Helms, K. R., J. H. Fewell, and S. W. Rissing. 2000. Sex ratio determination by queens and workers in the ant *Pheidole desertorum*. *Anim. Behav.* 59: 523–527.

Heming-Van Battum, K. E., and B. S. Heming. 1986. Structure, function and evolution of the reproductive system in females of *Hebrus pusillus* and *H. ruficeps* (Hemiptera, Gerromorpha, Hebridae). *J. Morph.* 190: 121–167.

Herbers, J. M., and R. L. Mouser. 1998. Microsatellite DNA markers reveal details of social structure in forest ants. *Mol. Ecol.* 7: 299–306.

Herman, W. S., and P. Peng. 1976. Juvenile hormone stimulation of sperm activator production in male monarch butterflies. *J. Insect Physiol.* 22: 579–581.

Herndon, L. A., and M. F. Wolfner. 1995. A *Drosophila* seminal fluid protein, Acp26Aa, stimulates egg laying in females for 1 day after mating. *Proc. Natl. Acad. Sci. USA* 92: 10114–10118.

Hewitt, G. M., P. Mason, and R. A. Nichols. 1989. Sperm precedence and homogamy across a hybrid zone in the alpine grasshopper *Podisma pedestris*. *Heredity* 62: 343–354.

Hinton, H. E. 1964. Sperm transfer in insects and the evolution of haemocoelic insemination. In K. C. Highnam, ed., *Insect Reproduction*, pp. 95–107. Symp. Roy. Entomol. Soc. Lond., no. 2.

Hockman, L. R., and K. Vahed. 1997. The function of mate guarding in a field cricket (Orthoptera: Gryllidae; *Teleogryllus natalensis* Otte and Cade). *J. Insect Behav.* 10: 247–256.

Holland, B., and W. R. Rice. 1998. Chase-away sexual selection: Antagonistic seduction versus resistance. *Evolution* 52: 1–7.

Holland, B., and W. R. Rice. 1999. Experimental removal of sexual selection reverses intersexual antagonistic coevolution and removes a reproductive load. *Proc. Natl. Acad. Sci. USA* 96: 5083–5088.

Holland, G. P. 1955. Primary and secondary sexual characteristics of some Ceratophyllinae, with notes on the mechanism of copulation. *Trans. Roy. Entomol. Soc. Lond.* 107: 233–248.

Hölldobler, B., and E. O. Wilson. 1990. *The Ants*. Belknap/Harvard University Press, Cambridge, MA.

Holmes, H. B. 1974. Patterns of sperm competition in *Nasonia vitripennis*. *Can. J. Genet. Cytol.* 16: 789–795.

Holt, G. G., and D. T. North. 1970. Effects of gamma irradiation on the mechanism of sperm transfer in *Trichoplusia ni*. *J. Insect Physiol.* 16: 2211–2222.

Hooper, R. E., and M. T. Siva-Jothy. 1996. Last male sperm precedence in a damselfly demonstrated by RAPD profiling. *Mol. Ecol.* 5: 449–452.

Hosken, D. J. 1997. Sperm competition in bats. *Proc. Roy. Soc. Lond. B* 264: 385–392.

Hosken, D. J. 1998. Testes mass in megachiropteran bats varies in accordance with sperm competition theory. *Behav. Ecol. Sociobiol.* 44: 169–177.

Hosken, D. J., and P. I. Ward. 2000. Copula in yellow dung flies (*Scathophaga stercoraria*): Investigating sperm competition models by direct observation. *J. Insect Physiol.* 46: 1355–1363.

Hosken, D. J., and P. I. Ward. 2001. Experimental evidence for testis size evolution via sperm competition. *Ecol. Let.* 4: 10–13.

Hosken, D. J., E. P. Meyer, and P. I. Ward. 1999. Internal female reproductive anatomy and genital interactions during copula in the yellow dung fly, *Scathophaga stercoraria* (Diptera: Scathophagidae). *Can. J. Zool.* 77: 1975–1983.

Houde, A. E. 1994. Effect of artificial selection on male patterns on mating preference of female guppies. *Proc. Roy. Soc. Lond. B* 256: 125–130.

Houde, A. E., and J. A. Endler. 1990. Correlated evolution of female mating preferences and male color patterns in the guppy *Poecilia reticulata*. *Science* 248: 1405–1408.

Howard, D. J. 1999. Conspecific sperm and pollen precedence and speciation. *Ann. Rev. Ecol. Syst.* 30: 109–132.

Howard, D. J., and P. G. Gregory. 1993. Post-insemination signalling systems and reinforcement. *Phil. Trans. Roy. Soc. Lond. B* 340: 231–236.

Howard, D. J., P. G. Gregory, J. Chu, and M. L. Cain. 1998. Conspecific sperm precedence is an effective barrier to hybridization between closely related species. *Evolution* 52: 511–516.

Huettel, M. D., C. O. Calkins, and A. J. Hill. 1972. Allozyme markers in the study of sperm precedence in the plum curculio, *Conotrachelus nenuphar*. *Ann. Entomol. Soc. Am.* 69: 465–468.

Hughes, A. L. 1981. Differential male mating success in the whitespotted sawyer, *Monochamus scutellatus* (Coleoptera: Cerambycidae). *Ann. Entomol. Soc. Am.* 74: 180–184.

Hughes, A. L., and M. K. Hughes. 1982. Male size, mating success, and breeding habit partitioning in the whitespotted sawyer, *Monochamus scutellatus* (Say) (Coleoptera: Cerambycidae). *Oecologia* 55: 258–263.

Hughes, A. L., and M. K. Hughes. 1985. Female choice of mates in a polygynous insect, the whitespotted sawyer, *Monachamus scutellatus*. *Behav. Ecol. Sociobiol.* 17: 385–388.

Hughes, K. A. 1997. Quantitative genetics of sperm precedence in *Drosophila melanogaster. Genetics* 145: 139–151.

Hughes, L., B. Sie-Woon Chang, D. Wagner, and N. E. Pierce. 2000. Effects of mating history on ejaculate size, fecundity, longevity, and copulation duration in the ant-tended lycaenid butterfly, *Jalmenus evagoras*. *Behav. Ecol. Sociobiol.* 47: 119–128.

Huignard, J. 1983. Transfer and fate of male secretions deposited in the spermatophore of females of *Acanthoscelides obtectus* Say (Coleoptera Bruchidae). *J. Insect Physiol.* 29: 55–63.

Hunt, J., and L. W. Simmons. 1997. Patterns of fluctuating asymmetry in beetle horns: An experimental examination of the honest signalling hypothesis. *Behav. Ecol. Sociobiol.* 41: 109–114.

Hunt, J., and L. W. Simmons. 1998. Patterns of parental provisioning covary with male morphology in a horned beetle (*Onthophagus taurus*) (Coleoptera: Scarabaeidae). *Behav. Ecol. Sociobiol.* 42: 447–451.

Hunter-Jones, P. 1960. Fertilization of eggs of the desert locust by spermatozoa from successive copulations. *Nature* 185: 336.

Hurst, L. D. 1990. Parasite diversity and the evolution of diploidy, multicellularity and anisogamy. *J. Theor. Biol.* 144: 429–443.

Hurst, L. D., and W. D. Hamilton. 1992. Cytoplasmic fusion and the nature of sexes. *Proc. Roy. Soc. Lond. B* 247: 189–194.

Huxley, J. S. 1938. Darwin's theory of sexual selection and the data subsumed by it, in the light of recent research. *Am. Nat.* 72: 416–433.

Imamura, M., K. Hainofukushima, T. Aigaki, and Y. Fuyama. 1998. Ovulation stimulating substances in *Drosophila biarmipes* males—their origin, genetic variation in the response of females, and molecular characterization. *Insect Biochem. Mol. Biol.* 28: 365–372.

Imhof, M., B. Harr, G. Brem, and C. Schlotterer. 1998. Multiple mating in wild *Drosophila melanogaster* revisited by microsatellite analysis. *Mol. Ecol.* 7: 915–917.

Iriki, S. 1941. The two sperm types in the silkworm and their functions. *Zool. Mag. (Tokyo)* 53: 123–124.

Ivy, T. M., J. C. Johnson, and S. K. Sakaluk. 1999. Hydration benefits to courtship feeding in crickets. *Proc. Roy. Soc. Lond. B* 266: 1523–1527.

Iwasa, Y., F. J. Odendaal, D. D. Murphy, P. R. Ehrlich, and A. E. Launer. 1983. Emergence patterns in male butterflies: A hypothesis and a test. *Theor. Pop. Biol.* 23: 363–379.

Iwasa, Y., A. Pomiankowski, and S. Nee. 1991. The evolution of costly mate preferences. II. The "handicap" principle. *Evolution* 45: 1431–1442.

Jabłoński, P., and S. Kaczanowski. 1994. Influence of mate guarding duration on male reproductive success: An experiment with irradiated water strider (*Gerris lacustris*) males. *Ethology* 98: 312–320.

Jabłoński, P., and K. Vepsäläinen. 1995. Conflict between sexes in the water strider, *Gerris lacustris*: A test of two hypotheses for male guarding behavior. *Behav. Ecol.* 6: 388–392.

Jacobs, M. E. 1955. Studies on territorialism and sexual selection in dragonflies. *Ecology* 36: 566–586.

James, H. C. 1937. The effect of delayed fertilization on the sex ratio of a species of insect in which the female is the heterogametic sex (*Ephestia kühniella* Zell. Lepid., Phycitidae). *Proc. Roy. Entomol. Soc. Lond. A* 12: 92–98.

Jamieson, B. G. M. 1987. *The Ultrastructure and Phylogeny of Insect Spermatozoa*. Cambridge University Press, Cambridge, UK.

Janz, N., and S. Nylin. 1998. Butterflies and plants: A phylogenetic study. *Evolution* 52: 486–502.

Johnson, C. 1962. Breeding behavior and oviposition in *Calopteryx maculatum* (Beauvais) (Odonata: Calopterygidae). *Am. Midl. Nat.* 68: 242–247.

Johnson, J. A., and M. Niedzlek-Feaver. 1998. A histological study of copulation duration, patterns of sperm transfer and organization inside the spermatheca of a grasshopper, *Dichromorpha viridis* (Scudder). *J. Orthop. Res.* 7: 139–146.

Johnson, J. C., T. M. Ivy, and S. K. Sakaluk. 1999. Female remating propensity contingent on sexual cannibalism in sagebrush crickets, *Cyphoderris strepitans*: A mechanism of cryptic female choice. *Behav. Ecol.* 10: 227–233.

Johnson, L. K. 1982. Sexual selection in a brentid weevil. *Evolution* 36: 251–262.

Johnson, L. K., and S. P. Hubbell. 1984. Male choice: Experimental demonstration in a brentid weevil. *Behav. Ecol. Sociobiol.* 15: 183–188.

Johnstone, R. A. 1995. Sexual selection, honest advertisement and the handicap principle: Reviewing the evidence. *Biol. Rev.* 70: 1–65.

Johnstone, R. A., J. D. Reynolds, and J. C. Deutsch. 1996. Mutual mate choice and sex differences in choosiness. *Evolution* 50: 1382–1391.

Joly, D., and D. Lachaise. 1994. Polymorphism in the sperm heteromorphic species of the *Drosophila obscura* group. *J. Insect Physiol.* 40: 933–938.

Joly, D., C. Bazin, L.-W. Zeng, and R. S. Singh. 1995. Genetic basis of sperm and testis length differences and epistatic effect on hybrid inviability and sperm motility between *Drosophila simulans* and *D. sechellia*. *Heredity* 78: 354–362.

Joly, D., C. Bressac, J. Devaux, and D. Lachaise. 1991a. Sperm length diversity in Drosophilidae. *Dros. Inf. Svc.* 70: 104–108.

Joly, D., M. L. Cariou, and D. Lachaise. 1991b. Can sperm competition explain sperm polymorphism in *Drosophila teissieri*? *Evolucion Biologica* 5: 25–44.

Jones, J. S. 1987. The heritability of fitness: Bad news for "good genes"? *TREE* 2: 35–38.

Kaitala, A., and M. Miettinen. 1997. Female egg dumping and the effect of sex ratio on male egg carrying in a coreid bug. *Behav. Ecol.* 8: 429–432.

Kaitala, A., and C. Wiklund. 1994. Polyandrous female butterflies forage for matings. *Behav. Ecol. Sociobiol.* 35: 385–388.

Kaitala, A., and C. Wiklund. 1995. Female mate choice and mating costs in the polyandrous butterfly *Pieris napi* (Lepidoptera: Pieridae). *J. Insect Behav.* 8: 355–363.

Kalb, J. M., A. J. D. DiBenedetto, and M. F. Wolfner. 1993. Probing the function of *Drosophila* accessory glands by directed cell ablation. *Proc. Natl. Acad. Sci. USA* 90: 8093–8097.

Kappeler, P. M. 1997. Intrasexual selection and testis size in strepsirhine primates. *Behav. Ecol.* 8: 10–19.

Karlsson, B. 1995. Resource allocation and mating systems in butterflies. *Evolution* 49: 955–961.

Karlsson, B. 1996. Male reproductive reserves in relation to mating system in butterflies: A comparative study. *Proc. Roy. Soc. Lond. B* 263: 187–192.

Karlsson, B., O. Leimar, and C. Wiklund. 1997. Unpredictable environments, nuptial gifts and the evolution of sexual size dimorphism in insects: An experiment. *Proc. Roy. Soc. Lond. B* 264: 475–479.

Karpenko, C. P., and D. T. North. 1973. Ovipositional response elicited by normal, irradiated, F1 male progeny, or castrated male *Trichoplusia ni* (Lepidoptera: Noctuidae). *Ann. Entomol. Soc. Am.* 66: 1278–1280.

Karr, T. L. 1991. Intracellular sperm/egg interactions in *Drosophila*: A three-dimensional structural analysis of a paternal product in the developing egg. *Mech. Dev.* 34: 101–112.

Karr, T. L., and S. Pitnick. 1996. The ins and outs of fertilization. *Nature* 379: 405–406.

Katakura, H. 1986. Evidence for the incapacitation of heterospecific sperm in the female genital tract in a pair of closely related ladybirds (Insecta, Coleoptera, Coccinellidae). *Zool. Sci.* 3: 115–121.

Katsuno, S. 1977a. Studies on eupyrene and apyrene spermatozoa in the silkworm, *Bombyx mori* L. (Lepidoptera: Bombycidae). V. The factor related to the separation of eupyrene sperm bundles. *Appl. Entomol. Zool.* 12: 370–371.

Katsuno, S. 1977b. Studies on eupyrene and apyrene spermatozoa in the silkworm, *Bombyx mori* L. (Lepidoptera: Bombycidae). I. The intratesticular behaviour of the spermatozoa at various stages from the 5th instar to the adult. *Appl. Entomol. Zool.* 12: 142–153.

Katsuno, S. 1978. Studies on eupyrene and apyrene spermatozoa in the silkworm, *Bombyx mori* L. (Lepidoptera: Bombycidae). VII. The motility of sperm bundles and spermatozoa in the reproductive organs of males and females. *Appl. Entomol. Zool.* 13: 91–96.

Kaufmann, T. 1996. Dynamics of sperm transfer, mixing and fertilization in *Cryptolaemus montrouzieri* (Coleoptera: Coccinellidae) in Kenya. *Ann. Entomol. Soc. Am.* 89: 238–242.

Keller, L., and L. Passera. 1992. Mating system, optimal number of matings, and sperm transfer in the Argentine ant *Iridomyrmex humilis*. *Behav. Ecol. Sociobiol.* 31: 359–366.

Keller, L., and H. K. Reeve. 1994. Genetic variability, queen number, and polyandry in social hymenoptera. *Evolution* 48: 694–704.

Keller, L., and H. K. Reeve. 1995. Why do females mate with multiple males? The sexually selected sperm hypothesis. *Adv. Stud. Behav.* 24: 291–315.

Keller, L., L. Sundström, and M. Chapuisat. 1997. Male reproductive success: Paternity contribution to queens and workers in *Formica* ants. *Behav. Ecol. Sociobiol.* 41: 11–15.

Kempenaers, B., and B. C. Sheldon. 1997. Studying paternity and paternal care: Pitfalls and problems. *Anim. Behav.* 53: 423–427.

Kempenaers, B., K. Foerster, S. Questiau, B. C. Robertson, and E. L. M. Vermeirssen. 2000. Distinguishing between female sperm choice versus male sperm competition: A comment on Birkhead (1998). *Evolution* 54: 1050–1052.

Kempenaers, B., G. R. Verheyen, M. Van den Broeck, T. Burke, C. Van Broekhoven, and A. A. Dhondt. 1992. Extra-pair paternity results from female preference for high-quality males in the blue tit. *Nature* 357: 494–496.

Kent, D. S., and J. A. Simpson. 1992. Eusociality in the beetle *Austroplatypus incompertus* (Coleoptera: Curculionidae). *Naturwissenschaften* 79: 86–87.

Khalifa, A. 1949. The mechanism of insemination and the mode of action of the spermatophore in *Gryllus domesticus*. *Q. J. Micr. Sci.* 90: 281–292.

Khalifa, A. 1950. Sexual behaviour in *Gryllus domesticus* L. *Behaviour* 2: 264–274.

King, B. H. 1996. Fitness effects of sex ratio response to host quality and size in the parasitoid wasp *Spalangia cameroni*. *Behav. Ecol.* 7: 35–42.

Kingan, T. G., W. M. Bodnar, A. K. Raina, J. Shabanowitz, and D. F. Hunt. 1995. The loss of female sex pheromone after mating in the corn earworm moth *Helicoverpa zea*: identification of a male pheromonostatic peptide. *Proc. Natl. Acad. Sci. USA* 92: 5082–5086.

Kingan, T. G., P. A. Thomaslaemont, and A. K. Raina. 1993. Male accessory gland factors elicit change from virgin to mated behaviour in the female corn earworm moth *Helicoverpa zea*. *J. Exp. Biol.* 183: 61–76.

Kirkendall, L. R. 1984. Long copulations and post-copulatory "escort" behaviour in the locust leaf miner, *Odontota dorsalis* (Coleoptera: Chrysomelidae). *J. Nat. Hist.* 18: 905–919.

Kirkpatrick, M. 1982. Sexual selection and the evolution of female choice. *Evolution* 36: 1–12.

Kirkpatrick, M., and M. J. Ryan. 1991. The evolution of mating preferences and the paradox of the lek. *Nature* 350: 33–38.

Klowden, M. J., and G. M. Chambers. 1991. Male accessory gland substances activate egg development in nutritionally stressed *Aedes aegypti* mosquitoes. *J. Insect Physiol.* 37: 721–726.

Knowlton, N., and S. R. Greenwell. 1984. Male sperm competition avoidance mechanisms: The influence of female interests. In R. L. Smith, ed., *Sperm Competition and the Evolution of Animal Mating Systems*, pp. 62–85. Academic Press, London.

Koenig, W. D. 1991. Levels of female choice in the white-tailed skimmer *Plathemis lydia* (Odonata: Libellulidae). *Behaviour* 119: 193–224.

Koeniger, G. 1990. The role of the mating sign in honey bees, *Apis mellifera* L.: Does it hinder or promote multiple mating? *Anim. Behav.* 39: 444–449.

Koga, T., and K. Hayashi. 1993. Territorial behavior of both sexes in the water strider *Metrocoris histrio* (Hemiptera: Gerridae) during the mating season. *J. Insect Behav.* 6: 65–77.

Komdeur, J., S. Daan, J. Tinbergen, and C. Mateman. 1997. Extreme adaptive modification in sex ratio of the Seychelles warbler's eggs. *Nature* 385: 522–525.

Kotrba, M. 1990. Sperm transfer by spermatophore in an acalyptrate fly (Diptera: Diopsidae). *Entomol. Gener.* 15: 181–183.

Kotrba, M. 1996. Sperm transfer by spermatophore in Diptera: New results from the Diopsidae. *Zool. J. Linn. Soc.* 117: 305–323.

Kraus, B., and R. C. Lederhouse. 1983. Contact guarding during courtship in the tiger beetle *Cicindela martha* Dow (Coleoptera: Cicindelidae). *Am. Midl. Nat.* 110: 208–211.

Kraus, B., and R. E. Page. 1998. Parasites, pathogens, and polyandry in social insects. *Am. Nat.* 151: 383–391.

Krebs, J. R., and N. B. Davies. 1993. *An Introduction to Behavioural Ecology*. Blackwell Scientific, Oxford.

Krebs, R. A. 1991. Function and genetics of long versus short copulations in the cactophilic fruit fly, *Drosophila mojavensis* (Diptera: Drosophilidae). *J. Insect Behav.* 4: 221–233.

Krieger, M. J. B., and L. Keller. 2000. Mating frequency and genetic structure of the argentine ant *Linepithema humile*. *Mol. Ecol.* 9: 119–126.

Krupa, J. J., and A. Sih. 1993. Experimental studies on water strider mating dynamics: Spatial variation in density and sex ratio. *Behav. Ecol. Sociobiol.* 33: 107–120.

Kubli, E. 1996. The *Drosophila* sex peptide: A peptide pheromone involved in reproduction. *Adv. Dev. Biochem.* 4: 99–128.

Kukuk, P. F., G. C. Eickwort, and B. May. 1987. Multiple maternity and multiple paternity in first generation brood from single foundress colonies of the sweat bee *Dialictus zephyrus* (Hymenoptera: Halictidae). *Insect. Soc.* 34: 131–135.

Kura, T., and Y. Nakashima. 2000. Conditions for the evolution of soldier sperm classes. *Evolution* 54: 72–80.

Labine, P. A. 1964. Population biology of the butterfly *Euphydryas editha*. I. Barriers to multiple inseminations. *Evolution* 18: 335–336.

Labine, P. A. 1966. The population biology of the butterfly, *Euphydryas editha*. IV. Sperm precedence—a preliminary report. *Evolution* 20: 580–586.

LaChance, L. E., F. I. Proshold, and R. L. Ruud. 1978. Pink bollworm: Effects of male irradiation and ejaculate sequence on female ovipositional response and sperm radiosensitivity. *J. Econ. Entomol.* 71: 361–365.

Lachmann, A. D. 1997. Sperm transfer during copulation in five *Coproica* species (Diptera: Sphaeroceridae). *Eur. J. Entomol.* 94: 271–286.

Lackie, A. M. 1986. Transplantation: The limits of recognition. In A. P. Gupta, ed., *Hemocytic and Humoral Immunity in Arthropods*, pp. 191–223. CRC Press, Boca Raton, FL.

Ladd, T. L. 1966. Egg viability and longevity of Japanese beetles treated with tepa, apholate, and metepa. *J. Econ. Entomol.* 59: 422–425.

Ladle, R. J., and E. Foster. 1992. Are giant sperm copulatory plugs? *Acta Oecologica* 13: 635–638.

Laidlow, H. H., and R. E. Page. 1984. Polyandry in honey bees (*Apis mellifera* L.): Sperm utilization and intracolony genetic relationships. *Genetics* 108: 985–997.

LaMunyon, C. W. 1994. Paternity in naturally-occurring *Utetheisa ornatrix* (Lepidoptera: Arctiidae) as estimated using enzyme polymorphism. *Behav. Ecol. Sociobiol.* 34: 403–408.

LaMunyon, C. W. 2000. Sperm storage by females of the polyandrous noctuid moth *Heliothis virescens*. *Anim. Behav*. 59: 395–402.

LaMunyon, C. W., and T. Eisner. 1993. Postcopulatory sexual selection in an arctiid moth (*Utetheisa ornatrix*). *Proc. Natl. Acad. Sci. USA* 90: 4689–4692.

LaMunyon, C. W., and T. Eisner. 1994. Spermatophore mass as determinant of paternity in an arctiid moth (*Utetheisa ornatrix*). *Proc. Natl. Acad. Sci. USA* 91: 7081–7084.

LaMunyon, C. W., and S. Ward. 1998. Larger sperm outcompete smaller sperm in the nematode *Caenorhabditis elegans*. *Proc. Roy. Soc. Lond. B* 265: 1997–2002.

LaMunyon, C. W., and S. Ward. 1999. Evolution of sperm size in nematodes: Sperm competition favours larger sperm. *Proc. Roy. Soc. Lond. B* 266: 263–267.

Lande, R. 1981. Models of speciation by sexual selection on polygenic traits. *Proc. Natl. Acad. Sci. USA* 78: 3721–3725.

Lange, A. B., and B. G. Loughton. 1985. An oviposition-stimulating factor in the male accessory reproductive gland of the locust, *Locusta migratoria*. *Gen. Comp. Endocrinol.* 57: 208–215.

Larsen, O. N., G. Gleffe, and J. Tengo. 1986. Vibration and sound communication in solitary bees *Colletes cunicularius* and wasps *Bembix rostrata*. *Physiol. Entomol.* 11: 287–296.

Lauer, M. J., A. Sih, and J. J. Krupa. 1996. Male density, female density and intersexual conflict in a stream-dwelling insect. *Anim. Behav*. 52: 929–939.

Lawrence, S. 1991. Sexual cannibalism in Praying Mantids. Ph.D. diss., University of Sheffield, Sheffield, U.K.

Leahy, S. M. G. 1973a. Oviposition of *Schistocerca gregaria* (Forskål) (Orthoptera: Acrididae) mated with males unable to transfer spermatophores. *J. Entomol. A* 48: 79–84.

Leahy, S. M. G. 1973b. Oviposition of virgin *Schistocerca gregaria* (Forskål) (Orthoptera: Acrididae) after implant of the male accessory gland complex. *J. Entomol. A* 48: 69–78.

Lederhouse, R. C. 1981. The effect of female mating frequency on egg fertility in the black swallowtail, *Papilio polyxenes asterius* (Papilionidae). *J. Lepid. Soc.* 35: 266–277.

Lee, P. E., and A. Wilkes. 1965. Polymorphic spermatozoa in the hymenopterous wasp *Dahlbominous*. *Science* 147: 1445–1446.

Lefevre, G. J., and U. B. Jonsson. 1962. Sperm transfer, storage, displacement, and utilization in *Drosophila melanogaster*. *Genetics* 47: 1719–1736.

Lehmann, G. U. C., and A. W. Lehmann. 2000. Spermatophore characteristics in bushcrickets vary with parasitism and remating interval. *Behav. Ecol. Sociobiol.* 47: 393–399.

Leimar, O., B. Karlsson, and C. Wicklund. 1994. Unpredictable food and sexual size dimorphism in insects. *Proc. Roy. Soc. Lond. B* 258: 121–125.

Leopold, R. A. 1976. The role of male accessory glands in insect reproduction. *Ann. Rev. Entomol.* 21: 199–221.

Lessells, C. M., and T. R. Birkhead. 1990. Mechanisms of sperm competition in birds: Mathematical models. *Behav. Ecol. Sociobiol.* 27: 325–337.

Letsinger, J. T., and M. H. Gromko. 1985. The role of sperm numbers in sperm competition and female remating in *Drosophila melanogaster*. *Genetica* 66: 195–202.

Leviatan, R., and M. Friedländer. 1997. The eupyrene-apyrene dichotomous spermatogenesis of Lepidoptera. I. The relationship with postembryonic development and the role of the decline in juvenile hormone titer towards pupation. *Dev. Biol.* 68: 515–524.

Levitan, D. R. 1995. Sperm limitation in the sea. *TREE* 10: 228–231.

Levitan, D. R. 1996. Effects of gamete traits on fertilization in the sea and the evolution of sexual dimorphism. *Nature* 382: 153–155.

Lewis, S. M., and J. Iannini. 1995. Fitness consequences of differences in male mating behaviour in relation to female reproductive status in flour beetles. *Anim. Behav.* 50: 1157–1160.

Lewis, S. M., and S. N. Austad. 1990. Sources of intraspecific variation in sperm precedence in red flour beetles. *Am. Nat.* 135: 351–359.

Lewis, S. M., and S. N. Austad. 1994. Sexual selection in flour beetles: The relationship between sperm precedence and male olfactory attractiveness. *Behav. Ecol.* 5: 219–224.

Lewis, S. M., and E. Jutkiewicz. 1998. Sperm precedence and sperm storage in multiply mated red flour beetles. *Behav. Ecol. Sociobiol.* 43: 365–370.

Lewis, T. 1984. *Insect Communication*. Academic Press, London.

Liebherr, J. K. 1992. Phylogeny and revision of the *Platynus degallieri* species group (Coleoptera: Carabidae: Platynini). *Bull. Am. Mus. Nat. Hist.* 214: 1–115.

Liersch, S., and P. Schmid-Hempel. 1997. Genetic variation within social insect colonies reduces parasite load. *Proc. Roy. Soc. Lond. B* 265: 221–225.

Liles, J. N. 1965. Effects of mating or association of the sexes on longevity in *Aedes aegypti* (L.). *Mosquito News* 25: 434–439.

Linley, J. R. 1975. Sperm supply and its utilization in doubly inseminated flies, *Culicoides melleus*. *J. Insect Physiol.* 21: 1785–1788.

Linley, J. R. 1981. Emptying of the spermatophore and spermathecal filling in *Culicoides melleus* (Coq.) (Diptera: Ceratopogonidae). *Can. J. Zool.* 59: 347–356.

Linley, J. R., and M. J. Hinds. 1975a. Sperm loss at copulation in *Culicoides melleus. J. Entomol. A* 50: 37–41.

Linley, J. R., and M. J. Hinds. 1975b. Quantity of the male ejaculate influenced by female unreceptivity in the fly, *Culicoides melleus. J. Insect Physiol.* 21: 281–285.

Linley, J. R., and M. S. Mook. 1975. Behavioural interaction between sexually experienced *Culicoides melleus* (Coquillett) (Diptera Ceratopogonidae). *Behaviour* 54: 7–110.

Lissemore, F. M. 1997. Frass clearing by male pine engraver beetles (*Ips pini*; Scolytidae): Paternal care or paternity assurance? *Behav. Ecol.* 8: 318–325.

Lloyd, J. E. 1979. Mating behavior and natural selection. *Fla. Entomol.* 62: 17–34.

Loher, W. 1979. The influence of prostaglandin E2 on oviposition in *Teleogryllus commodus. Entomol. Exp. Appl.* 25: 107–119.

Loher, W. 1981. The effect of mating on female sexual behavior in *Teleogryllus commodus* Walker. *Behav. Ecol. Sociobiol.* 9: 219–225.

Loher, W., and K. Edson. 1973. The effect of mating on egg production and release in the cricket *Teleogryllus commodus. Entomol. Exp. Appl.* 16: 483–490.

Loher, W., and B. Rence. 1978. The mating behavior of *Teleogryllus commodus* (Walker) and its central and peripheral control. *Z. Tierpsych.* 46: 225–259.

Loher, W., I. Ganjian, I. Kubo, D. Stanley-Samuelson, and S. Tobe. 1981. Prostaglandins: Their role in egg laying of the cricket *Teleogryllus commodus. Proc. Natl. Acad. Sci. USA* 78: 7835–7838.

Loher, W., T. Weber, and F. Huber. 1993. The effect of mating on phonotactic behaviour in *Gryllus bimaculatus* (De Geer). *Physiol. Entomol.* 18: 57–66.

López-León, M. D., J. Cabrero, M. C. Pardo, E. Viseras, and J. P. M. Camacho. 1993. Paternity displacement in the grasshopper *Eyprepocnemis plorans. Heredity* 71: 539–545.

López-León, M. D., M. C. Pardo, J. Cabrero, and J. P. M. Camacho. 1995. Evidence for multiple paternity in two natural populations of the grasshopper *Eyprepocnemis plorans. Hereditas* 123: 89–90.

Lorch, P. D., G. S. Wilkinson, and P. R. Reillo. 1993. Copulation duration and sperm precedence in the stalk-eyed fly *Cyrtodiopsis whitei* (Diptera: Diopsidae). *Behav. Ecol. Sociobiol.* 32: 303–311.

Lum, P. T. 1961. The reproductive system of some Florida mosquitoes. II. The male accessory glands and their roles. *Ann. Entomol. Soc. Am.* 54: 430–433.

Lum, P. T. M. 1977. High temperature inhibition of development of eupyrene sperm and of reproduction in *Plodia interpunctella* and *Ephestia cautella. J. Georgia Entomol. Soc.* 12: 199–203.

Lum, P. T. M., and U. E. Brady. 1973. Levels of pheromone in female *Plodia interpunctella* mating with males reared in different light regimes. *Ann. Entomol. Soc. Am.* 66: 821–823.

Lum, P. T. M., and B. R. Flaherty. 1970. Effect of continuous light on the potency of *Plodia interpunctella* males (Lepidoptera: Phycitidae). *Ann. Entomol. Soc. Am.* 63: 1470–1471.

Machado, M. F. P. S., E. P. B. Contel, and W. E. Kerr. 1984. Proportion of males sons-of-queen and sons-of-workers in *Plebeia droryana* (Hymenoptera, Apidae) estimated from data of an MDH isozymic polymorphic system. *Genetica* 65: 193–198.

Madsen, T., R. Shine, J. Loman, and T. Hakansson. 1992. Why do female adders copulate so frequently? *Nature* 355: 440–441.

Majerus, M. E. N., P. O'Donald, P. W. E. Kearns, and H. Ireland. 1986. Genetics and evolution of female choice. *Nature* 321: 164–167.

Majerus, M. E. N., P. O'Donald, and J. Weir. 1982. Female mating preference is genetic. *Nature* 300: 521–523.

Mane, S. D., L. Tomkins, and R. C. Richmond. 1983. Male esterase 6 catalyzes the synthesis of a sex pheromone in *Drosophila melanogaster* females. *Science* 222: 419–421.

Mangan, R. L. 1979. Reproductive behavior of the cactus fly, *Odontoloxozus longicornis*, male territoriality and female guarding as adaptive strategies. *Behav. Ecol. Sociobiol.* 4: 265–278.

Mange, A. P. 1970. Possible nonrandom utilization of X- and Y-bearing sperm in *Drosophila melanogaster*. *Genetics* 65: 95–106.

Mann, T. 1984. *Spermatophores*. Springer-Verlag, Heidelberg.

Mann, T., and C. Lutwak-Mann. 1981. *Male Reproductive Function and Semen*. Springer-Verlag, New York.

Manning, A. 1962. A sperm factor affecting the receptivity of *Drosophila melanogaster* females. *Nature* 194: 252–253.

Manning, A. 1967. The control of sexual receptivity in female *Drosophila*. *Anim. Behav.* 15: 239–250.

Manning, J. T., and A. T. Chamberlain. 1994. Sib competition and sperm competitiveness: An answer to "Why so many sperms?" and the recombination/sperm numbers correlation. *Proc. Roy. Soc. Lond. B* 256: 177–182.

Marden, J. H., and J. K. Waage. 1990. Escalated damselfly territorial contests are energetic wars of attrition. *Anim. Behav.* 39: 954–959.

Markow, T. A. 1982. Mating systems of cactophilic *Drosophila*. In J. S. Barker and W. T. Starmer, eds., *Ecological Genetics and Evolution*, pp. 273–287. Academic Press, London.

Markow, T. A. 1985. A comparative investigation of the mating system of *Drosophila hydei*. *Anim. Behav.* 33: 775–781.

Markow, T. A. 1988. *Drosophila* males provide a material contribution to offspring sired by other males. *Funct. Ecol.* 2: 77–79.

Markow, T. A., and P. F. Ankney. 1988. Insemination reaction in *Drosophila*: Found in species whose males contribute material to oocytes before fertilization. *Evolution* 42: 1097–1101.

Marks, R. W., R. D. Seager, and L. G. Barr. 1988. Local ecology and multiple mating in a natural population of *Drosophila melanogaster*. *Am. Nat.* 131: 918–923.

Marshall, L. D. 1985. Protein and lipid composition of *Colias philodice* and *C. eurytheme* spermatophores and their changes over time. *J. Res. Lepid.* 24: 21–30.

Martens, A. 1991. Plasticity of mate-guarding and oviposition behavior in *Zygonyx natalensis* (Martin) (Anisoptera: Libellulidae). *Odonatologica* 20: 293–302.

Martinez, L., J. L. Leyvavazquez, and H. B. Mojica. 1993. Sperm competition in the female *Diachasmimorpha longicaudata* (Hymenotera, Braconidae). *Southw. Entomol.* 18: 293–299.

Mason, L. G. 1980. Sexual selection and the evolution of pair-bonding in soldier beetles. *Evolution* 34: 174–180.

Mason, L. J., and D. P. Pashley. 1991. Sperm competition in the soybean looper (Lepidoptera: Noctuidae). *Ann. Entomol. Soc. Am.* 84: 268–271.

Matsumoto, K., and N. Suzuki. 1992. Effectiveness of the mating plug in *Atrophaneura alcinous* (Lepidoptera: Papilionidae). *Behav. Ecol. Sociobiol.* 30: 157–163.

Matsumoto, K., and N. Suzuki. 1995. The nature of mating plugs and the probability of reinsemination in Japanese Papilionidae. In J. M. Scriber, Y. Tsubaki, and R. C. Lederhouse, eds., *Swallowtail Butterflies: Their Ecology and Evolutionary Biology*, pp. 145–154. Scientific Publishers, Gainseville, FL.

Maynard Smith, J. 1977. Parental care: A prospective analysis. *Anim. Behav.* 25: 1–9.

Maynard Smith, J. 1978. *The Evolution of Sex*. Cambridge University Press, Cambridge, UK.

Maynard Smith, J. 1982. *Evolution and the Theory of Games*. Cambridge University Press, Cambridge, UK.

Maynard Smith, J. 1991. Theories of sexual selection. *TREE* 6: 146–151.

Mayr, E. 1963. *Animal Species and Evolution*. Harvard University Press, Cambridge, MA.

Mazzini, M. 1976. Giant spermatozoa in *Davales bipustulatus* F. (Coleoptera: Cleridae). *Int. J. Insect Morphol. Embryol.* 5: 107–115.

Mbata, G. N., and S. B. Ramaswamy. 1990. Rhythmicity of sex pheromone content in female *Heliothis virescens*: Impact of mating. *Physiol. Entomol.* 15: 423–432.

McCauley, D. E., and L. M. Reilly. 1984. Sperm storage and sperm precedence in the milkweed beetle *Tetraopes tetraophthalmus* (Forster) (Coleoptera: Cerambycidae). *Ann. Entomol. Soc. Am.* 77: 526–530.

McLain, D. K. 1980. Female choice and the adaptive significance of prolonged copulation in *Nezara viridula* (Hemiptera: Pentatomaidae). *Psyche* 87: 325–336.

McLain, D. K. 1981. Interspecific interference competition and mate choice in the soldier beetle *Chauliognathus pennsylvanicus*. *Behav. Ecol. Sociobiol.* 9: 65–66.

McLain, D. K. 1985. Male size, sperm competition, and the intensity of sexual selection in the southern green stink bug, *Nezara viridula* (Hemiptera: Pentatomidae). *Ann. Entomol. Soc. Am.* 78: 86–89.

McLain, D. K. 1989. Prolonged copulation as a post-insemination guarding tactic in a natural population of the ragwort seed bug. *Anim. Behav.* 38: 659–664.

McLain, D. K. 1991. Components of variance in male lifetime copulatory and reproductive success in a seed bug. *Behav. Ecol. Sociobiol.* 29: 121–126.

McMillan, V. E. 1991. Variable mate guarding behavior in the dragonfly *Plathemis lydia* (Odonata: Libellulidae). *Anim. Behav.* 41: 979–988.

McVey, M., and B. J. Smittle. 1984. Sperm precedence in the dragonfly *Erythemis simplicollis*. *J. Insect Physiol.* 30: 619–628.

Meek, S. B., and T. B. Herman. 1990. A comparison of the reproductive behaviours of three *Calopteryx* species (Odonata: Calopterygidae) in Nova Scotia. *Can. J. Zool.* 68: 10–16.

Meikle, D. B., K. B. Sheehan, D. M. Phillis, and R. C. Richmond. 1990. Localization and longevity of seminal-fluid esterase 6 in mated female *Drosophila melanogaster*. *J. Insect Physiol.* 36: 93–101.

Merle, J. 1968. Fonctionnement ovarien et réceptivité sexuelle de *Drosophila melanogaster* après implantation de fragments de l'appareil génital male. *J. Insect Physiol.* 14: 1159–1168.

Mesterton-Gibbons, M. 1999. On sperm competition games: Incomplete fertilization risk and the equity paradox. *Proc. Roy. Soc. Lond. B* 266: 269–274.

Metcalf, R. A. 1980. Sex ratios, parent-offspring conflict, and local competition for mates in the social wasps *Polistes metricus* and *Polistes variatus*. *Am. Nat.* 116: 642–654.

Metcalf, R. A., and G. S. Whitt. 1977. Intra-nest relatedness in the social wasp *Polistes metricus*: A genetic analysis. *Behav. Ecol. Sociobiol.* 2: 339–351.

Metz, E. C., and S. R. Palumbi. 1996. Positive selection and sequence rearrangements generate extensive polymorphism in the gamete recognition protein bindin. *Mol. Biol. Evol.* 13: 397–406.

Meves, F. 1902. Über oligopyrene und apyrene Spermien und über ihre Entstehung, nach Beobachtungen an *Paludina* und *Pygaera*. *Arch. Mikrosk. Anat. Entwicklungsgesch.* 61: 1–84.

Michelsen, A. 1963. Observations on the sexual behaviour of some longicorn beetles, subfamily Lepturinae (Coleoptera, Cerambycidae). *Behaviour* 22: 154–166.

Michiels, N. K. 1989. Morphology of male and female genitalia in *Sympetrum danae* (Sulzer), with special reference to the mechanism of sperm removal during copulation (Anisoptera: Libellulidae). *Odonatologica* 18: 21–32.

Michiels, N. K. 1992. Consequences and adaptive significance of variation in copulation duration in the dragonfly *Sympetrum danae*. *Behav. Ecol. Sociobiol.* 29: 429–435.

Michiels, N. K., and A. A. Dhondt. 1988. Direct and indirect estimates of sperm precedence and displacement in the dragonfly *Sympetrum danae* (Odonata: Libellulidae). *Behav. Ecol. Sociobiol.* 23: 257–264.

Miller, A. 1950. The internal anatomy and histology of the imago of *Drosophila melanogaster*. In M. Demerec, ed., *Biology of Drosophila*, pp. 420–534. Wiley, New York.

Miller, A. K., P. L. Miller, and M. T. Siva-Jothy. 1984. Pre-copulatory guarding and other aspects of reproductive behaviour in *Sympetrum depressiusculum* (Selys) at rice fields in southern France (Anisoptera: Libellulidae). *Odonatologica* 13: 407–414.

Miller, P. L. 1983. The duration of copulation correlates with other aspects of mating behaviour in *Orthetrum chrysostigma* (Burmeister) (Anisoptera: Libellulidae). *Odonatologica* 12: 227–238.

Miller, P. L. 1984. The structure of the genitalia and the volumes of sperm stored in male and female *Nesciothemis farinosa* (Foerster) and *Orthetrum chrysostigma* (Burmeister) (Anisoptera: Libellulidae). *Odonatologica* 13: 415–428.

Miller, P. L. 1987a. An examination of the prolonged copulations of *Ischnura elegans* (Vander Linden) (Zygoptera: Coenagrionidae). *Odonatologica* 16: 37–56.

Miller, P. L. 1987b. Sperm competition in *Ischnura elegans* (Vander Linden) (Zygoptera: Coenagrionidae). *Odonatologica* 16: 201–208.

Miller, P. L. 1990. Mechanisms of sperm removal and sperm transfer in *Orthetrum coerulescens* (Fabricus) (Odonata: Libellulidae). *Physiol. Entomol.* 15: 199–209.

Miller, P. L. 1991. The structure and function of the genitalia in the Libellulidae (Odonata). *Zool. J. Linn. Soc.* 102: 43–74.

Miller, P. L., and C. A. Miller. 1981. Field observations on copulatory behaviour in Zygoptera, with an examination of the structure and activity of male genitalia. *Odonatologica* 10: 201–218.

Miyano, S., and E. Hasegawa. 1998. Genetic structure of the first brood of workers and mating frequency of queens in a Japanese paper wasp, *Polistes chinensis antennalis*. *Ethol. Ecol. Evol.* 10: 79–85.

Miyatake, T., T. Chapman, and L. Partridge. 1999. Mating-induced inhibition of mating in female Mediterranean fruit flies *Ceratitis capitata*. *J. Insect Physiol.* 45: 1021–1028.

Moczek, A. P., and D. J. Emlen. 2000. Male horn dimorphism in the scarab beetle *Onthophagus taurus*: Do alternative reproductive tactics favor alternative phenotypes? *Anim. Behav.* 59: 459–466.

Møller, A. P. 1988. Female choice selects for male sexual tail ornaments in the monogamous swallow. *Nature* 332: 640–642.

Møller, A. P. 1991. Sperm competition, sperm depletion, paternal care, and relative testis size in birds. *Am. Nat.* 137: 882–906.

Møller, A. P. 1998. Sperm competition and sexual selection. In T. R. Birkhead and A. P. Møller, eds., *Sperm Competition and Sexual Selection*, pp. 55–90. Academic Press, London.

Møller, A. P., and R. V. Alatalo. 1999. Good-genes effects in sexual selection. *Proc. Roy. Soc. Lond. B* 266: 85–91.

Møller, A. P., and Birkhead, T. R. 1993. Certainty of paternity covaries with paternal care in birds. *Behav. Ecol. Sociobiol.* 33: 261–268.

Møller, A. P., and Birkhead, T. R. 1994. The evolution of plumage brightness in birds is related to extrapair paternity. *Evolution* 48: 1089–1100.

Møller, A.P., and Cuervo, J. J. 2000. The evolution of paternity and paternal care in birds. *Behav. Ecol.* 11: 472–485.

Møller, A. P., and A. Pomiankowski. 1993. Why have birds got multiple sexual ornaments? *Behav. Ecol. Sociobiol.* 32: 167–176.

Møller, A. P., and J. P. Swaddle. 1997. *Asymmetry, Developmental Stability, and Evolution*. Oxford University Press, Oxford.

Monsma, S. A., and M. F. Wolfner. 1988. Structure and expression of a *Drosophila* male accessory gland gene whose product resembles a peptide pheromone precursor. *Dev. Biol.* 2: 1063–1073.

Moore, A. J. 1989. The behavioral ecology of *Libellula luctuosa* (Burmeister) (Odonata: Libellulidae): III. Male density, OSR, and male and female mating behavior. *Ethology* 80: 120–136.

Moore, A. J. 1990. Sexual selection and the genetics of pheromonally mediated social behavior in *Nauphoeta cinerea* (Dictyoptera: Blaberidae). *Entomol. Gener.* 15: 133–147.

Moore, A. J. 1994. Genetic evidence for the "good genes" process of sexual selection. *Behav. Ecol. Sociobiol.* 35: 235–241.

Moore, H. D. M., M. Martin, and T. R. Birkhead. 1999. No evidence for killer sperm or other selective interactions between human spermatozoa in ejaculates of different males *in vitro*. *Proc. Roy. Soc. Lond. B* 266: 2343–2350.

Moritz, R. F. A. 1985. The effects of multiple mating on the worker-queen conflict in *Apis mellifera* L. *Behav. Ecol. Sociobiol.* 16: 375–377.

Moritz, R. F. A. 1986. Intracolonial worker relationship and sperm competition in the honeybee (*Apis mellifera* L.). *Experientia* 42: 445–448.

Moritz, R. F. A., P. Kryger, G. Koeniger, N. Koeniger, A. Estoup, and S. Tingek. 1995. High degree of polyandry in *Apis dorsata* queens detected by DNA microsatellite variability. *Behav. Ecol. Sociobiol.* 37: 357–363.

Morrow, E. H. 2000. Giant sperm in a neotropical moth (Lepidoptera: Arctiidae). *Eur. J. Entomol.* 97: 281–283.

Morrow, E. H., and M. J. G. Gage. 2000. The evolution of sperm length in moths. *Proc. Roy. Soc. Lond. B* 267: 307–313.

Moshitzky, P., I. Fleischmann, N. Chaimov, P. Saudan, S. Klauser, E. Kubli, and S. W. Applebaum. 1996. Sex-peptide activates juvenile hormone biosynthesis in the *Drosophila melanogaster* corpus allatum. *Arch. Insect Bioch. Physiol.* 32: 363–374.

Mousseau, T. A., and D. A. Roff. 1987. Natural selection and the heritability of fitness components. *Heredity* 59: 181–197.

Mueller, U. G. 1991. Haplodiploidy and the evolution of facultative sex ratios in a primitively eusocial bee. *Science* 254: 442–444.

Mueller, U. G., G. C. Eickwort, and C. F. Aquadro. 1994. DNA fingerprinting analysis of parent-offspring conflict in a bee. *Proc. Natl. Acad. Sci. USA* 91: 5143–5147.

Mühlhäuser, C., W. U. Blanckenhorn, and P. I. Ward. 1996. The genetic component of copula duration in the yellow dung fly. *Anim. Behav.* 51: 1401–1407.

Müller, J. K., and A.-K. Eggert. 1989. Paternity assurance by "helpful" males: Adaptations to sperm competition in burying beetles. *Behav. Ecol. Sociobiol.* 24: 245–249.

Murakami, T., S. Higashi, and D. Windsor. 2000. Mating frequency, colony size, polyethism and sex ratio in fungus-growing ants (Attini). *Behav. Ecol. Sociobiol.* 48: 276–284.

Muralidharan, K., M. S. Shaila, and R. Gadagkar. 1986. Evidence for multiple mating in the primitively eusocial wasp *Ropalidia marginata* (Lep.) (Hymenoptera: Vespidae). *J. Genet.* 65: 153–158.

Murdoch, W. W. 1966. Population stability and life-history phenomena. *Am. Nat.* 10: 5–11.

Murray, A. M., and P. S. Giller. 1990. The life-history of *Aquarius najas* De Geer (Hemiptera: Gerridae) in Southern Ireland. *Entomologist* 109: 53–64.

Muse, W. A., and T. Ono. 1996. Copulatory behavior and post copulatory mate guarding in a grasshopper *Atractomorpha lata* Motschulsky (Orthoptera: Tetrigidae) under laboratory conditions. *Appl. Entomol. Zool.* 31: 233–241.

Myers, H. S., B. D. Barry, J. A. Burnside, and R. H. Rhode. 1976. Sperm precedence in female apple maggots alternately mated to normal and irradiated males. *Ann. Entomol. Soc. Am.* 69: 39–41.

Nabours, R. K. 1927. Polyandry in the grouse locust, *Paratettix texanus* Hancock, with notes on inheritance of aquired characters and telegony. *Am. Nat.* 61: 531–538.

Nakano, S. 1985. Sperm displacement in *Henosepilachna pustulosa* (Coleoptera, Coccinellidae). *Kotyu* 53: 516–519.

Neems, R. M., J. Lazarus, and A. J. McLachlan. 1998. Lifetime reproductive success in a swarming midge: Trade-offs and stabilizing selection for male body size. *Behav. Ecol.* 9: 279–286.

Neubaum, D. M., and M. F. Wolfner. 1999. Mated *Drosophila melanogaster* females require a seminal fluid protein, Acp36DE, to store sperm efficiently. *Genetics* 153: 845–857.

Nielsen, E. S., and I. F. B. Common. 1991. Lepidoptera. In Division of Entomology, CSIRO, ed., *The Insects of Australia*, pp. 817–915. Melbourne University Press, Melbourne.

Nijhout, M. M., and L. M. Riddiford. 1974. The control of egg maturation by juvenile hormone in the tobacco hornworm moth, *Manduca sexta*. *Biol. Bull.* 146: 377–392.

Nijhout, M. M., and L. M. Riddiford. 1979. Juvenile hormone and ovarian growth in *Manduca sexta*. *Int. J. Invert. Reprod. Dev.* 1: 209–219.

Nilakhe, S. S., and E. J. Villavaso. 1979. Measuring sperm competition in the boll weevil by the use of females whose spermathecae have been surgically removed. *Ann. Entomol. Soc. Am.* 72: 500–502.

Nomakuchi, S., K. Higashi, M. Harada, and M. Maeda. 1984. An experimental study of the territoriality in *Mnais pruinosa pruinosa* Selys (Zygoptera: Calopterygidae). *Odonatologica* 13: 259–267.

Nonidez, J. F. 1920. The internal phenomena of reproduction in *Drosophila*. *Biol. Bull. Mar. Biol. Lab. Woods Hole* 39: 207–230.

Norris, K. 1993. Heritable variation in a plumage indicator of viability in male great tits *Parus major*. *Nature* 362: 537–539.

North, D. T., and G. G. Holt. 1968. Genetic and cytogenetic basis of radiation-induced sterility in the adult male cabbage looper *Trichoplusia ni*. In *IAEA/FAO Symposium, Isotopes and Radiation in Entomology (1967)*, pp. 391–403. IAEA, Vienna.

Nuhardiyata, M. 1998. The mating system of the Leafhopper *Balclutha incisa* (Matsumura) (Cicadellidae: Auchenorrhyncha: Hemiptera). Ph.D. diss., University of Western Australia.

Nummelin, M. 1987. Ripple signals of the waterstrider *Limnoporus rufoscutellatus* (Heteroptera, Gerridae). *Ann. Entomol. Fenn.* 53: 17–22.

Nummelin, M. 1988. The territorial behaviour of four Ugandan waterstrider species (Heteroptera, Gerridae): A comparative study. *Ann. Entomol. Fenn.* 54: 121–134.

Nuyts, E. 1994. On the copulation duration of the yellow dung fly (*Scathophaga stercoraria*): The effects of optimisation per day. *Acta Biotheoretica* 42: 271–279.

Nuyts, E., and N. K. Michiels. 1993. Integration of immediate and long term sperm precedence patterns and mating costs in an optimization model of insect copulation duration. *J. Theor. Biol.* 160: 271–295.

Nuyts, E., and N. K. Michiels. 1994. The influence of sperm precedence patterns and mating costs on copulation duration in odonates: Predictions and supporting data. *Belg. J. Zool.* 124: 11–19.

Nuyts, E., and N. K. Michiels. 1996. High last-male sperm precedence despite unfavourable positioning of sperm in the bursa copulatrix of *Sympetrum danae* (Sulzer) (Anisoptera: Libellulidae). *Odonatologica* 25: 79–82.

Nylin, S., and N. Wedell. 1994. Sexual size dimorphism and comparative methods. In P. Eggleton, and R. Vane-Wright, eds., *Phylogenetics and Ecology*, pp. 255–280. Academic Press, London.

Obara, Y. 1982. Mate refusal hormone in the cabbage white butterfly? *Naturwissenschaften* 69: 551–552.

Oberhauser, K. S. 1988. Male monarch butterfly spermatophore mass and mating strategies. *Anim. Behav.* 36: 1384–1388.

Oberhauser, K. S. 1989. Effects of spermatophores on male and female reproductive success. *Behav. Ecol. Sociobiol.* 25: 237–246.

Oberhauser, K. S. 1992. Rate of ejaculate breakdown and intermating intervals in monarch butterflies. *Behav. Ecol. Sociobiol.* 31: 367–373.

Oberhauser, K. S. 1997. Fecundity, lifespan and egg mass in butterflies: effects of male-derived nutrients and female size. *Funct. Ecol.* 11: 166–175.

Oberhauser, K. S., and R. Hampton. 1995. The relationship between mating and oogenesis in monarch butterflies (Lepidoptera: Danainae). *J. Insect Behav.* 8: 701–713.

Oberhauser, K. S., R. Hampton, B. Jenson, and S. Weisberg. Unpublished manuscript. Variation in sperm precedence patterns in monarch butterflies (*Danaus plexippus*): Effects of male size, mating history, and time.

Odiambo, T. R. 1959. An account of parental care in *Rhinocoris albopilosus* Signoret (Hemiptera: Heteroptera: Reduviidae), with notes on its life history. *Proc. Roy. Entomol. Soc. Lond. A* 34: 175–185.

O'Donald, P. 1980. *Genetic Models of Sexual Selection*. Cambridge University Press, Cambridge, UK.

O'Donald, P., and M. E. N. Majerus. 1992. Non-random mating in *Adelia bipunctata* (the two-spot ladybird). III. New evidence of genetic preference. *Heredity* 69: 521–526.

Okelo, O. 1979. Mechanisms of sperm release from the receptaculum seminis of *Schistocerca vaga* Scudder (Orthoptera: Acrididae). *Int. J. Invert. Reprod.* 1: 121–131.

Oldroyd, B. P., M. J. Clifton, S. Wongsiri, T. E. Rinderer, H. A. Sylvester, and R. H. Crozier. 1997. Polyandry in the genus *Apis*, particularly *Apis andreniformis*. *Behav. Ecol. Sociobiol.* 40: 17–26.

Oldroyd, B. P., T. E. Rinderer, S. M. Buco, and L. D. Beaman. 1993. Genetic variance in honey bees for preferred foraging distance. *Anim. Behav*. 45: 323–332.

Oldroyd, B. P., T. E. Rinderer, J. R. Harbo, and S. M. Buco. 1992. Effects of intracolonial genetic divesity on honey bee (Hymenoptera: Apidae) colony performance. *Ann. Entomol. Soc. Am.* 85: 335–343.

Oldroyd, B. P., A. Smolenski, J.-M. Cornuet, and R. H. Crozier. 1994a. Anarchy in the beehive. *Nature* 371: 749.

Oldroyd, B. P., A. J. Smolenski, J.-M. Cornuet, S. Wongsiri, A. Estoup, T. E. Rinderer, and R. H. Crozier. 1995. Levels of polyandry and intracolonial genetic relationships in *Apis florea*. *Behav. Ecol. Sociobiol.* 37: 329–335.

Oldroyd, B. P., H. A. Sylvester, S. Wongsiri, and T. E. Rinderer. 1994b. Task specialization in a wild bee, *Apis florea* (Hymenoptera: Apidae), revealed by RFLP banding. *Behav. Ecol. Sociobiol.* 34: 25–30.

Olsson, M., R. Shine, T. Madsen, A. Gullberg, and H. Tegelström. 1996. Sperm selection by females. *Nature* 383: 585.

Olsson, M., R. Shine, T. Madsen, A. Gullberg, and H. Tegelström. 1997. Sperm choice by females. *TREE* 12: 445–446.

Omura, S. 1939. Selective fertilization in *Bombyx mori*. *Jap. J. Genet.* 15: 29–35.

Ono, T., M. T. Siva-Jothy, and A. Kato. 1989. Removal and subsequent ingestion of rivals' semen during copulation in a tree cricket. *Physiol. Entomol.* 14: 195–202.

Opp, S. B., J. Ziegner, N. Bui, and R. J. Prokopy. 1990. Factors influencing estimates of sperm competition in *Rhagoletis pomonella* (Walsh) (Diptera: Tephritidae). *Ann. Entomol. Soc. Am.* 83: 521–526.

Orr, A. G. 1988. Mate conflict and the evolution of the sphragis in butterflies. Ph.D. diss., Griffith University, Queensland.

Orr, A. G. 1995. The evolution of the sphragis in the Papilionidae and other butterflies. In J. M. Scriber, Y. Tsubaki, and R. C. Lederhouse, eds., *Swallowtail Butterflies: Their Ecology and Evolutionary Biology*, pp. 155–164. Scientific Publishers, Gainseville, FL.

Orr, A. G., and R. L. Rutowski. 1991. The function of the sphragis in *Cressida cressida* (Fab) (Lepidoptera, Papilionidae): A visual deterrent to copulation attempts. *J. Nat. Hist.* 25: 703–710.

Otronen, M. 1984. Male contests for territories and females in the fly *Dryomyza anilis*. *Anim. Behav*. 32: 891–898.

Otronen, M. 1990. Mating behavior and sperm competition in the fly, *Dryomyza anilis*. *Behav. Ecol. Sociobiol.* 26: 349–356.

Otronen, M. 1994a. Fertilisation success in the fly *Dryomyza anilis* (Dryomyzidae): Effects of male size and the mating situation. *Behav. Ecol. Sociobiol.* 35: 33–38.

Otronen, M. 1994b. Repeated copulations as a strategy to maximize fertilization in the fly, *Dryomyza anilis* (Dryomyzidae). *Behav. Ecol.* 5: 51–56.

Otronen, M. 1997a. Sperm numbers, their storage and usage in the fly *Dryomyza anilis*. *Proc. Roy. Soc. Lond. B* 264: 777–782.

Otronen, M. 1997b. Variation in sperm precedence during mating in male flies, *Dryomyza anilis*. *Anim. Behav*. 53: 1233–1240.

Otronen, M. 1998. Male asymmetry and postcopulatory sexual selection in the fly *Dryomyza anilis*. *Behav. Ecol. Sociobiol.* 42: 185–192.

Otronen, M., and M. T. Siva-Jothy. 1991. The effect of postcopulatory male behaviour on ejaculate distribution within the female sperm storage organs of the fly, *Dryomyza anilis* (Diptera: Dryomyzidae). *Behav. Ecol. Sociobiol.* 29: 33–37.

Otronen, M., P. Reguera, and P. I. Ward. 1997. Sperm storage in the yellow dung fly *Scathophaga stercoraria*: Identifying the sperm of competing males in separate female spermathecae. *Ethology* 103: 844–854.

Otte, D. 1979. Historical development of sexual selection theory. In M. S. Blum and N. A. Blum, eds., *Sexual Selection and Reproductive Competition in Insects*, pp. 1–18. Academic Press, London.

Owen, D. F., J. Owen, and D. O. Chanter. 1973. Low mating frequency in an African butterfly. *Nature* 244: 116–117.

Packer, L., and R. E. Owen. 1994. Relatedness and sex ratio in a primitively eusocial halictine bee. *Behav. Ecol. Sociobiol.* 34: 1–10.

Page, R. E. 1986. Sperm utilization in social insects. *Ann. Rev. Entomol.* 31: 297–320.

Page, R. E., and R. A. Metcalf. 1982. Multiple mating, sperm utilization, and social evolution. *Am. Nat.* 119: 263–281.

Page, R. E., and G. E. Robinson. 1991. The genetics of division of labor in honey bee colonies. *Adv. Insect Physiol.* 23: 117–169.

Page, R. E., G. E. Robinson, M. K. Fondrk, and M. E. Nasr. 1995. Effects of worker genotypic diversity on honey bee colony development and behavior (*Apis mellifera* L.). *Behav. Ecol. Sociobiol.* 36: 387–396.

Pair, S. D., M. L. Laster, and D. F. Martin. 1977. Hybrid sterility of the tobacco budworm: Effects of alternate sterile and normal matings on fecundity and fertility. *Ann. Entomol. Soc. Am.* 70: 952–954.

Pajunen, V. L. 1963. Reproductive behaviour in *Leucorrhinia dubia* v. d. Lind. and *L. rubicunda* L. (Odon., Libellulidae). *Ann. Entomol. Fenn.* 29: 106–118.

Palumbi, S. R. 1999. All males are not created equal: Fertility differences depend on gamete recognition polymorphisms in sea urchins. *Proc. Natl. Acad. Sci. USA* 96: 12632–12637.

Palumbi, S. R., and E. C. Metz. 1991. Strong reproductive isolation between closely related tropical sea urchins (genus *Echinometra*). *Mol. Biol. Evol.* 8: 227–239.

Pamilo, P. 1982a. Genetic evolution of sex ratios in eusocial Hymenoptera: Allele frequency simulations. *Am. Nat.* 119: 638–656.

Pamilo, P. 1982b. Multiple mating in *Formica* ants. *Hereditas* 97: 37–45.

Pamilo, P. 1991a. Evolution of colony characteristics in social insects. I. Sex allocation. *Am. Nat.* 137: 83–107.

Pamilo, P. 1991b. Evolution of colony characteristics in social insects. II. Number of reproductive individuals. *Am. Nat.* 138: 412–433.

Pamilo, P. 1991c. Evolution of the sterile caste. *J. Theor. Biol.* 149: 75–95.

Pamilo, P. 1991d. Lifespan of queens in the ant *Formica exsecta. Insect. Soc.* 38: 111–120.

Pamilo, P. 1993. Polyandry and allele frequency differences between the sexes in the ant *Formica aquilonia. Heredity* 70: 472–480.

Pamilo, P., and R. Rosengren. 1984. Evolution of nesting strategies of ants: Genetic evidence from different population types of *Formica* ants. *Biol. J. Linn. Soc.* 21: 331–348.

Pamilo, P., and P. Seppä. 1994. Reproductive competition and conflicts in colonies of the ant *Formica sanguinea. Anim. Behav.* 48: 1201–1206.

Pamilo, P., L. Sundström, F. W., and R. Rosengren. 1994. Diploid males and colony-level selection in *Formica* ants. *Ethol. Ecol. Evol.* 6: 221–235.

Pamilo, P., and S.-L. Varvio-Aho. 1979. Genetic structure of nests in the ant *Formica sanguinea. Behav. Ecol. Sociobiol.* 6: 91–98.

Park, M., and M. F. Wolfner. 1995. Male and female cooperate in the prohormone-like processing of a *Drosophila melanogaster* seminal fluid protein. *Dev. Biol.* 171: 694–702.

Park, Y. I., S. B. Ramaswamy, and A. Srinivasan. 1998. Spermatophore formation and regulation of egg maturation and oviposition in female *Heliothis virescens* by the male. *J. Insect Physiol.* 44: 903–908.

Parker, G. A. 1969. The reproductive behaviour and the nature of sexual selection in *Scatophaga stercoraria* L. (Diptera: Scatophagidae). III. Apparent intersex individuals and their evolutionary cost to normal, searching males. *Trans. Roy. Entomol. Soc. Lond.* 121: 305–323.

Parker, G. A. 1970a. The reproductive behaviour and the nature of sexual selection in *Scatophaga stercoraria* L. (Diptera: Scatophagidae). II. The fertilization rate and the spatial and temporal relationships of each sex around the site of mating and oviposition. *J. Anim. Ecol.* 39: 205–228.

Parker, G. A. 1970b. The reproductive behaviour and the nature of sexual selection in *Scatophaga stercoraria* L. (Diptera: Scatophagidae). IV. Epigamic recognition and competition between males for the possession of females. *Behaviour* 37: 114–139.

Parker, G. A. 1970c. The reproductive behaviour and the nature of sexual selection in *Scatophaga stercoraria* L. (Diptera: Scatophagidae). V. The female's behaviour at the oviposition site. *Behaviour* 37: 140–168.

Parker, G. A. 1970d. The reproductive behaviour and the nature of sexual selection in *Scatophaga stercoraria* L. (Diptera: Scatophagidae). VII. The origin and evolution of the passive phase. *Evolution* 24: 774–788.

Parker, G. A. 1970e. Sperm competition and its evolutionary consequences in the insects. *Biol. Rev.* 45: 525–567.

Parker, G. A. 1970f. Sperm competition and its evolutionary effect on copula duration in the fly *Scatophaga stercoraria. J. Insect Physiol.* 16: 1301–1328.

Parker, G. A. 1971. The reproductive behaviour and the nature of sexual selection in *Scatophaga stercoraria* L. (Diptera: Scatophagidae). VI. The adaptive significance of emigration from the oviposition site during the phase of genital contact. *J. Anim. Ecol.* 40: 215–233.

Parker, G. A. 1972a. Reproductive behaviour of *Sepsis cynipsea* (L.) (Diptera: Sepsidae). I. A preliminary analysis of the reproductive strategy and its associated behaviour patterns. *Behaviour* 41: 172–206.

Parker, G. A. 1972b. Reproductive behaviour of *Sepsis cynipsea* (L.) (Diptera: Sepsidae). II. The significance of the precopulatory passive phase and emigration. *Behaviour* 41: 242–250.

Parker, G. A. 1974. Courtship persistence and female-guarding as male time investment strategies. *Behaviour* 48: 157–184.

Parker, G. A. 1978. Searching for mates. In J. R. Krebs and N. B. Davies, eds., *Behavioural Ecology: An Evolutionary Approach*, pp. 214–244. Blackwell Scientific, Oxford.

Parker, G. A. 1979. Sexual selection and sexual conflict. In M. S. Blum and N. A. Blum, eds., *Sexual Selection and Reproductive Competition in Insects*, pp. 123–166. Academic Press, London.

Parker, G. A. 1982. Why are there so many tiny sperm? Sperm competition and the maintenance of two sexes. *J. Theor. Biol.* 96: 281–294.

Parker, G. A. 1984. Sperm competition and the evolution of animal mating strategies. In R. L. Smith, ed., *Sperm Competition and the Evolution of Animal Mating Systems*, pp. 2–60. Academic Press, London.

Parker, G. A. 1990a. Sperm competition games: Raffles and roles. *Proc. Roy. Soc. Lond. B* 242: 120–126.

Parker, G. A. 1990b. Sperm competition games: Sneaks and extra-pair copulations. *Proc. Roy. Soc. Lond. B* 242: 127–133.

Parker, G. A. 1992a. Marginal value theorem with exploitation time costs: Diet, sperm reserves, and optimal copula duration in dung flies. *Am. Nat.* 139: 1237–1256.

Parker, G. A. 1992b. Snakes and female sexuality. *Nature* 355: 395–396.

Parker, G. A. 1992c. The evolution of sexual size dimorphism in fish. *J. Fish Biol.* 41 (Suppl. B): 1–20.

Parker, G. A. 1993. Sperm competition games: Sperm size and sperm number under adult control. *Proc. Roy. Soc. Lond. B* 253: 245–254.

Parker, G. A. 1998. Sperm competition and the evolution of ejaculates: Towards a theory base. In T. R. Birkhead and A. P. Møller, eds., *Sperm Competition and Sexual Selection*, pp. 3–54. Academic Press, London.

Parker, G. A. 2001. Golden flies, sunlit meadows: A tribute to the yellow dungfly. In L. A. Dugatkin, ed., *Model Systems in Behavioral Ecology*. Monographs in Behavior and Ecology. Princeton University Press, Princeton, NJ.

Parker, G. A., and M. Begon. 1993. Sperm competition games: Sperm size and number under gametic control. *Proc. Roy. Soc. Lond. B* 253: 255–262.

Parker, G. A., and S. P. Courtney. 1983. Seasonal incidence: Adaptive variation in the timing of life history stages. *J. Theor. Biol.* 105: 147–155.

Parker, G. A., and L. Partridge. 1998. Sexual conflict and speciation. *Phil. Trans. Roy. Soc. Lond. B* 353: 261–274.

Parker, G. A., and L. W. Simmons. 1989. Nuptial feeding in insects: Theoretical models of male and female interests. *Ethology* 82: 3–26.

Parker, G. A., and L. W. Simmons. 1991. A model of constant random sperm displacement during mating: Evidence from *Scatophaga*. *Proc. Roy. Soc. Lond. B* 246: 107–115.

Parker, G. A., and L. W. Simmons. 1994. Evolution of phenotypic optima and copula duration in dungflies. *Nature* 370: 53–56.

Parker, G. A., and L. W. Simmons. 1996. Parental investment and the control of sexual selection: Predicting the direction of sexual competition. *Proc. Roy. Soc. Lond. B* 263: 315–321.

Parker, G. A., and L. W. Simmons. 2000. Optimal copula duration in yellow dung flies: Ejaculatory duct dimensions and size-dependent sperm displacement. *Evolution* 54: 924–935.

Parker, G. A., and J. L. Smith. 1975. Sperm competition and the evolution of the precopulatory passive phase behaviour in *Locusta migratoria migratorioides*. *J. Entomol. A* 49: 155–171.

Parker, G. A., and R. A. Stuart. 1976. Animal behavior as a strategy optimizer: Evolution of resource assessment strategies and optimal emigration thresholds. *Am. Nat.* 110: 1055–1076.

Parker, G. A., R. R. Baker, and V. G. F. Smith. 1972. The origin and evolution of gamete dimorphism and the male-female phenomenon. *J. Theor. Biol.* 36: 529–533.

Parker, G. A., M. A. Ball, P. Stockley, and M. J. G. Gage. 1996. Sperm competition games: Individual assessment of sperm competition intensity by group spawners. *Proc. Roy. Soc. Lond. B* 263: 1291–1297.

Parker, G. A., M. A. Ball, P. Stockley, and M. J. G. Gage. 1997. Sperm competition games: A prospective analysis of risk assessment. *Proc. Roy. Soc. Lond. B* 264: 1793–1802.

Parker, G. A., L. W. Simmons, and H. Kirk. 1990. Analysing sperm competition data: Simple models for predicting mechanisms. *Behav. Ecol. Sociobiol.* 27: 55–65.

Parker, G. A., L. W. Simmons, P. Stockley, D. M. McChristy, and E. L. Charnov. 1999. Optimal copula duration in yellow dung flies: Effects of female size and egg content. *Anim. Behav.* 57: 795–805.

Parker, G. A., L. W. Simmons, and P. I. Ward. 1993. Optimal copula duration in dungflies: Effects of frequency dependence and female mating status. *Behav. Ecol. Sociobiol.* 32: 157–166.

Partridge, L. 1994. Genetic and nongenetic approaches to questions about sexual selection. In R. B. Boake, ed., *Quantitative Genetic Studies of Behavioural Evolution*, pp. 126–141. University of Chicago Press, Chicago.

Partridge, L., and Endler, J. A. 1987. Life history constraints on sexual selection. In J. W. Bradbury and M. B. Andersson, eds., *Sexual Selection: Testing the Alternatives*, pp. 265–277. Wiley, Chichister, UK.

Partridge, L., and K. Fowler. 1990. Non-mating costs of exposure to males in female *Drosophila melanogaster*. *J. Insect Physiol.* 36: 419–425.

Partridge, L., and T. Halliday. 1984. Mating patterns and mate choice. In J. R. Krebs and N. B. Davies, eds., *Behavioural Ecology: An Evolutionary Approach*, pp. 222–250. Blackwell Scientific, Oxford.

Partridge, L., A. Green, and K. Fowler. 1987. Effects of egg-production and of exposure to males on female survival in *Drosophila melanogaster*. *J. Insect Physiol.* 33: 745–749.

Patterson, J. T. 1943. Studies in the genetics of *Drosophila*. III. The Drosophilidae of the Southwest. *U. Texas Publ.* 4313: 7–203.

Patterson, J. T. 1946. A new type of isolating mechanism. *Proc. Natl. Acad. Sci. USA* 32: 202–208.

Pease, R. W. 1968. The evolution and biological significance of multiple pairing in Lepidoptera. *J. Lepid. Soc.* 22: 197–209.

Pedersen, J. S., and J. J. Boomsma. 1999. Positive association of queen number and queen-mating frequency in Myrmica ants: A challange to the genetic-variability hypothesis. *Behav. Ecol. Sociobiol.* 45: 185–194.

Perotti, M. E. 1973. The mitochondrial derivative of the spermatozoon of *Drosophila* before and after fertilization. *J. Ultra. Res.* 44: 181–198.

Peters, J. M., D. C. Queller, J. E. Strassmann, and C. R. Solís. 1995. Maternity assignment and queen replacement in a social wasp. *Proc. Roy. Soc. Lond. B* 260: 7–12.

Petrie, M. 1994. Improved growth and survival of offspring of peacocks with more elaborate trains. *Nature* 371: 598–599.

Pezalla, V. M. 1979. Behavioral ecology of the dragonfly *Libellula pulchella* Drury (Odonata: Anisoptera). *Am. Midl. Nat.* 102: 1–22.

Pickford, R., and C. Gillot. 1972. Courtship behaviour of the migratory grasshopper *Melanoplus sanguinipes* (Orthoptera: Acrididae). *Can. Entomol.* 104: 715–722.

Pickford, R., A. B. Ewen, and C. Gillot. 1969. Male accessory gland substance and egg-laying stimulant in *Melanoplus sanguinipes* (F.) (Orthoptera: Acrididae). *Can. J. Zool.* 47: 1199–1203.

Pierre, J. 1986. Morphologie comparée de l'appareil génital femelle des Acraeinae (Lepidoptera, Nymphalidae). *Ann. Soc. Entomol. Fr.* 22: 53–65.

Pitnick, S. 1993. Operational sex ratios and sperm limitation in populations of *Drosophila pachea*. *Behav. Ecol. Sociobiol.* 33: 383–391.

Pitnick, S. 1996. Investment in testes and the cost of making long sperm in *Drosophila*. *Am. Nat.* 57–80.

Pitnick, S., and W. D. Brown. 2000. Criteria for demonstrating female sperm choice. *Evolution* 54: 1052–1056.

Pitnick, S., and T. L. Karr. 1998. Paternal products and by-products in *Drosophila* development. *Proc. Roy. Soc. Lond. B* 265: 821–826.

Pitnick, S., and T. A. Markow. 1994a. Large-male advantages associated with costs of sperm production in *Drosophila hydei*, a species with giant sperm. *Proc. Natl. Acad. Sci. USA* 91: 9277–9281.

Pitnick, S., and T. A. Markow. 1994b. Male gametic strategies: Sperm size, testes size, and the allocation of ejaculate among successive mates by the sperm-limited fly *Drosophila pachea* and its relatives. *Am. Nat.* 143: 785–819.

Pitnick, S., and G. T. Miller. 2000. Correlated response in reproductive and life history traits to selection on testis length in *Drosophila hydei*. *Heredity* 84: 416–426.

Pitnick, S., T. A. Markow, and M. F. Riedy. 1991. Transfer of ejaculate and incorporation of male-derived substances by females in the Nannoptera species group (Diptera: Drosophilidae). *Evolution* 45: 774–780.

Pitnick, S., T. A. Markow, and G. S. Spicer. 1995a. Delayed male maturity is a cost of producing large sperm in *Drosophila. Proc. Natl. Acad. Sci. USA* 92: 10614–10618.

Pitnick, S., T. A. Markow, and G. S. Spicer. 1999. Evolution of multiple kinds of female sperm-storage organs in *Drosophila. Evolution* 53: 1804–1822.

Pitnick, S., G. T. Miller, J. Reagan, and B. Holland. 2001. Males' evolutionary responses to experimental removal of sexual selection. *Proc. Roy. Soc. Lond. B* 268: 1071–1080.

Pitnick, S., G. S. Spicer, and T. A. Markow. 1995b. How long is a giant sperm? *Nature* 375: 109.

Pitnick, S., G. S. Spicer, and T. A. Markow. 1997. Phylogenetic examination of female incorporation of ejaculate in *Drosophila. Evolution* 51: 833–845.

Plaistow, S., and M. T. Siva-Jothy. 1996. Energetic constraints and male mate-securing tactics in the damselfly *Calopteryx splendens xanthostoma* (Charpentier). *Proc. Roy. Soc. Lond. B* 263: 1233–1238.

Pliske, T. E. 1973. Factors determining mating frequencies in some new world butterflies and skippers. *Ann. Entomol. Soc. Am.* 66: 164–169.

Pliske, T. E. 1975. Courtship behavior of the monarch butterfly, *Danaus plexippus* L. *Ann. Entomol. Soc. Am.* 68: 143–151.

Polak, M. 1998. Effects of ectoparasitism on host condition in the *Drosophila-Macrocheles* system. *Ecology* 79: 1807–1817.

Polak, M., W. T. Starmer, and J. S. F. Barker. 1998. A mating plug and male mate choice in *Drosophila hibisci* Bock. *Anim. Behav.* 56: 919–926.

Polak, M., L. L. Wolf, W. T. Starmer, and J. S. F. Barker. 2001. Function of the mating plug in *Drosophila hibisci* Bock. *Behav. Ecol. Sociobiol.* 49: 196–205.

Policansky, D. 1970. Three sperm sizes in *D. pseudoobscura* and *D. persimilis*. *Dros. Inf. Serv.* 45: 119–120.

Pomiankowski, A. 1987. Sexual selection: The handicap principle does work—sometimes. *Proc. R. Soc. Lond. B* 231: 123–145.

Pomiankowski, A. 1988. The evolution of female mate preferences for male genetic quality. *Oxford Surv. Evol. Biol.* 5: 136–184.

Pomiankowski, A., Y. Iwasa, and S. Nee. 1991. The evolution of costly mate preferences I. Fisher and biased mutation. *Evolution* 45: 1422–1430.

Pomiankowski, A., and A. P. Møller. 1995. A resolution of the lek paradox. *Proc. Roy. Soc. Lond. B* 260: 21–29.

Popham, E. J. 1965. The functional morphology of the reproductive organs of the common earwig (*Forficula auricularia*) and other Dermaptera with reference to the natural classification of the order. *J. Zool.* 146: 1–43.

Popov, G. B. 1958. Ecological studies on oviposition by swarms of the desert locust (*Schistocerca gregaria* Forskål) in eastern Africa. *Anti-Locust Bull.* 31: 1–70.

Presgraves, D. C., R. H. Baker, and G. S. Wilkinson. 1999. Coevolution of sperm and female reproductive tract morphology in stalk-eyed flies. *Proc. Roy. Soc. Lond. B* 266: 1041–1047.

Presgraves, D. C., E. Severance, and G. S. Wilkinson. 1997. Sex chromosome meiotic drive in stalk-eyed flies. *Genetics* 147: 1169–1180.

Price, C. S. C. 1997. Conspecific sperm precedence in *Drosophila. Nature* 388: 663–666.

Price, C. S. C., K. A. Dyer, and J. A. Coyne. 1999. Sperm competition between *Drosophila* males involves both displacement and incapacitation. *Nature* 400: 449–452.

Price, C. S. C., C. H. Kim, J. Posluszny, and J. A. Coyne. 2000. Mechanisms of conspecific sperm precedence in *Drosophila. Evolution* 54: 2028–2037.

Price, T. 1998. Sexual selection and natural selection in bird speciation. *Phil. Trans. Roy. Soc. Lond. B* 353: 251–260.

Proctor, H. C. 1992. Sensory exploitation and the evolution of male mating behaviour: A cladistic test using water mites (Acari: Parasitengona). *Anim. Behav.* 44: 745–752.

Proshold, F. I., and L. E. LaChance. 1974. Analysis of sterility in hybrids from interspecific crosses between *Heliothis virescens* and *H. subflexa. Ann. Entomol. Soc. Am.* 67: 445–449.

Proshold, F. I., L. E. LaChance, and R. D. Richard. 1975. Sperm production and transfer by *Heliothis virescens, H. subflexa* and the sterile hybrid males. Ann. Entomol. Soc. Am. 68: 31–34.

Prout, T., and J. Bundgaard. 1977. The population genetics of sperm displacement. *Genetics* 85: 95–124.

Prout, T., and A. G. Clark. 1996. Polymorphism in genes that influence sperm displacement. *Genetics* 144: 401–408.

Prout, T., and A. G. Clark. 2000. Seminal fluid causes temporary reduced egg hatch in previously mated females. *Proc. Roy. Soc. Lond. B* 267: 201–203.

Proverbs, M. D., and J. R. Newton. 1962. Some effects of gamma radiation on the potential of the codling moth, *Carpocapsa pomonella* (L.) (Lepidoptera: Olethreutidae). *Can. J. Zool.* 94: 1162–1170.

Prowse, N., and L. Partridge. 1997. The effects of reproduction on longevity and fertility in male *Drosophila melanogaster. J. Insect Physiol.* 43: 501–511.

Purvis, A. 1991. Comparative analysis by independent contrasts. Oxford University, Oxford.

Pyle, D. W., and M. H. Gromko. 1978. Repeated mating by female *Drosophila melanogaster*: The adaptive significance. *Experientia* 34: 449–450.

Queller, D. C. 1993. Worker control of sex ratios and selection for extreme multiple mating by queens. *Am. Nat.* 142: 346–351.

Queller, D. C., and J. E. Strassmann. 1998. Kin selection and social insects. *Bioscience* 48: 164–175.

Queller, D. C., J. E. Strassmann, and C. R. Hughes. 1993a. Microsatellites and kinship. *TREE* 8: 285–288.

Queller, D. C., J. E. Strassmann, C. R. Solís, C. R. Hughes, and D. M. DeLoach. 1993b. A selfish strategy of social insect workers that promotes social cohesion. *Nature* 365: 639–641.

Radwan, J. 1996. Intraspecific variation in sperm competition success in the bulb mite: A role for sperm size. *Proc. Roy. Soc. Lond. B* 263: 855–859.

Radwan, J. 1998. Heritability of sperm competition success in the bulb mite, *Rhizoglyphus robini. J. Evol. Biol.* 11: 321–328.

Raina, A. K. 1989. Male induced termination of sex pheromone production and receptivity in mated females of *Heliothis zea. J. Insect Physiol.* 35: 821–826.

Raina, A. K., and J. A. Klun. 1984. Brain factor control of sex pheromone production in the female corn earworm moth. *Science* 225: 531.

Raina, A. K., and E. A. Stadelbacher. 1990. Pheromone titer and calling in *Heliothis virescens* (Lepidoptera: Noctuidae): Effect of mating with normal and sterile backcross males. *Ann. Entomol. Soc. Am.* 83: 987–990.

Raina, A. K., T. G. Kingan, and J. M. Giebultowicz. 1994. Mating-induced loss of sex pheromone and sexual receptivity in insects with emphasis on *Helicoverpa zea* and *Lymantria dispar. Arch. Insect Biochem. Physiol.* 25: 317–327.

Ramaswamy, S. B., and N. E. Cohen. 1992. Ecdysone: An inhibitor of receptivity in the moth, *Heliothis virescens? Naturwissenschaften* 79: 29–31.

Ramaswamy, S. B., G. N. Mbata, N. E. Cohen, A. Moore, and N. M. Cox. 1994. Pheromonotropic and pheromonostatic activity in moths. *Arch. Insect Biochem. Physiol.* 25: 301–315.

Ramaswamy, S. B., S. Shu, Y. I. Park, and F. Zeng. 1997. Dynamics of juvenile hormone-mediated gonadotropism in the Lepidoptera. *Arch. Insect Biochem. Physiol.* 35: 539–558.

Rananavare, H. D., M. R. Harwalkar, and G. W. Rahalkar. 1990. Studies on the mating behavior of radiosterilized males of potato tuberworm *Phthorimaea operculella* Zeller. *J. Nucl. Agr. Biol.* 19: 47–53.

Randerson, J. P., and L. D. Hurst. 1999. Small sperm, uniparental inheritance and selfish cytoplasmic elements: A comparison of two models. *J. Evol. Biol.* 12: 1110–1124.

Ransford, M. 1997. Sperm competition in the 2-Spot Ladybird, *Adelia bipuntata.* Ph.D. Diss. University of Cambridge, Cambridge, UK.

Ratnieks, F. L. W. 1988. Reproductive harmony via mutual policing by workers in eusocial Hymenoptera. *Am. Nat.* 132: 217–236.

Ratnieks, F. L. W. 1990a. Assessment of queen mating frequency by workers in social Hymenoptera. *J. Theor. Biol.* 142: 87–93.

Ratnieks, F. L. W. 1990b. The evolution of polyandry by queens in social hymenoptera: The significance of the timing of removal of diploid males. *Behav. Ecol. Sociobiol.* 26: 343–348.

Ratnieks, F. L. W. 1991. Facultative sex allocation biasing by workers in social hymenoptera. *Evolution* 45: 281–292.

Ratnieks, F. L. W., and J. J. Boomsma. 1995. Facultative sex allocation by workers and the evolution of polyandry by queens in social hymenoptera. *Am. Nat.* 145: 969–993.

Ratnieks, F. W. L., and H. K. Reeve. 1992. Conflict in single-queen Hymenopteran societies: The structure of conflict and processes that reduce conflict in advanced eusocial species. *J. Theor. Biol.* 158: 33–65.

Ratnieks, F. W. L., and P. K. Visscher. 1989. Worker policing in the honeybee. *Nature* 342: 796–797.

Reichardt, A. K., and D. E. Wheeler. 1996. Multiple mating in the ant *Acromyrmex versicolor*: A case of female control. *Behav. Ecol. Sociobiol.* 38: 219–225.

Reinecke, J. P. 1985. Nutrition: Artificial diets. In G. A. Kerkut and L. I. Gilbert, eds., *Comprehensive Insect Physiology, Biochemistry and Pharmacology*, pp. 391–420. Pergamon, Oxford.

Reinhardt, K., G. Köhler, and J. Schumacher. 1999. Females of the grasshopper *Chorthippus parallelus* (Zett.) do not remate for fresh sperm. *Proc. Roy. Soc. Lond. B* 266: 2003–2009.

Reinhold, K. 1994. Inheritance of body and testis size in the bushcricket *Poecilimon veluchianus* Ramme (Orthoptera; Tettigoniidae) examined by means of subspecies hybrids. *Biol. J. Linn. Soc.* 52: 305–316.

Reinhold, K. 1996a. Biased primary sex ratio in the bushcricket *Poecilimon veluchianus*, an insect with sex chromosomes. *Behav. Ecol. Sociobiol.* 39: 189–194.

Reinhold, K. 1996b. Sex-ratio selection with asymmetrical migration of the sexes can lead to an uneven sex ratio. *Oikos* 75: 15–19.

Reinhold, K. 1999. Paternal investment in *Poecilimon veluchianus* bushcrickets: Beneficial effects of nuptial feeding on offspring viability. *Behav. Ecol. Sociobiol.* 45: 293–300.

Reinhold, K., and K.-G. Heller. 1993. The ultimate function of nuptial feeding in the bushcricket *Poecilimon veluchianus* (Orthoptera: Tettigoniidae: Phaneropterinae). *Behav. Ecol. Sociobiol.* 32: 55–60.

Retnakaran, A. 1974. The mechanism of sperm precedence in the spruce budworm, *Choristoneura fumiferana* (Lepidoptera: Tortricidae). *Can. Entomol.* 106: 1189–1194.

Reynolds, J. D., and M. R. Gross. 1992. Female mate preference enhances offspring growth and reproduction in a fish, *Poecilia reticulata*. *Proc. Roy. Soc. Lond. B* 250: 57–62.

Rice, W. R. 1984. Sex chromosomes and the evolution of sexual dimorphism. *Evolution* 38: 735–742.

Rice, W. R. 1988. Heritable variation in fitness as a prerequisite for adaptive female choice: The effect of mutation-selection balance. *Evolution* 42: 817–820.

Rice, W. R. 1996. Sexually antagonistic male adaptation triggered by experimental arrest of female evolution. *Nature* 381: 232–234.

Rice, W. R. 1998a. Male fitness increases when females are eliminated from the gene pool: Implications for the Y chromosome. *Proc. Natl. Acad. Sci. USA* 95: 6217–6221.

Rice, W. R. 1998b. Intergenomic conflict, interlocus antagonistic coevolution, and the evolution of reproductive isolation. In D. J. Howard and S. H. Berlocher, eds., *Endless Forms: Species and Speciation*, pp. 261–270. Oxford University Press, Oxford.

Rice, W. R., and B. Holland. 1999. Reply to comments on the chase-away model of sexual selection. *Evolution* 53: 302–306.

Richard, R. D., L. E. LaChance, and F. I. Proshold. 1975. An ultrastructural study of sperm in sterile-hybrids from crosses of *Heleothis virescens* and *Heleothis subflexa*. *Ann. Entomol. Soc. Am.* 68: 35–39.

Richards, M. H., L. Packer, and J. Seger. 1995. Unexpected patterns of parentage and relatedness in a primitively eusocial bee. *Nature* 373: 239–241.

Richards, O. W. 1927. Sexual selection and related problems in the insects. *Biol. Rev.* 2: 298–364.

Richmond, R. C., D. G. Gilbert, K. B. Sheehan, M. H. Gromko, and F. M. Butterworth. 1980. Esterase 6 and reproduction in *Drosophila melanogaster*. *Science* 207: 1483–1485.

Riddiford, L. M. 1993. Hormones and *Drosophila* development. In M. Bate and A. Martinez Arias, eds., *The Development of Drosophila melanogaster*, pp. 899–940. Cold Spring Harbor Press, New York.

Riddiford, L. M., and J. Ashenhurst. 1973. The switchover from virgin to mated behaviour in female *Cercropia* moths: The role of the bursa copulatrix. *Biol. Bull.* 144: 162–171.

Ridley, M. 1988. Mating frequency and fecundity in insects. *Biol. Rev.* 63: 509–549.

Ridley, M. 1989. The incidence of sperm displacement in insects: Four conjectures, one corroboration. *Biol. J. Linn. Soc.* 38: 349–367.

Riemann, J. G. 1973. Ultrastructure of sperm in F1 progeny of irradiated males of the Mediterranean flour moth, *Anagasta kuehniella*. *Ann. Entomol. Soc. Am.* 66: 147–153.

Riemann, J. G., and G. Gassner. 1973. Ultrastructure of lepidopteran sperm within spermathecae. *Ann. Entomol. Soc. Am.* 66: 154–159.

Riemann, J. G., D. O. Moen, and B. J. Thorson. 1967. Female monogamy and its control in the housefly, *Musca domestica* L. *J. Insect Physiol.* 13: 407–418.

Robert, P.-A. 1958. *Les libellules (Odonates)*. Delachaux et Niestlé, Paris.

Robertson, H. G. 1995. Sperm transfer in the ant *Carebara vidua* F. Smith (Hymenoptera: Formicidae). *Insect. Soc.* 42: 411–418.

Robertson, H. M. 1985. Female dimorphism and mating behaviour in a damselfly, *Ischnura ramburi*: Females mimicking males. *Anim. Behav.* 33: 805–809.

Robinson, G. E., and R. E. Page. 1988. Genetic determination of guarding and undertaking in honey-bee colonies. *Nature* 333: 356–358.

Robinson, G. E., and R. E. Page. 1989. Genetic determination of nectar foraging, pollen foraging, and nest site scouting in honey bee colonies. *Behav. Ecol. Sociobiol.* 24: 317–323.

Robinson, J. V., and R. Allgeyer. 1996. Covariation in life-history traits, demographics and behaviour in ischnuran damselflies: The evolution of monandry. *Biol. J. Linn. Soc.* 58: 85–98.

Robinson, J. V., and K. L. Novak. 1997. The relationship between mating system and penis morphology in ischnuran damselflies (Odonata: Coenagrionidae). *Biol. J. Linn. Soc.* 60: 187–200.

Robinson, T., N. A. Johnson, and M. J. Wade. 1994. Postcopulatory, prezygotic isolation: Intraspecific and interspecific sperm precedence in *Tribolium* spp., flour beetles. *Heredity* 73: 155–159.

Rodriguez, V. 1994a. Fuentes de variación en la precedencia de espematozoides de *Chelymorpha alternans* Boheman 1854 (Coleoptera: Chrysomelidae: Cassidinae). Ph.D. diss., University of Costa Rica.

Rodriguez, V. 1994b. Function of the spermathecal muscle in *Chelymorpha alternans* Boheman (Coleoptera: Chrysomelidae: Cassidinae). *Physiol. Entomol.* 19: 198–202.

Rodriguez, V. 1995. Relation of flagellum length to reproductive success in male *Chelymorpha alternans* Boheman (Coleoptera: Chrysomelidae: Cassidinae). *Coleopt. Bull.* 49: 201–205.

Roff, D. A. 1984. The cost of being able to fly: A study of wing polymorphism in two species of crickets. *Oecologia* 63: 30–37.

Roff, D. A., and D. J. Fairbairn. 1993. The evolution of alternate morphologies: Fitness and wing morphology in male sand crickets. *Evolution* 47: 1572–1584.

Roig-Alsina, A. 1993. The evolution of the apoid endophallus, its phylogenetic implications, and functional significance of the genital capsule (Hymenoptera, Apoidea). *Boll. Zool.* 60: 169–183.

Rose, R. W., C. M. Nevison, and A. F. Dixson. 1997. Testes weight, body weight and mating systems in marsupials and monotremes. *J. Zool.* 243: 523–531.

Rosenthal, G. G., and M. R. Servedio. 1999. Chase-away sexual selection: Resistance to "resistance." *Evolution* 53: 296–299.

Ross, K. G. 1986. Kin selection and the problem of sperm utilization in social insects. *Nature* 323: 798–800.

Ross, K. G., and J. M. Carpenter. 1991. Population genetic structure, relatedness, and breeding systems. In K. G. Ross and R. W. Matthews, eds., *The Social Biology of Wasps*, pp. 451–479. Cornell University Press, Ithaca, NY.

Ross, K. G., and D. J. C. Fletcher. 1985. Comparative study of genetic and social structure in two forms of the fire ant *Solenopsis invicta* (Hymenoptera: Formicidae). *Behav. Ecol. Sociobiol.* 17: 349–356.

Ross, K. G., and D. J. C. Fletcher. 1986. Diploid male production—a significant colony mortality factor in the fire ant *Solenopsis invicta* (Hymenoptera: Formicidae). *Behav. Ecol. Sociobiol.* 19: 283–291.

Ross, K. G., E. L. Vargo, and D. J. Fletcher. 1987. Comparative biochemical genetics of three fire ant species in North America, with special reference to the two social forms of *Solenopsis invicta* (Hymenoptera: Formicidae). *Evolution* 41: 979–990.

Ross, K. G., E. L. Vargo, and D. J. C. Fletcher. 1988. Colony genetic structure and queen mating frequency in fire ants of the subgenus *Solenopsis* (Hymenoptera: Formicidae). *Biol. J. Linn. Soc.* 34: 105–117.

Roth, L. M. 1962. Hypersexual activity induced in females of the cockroach *Nauphoeta cinerea*. *Science* 138: 1267–1269.

Rothschild, M. L. 1991. Arrangement of sperm within the spermatheca of fleas, with remarks on sperm displacement. *Biol. J. Linn. Soc.* 43: 313–323.

Rowe, L. 1992. Convenience polyandry in a water strider: Foraging conflicts and female control of copulation frequency and guarding duration. *Anim. Behav.* 44: 189–202.

Rowe, L. 1994. The costs of mating and mate choice in water striders. *Anim. Behav.* 48: 1049–1056.

Rowe, L., and G. Arnqvist. 1996. Analysis of the causal components of assortative mating in water striders. *Behav. Ecol. Sociobiol.* 38: 279–286.

Rowe, L., and D. Houle. 1996. The lek paradox and the capture of genetic variance by condition dependent traits. *Proc. Roy. Soc. Lond. B* 263: 1415–1421.

Rowe, L., G. Arnqvist, A. Sih, and J. J. Krupa. 1994. Sexual conflict and the evolutionary ecology of mating patterns: Water striders as a model system. *TREE* 9: 289–293.

Rubenstein, D. I. 1984. Resource acquisition and alternative mating strategies in water striders. *Am. Zool.* 24: 345–353.

Rubenstein, D. I. 1989. Sperm competition in the water strider, *Gerris remigis*. *Anim. Behav.* 38: 631–636.

Rutowski, R. L. 1978. The courtship behaviour of the small sulphur butterfly *Eurema lisa* (Lepidoptera, Pieridae). *Anim. Behav.* 26: 892–903.

Rutowski, R. L. 1979. Courtship behavior of the checkered white, *Pieris protodice* (Pieridae). *J. Lepid. Soc.* 33: 42–49.

Rutowski, R. L. 1982. Epigamic selection by males as evidenced by courtship partner preferences in the checkered white butterfly (*Pieris protodice*). *Anim. Behav.* 30: 108–112.

Rutowski, R. L., and J. Alcock. 1980. Temporal variation in male copulatory behavior in the solitary bee *Nomadopsis puellae* (Hymenoptera: Andrenidae). *Behaviour* 73: 175–188.

Rutowski, R. L., and G. W. Gilchrist. 1987. Courtship, copulation and oviposition in the chalcedon checkerspot, *Euphydryas chalcedona* (Lepidoptera: Nymphalidae). *J. Nat. Hist.* 21: 1109–1118.

Rutowski, R. L., J. L. Dickinson, and B. Terkanian. 1989. The structure of the mating plug in the checkerspot butterfly, *Euphydryas chalcedona*. *Psyche* 96: 3–4.

Rutowski, R. L., G. W. Gilchrist, and B. Terkanian. 1987. Female butterflies mated with recently mated males show reduced reproductive output. *Behav. Ecol. Sociobiol.* 20: 319–322.

Ryan, M. J., and A. S. Rand. 1993. Sexual selection and signal evolution: The ghost of biases past. *Phil. Trans. Roy. Soc. Lond. B* 340: 187–195.

Ryan, M. J., J. H. Fox, W. Wilczynski, and A. S. Rand. 1990. Sexual selection for sensory exploitation in the frog *Physalaemus pustulosus*. *Nature* 343: 66–67.

Sadowski, J. A., A. J. Moore, and E. D. I. Brodie. 1999. The evolution of empty nuptial gifts in a dance fly, *Empis snoddyi* (Diptera: Empidae): Bigger isn't always better. *Behav. Ecol. Sociobiol.* 45: 161–166.

Sætre, G.-P., T. Moum, S. Bures, M. Král, M. Adamjan, and J. Moreno. 1997. A sexually selected character displacement in flycatchers reinforces premating isolation. *Nature* 387: 589–592.

Sait, S., M. Begon, and D. Thompson. 1994. The influence of larval age on the response of *Plodia interpunctella* to granulosis virus. *J. Invert. Pathol.* 63: 107–110.

Sait, S. M., M. J. G. Gage, and P. A. Cook. 1998. Effects of fertility reducing baculovirus on sperm numbers and sizes in the Indian meal moth, *Plodia interpunctella*. *Funct. Ecol.* 12: 56–62.

Sakaluk, S. K. 1984. Male crickets feed females to ensure complete sperm transfer. *Science* 223: 609–610.

Sakaluk, S. K. 1986. Sperm competition and the evolution of nuptial feeding behavior in the cricket, *Grylloides supplicans* (Walker). *Evolution* 40: 584–593.

Sakaluk, S. K. 1991. Post-copulatory mate guarding in decorated crickets. *Anim. Behav.* 41: 207–216.

Sakaluk, S. K. 1997. Cryptic female choice predicted on wing dimorphism in decorated crickets. *Behav. Ecol.* 8: 326–331.

Sakaluk, S. K. 2000. Sensory exploitation as an evolutionary origin to nuptial food gifts in insects. *Proc. Roy. Soc. Lond. B* 267: 339–343.

Sakaluk, S. K., and W. H. Cade. 1980. Female mating frequency and progeny production in singly and doubly mated house and field crickets. *Can. J. Zool.* 58: 404–411.

Sakaluk, S. K., and W. H. Cade. 1983. The adaptive significance of female multiple matings in house and field crickets. In D. T. Gwynne and G. K. Morris, eds., *Orthopteran Mating Systems: Sexual Competition in a Diverse Group of Insects*, pp. 319–336. Westview Press, Boulder, CO.

Sakaluk, S. K., and A.-K. Eggert. 1996. Female control of sperm transfer and intraspecific variation in sperm precedence: Antecedents to the evolution of a courtship food gift. *Evolution* 50: 694–703.

Sakaluk, S. K., P. J. Bangert, A.-K. Eggert, C. Gack, and L. V. Swanson. 1995. The gin trap as a device facilitating coercive mating in sagebrush crickets. *Proc. Roy. Soc. Lond. B* 261: 65–71.

Sander, K. 1985. Fertilization and egg cell activation in insects. In C. B. Metz and A.

Monroy, eds., *Biology of Fertilization, Vol. 2, Biology of the Sperm*, pp. 409–430. Academic Press, London.

Sang, J. H., and R. C. King. 1961. Nutritional requirements of axenically cultured *Drosophila melanogaster*. *J. Exp. Biol.* 38: 793–809.

Santhosh-Babu, P. B., and V. K. K. Prabhu. 1987. Oviposition-inducing and mating-inhibiting factor in the male accessory glands of *Opisina arenosella* Walker (Lepidoptera: Cryptophasidae). *Current Science* 56: 967–968.

Sauer, K. P., C. Epplen, I. Over, T. Lubjuhn, A. Schmidt, T. Gerken, and J. T. Epplen. 1999. Molecular genetic analysis of remating frequencies and sperm competition in the scorpionfly *Panorpa vulgaris* (Imhoff and Labram). *Behaviour* 136: 1107–1121.

Saul, S. H., and S. D. McCombs. 1993. Dynamics of sperm use in the Mediterranean fruit fly (Diptera: Tephritidae): Reproductive fitness of multiple-mated females and sequentially mated males. *Ann. Entomol. Soc. Am.* 86: 198–202.

Saul, S. H., S. Y. T. Tam, and D. O. McInnis. 1988. Relationship between sperm competition and copulation duration in the Mediterranean fruit fly (Diptera: Tephritidae). *Ann. Entomol. Soc. Am.* 81: 498–502.

Savalli, U. M., and C. W. Fox. 1998a. Genetic variation in paternal investment in a seed beetle. *Anim. Behav.* 56: 953–961.

Savalli, U. M., and C. W. Fox. 1998b. Sexual selection and the fitness consequences of male body size in the seed beetle *Stator limbatus*. *Anim. Behav.* 55: 473–483.

Savalli, U. M., and C. W. Fox. 1999. The effect of male mating history on paternal investment, fecundity and female remating in the seed beetle *Callosobruchus maculatus*. *Funct. Ecol.* 13: 169–177.

Sawada, K. 1995. Male's ability of sperm displacement during prolonged copulations in *Ischnura senegalensis* (Rambur) (Zygoptera: Coenagrionidae). *Odonatologica* 24: 237–244.

Schlager, G. 1960. Sperm precedence in the fertilization of eggs in *Tribolium castaneum*. *Ann. Entomol. Soc. Am.* 53: 557–560.

Schmid-Hempel, P. 1998. *Parasites in Social Insects*. Princeton University Press, Princeton, NJ.

Schmid-Hempel, P., and R. H. Crozier. 1999. Polyandry versus polygyny versus parasites. *Phil. Trans. Roy. Soc. Lond. B* 354: 507–515.

Schmid-Hempel, P., and R. Schmid-Hempel. 1993. Transmission of a pathogen in *Bombus terrestris*, with a note on division of labour in social insects. *Behav. Ecol. Sociobiol.* 33: 319–328.

Schmid-Hempel, P., and R. Schmid-Hempel. 2000. Female mating frequency in *Bombus* spp. from Central Europe. *Insect. Soc.* 47: 36–41.

Schmidt, T., Y. Choffat, S. Klauser, and E. Kubli. 1993. *Drosophila melanogaster* sex peptide: A molecular analysis of structure-function relationships. *J. Insect Physiol.* 39: 361–368.

Schrader, F., and C. Leuchtenberger. 1950. A cytochemical analysis of the functional interrelationships of various cell structures in *Arvelius albopunctatus* (De Geer). *Exp. Cell Res.* 1: 421–452.

Schwagmeyer, P. L., and D. W. Mock. 1993. Shaken confidence of paternity. *Anim. Behav.* 46: 1020–1022.

Schwagmeyer, P. L., and G. A. Parker. 1990. Male mate choice as predicted by sperm competition in thirteen-lined ground squirrels. *Nature* 348: 62–64.

Scott, D. 1986a. Inhibition of female *Drosophila melanogaster* remating by a seminal fluid protein esterase 6. *Evolution* 40: 1084–1091.

Scott, D. 1986b. Sexual mimicry regulates the attractiveness of mated *Drosophila melanogaster* females. *Proc. Natl. Acad. Sci. USA* 83: 8429–8433.

Scott, D. 1987. The timing of the sperm effect on female *Drosophila melanogaster* receptivity. *Anim. Behav.* 35: 142–149.

Scott, D., and L. L. Jackson. 1988. Interstrain comparison of male predominant anti-aphrodisiacs in *Drosophila melanogaster*. *J. Insect Physiol.* 34: 863–872.

Scott, D., and R. C. Richmond. 1985. An effect of mate fertility on the attractiveness and oviposition rates of mated *Drosophila melanogaster* females. *Anim. Behav.* 33: 817–824.

Scott, D., and R. C. Richmond. 1987. Evidence against an antiaphrodisiac role for cis-vaccenyl acetate in *Drosophila melanogaster*. *J. Insect Physiol.* 33: 363–369.

Scott, D., and R. C. Richmond. 1990. Sperm loss by remating *Drosophila melanogaster* females. *J. Insect Physiol.* 36: 451–456.

Seppä, P. 1994. Sociogenetic organization of the ants *Myrmica ruginodis* and *Myrmica lobicornis*: Number, relatedness and longevity of reproducing individuals. *J. Evol. Biol.* 7: 71–95.

Service, P. M., and A. J. Fales. 1993. Evolution of delayed reproductive senescence in male fruit flies: Sperm competition. *Genetica* 91: 111–125.

Service, P. M., and R. E. Vossbrink. 1996. Genetic variation in "first" male effects on egg laying and remating by female *Drosophila melanogaster*. *Behav. Gen.* 26: 39–48.

Shapiro, A. M., and A. H. Porter. 1989. The lock-and-key hypothesis: Evolutionary and biosystematic interpretation of insect genitalia. *Ann. Rev. Entomol.* 34: 231–245.

Sheldon, B. C. 1994. Male phenotype, fertility, and the pursuit of extra-pair copulations by female birds. *Proc. Roy. Soc. Lond. B* 257: 25–30.

Shellman-Reeve, J. S. 1997. The spectrum of eusociality in termites. In J. C. Choe and B. J. Crespi, eds., *The Evolution of Social Behavior in Insects and Arachnids*, pp. 52–93. Cambridge University Press, Cambridge, UK.

Sheperd, J. G. 1975. A polypeptide sperm activator from male saturniid moths. *J. Insect Physiol.* 21: 9–22.

Sherman, K. J. 1983. The adaptive significance of postcopulatory mate guarding in a dragonfly, *Pachydiplax longipennis*. *Anim. Behav.* 31: 1107–1115.

Sherman, P. W., T. D. Seeley, and H. K. Reeve. 1988. Parasites, pathogens and polyandry in social hymenoptera. *Am. Nat.* 131: 602–610.

Sherman, P. W., T. D. Seeley, and H. K. Reeve. 1998. Parasites, pathogens, and polyandry in honey bees. *Am. Nat.* 151: 392–396.

Sherwood, D. R., and E. Levine. 1993. Copulation and its duration affects female weight, oviposition, hatching patterns, and ovarian development in the western corn rootworm (Coleoptera: Chrysomelidae). *J. Econ. Entomol.* 86: 1664–1671.

Shields, O. 1967. Hilltopping. *J. Res. Lepid.* 6: 69–178.

Shirk, P. D., G. Bhaskaran, and H. Röller. 1980. The transfer of juvenile hormone from male to female during mating in the cecropia silkmoth. *Experientia* 36: 682–683.

Shivashankar, T., and D. L. Pearson. 1994. A comparison of mate guarding among five syntopic tiger beetle species from Peninsular India (Coleoptera: Cicindelidae). *Biotropica* 26: 436–442.

Short, R. V. 1979. Sexual selection and component parts, somatic and genital selection, as illustrated by man and the great apes. *Adv. Stud. Behav.* 9: 131–158.

Shykoff, J. A., and P. Schmid-Hempel. 1991a. Gentic relatedness and eusociality: Parasite-mediated selection on the genetic composition of groups. *Behav. Ecol. Sociobiol.* 28: 371–376.

Shykoff, J. A., and P. Schmid-Hempel. 1991b. Parasites and the advantage of genetic variability within social insect colonies. *Proc. Roy. Soc. Lond. B* 243: 55–58.

Sigurjónsdóttir, H., and G. A. Parker. 1981. Dung fly struggles: Evidence for assessment strategy. *Behav. Ecol. Sociobiol.* 8: 219–230.

Sih, A., J. Krupa, and S. Travers. 1990. An experimental study on the effects of predation risk and feeding regime on the mating behavior of the water strider. *Am. Nat.* 135: 284–290.

Silberglied, R. E., J. G. Sheperd, and J. L. Dickinson. 1984. Eunuchs: The role of apyrene sperm in lepidoptera? *Am. Nat.* 123: 255–265.

Sillen-Tullberg, B. 1981. Prolonged copulation: A male "postcopulatory" strategy in a promiscuous species, *Lygaeus equestris* (Heteroptera: Lygaenidae). *Behav. Ecol. Sociobiol.* 9: 283–289.

Simmons, L. W. 1986. Female choice in the field cricket, *Gryllus bimaculatus* (De Geer). *Anim. Behav.* 34: 1463–1470.

Simmons, L. W. 1987a. Female choice contributes to offspring fitness in the field cricket, *Gryllus bimaculatus* (De Geer). *Behav. Ecol. Sociobiol.* 21: 313–321.

Simmons, L. W. 1987b. Sperm competition as a mechanism of female choice in the field cricket, *Gryllus bimaculatus*. *Behav. Ecol. Sociobiol.* 21: 197–202.

Simmons, L. W. 1988a. The contribution of multiple mating and spermatophore consumption to the lifetime reproductive success of female field crickets (*Gryllus bimaculatus*). *Ecol. Entomol.* 13: 57–69.

Simmons, L. W. 1988b. Male size, mating potential and lifetime reproductive success in the field cricket, *Gryllus bimaculatus* (De Geer). *Anim. Behav.* 36: 372–379.

Simmons, L. W. 1989. Kin recognition and its influence on mating preferences of the field cricket, *Gryllus bimaculatus* (De Geer). *Anim. Behav.* 38: 68–77.

Simmons, L. W. 1990a. Nuptial feeding in tettigoniids: Male costs and the rates of fecundity increase. *Behav. Ecol. Sociobiol.* 27: 43–47.

Simmons, L. W. 1990b. Pheromonal cues for the recognition of kin by female field crickets, *Gryllus bimaculatus*. *Anim. Behav.* 40: 192–195.

Simmons, L. W. 1990c. Post-copulatory guarding, female choice and the levels of gregarine infections in the field cricket, *Gryllus bimaculatus*. *Behav. Ecol. Sociobiol.* 26: 403–407.

Simmons, L. W. 1991a. Female choice and the relatedness of mates in the field cricket, *Gryllus bimaculatus*. *Anim. Behav.* 41: 493–501.

Simmons, L. W. 1991b. On the post-copulatory guarding behaviour of male field crickets. *Anim. Behav.* 42: 504–505.

Simmons, L. W. 1993. Some constraints on reproduction for male bushcrickets, *Requena verticalis* (Orthoptera: Tettigoniidae): Diet, size and parasite load. *Behav. Ecol. Sociobiol.* 32: 135–140.

Simmons, L. W. 1995a. Courtship feeding in katydids (Orthoptera: Tettigoniidae): Investment in offspring and in obtaining fertilizations. *Am. Nat.* 146: 307–315.

Simmons, L. W. 1995b. Male bushcrickets tailor spermatophores in relation to their remating intervals. *Funct. Ecol.* 9: 881–886.

Simmons, L. W. 1995c. Relative parental investment, potential reproductive rates, and the control of sexual selection in katydids. *Am. Nat.* 145: 797–808.

Simmons, L. W. In press. The evolution of polyandry: an examination of the genetic incompatibility and good-sperm hypotheses. *J. Evol. Biol.*

Simmons, L. W., and R. Achmann. 2000. Microsatellite analysis of sperm utilisation patterns in the bushcricket, *Requena verticalis*. *Evolution* 54: 942–952.

Simmons, L. W., and W. J. Bailey. 1990. Resource influenced sex roles of zaprochiline tettigoniids (Orthoptera: Tettigoniidae). *Evolution* 44: 1853–1868.

Simmons, L. W., and D. T. Gwynne. 1991. The refractory period of female katydids (Orthoptera: Tettigoniidae): Sexual conflict over the remating interval? *Behav. Ecol.* 2: 276–282.

Simmons, L. W., and C. Kvarnemo. 1997. Ejaculate expenditure by male bushcrickets decreases with sperm competition intensity. *Proc. Roy. Soc. Lond. B* 264: 1203–1208.

Simmons, L. W., and G. A. Parker. 1989. Nuptial feeding in insects: Mating effort versus paternal investment. *Ethology* 81: 332–343.

Simmons, L. W., and G. A. Parker. 1992. Individual variation in sperm competition success of yellow dung flies, *Scatophaga stercoraria*. *Evolution* 46: 366–375.

Simmons, L. W., and G. A. Parker. 1996. Parental investment and the control of sexual selection: Can sperm competition affect the direction of sexual competition? *Proc. Roy. Soc. Lond. B* 263: 515–519.

Simmons, L. W., and M. J. Siva-Jothy. 1998. Sperm competition in insects: Mechanisms and the potential for selection. In T. R. Birkhead and A. P. Møller, eds., *Sperm Competition and Sexual Selection*, pp. 341–434. Academic Press, London.

Simmons, L. W., L. Beesley, P. Lindhjem, D. Newbound, J. Norris, and A. Wayne. 1999a. Nuptial feeding by male bushcrickets: An indicator of male quality? *Behav. Ecol.* 10: 263–269.

Simmons, L. W., M. Craig, T. Llorens, M. Schinzig, and D. Hosken. 1993. Bushcricket spermatophores vary in accord with sperm competition and parental investment theory. *Proc. Roy. Soc. Lond. B* 251: 183–186.

Simmons, L. W., T. Llorens, M. Schinzig, D. Hosken, and M. Craig. 1994. Sperm competition selects for male mate choice and protandry in the bushcricket, *Requena verticalis* (Orthoptera: Tettigoniidae). *Anim. Behav.* 47: 117–122.

Simmons, L. W., G. A. Parker, and P. Stockley. 1999b. Sperm displacement in the yellow dung fly, *Scatophaga stercoraria:* An investigation of male and female processes. *Am. Nat.* 153: 302–314.

Simmons, L. W., P. Stockley, R. L. Jackson, and G. A. Parker. 1996. Sperm competition or sperm selection: No evidence for female influence over paternity in yellow dung flies *Scatophaga stercoraria*. *Behav. Ecol. Sociobiol.* 38: 199–206.

Simmons, L. W., R. J. Teale, M. Maier, R. J. Standish, W. J. Bailey, and P. C. Withers. 1992. Some costs of reproduction for male bushcrickets, *Requena verticalis* (Orthoptera: Tettigoniidae): Allocating resources to mate attraction and nuptial feeding. *Behav. Ecol. Sociobiol.* 31: 57–62.

Simmons, L. W., J. L. Tomkins, and J. Alcock. 2000. Can minor males of Dawson's burrowing bee, *Amegilla dawsoni* (Hymenoptera: Anthophorini) compensate for reduced access to virgin females through sperm competition? *Behav. Ecol.* 11: 319–325.

Simmons, L. W., J. L. Tomkins, and J. Hunt. 1999c. Sperm competition games played by dimorphic male beetles. *Proc. Roy. Soc. Lond. B* 266: 145–150.

Singer, F. 1987. A physiological basis of variation in postcopulatory behaviour in a dragonfly *Sympetrum obtrusum*. *Anim. Behav*. 35: 1575–1577.

Siva-Jothy, M. T. 1984. Sperm competition in the family Libellulidae (Anisoptera) with special reference to *Crocothemis erythraea* (Brulle) and *Orthetrum cancellatum* (L.). *Adv. Odonatol.* 2: 195–207.

Siva-Jothy, M. T. 1987a. The structure and function of the female sperm-storage organs in libellulid dragonflies. *J. Insect Physiol.* 33: 559–568.

Siva-Jothy, M. T. 1987b. Variation in copulation duration and the resultant degree of sperm removal in *Orthetrum cancellatum* (L.) (Libellulidae: Odonata). *Behav. Ecol. Sociobiol.* 20: 147–151.

Siva-Jothy, M. T. 1988. Sperm "repositioning" in *Crocothemis erythraea*, a libellulid dragonfly with a brief copulation. *J. Insect Behav.* 1: 235–245.

Siva-Jothy, M. T., and R. E. Hooper. 1995. The disposition and genetic diversity of stored sperm in females of the damselfly *Calopteryx splendens xanthostoma* (Charpentier). *Proc. Roy. Soc. Lond. B* 259: 313–318.

Siva-Jothy, M. T., and R. E. Hooper. 1996. Differential use of stored sperm during oviposition in the damselfly *Calopteryx splendens xanthostoma* (Charpentier). *Behav. Ecol. Sociobiol.* 39: 389–394.

Siva-Jothy, M. T., and Y. Tsubaki. 1989a. Variation in copulation duration in *Mnais pruinosa pruinosa* Selys (Odonata: Calopterygidae). 1. Alternative mate-securing tactics and sperm precedence. *Behav. Ecol. Sociobiol.* 24: 39–45.

Siva-Jothy, M. T., and Y. Tsubaki. 1989b. Variation in copulation duration in *Mnais pruinosa pruinosa* Selys (Odonata: Calopterygidae). 2. Causal factors. *Behav. Ecol. Sociobiol.* 25: 261–267.

Siva-Jothy, M. T., and Y. Tsubaki. 1994. Sperm competition and sperm precedence in the dragonfly *Nanophya pygmaea*. *Physiol. Entomol.* 19: 363–366.

Siva-Jothy, M. T., D. Earle Blake, J. Thompson, and J. J. Ryder. 1996. Short- and long-term sperm precedence in the beetle *Tenebrio molitor*: A test of the "adaptive sperm removal" hypothesis. *Physiol. Entomol.* 21: 313–316.

Sivinski, J. 1979. Intrasexual aggression in the stick insects *Diapheromera veliei* and *D. covilleae* and sexual dimorphism in the Phasmatodea. *Psyche* 85: 395–405.

Sivinski, J. 1980. Sexual selection and insect sperm. *Fla. Entomol.* 63: 99–111.

Sivinski, J. 1983. Predation and sperm competition in the evolution of coupling durations, particularly in the stick insect *Diapheromera veliei*. In D. T. Gwynne and G. K. Morris, eds., *Orthopteran Mating Systems: Sexual Competition in a Diverse Group of Insects*, pp. 147–162. Westview Press, Boulder, CO.

Sivinski, J. 1984. Sperm in competition. In R. L. Smith, ed., *Sperm Competition and the Evolution of Animal Mating Systems*, pp. 86–115. Academic Press, London.

Sivinski, J. M., and G. Dodson. 1992. Sexual dimorphism in *Anastrepha suspensa* (Loew) and other tephritid fruit flies (Diptera: Tephrididae): Possible roles for development rate, fecundity, and dispersal. *J. Insect Behav.* 5: 491–506.

Smid, H. M. 1997. Chemical mate guarding and oviposition stimulation in insects—a model mechanism alternative to the *Drosophila* sex-peptide paradigm. *Proc. Koninklij. Ned. Akad. Wetensch.* 100: 269–278.

Smid, H. M. 1998. Transfer of a male accessory gland peptide to the female during mating in *Leptinotarsa decemlineata*. *Invert. Reprod. Dev.* 34: 47–53.

Smith, A. F., and C. Schal. 1990. The physiological basis for the termination of pheromone-releasing behaviour in the female brown-banded cockroach, *Supell longipalpa* (F.) (Dictyoptera: Blattellidae). *J. Insect Physiol.* 36: 369–373.

Smith, P. H., L. B. Browne, and A. C. M. van Gerwen. 1989. Causes and correlates of loss and recovery of sexual receptivity in *Lucilia cuprina* females after their first mating. *J. Insect Behav.* 2: 325–338.

Smith, P. H., C. Gillott, L. Barton Browne, and A. C. M. van Gerwen. 1990. The mating-induced refractoriness of *Lucilia cuprina* females: Manipulating the male contribution. *Physiol. Entomol.* 15: 469–481.

Smith, R. L. 1976. Male brooding behavior of the water bug *Abedus herberti* (Heteroptera: Belostomatidae). *Ann. Entomol. Soc. Am.* 69: 740–747.

Smith, R. L. 1979. Repeated copulation and sperm precedence: Paternity assurance for a male brooding water bug. *Science* 205: 1029–1031.

Smith, R. L. 1984. *Sperm Competition and the Evolution of Animal Mating Systems.* Academic Press, London.

Snook, R. R. 1997. Is the production of multiple sperm types adaptive? *Evolution* 51: 797–808.

Snook, R. R. 1998. The risk of sperm competition and the evolution of sperm heteromorphism. *Anim. Behav.* 56: 1497–1507.

Snook, R. R., and T. L. Karr. 1998. Only long sperm are fertilization-competent in six sperm-heteromorphic *Drosophila* species. *Current Biology* 8: 291–294.

Snook, R. R., and T. A. Markow. 1996. Possible role of nonfertilizing sperm as a nutrient source for female *Drosophila pseudoobscura* Frolova (Diptera: Drosophilidae). *Pan-Pacific Entomol.* 72: 121–129.

Snook, R. R., T. A. Markow, and T. L. Karr. 1994. Functional nonequivalence of sperm in *Drosophila pseudoobscura*. *Proc. Natl. Acad. Sci. USA* 91: 11222–11226.

Snow, J. W., J. R. Young, and R. L. Jones. 1970. Competitiveness of sperm in female fall armyworms mating with normal and chemosterilized males. *J. Econ. Entomol.* 63: 1799–1802.

Snyder, L. E. 1992. The genetics of social behavior in a polygynous ant. *Naturwissenschaften* 79: 525–527.

Snyder, L. E. 1993. Non-random behavioural interactions among genetic subgroups in a polygynous ant. *Anim. Behav.* 46: 431–439.

Snyder, L. E., and J. M. Herbers. 1991. Polydomy and sexual allocation ratios in the ant *Myrmica punctiventris*. *Behav. Ecol. Sociobiol.* 28: 409–415.

Solymar, B. D., and W. H. Cade. 1990. Heritable variation for female mating frequency in field crickets, *Gryllus integer*. *Behav. Ecol. Sociobiol.* 26: 73–76.

Souroukis, K., and A.-M. Murray. 1995. Female mating behavior in the field cricket, *Gryllus pennsylvanicus* (Orthoptera: Gryllidae) at different operational sex ratios. *J. Insect Behav.* 8: 269–279.

Spence, J. R., and R. S. Wilcox. 1986. The mating system of two hybridizing species of water striders (Gerridae). II. Alternative tactics of males and females. *Behav. Ecol. Sociobiol.* 19: 87–95.

Sridevi, R., A. Ray, and P. S. Ramamurty. 1987. An oviposition stimulant in the male accessory gland extract of *Spodoptera litura* (Lepidoptera, Noctuidae). *Entomon.* 12: 231–234.

Srivastava, U. S., and B. P. Srivastava. 1957. Notes on the spermatophore formation and transference of sperms in the female reproductive organs of *Leucinodes orbonalis* Guen. (Lepidoptera, Pyraustidae). *Zool. Anz.* 158: 258–266.

Stanley-Samuelson, D. W., and W. Loher. 1983. Arachidonic and other long-chain polyunsaturated fatty acids in spermatophores and spermathecae of *Teleogryllus commodus*: Significance in prostaglandin-mediated reproductive behaviour. *J. Insect Physiol.* 29: 41–45.

Stanley-Samuelson, D. W., and W. Loher. 1985. The disappearance of injected prostaglandins from the circulation of adult female Australian field crickets, *Teleogryllus commodus*. *Arch. Insect Biochem. Physiol.* 2: 367–374.

Stanley-Samuelson, D. W., and W. Loher. 1986. Prostaglandins in insect reproduction. *Ann. Entomol. Soc. Am.* 79: 841–853.

Stanley-Samuelson, D. W., J. J. Peloquin, and W. Loher. 1986. Egg-laying in response to prostaglandin injections in the Australian field cricket, *Teleogryllus commodus*. *Physiol. Entomol.* 11: 213–219.

Starr, C. K. 1984. Sperm competition, kinship, and sociality in the aculeate Hymenoptera. In R. L. Smith, ed., *Sperm Competition and the Evolution of Animal Mating Systems*, pp. 427–464. Academic Press, London.

Stearns, S. C. 1983. The influence of size and phylogeny on patterns of covariation among life-history traits in mammals. *Oikos* 41: 173–187.

Steele, R. H. 1986. Courtship feeding in *Drosophila subobscura*. II. Courtship feeding by males influences female mate choice. *Anim. Behav.* 34: 1099–1108.

Stephens, D. W., and S. R. Dunbar. 1993. Dimensional analysis in behavioral ecology. *Behav. Ecol.* 4: 172–183.

Stockley, P. 1999. Sperm selection and genetic incompatibility: Does relatedness of mates affect male success in sperm competition? *Proc. Roy. Soc. Lond. B* 266: 1663–1669.

Stockley, P., and L. W. Simmons. 1998. Consequences of sperm displacement for female dung flies, *Scatophaga stercoraria*. *Proc. Roy. Soc. Lond. B* 265: 1755–1760.

Stockley, P., M. J. G. Gage, G. A. Parker, and A. P. Møller. 1997. Sperm competition in fishes: The evolution of testis size and ejaculate characteristics. *Am. Nat.* 149: 933–954.

Stoks, R., L. De Bruyn, and E. Matthysen. 1997. The adaptiveness of intense contact mate guarding by males of the emerald damselfly, *Lestes sponsa* (Odonata: Lestidae): The male's perspective. *J. Insect Behav.* 10: 289–298.

Stuart, R. J., and R. E. Page. 1991. Genetic component to division of labor among workers of a leptothoracine ant. *Naturwissenschaften* 78: 375–377.

Stumm-Zollinger, E., and P. S. Chen. 1985. Protein metabolism of *Drosophila* male accessory glands. I. Characterization of secretory proteins. *Insect Biochem.* 15: 375–383.

Stumm-Zollinger, E., and P. S. Chen. 1988. Gene expression of male accessory glands of interspecific hybrids of *Drosophila*. *J. Insect Physiol.* 34: 59–74.

Sugawara, T. 1979. Stretch reception in the bursa copulatrix of the butterfly, *Pieris rapae crucivora*, and its role in behaviour. *J. Comp. Physiol.* 130: 191–199.

Sugawara, T. 1993. Oviposition behavior of the cricket *Teleogryllus commodus*: Mechanosensory cells in the genital chamber and their role in the switch-over of steps. *J. Insect Physiol.* 39: 335–346.

Sundberg, J., and A. Dixon. 1996. Old, colourful male yellowhammers, *Emberiza citrinella*, benefit from extra-pair copulations. *Anim. Behav.* 52: 113–122.

Sundström, L. 1993. Genetic population structure and sociogenetic organisation in *Formica truncorum* (Hymenoptera; Formicidae). *Behav. Ecol. Sociobiol.* 33: 345–354.

Sundström, L. 1994. Sex ratio bias, relatedness asymmetry and queen mating frequency in ants. *Nature* 367: 266–268.

Sundström, L., and F. L. W. Ratnieks. 1998. Sex ratio conflicts, mating frequency, and queen fitness in the ant *Formica truncorum*. *Behav. Ecol.* 9: 116–121.

Sundström, L., M. Chapuisat, and L. Keller. 1996. Conditional manipulation of sex ratios by ant workers: A test of kin selection theory. *Science* 274: 993–995.

Suzuki, N., T. Okuda, and H. Shinbo. 1996. Sperm precedence and sperm movement under different copulation intervals in the silkworm, *Bombyx mori*. *J. Insect Physiol.* 42: 199–204.

Svärd, L., and J. N. McNeil. 1994. Female benefit, male risk: Polyandry in the true armyworm *Pseudaletia unipuncta*. *Behav. Ecol. Sociobiol.* 35: 319–326.

Svärd, L., and C. Wiklund. 1988a. Fecundity, egg weight and longevity in relation to multiple matings in females of the monarch butterfly. *Behav. Ecol. Sociobiol.* 23: 39–43.

Svärd, L., and C. Wiklund. 1988b. Prolonged mating in the monarch butterfly *Danaus plexippus* and nightfall as a cue for sperm transfer. *Oikos* 52: 351–354.

Svärd, L., and C. Wiklund. 1989. Mass and production rate of ejaculates in relation to monandry/polyandry in butterflies. *Behav. Ecol. Sociobiol.* 24: 395–402.

Svensson, B. G., and E. Petersson. 1987. Sex-role reversed courtship behaviour, sexual dimorphism and nuptial gifts in the dance fly, *Empis borealis* (L.). *Ann. Zool. Fenn.* 24: 323–334.

Svensson, B. G., E. Petersson, and M. Frisk. 1990. Nuptial gift size prolongs copulation duration in the dance fly *Empis borealis*. *Ecol. Entomol.* 15: 225–229.

Svensson, E., and J. Nilsson. 1996. Mate quality affects offspring sex ratio in blue tits. *Proc. Roy. Soc. Lond. B* 263: 357–361.

Szöllösi, A. 1974. Ultrastructural study of the spermatodesm of *Locusta migratoria migratorioides* (R.F.): Acrosome and cap formation. *Acrida* 3: 175–192.

Taber, S. 1955. Sperm distribution in the spermathecae of multiple mated queen honey bees. *J. Econ. Entomol.* 48: 522–525.

Taber, S., and J. Wendel. 1958. Concerning the number of times queen bees mate. *J. Econ. Entomol.* 48: 522–525.

Tadler, A. 1999. Selection of a conspicuous male genitalic trait in seedbug *Lygaeus simulans*. *Proc. Roy. Soc. Lond. B* 266: 1773–1777.

Tadler, A., and H. L. Nemeschkal. 1999. Selection of male traits during and after copulation in the seedbug *Lygaeus simulans* (Heteroptera, Lygaeidae). *Biol. J. Linn. Soc.* 68: 471–483.

Tang, J. D., R. E. Charlton, R. A. Jurenka, W. A. Wolf, P. L. Phelan, L. Sreng, and W. L. Roelofs. 1989. Regulation of pheromone biosynthesis by a brain hormone in two moth species. *Proc. Natl. Acad. Sci. USA* 86: 1806–1810.

Taylor, G. 1952. The action of waving cylindrical tails in propelling microscopic organisms. *Proc. Roy. Soc. Lond. A* 211: 225–239.

Taylor, O. R. 1967. Relationship of multiple mating to fertility in *Atteva punctella* (Lepidoptera: Yponomeutidae). *Ann. Entomol. Soc. Am.* 60: 583–590.

Taylor, P. W., and B. Yuval. 1999. Postcopulatory sexual selection in Mediterranean fruit flies: Advantages for large and protein-fed males. *Anim. Behav.* 58: 247–254.

Taylor, V. A. 1982. The giant sperm of a minute beetle. *Tissue and Cell* 14: 113–123.

Thibout, E. 1975. Analyse des causes de l'inhibition de la réceptivité sexualle et de l'influence d'une eventualle seconde copulation sur la reproduction chez la Teigne du poireau, *Acrolepia assectella* (Lepidoptera: Plutellidae). *Entomol. Expl. Appl.* 18: 105–116.

Thompson, D. J. 1990. On the biology of the damselfly *Nososticta kalumbaru* Watson & Theischinger (Zygoptera: Protoneuridae). *Biol. J. Linn. Soc.* 40: 347–356.

Thorén, P. A., R. J. Paxton, and A. Estoup. 1995. Unusually high frequency of (CT)n and (GT)n microsatellite loci in a yellowjacket wasp, *Vespula rufa* (L.) (Hymenoptera: Vespidae). *Insect Mol. Biol.* 4: 141–148.

Thornhill, R. 1976a. Reproductive behavior of the lovebug, *Plecia nearctica* (Diptera: Bibionidae). *Ann. Entomol. Soc. Am.* 69: 843–847.

Thornhill, R. 1976b. Sexual selection and nuptial feeding behavior in *Bittacus apicalis* (Insecta: Mecoptera). *Am. Nat.* 110: 529–548.

Thornhill, R. 1976c. Sexual selection and paternal investment in insects. *Am. Nat.* 110: 153–163.

Thornhill, R. 1979. Male and female sexual selection and the evolution of mating systems in insects. In M. S. Blum and N. A. Blum, eds., *Sexual Selection and Reproductive Competition in Insects*, pp. 81–122. Academic Press, New York.

Thornhill, R. 1980a. Rape in *Panorpa* scorpionflies and a general rape hypothesis. *Anim. Behav.* 28: 52–59.

Thornhill, R. 1980b. Sexual selection within mating swarms of the lovebug, *Plecia nearctica* (Diptera: Bibionidae). *Anim. Behav.* 28: 405–412.

Thornhill, R. 1983. Cryptic female choice and its implications in the scorpionfly *Harpobittacus nigriceps*. *Am. Nat.* 122: 765–788.

Thornhill, R. 1984a. Alternative female choice tactics in the scorpionfly *Hylobittacus apicalis* (Mecoptera) and their implications. *Am. Zool.* 24: 367–383.

Thornhill, R. 1984b. Alternative hypotheses for traits believed to have evolved by sperm competition. In R. L. Smith, ed., *Sperm Competition and the Evolution of Animal Mating Systems*, pp. 151–178. Academic Press, London.

Thornhill, R., and J. Alcock. 1983. *The Evolution of Insect Mating Systems*. Harvard University Press, Cambridge, MA.

Thornhill, R., and K. P. Sauer. 1991. The notal organ of the scorpionfly (*Panorpa vulgaris*): An adaptation to coerce mating duration. *Behav. Ecol.* 2: 156–164.

Tomkins, J. L., and L. W. Simmons. 1996. Dimorphisms and fluctuating asymmetry in the forceps of male earwigs. *J. Evol. Biol.* 9: 753–770.

Tomkins, J. L., and L. W. Simmons. 2000. Sperm competition games played by dimorphic male beetles: fertilisation gains with equal mating access. *Proc. Roy. Soc. Lond. B* 267: 1547–1553.

Tomkins, L., and J. C. Hall. 1981. The different effects on courtship of volatile compounds from mated and virgin *Drosophila* females. *J. Insect Physiol.* 26: 689–697.

Tomkins, L., and J. C. Hall. 1983. Identification of brain sites controlling female receptivity in mosaics of *Drosophila melanogaster*. *Genetics* 103: 179–195.

Tram, U., and M. F. Wolfner. 1998. Seminal fluid regulation of female sexual attractiveness in *Drosophila melanogaster*. *Proc. Natl. Acad. Sci. USA* 95: 4051–4054.

Tram, U., and M. F. Wolfner. 1999. Male seminal fluid proteins are essential for sperm storage in *Drosophila melanogaster*. *Genetics* 153: 837–844.

Travers, S. E., and A. Sih. 1991. The influence of starvation and predators on the mating behavior of a semiaquatic insect. *Ecology* 72: 2123–2136.

Tregenza, T., and N. Wedell. 1998. Benefits of multiple mates in the cricket *Gryllus bimaculatus*. *Evolution* 52: 1726–1730.

Tregenza, T., and N. Wedell. 2000. Genetic compatibility, mate choice and patterns of parentage. *Mol. Ecol.* 9: 1013–1027.

Trivers, R. L. 1972. Parental investment and sexual selection. In B. Campbell, ed., *Sexual Selection and the Descent of Man, 1871–1971*, pp. 136–172. Aldine-Atherton, Chicago.

Trivers, R. L., and H. Hare. 1976. Haplodiploidy and the evolution of the social insects. *Science* 191: 249–263.

Trumbo, S. T., and A. J. Fiore. 1991. A genetic marker for investigating paternity and maternity in the burying beetle *Nicrophorus orbicollis* (Coleoptera: Silphidae). *J. N.Y. Entomol. Soc.* 99: 637–642.

Tsaur, S.-C., C.-T. Tinb, and C.-I. Wu. 1998. Positive selection driving the evolution of a gene of male reproduction, *Acp26Aa*, of *Drosophila*: II. Divergence versus polymorphism. *Mol. Biol. Evol.* 15: 1040–1046.

Tschinkel, W. R. 1987. Relationship between ovariole number and spermathecal sperm count in ant queens: A new allometry. *Ann. Entomol. Soc. Am.* 80: 208–211.

Tschudi-Rein, K., and G. Benz. 1990. Mechanisms of sperm transfer in female *Pieris brassicae* (Lepidoptera: Pieridae). *Ann. Entomol. Soc. Am.* 83: 1158–1164.

Tsubaki, Y., and T. Ono. 1985. The adaptive significance of non-contact mate guarding by males of the dragonfly, *Nannophya pygmaea* Rambur (Odonata: Libellulidae). *J. Ethol.* 3: 135–141.

Tsubaki, Y., and Y. Sokei. 1988. Prolonged mating in the melon fly, *Dacus cucurbitae* (Diptera: Tephritidae): Competition for fertilization by sperm-loading. *Res. Pop. Biol.* 30: 343–352.

Tsubaki, Y., and M. Yamagishi. 1991. "Longevity" of sperm within the female of the melon fly, *Dacus cucurbitae* (Diptera: Tephritidae), and its relevence to sperm competition. *J. Insect Behav.* 4: 243–250.

Tsubaki, Y., M. T. Siva-Jothy, and T. Ono. 1994. Re-copulation and post-copulatory mate guarding increase immediate female reproductive output in the dragonfly *Nannophya pygmaea* Rambur. *Behav. Ecol. Sociobiol.* 35: 219–225.

Tsuchida, K. 1994. Genetic relatedness and the breeding structure of the Japanese wasp, *Polistes jadwigae*. *Ethol. Ecol. Evol.* 6: 237–242.

Tsukamoto, L., K. Kuki, and S. Tojo. 1994. Mating tactics and constraints in the gregarious insect *Parastrachia japonensis* (Hemiptera: Cydnidae). *Ann. Entomol. Soc. Am.* 87: 962–971.

Turner, M. E., and W. W. Anderson. 1984. Sperm predominance among *Drosophila pseudoobscura* karyotypes. *Evolution* 38: 983–995.

Ueda, T. 1979. Plasticity of the reproductive behaviour of dragonfly, *Sympetrum parvulum* Bartneff, with reference to the social relationships of males and the density of territories. *Res. Pop. Ecol.* 21: 135–152.

Ueno, H. 1994. Intraspecific variation of P2 value in a coccinellid beetle, *Harmonia axyridis*. *J. Ethol.* 12: 169–174.

Ueno, H., and Y. Ito. 1992. Sperm precedence in *Eysarcoris lewisi* (Heteroptera: Pentatomidae) in relation to duration between oviposition and the last copulation. *Appl. Entomol. Zool.* 27: 421–426.

Utzeri, C., and L. Dell'Anna. 1989. Wandering and territoriality in *Libellula depressa* L. (Anisoptera: Libellulidae). *Adv. Odonatol.* 4: 133–147.

Vacquier, V. D. 1998. Evolution of gamete recognition proteins. *Science* 281: 1995–1998.

Vacquier, V. D., and Y.-H. Lee. 1993. Abolone sperm lysin: unusual mode of evolution of a gamete recognition protein. *Zygote* 1: 181–196.

Vahed, K. 1996. Prolonged copulation in Oak Bushcrickets (Tettigoniidae: Meconematinae: *Meconema thalassinium* and *M. meridionale*). *J. Orth. Res.* 5: 199–204.

Vahed, K. 1998a. Copulation and spermatophores in the Ephippigerinae (Orthoptera: Tettigoniidae): Prolonged copulation is associated with a smaller nuptial gift in *Uromenus rugosicollis* Serville. *J. Orth. Res.* 6: 83–89.

Vahed, K. 1998b. The function of nuptial feeding in insects: A review of empirical studies. *Biol. Rev.* 73: 43–78.

Vahed, K. 1998c. Sperm precedence and the potential of the nuptial gift to function as paternal investment in the tettigoniid *Steropleurus stali* Bolivar (Orthoptera: Tettigoniidae: Ephippigerinae). *J. Orthop. Res.* 7: 223–226.

Vahed, K., and F. S. Gilbert. 1996. Differences across taxa in nuptial gift size correlate with differences in sperm number and ejaculate volume in bushcrickets (Orthoptera: Tettigoniidae). *Proc. Roy. Soc. Lond. B* 263: 1257–1265.

van den Assem, J., J. J. A. van Iersel, and R. L. Los-Den Hartogh. 1989. Is being large more important for female than for male parasitic wasps? *Behaviour* 108: 160–195.

van der Have, T. M., J. J. Boomsma, and S. B. J. Menken. 1988. Sex-investment ratios and relatedness in the monogynous ant *Lasius niger* (L.). *Evolution* 42: 160–172.

Vander Meer, R. K., M. S. Obin, S. Zawistowski, K. B. Sheehan, and R. C. Richmond. 1986. A reevaluation of the role of *cis*-vaccenyl acetate, *cis*-vaccenol and esterase 6 in the regulation of mated female sexual attractiveness in *Drosophila melanogaster*. *J. Insect Physiol.* 32: 681–686.

Vardell, H. H., and J. H. Brower. 1978. Sperm precedence in *Tribolium confusum* (Coleoptera: Tenebrionidae). *J. Kan. Entomol. Soc.* 51: 187–190.

Vepsäläinen, K., and M. Nummelin. 1985. Male territoriality in the waterstrider *Limnoporus rufoscutellatus*. *Ann. Zool. Fenn.* 22: 441–448.

Vepsäläinen, K., and R. Savolainen. 1995. Operational sex ratios and mating conflict between the sexes in the water strider *Gerris lacustris*. *Am. Nat.* 146: 869–880.

Vick, K. W., W. E. Burkholder, and B. J. Smittle. 1972. Duration of mating refractory period and frequency of second matings in female *Trogoderma inclusum* (Coleoptera: Dermestidae). *Ann. Entomol. Soc. Am.* 65: 790–793.

Villarreal, C., G. Fuentes-Maldonado, M. H. Rodriguez, and B. Yuval. 1994. Low rates of multiple fertilization in parous *Anopheles albimanus*. *J. Am. Mosquito Control Assoc.* 10: 67–69.

Villavaso, E. J. 1975a. Functions of the spermathecal muscle of the boll weevil, *Anthonomis grandis*. *J. Insect Physiol.* 21: 1275–1278.

Villavaso, E. J. 1975b. The role of the spermathecal gland of the boll weevil, *Anthonomus grandis*. *J. Insect Physiol.* 21: 1457–1462.

Villesen, P., P. J. Gertsch, J. Frydenberg, U. G. Mueller, and J. J. Boomsma. 1999. Evolutionary transition from single to multiple mating in fungus-growing ants. *Mol. Ecol.* 8: 1819–1825.

von Helversen, D., and O. von Helversen. 1991. Pre-mating sperm removal in the bushcricket *Metaplastes ornatus* Ramme 1931 (Orthoptera, Tettigoniidae, Phaneropteridae). *Behav. Ecol. Sociobiol.* 28: 391–396.

Waage, J. K. 1973. Reproductive behavior and its relation to territoriality in *Calopteryx maculata* (Beauvois) (Odonata: Calopterygidae). *Behaviour* 47: 240–256.

Waage, J. K. 1978. Oviposition duration and egg deposition rates in *Calopteryx maculata* (P. de Beauvois) (Zygoptera: Calopterygidae). *Odonatologica* 7: 77–88.

Waage, J. K. 1979a. Adaptive significance of postcopulatory guarding of mates and non-mates by male *Calopteryx maculata* (Odonata). *Behav. Ecol. Sociobiol.* 6: 147–154.

Waage, J. K. 1979b. Dual function of the damselfly penis: Sperm removal and transfer. *Science* 203: 916–918.

Waage, J. K. 1982. Sperm displacement by male *Lestes vigilax* Hagen (Odonata: Zygoptera). *Odonatologica* 11: 201–209.

Waage, J. K. 1984. Sperm competition and the evolution of odonate mating systems. In R. L. Smith, ed., *Sperm Competition and the Evolution of Animal Mating Systems*, pp. 251–290. Academic Press, London.

Waage, J. K. 1986. Evidence for widespread sperm displacement ability among Zygoptera (Odonata) and the means for predicting its presence. *Biol. J. Linn. Soc.* 28: 285–300.

Wade, M. J., and N. W. Chang. 1995. Increased male fertility in *Tribolium confusum* beetles after infection with the intracellular parasite *Wolbachia*. *Nature* 373: 72–74.

Wade, M. J., H. Petterson, N. W. Chang, and N. A. Johnson. 1994. Postcopulatory, prezygotic isolation in flour beetles. *Heredity* 72: 163–167.

Waldman, B., P. C. Frumhoff, and P. W. Sherman. 1988. Problems of kin recognition. *TREE* 3: 8–13.

Walker, W. F. 1979. Mating behaviour in *Oncopeltus fasciatus*: Circadian rhythms of coupling, copula duration and "rocking" behaviour. *Physiol. Entomol.* 4: 275–283.

Walker, W. F. 1980. Sperm utilization strategies in nonsocial insects. *Am. Nat.* 115: 780–799.

Wallace, A. R. 1889. *Darwinism: An Exposition of the Theory of Natural Selection*. Macmillan, London.

Wang, Q., and J. G. Millar. 1997. Reproductive behavior of *Thyanta pallidovirens* (Heteroptera: Pentatomidae). *Ann. Entomol. Soc. Am.* 90: 380–388.

Ward, K. E., and P. J. Landolt. 1995. Influence of multiple matings on fecundity and longevity of female cabbage looper moths (Lepidoptera, Noctuidae). *Ann. Entomol. Soc. Am.* 88: 768–772.

Ward, P. I. 1988. Sexual selection, natural selection, and body size in *Gammarus pulex* (Amphipoda). *Am. Nat.* 131: 348–359.

Ward, P. I. 1993. Females influence sperm storage and use in the yellow dung fly *Scathophaga stercoraria* (L.). *Behav. Ecol. Sociobiol.* 32: 313–319.

Ward, P. I. 1998a. Intraspecific variation in sperm size characters. *Heredity* 80: 655–659.

Ward, P. I. 1998b. A possible explanation for cryptic female choice in the yellow dung fly, *Scathophaga stercoraria* (L.). *Ethology* 104: 97–110.

Ward, P. I. 2000. Cryptic female choice in the yellow dung fly *Scathophaga stercoraria* (L.). *Evolution* 54: 1680–1686.

Ward, P. I., and E. Hauschteck-Jungen. 1993. Variation in sperm length in the yellow dung fly *Scathophaga stercoraria* (L.). *J. Insect Physiol.* 39: 545–547.

Ward, P. I., and L. W. Simmons. 1991. Copula duration and testes size in the yellow dung fly, *Scathophaga stercoraria* (L.): The effects of diet, body size, and mating history. *Behav. Ecol. Sociobiol.* 29: 77–85.

Warwick, S. 1999. Nutritional regulation and spermatophylax donation in the mating system of *Gryllodes sigillatus* (Orthoptera: Gryllidae). D. Phil. diss. University of Oxford, UK.

Watanabe, M., and M. Taguchi. 1990. Mating tactics and male wing dimorphism in the damselfly, *Mnais pruinosa costalis* Selys (Odonata: Calopterygidae). *J. Ethol.* 8: 129–137.

Watanabe, M., and M. Taguchi. 1997. Competition for perching sites in the hyaline-winged males of the damselfly *Mnais pruinosa costalis* Selys that use sneaky mate-securing tactics (Zygoptera: Calopterygidae). *Odonatologica* 26: 183–191.

Watanabe, M., C. Wiklund, and M. Bon'no. 1998. The effect of repeated matings on sperm numbers in successive ejaculates of the cabbage white butterfly *Pieris rapae* (Lepidoptera: Pieridae). *J. Insect Behav.* 11: 559–569.

Watson, P. J. 1998. Multi-male mating and female choice increase offspring growth in the spider *Neriene litigiosa* (Linyphiidae). *Anim. Behav.* 55: 387–403.

Watson, P. J., and J. R. B. Lighton. 1994. Sexual selection and the energetics of copulatory courtship in the Sierra dome spider, *Linyphila litigiosa. Anim. Behav.* 48: 615–626.

Watson, P. J., G. Arnqvist, and R. R. Stallmann. 1998. Sexual conflict and the energetic costs of mating and mate choice in water striders. *Am. Nat.* 151: 46–58.

Watt, W. B., P. A. Carter, and K. Donohue. 1986. Females' choice of "good genotypes" as mates is promoted by an insect mating system. *Science* 233: 1187–1190.

Webb, R. E., and F. F. Smith. 1968. Fertility of eggs of Mexican bean beetles from females mated alternately with normal and Apholate-treated males. *J. Econ. Entomol.* 61: 521–523.

Weber, H. 1930. *Biologie der Hemipteren.* Springer, Berlin.

Webster, R. P., and R. T. Cardé. 1984. The effects of mating, exogenous juvenile hormone and juvenile hormone analogue on pheromone titre, calling and oviposition in the omnivorous leafroller moth (*Platynota stultana*). *J. Insect Physiol.* 30: 113–118.

Wedell, N. 1991. Sperm competition selects for nuptial feeding in a bushcricket. *Evolution* 45: 1975–1978.

Wedell, N. 1992. Protandry and mate assessment in the wartbiter *Decticus verrucivorus* (Orthoptera: Tettigoniidae). *Behav. Ecol. Sociobiol.* 31: 301–308.

Wedell, N. 1993a. Mating effort or paternal investment? Incorporation rate and cost of male donations in the wartbiter. *Behav. Ecol. Sociobiol.* 32: 239–246.

Wedell, N. 1993b. Spermatophore size in bushcrickets: Comparative evidence for nuptial gifts as a sperm protection device. *Evolution* 47: 1203–1212.

Wedell, N. 1994. Dual function of the bushcricket spermatophore. *Proc. Roy. Soc. Lond. B* 258: 181–185.

Wedell, N. 1996. Mate quality affects reproductive effort in a paternally investing species. *Am. Nat.* 148: 1075–1088.

Wedell, N. 1997. Ejaculate size in bushcrickets: The importance of being large. *J. Evol. Biol.* 10: 315–325.

Wedell, N. 1998. Sperm protection and mate assessment in the bushcricket *Coptaspis* sp. 2. *Anim. Behav.* 56: 357–363.

Wedell, N., and A. Arak. 1989. The wartbiter spermatophore and its effect on female reproductive output (Orthoptera: Tettigoniidae, *Decticus verrucivorus*). *Behav. Ecol. Sociobiol.* 24: 117–125.

Wedell, N., and P. A. Cook. 1998. Determinants of paternity in a butterfly. *Proc. Roy. Soc. Lond. B* 265: 625–630.

Wedell, N., and P. A. Cook. 1999a. Butterflies tailor their ejaculate in response to sperm competition risk and intensity. *Proc. Roy. Soc. Lond. B* 266: 1033–1039.

Wedell, N., and P. A. Cook. 1999b. Strategic sperm allocation in the small white butterfly, *Pieris rapae* (Lepidoptera: Pieridae). *Funct. Ecol.* 13: 85–93.

Wedell, N., and T. Tregenza. 1999. Successful fathers sire successful sons. *Evolution* 53: 620–625.

Weigensberg, I., and D. J. Fairbairn. 1994. Conflicts of interest between the sexes: A study of mating interactions in a semiaquatic bug. *Anim. Behav.* 48: 893–901.

Wellings, K., J. Field, A. M. Johnson, and J. Wadsworth. 1994. *Sexual Behaviour in Britain.* Penguin, London.

Werren, J. H., and E. L. Charnov. 1978. Facultative sex ratios and population dynamics. *Nature* 272: 349–350.

West-Eberhard, M. J. 1983. Sexual selection, social competition, and speciation. *Q. Rev. Biol.* 58: 155–183.

West-Eberhard, M. J. 1984. Sexual selection, competitive communication and species-specific signals in insects. In T. Lewis, ed., *Insect Communication*, pp. 283–324. Academic Press, London.

Westneat, D. F., and P. W. Sherman. 1993. Parentage and the evolution of parental behavior. *Behav. Ecol.* 4: 66–77.

Wheeler, D. E. 1986. Developmental and physiological determinants of caste in social Hymenoptera: Evolutionary implications. *Am. Nat.* 128: 13–34.

Wheelwright, N. T., and G. S. Wilkinson. 1985. Space use by a neotropical water strider (Hemiptera: Gerridae): Sex and age-class differences. *Biotropica* 17: 165–169.

Wiklund, C., and T. Fagerstrom. 1977. Why do males emerge before females? A hypothesis to explain the incidence of protandry in butterflies. *Oecologia* 31: 153–158.

Wiklund, C., and J. Forsberg. 1985. Courtship and male discrimination between virgin and mated females in the orange tip butterfly *Anthocharis cardamines. Anim. Behav.* 34: 328–332.

Wiklund, C., and J. Forsberg. 1991. Sexual size dimorphism in relation to female polygamy and protandry in butterflies: A comparative study of Swedish Pieridae and Satyridae. *Oikos* 60: 373–381.

Wiklund, C., and A. Kaitala. 1995. Sexual selection for large male size in a polyandrous butterfly: The effect of body size on male versus female reproductive success in *Pieris napi. Behav. Ecol.* 6: 6–13.

Wiklund, C., and B. Karlsson. 1988. Sexual size dimorphism in relation to fecundity in some Swedish satyrid butterflies. *Am. Nat.* 131: 132–138.

Wiklund, C., A. Kaitala, V. Lindfors, and J. Abenius. 1993. Polyandry and its effects on female reproduction in the green-veined white butterfly (*Pieris napi* L.). *Behav. Ecol. Sociobiol.* 33: 25–33.

Wiklund, C., A. Kaitala, and N. Wedell. 1998. Decoupling of reproductive rates and parental expenditure in a polyandrous butterfly. *Behav. Ecol.* 9: 20–25.

Wilcox, R. S. 1972. Communication by surface waves: Mating behaviour of a water strider (Gerridae). *J. Comp. Physiol.* 80: 255–266.

Wilcox, R. S. 1984. Male copulatory guarding enhances female foraging in a water strider. *Behav. Ecol. Sociobiol.* 15: 171–174.

Wilcox, R. S., and J. Di Stefano. 1991. Vibratory signals enhance mate-guarding in a water strider (Hemiptera: Gerridae). *J. Insect Behav.* 4: 43–50.

Wilcox, R. S., and J. R. Spence. 1986. The mating system of two hybridizing species of water striders (Gerridae). I. Ripple signal functions. *Behav. Ecol. Sociobiol.* 19: 79–85.

Wilkes, A. 1965. Sperm transfer and utilization by the arrhenotokous wasp *Dahlbominus fuscipennis* (Zett.) (Hymenoptera: Eulophidae). *Can. Entomol.* 97: 647–657.

Wilkes, A. 1966. Sperm utilization following multiple insemination in the wasp *Dahlbominus fuscipennis*. *Can. J. Genet. Cytol.* 8: 451–461.

Wilkinson, G. S., and P. R. Reillo. 1994. Female choice response to artificial selection on an exaggerated male trait in a stalk-eyed fly. *Proc. Roy. Soc. Lond. B* 255: 1–6.

Williams, C. M. 1963. The juvenile hormone. III. Its accumulation and storage in abdomens of certain male moths. *Biol. Bull.* 124: 355–367.

Williams, G. C. 1966. *Adaptation and Natural Selection*. Princeton University Press, Princeton, NJ.

Williams, G. C. 1978. *Sex and Evolution*. Princeton University Press, Princeton, NJ.

Wilson, E. O. 1971. *The Insect Societies*. Harvard University Press, Cambridge, MA.

Wilson, N., S. C. Tubman, P. E. Eady, and G. Robertson. 1997. Female genotype affects male success in sperm competition. *Proc. Roy. Soc. Lond. B* 264: 1491–1495.

Wolf, L. L., E. C. Waltz, K. Wakeley, and D. Klockowski. 1989. Copulation duration and sperm competition in white-faced dragonflies (*Leucorrhinia intacta*; Odonata: Libellulidae). *Behav. Ecol. Sociobiol.* 24: 63–68.

Wolfner, M. F. 1997. Tokens of love: Functions and regulation of *Drosophila* male accessory gland products. *Insect Biochem. Mol. Biol.* 27: 179–192.

Wolfner, M. F., H. A. Harada, M. J. Bertram, T. J. Stelick, K. W. Kraus, J. M. Kalb, Y. O. Lung, D. M. Neubaum, M. Park, and U. Tram. 1997. New genes for male accessory gland proteins in *Drosophila melanogaster*. *Insect Biochem. Mol. Biol.* 27: 825–834.

Woodhead, A. P. 1984. Effect of duration of larval development on sexual competence in young adult male *Diploptera punctata*. *Physiol. Entomol.* 9: 473–477.

Woodhead, A. P. 1985. Sperm mixing in the cockroach *Diploptera punctata*. *Evolution* 39: 159–164.

Woyciechowski, M., and A. Lomnicki. 1987. Multiple mating of queens and the sterility of workers among eusocial Hymenoptera. *J. Theor. Biol.* 128: 317–327.

Wright, J. 1998. Paternity and paternal care. In T. R. Birkhead and A. P. Møller, eds., *Sperm Competition and Sexual Selection*, pp. 117–145. Academic Press, London.

Wu, C.-I. 1983. Virility deficiency and the sex-ratio trait in *Drosophila pseudoobscura*. I. Sperm displacement and sexual selection. *Genetics* 105: 651–662.

Wyckoff, G. J., W. Wang, and C.-I. Wu. 2000. Rapid evolution of male reproductive genes in the descent of man. *Nature* 403: 304–309.

Xia, X. 1992. Uncertainty of paternity can select against paternal care. *Am. Nat.* 139: 1126–1129.

Yamagishi, M., Y. Ito, and Y. Tsubaki. 1992. Sperm competition in the melon fly,

Bactrocera cucurbitae (Diptera: Tephritidae): Effects of sperm "longevity" on sperm precedence. *J. Insect Behav.* 5: 599–608.

Yamamura, N. 1986. An evolutionarily stable strategy (ESS) model of postcopulatory guarding in insects. *Theor. Pop. Biol.* 29: 438–455.

Yamamura, N., and N. Tsuji. 1989. Postcopulatory guarding strategy in a finite mating period. *Theor. Pop. Biol.* 35: 36–50.

Yamaoka, K., and T. Hirao. 1977. Stimulation of vaginal oviposition by male factor and its effect on spontaneous nervous activity in *Bombyx mori*. *J. Insect Physiol.* 23: 57–63.

Yan, G., and L. Stevens. 1995. Selection by parasites on components of fitness in *Tribolium* beetles: The effect of intraspecific competition. *Am. Nat.* 146: 795–813.

Yasui, Y. 1997. A "good-sperm" model can explain the evolution of costly multiple mating by females. *Am. Nat.* 149: 573–584.

Yokoi, N. 1990. The sperm removal behaviour of the yellow spotted longicorn beetle *Psacothea hilaris* (Coleoptera: Cerambycidae). *Appl. Entomol. Zool.* 25: 383–388.

Young, A. D. M., and A. E. R. Downe. 1982. Renewal of sexual receptivity in mated female mosquitoes, *Aedes aegypti*. *Physiol. Entomol.* 7: 467–471.

Young, A. D. M., and A. E. R. Downe. 1983. Influence of mating on sexual receptivity and oviposition in mosquito, *Culex tarsalis*. *Physiol. Entomol.* 8: 213–217.

Young, A. D. M., and A. E. R. Downe. 1987. Male accessory gland substances and the control of sexual receptivity in female *Culex tarsalis*. *Physiol. Entomol.* 12: 233–239.

Yund, P. O. 2000. How severe is sperm limitation in natural populations of marine free-spawners? *TREE* 15: 10–13.

Yuval, B., and G. N. Fritz. 1994. Multiple mating in female mosquitoes: Evidence from a field population of *Anopheles freeborni* (Diptera: Culicidae). *Bull. Entomol. Res.* 84: 137–139.

Zahavi, A. 1975. Mate selection—a selection for a handicap. *J. Theor. Biol.* 53: 205–214.

Zahavi, A. 1977. The cost of honesty (further remarks on the Handicap Principal). *J. Theor. Biol.* 67: 603–605.

Zeh, D. W., and R. L. Smith. 1985. Parental investment by terrestrial arthropods. *Am. Zool.* 25: 785–805.

Zeh, J. A. 1997. Polyandry and enhanced reproductive success in the harlequin-beetle-riding pseudoscorpion. *Behav. Ecol. Sociobiol.* 40: 111–118.

Zeh, J. A., and D. W. Zeh. 1994. Last-male sperm precedence breaks down when females mate with three males. *Proc. Roy. Soc. Lond. B* 257: 287–292.

Zeh, J. A., and D. W. Zeh. 1996. The evolution of polyandry I: intragenomic conflict and genetic incompatibility. *Proc. Roy. Soc. Lond. B* 263: 1711–1717.

Zeh, J. A., and D. W. Zeh. 1997. The evolution of polyandry. II: Post-copulatory defences against genetic incompatibility. *Proc. Roy. Soc. Lond. B* 264: 69–75.

Zeh, J. A., S. D. Newcomer, and D. W. Zeh. 1998. Polyandrous females discriminate against previous mates. *Proc. Natl. Acad. Sci. USA* 95: 13732–13736.

Zeigler, D. D. 1991. Passive choice and possible mate guarding in the stonefly *Pteronarcella badia* (Plecoptera: Pteronarcyidae). *Fla. Entomol.* 74: 335–340.

Zonneveld, C. 1992. Polyandry and protandry in butterflies. *Bull. Math. Biol.* 54: 957–976.

Zonneveld, C. 1996. Sperm competition cannot eliminate protandry. *J. Theor. Biol.* 178: 105–112.

Zuk, M. 1987. The effects of gregarine parasites, body size, and time of day on spermatophore production and sexual selection in field crickets. *Behav. Ecol. Sociobiol.* 21: 65–72.

Zuk, M., and L. W. Simmons. 1997. Reproductive strategies of the crickets (Orthoptera: Gryllidae). In J. C. Choe and B. J. Crespi, eds., *The Evolution of Mating Systems in Insects and Arachnids*, pp. 89–109. Cambridge University Press, Cambridge, UK.

Taxonomic Index

Subject Index